Understanding Pancreatic Cancer

This book provides a comprehensive examination of pancreatic cancer, integrating global insights with an African-focused perspective. *Understanding Pancreatic Cancer: Global Strategies and African Perspectives* is structured into five parts, each addressing key aspects of the disease, from epidemiology to future innovations in care.

Part I (Chapters 1 and 2) explores the global and African contexts, highlighting disparities in prevalence, diagnosis, and healthcare infrastructure. Part II (Chapter 3) delves into the biological and molecular mechanisms of pancreatic cancer, including genetic mutations and emerging therapeutic targets. Part III (Chapters 4 and 5) focuses on precision medicine and early detection, discussing advances such as artificial intelligence (AI) and biomarker-driven approaches. Part IV (Chapters 6–8) covers treatment strategies, from conventional therapies to novel interventions like immunotherapy and targeted treatments. Finally, Part V (Chapters 9–11) emphasizes the importance of advocacy, research, and policy reforms, outlining strategies for improving patient outcomes and shaping the future of pancreatic cancer care.

By bridging scientific advancements with real-world challenges, this book serves as an essential resource for researchers, clinicians, and policymakers working to enhance pancreatic cancer diagnosis, treatment, and prevention—both in Africa and globally.

Key Features

- Examines pancreatic cancer worldwide, with a focus on Africa's unique challenges and healthcare disparities.
- Explores molecular mechanisms, precision medicine, and AI-driven early detection.
- Covers surgery, chemotherapy, immunotherapy, and targeted therapies.
- Highlights research, healthcare policies, and strategies to improve patient care.

Understanding Pancreatic Cancer

Global Strategies and African Perspectives

Edited by
Zodwa Dlamini, PhD

CRC Press
Taylor & Francis Group
Boca Raton London New York

CRC Press is an imprint of the
Taylor & Francis Group, an **informa** business

Designed cover image: Shutterstock

First edition published 2026
by CRC Press
2385 NW Executive Center Drive, Suite 320, Boca Raton FL 33431

and by CRC Press
4 Park Square, Milton Park, Abingdon, Oxon, OX14 4RN

CRC Press is an imprint of Taylor & Francis Group, LLC

ISBN: 9781032897479 (hbk)
ISBN: 9781032897486 (pbk)
ISBN: 9781003544401 (ebk)

DOI: 10.1201/9781003544401

Typeset in Times
by Deanta Global Publishing Services, Chennai, India

Contents

Preface

Pancreatic cancer remains one of the most formidable challenges in oncology, with high mortality rates and limited treatment options. Despite significant advancements in molecular oncology and precision medicine, disparities in access to early diagnosis and effective therapies continue to affect patient outcomes—particularly in resource-limited settings such as those across Africa. This book seeks to bridge gaps in information by providing a comprehensive analysis of pancreatic cancer from both global and African perspectives, recognizing the unique epidemiological, genetic, and healthcare system challenges, which shape patient care on the continent.

Our approach is twofold: first, to present the latest research on pancreatic cancer, including its molecular drivers, potential therapeutic targets, and emerging diagnostic technologies; and second, to contextualize these advancements within the realities of healthcare delivery in Africa. By integrating discussions on artificial intelligence, precision medicine, and innovative treatment strategies, we aim to highlight opportunities for improving early detection and personalized care in diverse populations.

This book also takes a patient-centered perspective, emphasizing the role of advocacy, community engagement, and healthcare policy reform in addressing barriers to care. We explore the critical role of multidisciplinary collaborations between governments, academic institutions, and the private sector in transforming pancreatic cancer outcomes through research, investment, and knowledge-sharing.

Understanding Pancreatic Cancer: Global Strategies and African Perspectives is intended for oncologists, researchers, healthcare policymakers, and public health specialists who seek to advance cancer care on the African continent and beyond. It is our hope that this work will contribute to a more inclusive and equitable global oncology landscape, where innovative solutions can benefit all populations, regardless of geographic or economic constraints.

About the Editor

Zodwa Dlamini is a professor of molecular oncology, the founding director, and the executive head of the Pan African Cancer Research Institute (PACRI) in South Africa. She is the director of the South African Medical Research Council (SAMRC) Precision Oncology Research Unit (PORU). She is the DSTI/NRF SARCHI Chair in Precision Oncology and Cancer Prevention (POCP). She is a member of the American Association for Cancer Research (AACR) Advisory Group on sub-Saharan Africa and a member of the AACR Pathology Resources in the Africa Advisory Group. Prof. Dlamini is also a member of the African Organisation for Research and Training in Cancer (AORTIC) Research Committee Scientific Advisory Board. She is an elected member of the Academy of Science of South Africa, a member of the New York Academy of Sciences, and an overseas fellow of the Royal Society of Medicine (London).

Part I

Understanding the Global and African Contexts

1 The Pancreatic Cancer Landscape

Zodwa Dlamini[1], Rodney Hull[1], Fusi Madela[2], and Ravi Mehrotra[1,3,4]

[1]SAMRC Precision Oncology Research Unit (PORU), DSI/NRF SARChI Chair in Precision Oncology and Cancer Prevention (POCP), Pan African Cancer Research Institute (PACRI), University of Pretoria, Pretoria, South Africa.

[2]Department of Surgery, School of Clinical Medicine, College of Health Sciences, University of KwaZulu-Natal, Durban, South Africa.

[3]Cancer Genome Atlas, Bhageerath, Vetal Chowk, 402E, Senapati Bapat Road, Pune, Maharashtra, India.

[4]Centre for Health Innovation and Policy (CHIP) Foundation, Noida, Uttar Pradesh, India.

INTRODUCTION

The pancreas functions as both a digestive organ and a glandular organ. As an endocrine gland, it is responsible for producing insulin, glucagon, somatostatin, and pancreatic polypeptide. As a digestive organ, it secretes enzymes that assist in the digestive process in the small intestine. The pancreas consists of two types of cells, exocrine and endocrine, which secrete digestive enzymes and hormones, respectively (Karpińska and Czauderna, 2022). The endocrine tissues consist of cells known as the islets of Langerhans, which consist of three different types of cells, each secreting a different hormone (Da Silva Xavier, 2018). The pancreas is located between the stomach and the spine, nestled in the curvature of the duodenum (Figure 1.1). The pancreas can be divided into two regions: the larger head, nestled in the curve of the duodenum, and the uncinate process. The uncinate process can be further divided into the neck and the tail (Karpińska and Czauderna, 2022). The initial precancerous lesions on the pancreas are referred to as pancreatic intraepithelial neoplasms (PanINs), and it is worthwhile noting that only a small number of these lesions progress to high-grade dysplasia and pancreatic ductal adenocarcinoma (PDAC) (Terhune et al., 1998). The other more common precancerous lesions in pancreatic cancers are precancerous cysts that form in the main or branch pancreatic duct, known as intraductal papillary mucinous neoplasms (Oyama et al., 2020).

There are different subtypes of pancreatic cancer. These include PDAC, which can be distinguished through morphological differences, acinar carcinoma, pancreatic blastoma, and neuroendocrine tumors. Of these, PDAC accounts for 90% of all cases of pancreatic cancer. PDAC generally originates from non-invasive precursor lesions such as pancreatic intraepithelial neoplasia. Apart from these precursor lesions, pancreatic cancer can also develop from intraductal papillary mucinous neoplasms or mucinous cystic neoplasms (Kim and Hong, 2018). These subtypes can be distinguished based on molecular differences. Molecular subtypes also exist in pancreatic cancer, which are morphologically similar, leading to these subtypes having different clinical features that result in them responding differently to various treatments and having various outcomes (Collisson et al., 2019). PDAC can be classified into at least three subtypes using microarray analysis of gene

DOI: 10.1201/9781003544401-2

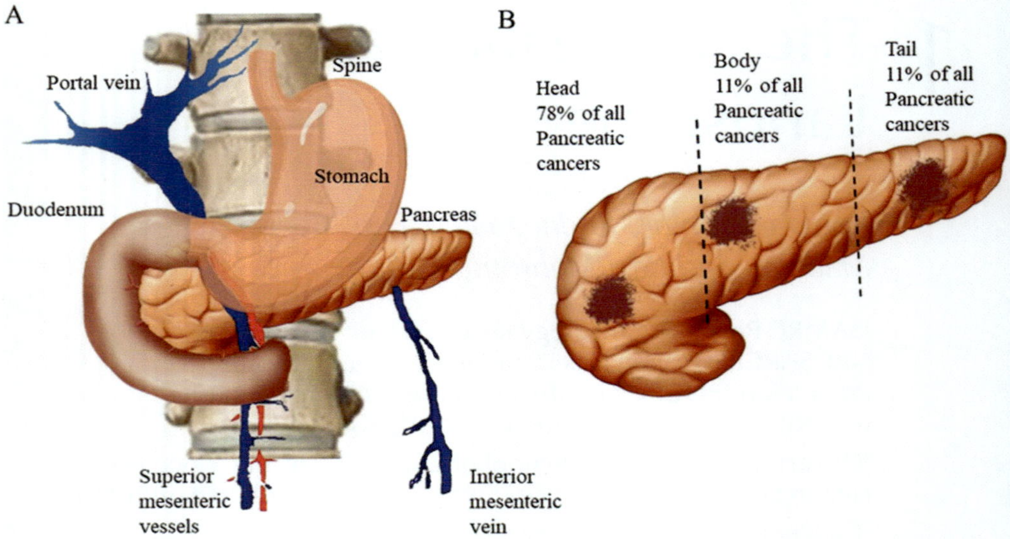

FIGURE 1.1 The pancreas. (A) The pancreas is positioned between the stomach and the spine, nestled in the curve of the duodenum. (B) The location of tumors on the pancreas. The pancreas is divided into three regions: the head, the body, and the tail. Most tumors are found in the head (78%).

expression. The first is the most aggressive quasi-mesenchymal, while the least aggressive with the best outcome is the classical subtype. The third subtype is known as the exocrine-like subtype (Collisson et al., 2011). A larger-scale classification technique using genome-wide expression patterns in a large cohort of pancreatic cancer patients identified four subtypes of PDAC. These are the pancreatic progenitor, squamous, aberrantly differentiated endocrine exocrine (ADEX), and immunogenic subtypes (Bailey et al., 2016). The most recent classification was created through the use of whole genome and transcriptome sequencing of single cells to identify Basal-like A and B, hybrid, and classical A and B subtypes (Chan-Seng-Yue et al., 2020). Classical subtypes are defined by the increased expression of genes such as GATA binding protein 6 (*gata6*) *kras*, KRAS G12V and *smad4* (Collisson et al., 2011). The progenitor subtype is characterized by the increased expression of genes associated with early tumor development. The squamous subtype is associated with poor outcomes, very likely due to the metabolic reprogramming, inflammation, and hypoxia associated with the subtype (Bailey et al., 2016). Squamous basal-like tumors are defined by *tp53* mutations and other genetic changes that drive epithelial-to-mesenchymal transition (EMT) and metastasis (Cancer Genome Atlas Research Network, 2017).

Additionally, pancreatic cancers can be classified based on the region of the pancreas where they arise. The three regions of the pancreas differ in the frequency at which cancers arise. The head is to the right of the superior mesenteric vessels, while those that arise in the tail are to the left of the inferior mesenteric vein, and in both the tail and the body are between the superior mesenteric vessels and the inferior mesenteric vein. By determining the location of tumors in the various regions of the pancreas in 100,313 patients, it was found that the majority of tumors were located in the head (78%), while both the body and the tail had an equal likelihood of being the location of the tumor (11%) (Figure 1.1) (Sener et al., 1999). Pancreatic cancers have also been classified using a combination of molecular biology and the identification of different pathologies in samples. This new classification is summarized in Table 1.1 (Sharma et al., 2011).

TABLE 1.1
Molecular and Pathological Classification of Pancreatic Cancer

Pancreatic Cancer Type	Frequency	Features	Genetic Changes	References
Ductal infiltrating adenocarcinoma	85–90%	Whitish mass that is poorly defined from surrounding tissues	Mutations in *kras2*, *tp53*, *p16/ kdkn2aA*, *madh4/dpc4*	Shi et al. (2008); Liu, (2001)
Solid pseudopapillary neoplasm	3%	Solid and cystic masses	No mutations in the *kras2*, *p16/* cdkn2a, *tp53*, *madh44*/dpc4 genes. Alteration in β-catenin protein expression	Abraham et al. (2002)
Intraductal papillary mucinous neoplasms (IPMNs)	5%	Papillary epithelial mucin-producing neoplasms arising in the pancreatic ducts	Loss of heterozygosity of the *stk/11/lb* gene is another frequent gene. Mutations in *tp53*, *kras2*, *and p16/cdkn2a*	Schönleben et al. (2006); Iacobuzio-Donahue et al. (2000)
Pancreatic intraepithelial neoplasia (PanIN)	The precursor to invasive ductal carcinoma. PanIN-1, PanIN-2, and PanIN-3	Neoplastic proliferation of mucin-producing epithelial cells	Telomerase shortening and activating point mutations in the *kras2* gene, activation of the p16/ cdkn2a gene, and alterations in the *tp53*, madh4/dpc*44*, and brca2 genes	Hruban et al. (2001); Shi et al. (2008)
Pancreatoblastoma	0.5% affects children.	Soft and well-demarcated mass with epithelial or acinar differentiation, endocrine, and mesenchymal characteristics	Loss of heterozygosity of chromosome 11p. Genetic alterations in the APC)/β-catenin pathway	Saif (2007)
Undifferentiated carcinoma	Arise from pre-existing ductal adenocarcinomas	Undifferentiated aggressive and tend to metastasize and infiltrate surrounding organs	*kras2* gene mutation	Hoorens et al. (1998)

(*Continued*)

TABLE 1.1 (CONTINUED)
Molecular and Pathological Classification of Pancreatic Cancer

Pancreatic Cancer Type	Frequency	Features	Genetic Changes	References
Medullary carcinoma		Well-demarcated soft masses that are usually poorly differentiated	Loss of expression of one of the DNA mismatch repair proteins (M1h1 and Msh2) mutation in the *brief* gene	Hoorens et al. (1998)
Mucinous cystadenocarcinoma	1%	Exocrine pancreas with different biological behaviors	Mutations in *kras, cdkn2a, tp53*, and *SMAD4*	Barbee et al. (1976); Fischer and Wood (2018)
Serous cystadenomas		Benign tumors	Mutations in , *kras, cdkn2a, tp53*, and *smad4*	King et al. (2009); Roggin et al. (2010)
Adenosquamous carcinoma	4%	Mixed populations of adenomatous cells and cells with a varying amount of keratinized squamous features	Mutations in *dnhd1*, *kprp*, *mdm4*, and *nuc6*	Del Arco et al. (2019)
Acinar cell carcinoma	2%	Presence of exocrine enzymes such as trypsin, chymotrypsin	Rare *kras. tp53*abd *cdkn2a* mutations. More prevalent homologous recombination-related genes, such as *brca1/2* and *atm*	Stelow et al. (2010)

EPIDEMIOLOGY

Global Prevalence and Mortality Rates

Worldwide, pancreatic cancer is the 14th most common cancer and the 9th leading cause of cancer-related death. In Africa, its overall incidence is slightly lower, being the 15th most common cancer and the 13th leading cause of cancer-related mortality. However, in South Africa, it is both the seventh most common and seventh leading cause of cancer-related death (Globocan) (Bray et al., 2024). In the United States, the median age at diagnosis is 71 years (Oyama et al., 2020), compared to 55.7 years in Zambia (Hu et al., 2021) and 64.6 years in Morocco (Amri et al., 2023). In South Africa, the median age of diagnosis was approximately 60 years in 2022 (National Cancer Registry, 2024). At the time of diagnosis of PDAC, most patients have metastatic disease (Siegel et al., 2021). Dividing the world's population by continent, Europe and North America have the highest incidence and mortality rates, while Africa has the lowest mortality and incidence rates amongst all the continents. However, in Africa, the mortality rate is comparatively the highest in comparison to the incidence rate (Figure 1.2(A)). Those regions with a higher incidence of pancreatic cancer are generally occupied by countries with higher human development indices (HDI). In these high HDI countries, it has been predicted that pancreatic cancer will soon overtake breast cancer as the third most common cause of cancer-related death (Bray et al., 2024).

South Africa, Egypt, Libya, Niger, and Burkina Faso all have the highest incidence and mortality rates reported on the continent (Figure 1.2(B)). However, currently, only 22 countries in Africa have national cancer registries, including these countries, and as such, these countries having the highest incidence and mortality rates may be the result of biased recording, as these countries are also some of the countries with cancer registries. As such, the higher rates recorded in these countries may be the result of these countries having more accurate records of pancreatic cancer cases. This means that the incidence of the disease may be much higher throughout the rest of Africa (World Health Organization, 2024). This is also the likely reason for the very low incidence of pancreatic cancer recorded in countries like Burundi, where cancer surveillance has been neglected in light of the civil war (1992–2005) (Manirakiza et al., 2020). In the North African region, there has been a noticeable increase in the incidence of pancreatic cancer in the last 20 years (Nejadghaderi et al., 2023). Data from the South African National Cancer Registry show a drastic increase in the number of recorded cases of pancreatic cancer over the period 2001–2022, with the number of recorded cases doubling over this time period (Figure 1.2(C)) (National Cancer Registry, 2024). A study of 23,581 pancreatic cancer cases in South Africa was conducted to identify factors that increased mortality rates in these patients. Certain types of occupations were linked to increased mortality. These included management jobs for both sexes, as well as craft and related trade work in males, and professional, elementary occupation, and clerical support work in females. This study also identified smoking as a primary risk factor in South Africa (Nhleko et al., 2023).

Socioeconomic Impacts of Pancreatic Cancer

Even though surgical resection is the first line of treatment and is the only curative therapy for pancreatic cancer, improving the chance of survival drastically, only 15–20% of pancreatic cancers are resectable (Vreeland and Katz, 2018). This is because, as with most health problems, socioeconomic status (SES) dictates both the presentation and therapeutic management of pancreatic cancer. Low SES individuals tend to present at a later stage and have lower access to treatment. SES also influences the quality of care as well as survival and recurrence (Cheung et al., 2010). SES also influences where a patient resides and their access to healthcare, while SES is in turn influenced by race and ethnicity. Finally, education can influence SES and be influenced by SES. All these factors play a role in contributing to pancreatic cancer outcomes (Fabregas et al., 2022). In South Africa, SES

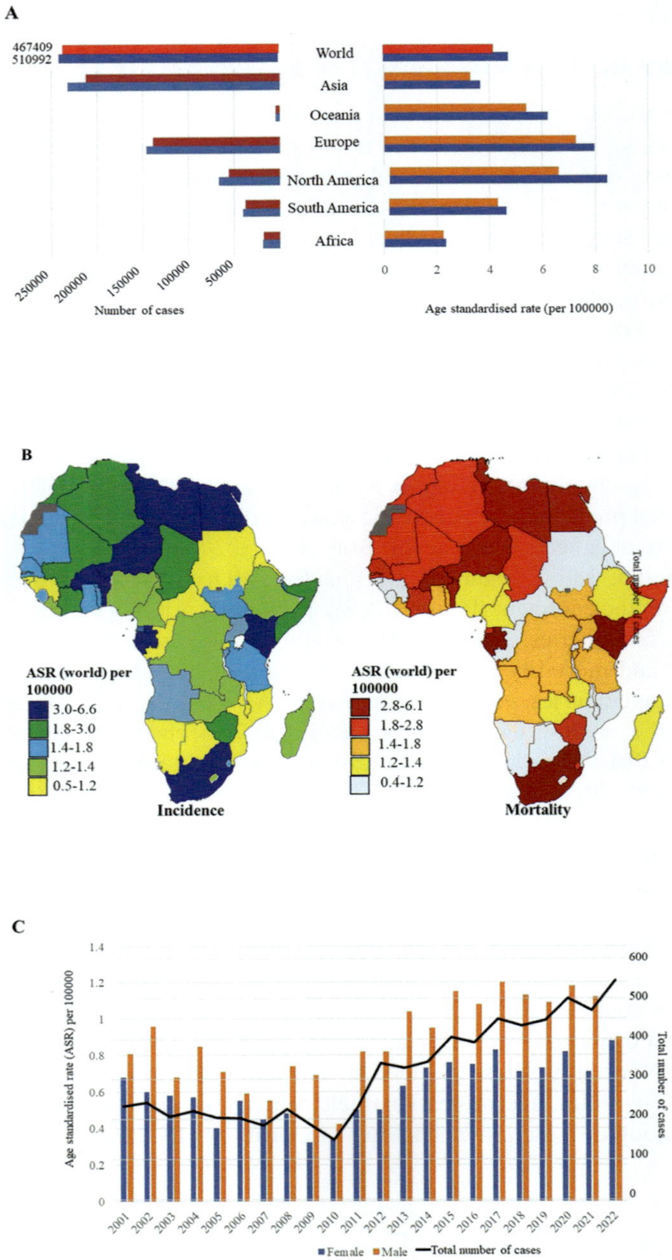

FIGURE 1.2 Epidemiology of pancreatic cancer. (A) Global incidence and mortality of pancreatic cancer by continent. This figure shows that the recorded incidence of the disease is highest in Europe and North America, with Africa having the lowest incidence rate. Due to its large population size, Asia has the highest number of cases but the second lowest age-standardized rate of pancreatic cancer. The mortality rates follow a similar pattern, but it is important to note that although most regions show a noticeable difference between the incidence and mortality rates, the mortality rate in Africa is comparatively high compared to the incidence rate. (B) Incidence and mortality rates in Africa showing certain select countries having a higher incidence and mortality rate. (C) Mortality and incidence rates of pancreatic cancer in South Africa by gender and the total number of cases for both genders. These data show a steady increase in the number of cases over the 20 years from 2001 to 2022.

is also a primary factor dictating access to the private health sector, where treatment is more rapid and the patient load is lighter, allowing more intensive patient care (Maphumulo and Bhengu, 2019). Any assessment of the role played by a difference in SES in contributing to health disparities is difficult to assess, as these various factors dictating SES cannot be directly compared or assigned a score on the same scale. The effect of socioeconomic factors also varies by country as their effects vary based on a community or population's historical background and culture (Fabregas et al., 2022).

Risk Factors for Pancreatic Cancer

Risk factors for pancreatic cancer include lifestyle-related factors such as tobacco use, with current active cigarette smoking being the risk factor with the strongest association with developing pancreatic cancer (Bosetti et al., 2012). It is estimated that up to 20% of all pancreatic tumors are the result of smoking cigarettes. These smoking-related pancreatic cancers have more genetic mutations compared to other pancreatic cancers (Blackford et al., 2009). Alcohol use is also a risk factor but has a weaker association with the development of PDAC and is normally associated with an intake of greater than three drinks a day (30 g per day) (Wang et al., 2016). Obesity is emerging as another lifestyle factor associated with the development of pancreatic cancer, with individuals with body weight in the top 20% percentile having an increased risk of PDAC (Johansen et al., 2010). Yet another lifestyle factor that can influence the development of pancreatic cancer is diet. The diets that are associated with pancreatic cancer include those rich in processed meat and saturated fat as well as low in vegetables and folate (Raimondi et al., 2009). Additionally, a high intake of high-fructose beverages is also associated with an increased risk of developing pancreatic cancer. These diets are also associated with both obesity and diabetes (Sung et al., 2019). The increased rates of obesity and diabetes may be contributing to the observation that pancreatic cancer is being diagnosed at an earlier age, with the annual incidence in individuals younger than 49 years steadily increasing over the last 30 years (Sung et al., 2019). These findings led to preliminary studies using the drug metformin, usually used to treat diabetes, as a preventative agent to lower the risk of developing pancreatic cancer (Conroy et al., 2018). Conditions such as chronic pancreatitis are also risk factors, with chronic pancreatitis or a family history of chronic pancreatitis increasing the chances of developing pancreatic cancer by about 13-fold (Genkinger et al., 2009). Other conditions associated with an increased risk for pancreatic cancer include periodontal disease and possibly *Helicobacter pylori* infection (Raimondi et al., 2009). Biological factors that increase the risk of developing pancreatic cancer include being male, older, having a non-O blood group (Amundadottir et al., 2009) and being of African-American ethnic origin (Raimondi et al., 2009).

A family history of pancreatic cancer is an important risk factor, as this suggests the presence of inherited genetic changes that predispose an individual to develop prostate cancer. It is thought that inherited genetic changes are responsible for up to 10% of all cases of pancreatic cancer (Petersen et al., 2006). Families in which a pair of first-degree relatives develop pancreatic cancer are nine times more likely to develop the disease (Klein et al., 2004). Familial pancreatic cancer is thought to account for many cases of younger-onset pancreatic cancer (diagnosis in patients under the age of 50) (Brune et al., 2010). Another difference between familial and other types of pancreatic cancer is the number of precancerous lesions. These are much higher in cases of familial pancreatic cancer (Shi et al., 2009).

THE ROLE OF EARLY DETECTION AND PRECISION MEDICINE IN COMBATING PANCREATIC CANCER

High-throughput technologies such as next-generation sequencing offer the promise of advanced diagnosis and treatment, where each patient's genetic sequence will provide the clinician or oncologist with information that can be used to predict the risks of the patient developing pancreatic cancer as well as the most appropriate treatment options when cancer has been diagnosed. Effective screening and prevention of pancreatic cancer involve the identification of individuals predisposed to developing the disease. One of the leading contributors to the high mortality observed in pancreatic cancer is the fact that it is commonly diagnosed at a very late stage. Only a small percentage of patients are diagnosed when the cancer is still at an early localized stage, while over half of the patients who present with PDAC already have metastatic PDAC (Siegel et al., 2021). As such, treatment has very little chance of success, and the best strategy to improve pancreatic cancer outcomes is the early screening and detection of pancreatic cancers when they are still treatable (Hur et al., 2016). If an individual is identified as having a genetic predisposition to pancreatic cancer, their family members can undergo genetic screening to assess their underlying risk of predisposition. The implementation of genetic testing is thought to be underused at present, allowing many high-risk patients to go unnoticed, contributing to the incidence of pancreatic cancer (Grover et al., 2004). The decision as to when genetic testing should be conducted can be informed by the presence of specific risk factors or indications, the patterns referred to above. For instance, the presence of *brca2* mutations should be considered in individuals of Jewish ancestry and individuals with a familial history of either pancreatic or breast cancer. Germline testing for *cdkn2a* mutations should be conducted in patients with a family history of certain melanoma subtypes. Genetic counseling should be the first step in any genetic testing to establish the predisposition of an individual or their families to developing pancreatic cancer. Apart from genetic testing, having good information concerning a detailed family history of cancer occurrence should allow for an informed and fairly accurate risk prediction.

The Molecular Basis of Pancreatic Cancer

As pancreatic cancer develops, genetic and epigenetic alterations occur. These genetic changes not only drive disease development and progression but also influence disease aggression and response to treatment. Some of the genes where mutations commonly occur, or the genes whose expression is altered in pancreatic cancer, include oncogenes such as *kras* and tumor-suppressor genes such as *cdkn2a*, *tp53*, *smad4* and *brca2* (Rozenblum et al., 1997). Apart from alterations related to these genes, pancreatic cancers also contain genetic alterations such as chromosomal losses, gene amplifications, and telomere shortening (Jones et al., 2008; van Heek et al., 2002). Due to their presence in the early stages of pancreatic cancer, *kras* mutations and telomere shortening are viable suspects for driving the initial development and progression of the disease (Hong et al., 2011). Mutations and alterations that result in the inactivation of tumor-suppressor genes normally occur in the later stages of pancreatic cancer, intraepithelial neoplasia, and invasive carcinomas (Hruban et al., 2008). Less than 20% of all pancreatic cancers contain a mutation in *braf*, *myb*, *akt2*, *rgfr*, *egfr*, *map3k4*, *stk11*, *tgfbr1*, *tgfbr2*, *acvr2a*, *acvr1b*, *fbxw7*, and *ep300*. (Jones et al., 2008). Mutations in genes such as *pik3cg*, *dgka*, *stk22*, and *prkcg* occur at an even lower frequency than those mentioned previously but are still able to drive pancreatic cancer development and progression (Carter et al., 2010).

Pancreatic cancer is also characterized by specific epigenetic changes which alter gene expression and function (Omura and Goggins, 2009). Alterations in the methylation of gene promoter regions and gene silencing promote pancreatic cancer by silencing tumor-suppressor genes such as *cdkn2a* (Schutte et al., 1997). Serine protease 1 (PRSS1) is a trypsinogen which is secreted by the pancreas into the small intestine. PRSS1 is cleaved multiple times to form trypsin, which digests protein. Mutations in PRSS1 lead to the creation of a protein that is activated within the pancreas, resulting in damage to the pancreas. This mutation is a common characteristic of pancreatitis and drastically increases the chance of developing PDAC (Rebours et al., 2008).

In a small percentage of pancreatic cancers, epigenetic modifications are found to occur in DNA repair genes such as *MLH1* (Li and Zhao, 2012). Hypomethylation of *sfn*, *msln*, and *s100a4*, resulting in the overexpression of genes, was found to occur in pancreatic cancers (Vincent et al., 2008). Non-coding RNAs also play a role in the development and progression of pancreatic cancers, where both microRNA (miRNA) and long-noncoding RNA (lncRNA) expression play a role. This has led to the proposed use of multiple miRNAs as diagnostic markers (Li et al., 2010). These include miRNAs such as miR-21, miR-34, miR-155, and miR-200 (Bloomston et al., 2007).

The Tumor Microenvironment

The above-mentioned mutations that drive pancreatic cancer lead to histological and morphological changes. This, in turn, leads to changes in the surrounding tissue stroma. These changes then promote further development of cancer as the modified tumor microenvironment (TME) has altered immune, vascular, and connective tissue components. The alterations in the TME can be targeted for therapeutic purposes (Foster et al., 2018). These alterations include increased levels of fibrosis (desmoplasia) resulting from pancreatic stellate cells becoming active (Vonlaufen et al., 2008). Desmoplasia can limit the exposure of the tumor to drugs such as chemotherapy by forming a barrier around the tumor (Provenzano et al., 2012). However, this process is not always cancer-promoting and its role in pancreatic cancer is more complex; thus, it requires a multi-faceted approach. Hyaluronan is a non-sulfated glycosaminoglycan and is a highly expressed component of the extracellular matrix (ECM) within the desmoplastic barrier. The higher the expression of hyaluronan, the poorer the prognosis of PDAC (Jacobetz et al., 2013). Recombinant human hyaluronidase (PEGPH20) has been used to target the fibrous barrier by catalyzing the degradation of hyaluronan. This strategy has been shown in previous studies to improve drug penetration and delivery in mouse models (Jacobetz et al., 2013).

The hedgehog signaling pathway is required for the development of the pancreas and plays a vital role in the creation of the TME. Expression of endodermal sonic hedgehog inhibits pancreatic differentiation, and during development, the pathway is downregulated to allow for the formation of pancreatic tissue (Kim and Melton, 1998). Perturbations in the regulation of the SHH pathway or alterations in the expression of its components result in pancreatic carcinogenesis, with increased SHH activity promoting desmoplasia (Strobel et al., 2010; Bailey et al., 2008).

The ECM regulates migration and invasion through the presence of various ECM protein components. These proteins remodel the ECM. One of the main classes of these proteins is the matrix metalloproteinases (MMPs) and proteins that inhibit their function. Increased expression of MMPs stimulates metastasis and is associated with poorer outcomes (Jones et al., 2004). MMPs that are overexpressed in pancreatic cancer include MMP2, MMP7, MMP9, and EMMPRIN (Schneiderhan et al., 2007). This has led to the development of multi-MMP inhibitors to suppress metastasis and tissue remodeling. These include marimastat and tanomastat, both of which are effective in various cancers, but not currently in pancreatic cancer (Moore et al., 2003).

Advances in Diagnostic Technologies

The primary aim of screening is to identify pancreatic lesions at early, curable stages where the disease is still in the initial stages or pre-invasive lesions. These include identifying pancreatic intraepithelial neoplasia, non-invasive intraductal papillary mucinous neoplasms, and mucinous cystic neoplasms (Klapman and Malafa, 2008). Ideally, a screening test would be non-invasive and rapid, such as an accurate blood test looking for biomarkers for pancreatic cancer. Currently, such a test does not exist as none have proven to be reliable or accurate enough (Harsha et al., 2009).

Some potential serum biomarkers include carbohydrate antigen 19-9 (CA19-9), carcinoembryonic antigen (CEA) (Meng et al., 2017), and carbohydrate antigens 242 (CA242) and 494 (CA494), carcinoembryonic antigen-related cell adhesion molecule 1 (CEACAM1), parathyroid hormone-related protein (PTHrP), TuM2-PK, CAM 17.1, and HC (Yang and Zhang, 2021). CA19-9 is used as a marker for various cancers (reviewed in Lee et al., 2020), where its overexpression can be detected through the 1116NS19-9 monoclonal antibody. The levels of CA19-9 can be affected by other factors such as bilirubin level, which can increase its serum levels. As such, patients with jaundice can receive false positive results (Luo et al., 2021). CEA is an intracellular adhesion molecule that is present in pancreatic secretions and is expressed at higher levels in the pancreatic juice of individuals with pancreatic cancer (Meng et al., 2017). The discovery and use of new biomarkers can be quite complex as it relies on identifying molecular patterns in large subsets of biological data such as whole genomic, epigenomic, transcriptomic, or proteomic data. As such, the development of artificial intelligence algorithms that use machines, or deep learning, is to identify these patterns and associate them with subtypes, stages, or outcomes of pancreatic cancer.

Currently, diagnosis through histological examination is the gold standard for the diagnosis of pancreatic cancer. This involves the cytological examination of either images or biopsied tissues. Abdominal ultrasound is a commonly used method to obtain medical images of the pancreas. This can be done through trans-abdominal ultrasound (Gandolfi et al., 2003). The tissue contrast, accuracy, and sensitivity of ultrasound are continually improving with the development of new ultrasound machines (Rezaei-Dalouei et al., 2018). The evaluation of the vascularity of pancreatic lesions allows for the identification of malignant pancreatic cancers and improved staging. Ultrasound is traditionally not able to clearly distinguish vascularization. However, the recent introduction of intravenous contrast has been shown to improve the effectiveness of ultrasound in this regard (Dietrich et al., 2008). Endoscopic ultrasonography (EUS) uses a sound emitting endoscopic tube to identify small pre-invasive lesions, where it has been shown to be more accurate than either magnetic resonance imaging (MRI) or computed tomography (CT) scans in detecting pancreatic cystic lesions (Canto et al., 2010). Unfortunately, EUS is invasive and is highly dependent on the skill of the operator. It is also expensive and can have negative side effects such as pancreatitis, bleeding, and duodenal perforation (Eloubeidi et al., 2006).

One of the most used diagnostic tools for pancreatic cancer is CT by using chest and pelvis CT scans. The diagnosis is performed by assessing vascular anatomy. This can also be used to stage the disease by determining the proximity between the tumor and any blood vessels near it. The two largest blood vessels which can be influenced by angiogenesis to supply the tumor are the celiac, hepatic, and superior mesenteric arteries (Conroy et al., 2011). This can result in the classification of the interaction between the tumor and blood vessels as uninvolved, abutment (blood vessels surrounding the tumor by 180° or less), and finally encasement which involves greater than 180° interaction between the tumor and surrounding blood vessels (Ribatti and Pezzella, 2021). Multi-detector CT scanners (MDCT) allow an increase in spatial and temporal resolution (Gangi et al., 2004; Ohwada et al., 1998).

MRI is used to perform cholangiopancreatography, an imaging procedure that can be used to detect obstructions in the biliary system. This is known as magnetic resonance cholangiopancreatography (MRCP) and can be used to determine the presence of pancreatic lesions, tumors, and the presence of metastatic tumors as well as provide more detailed information regarding tumors initially detected by CT scans (Vargas et al., 2004). Studies have been performed to determine the effectiveness of MRI and MRCP compared to CT, with many of these studies providing evidence that MRI is more effective at diagnosing and staging pancreatic cancers (Grenacher et al., 2004; Bipat et al., 2005).

Positron emission tomography (PET)/CT is another imaging method commonly used to diagnose pancreatic cancer. One of its strengths is its ability to detect elevated glucose metabolism in tumors by using a fluorodeoxyglucose (FDG) tracer. This different absorption and use of FDG by a tumor can be used to determine whether the tumor is benign or malignant and characterize its metastatic potential. Traditionally, however, it is unable to provide spatial resolution of the tumor, and results can be confusing due to increased glucose use and uptake occurring as a result of infection and inflammation (Wang et al., 2014). As such, PET/CT scans are not normally used as a routine method for the staging of pancreatic cancers. However, recent advances in technology have allowed the procedure to be used to detect small pancreatic cancers and diagnose up to 40% of metastatic tumors (Maemura et al., 2006). Since pancreatitis is characterized by a more diffuse uptake of FDG, PET/CT scans are useful tools to distinguish between pancreatitis and pancreatic cancer (Lee et al., 2009). 11C-Acetate-PET relies on the fact that acetate is incorporated and converted into acetyl-CoA, where it enters the tricarboxylic acid cycle. However, in cancer cells, this is not used to produce energy but are used to take part in the synthesis of lipids in the cell membrane of tumor cells. This technique appears to be better than FDG PET in the early detection and diagnosis of pancreatic cancer (Zhao et al., 2009).

THE TREATMENT AND COMPREHENSIVE CARE OF PANCREATIC CANCER

Currently, pancreatic cancer is mostly treated with combination chemotherapy, which has a high toxicity for patients while offering mild advantages. Adjuvant therapy using 5-fluorouracil, leucovorin, oxaliplatin, and irinotecan has resulted in improved patient survival (Conroy et al., 2018), but at the same time, this treatment is too harsh for many patients, especially those recovering from surgical intervention. For patients with advanced disease, the best option is thought to be gemcitabine/nab-paclitaxel. However, these treatments are still ineffective in treating late-stage pancreatic cancer (Von Hoff et al., 2013). Following diagnosis, the disease needs to be staged and characterized for patients to receive the appropriate treatment. For instance, patients with stage I or II pancreatic cancer can be treated through resection while stages IIa and IIb can be considered to be borderline resectable (Wood et al., 2022).

Surgery

Currently, surgery is the only treatment method that offers a cure for pancreatic cancer (Ozawa et al., 2001). Pancreatic cancers are generally classified for treatment purposes based on the resectability of the cancer. Resectable pancreatic cancer (RPC) is stage I or stage IIa cancer where the disease is still confined to a small region of the pancreas. Borderline resectable pancreatic cancers (BRPC) are stage IIb and some stage II cancers. These tumors have increased in size or have spread to the lymphatic system. As such, they may still be effectively surgically removed but may require both preoperative and postoperative treatment to ensure the complete removal of the tumor. Finally, unresectable pancreatic cancer (UPC) includes tumors that have spread to organs such as the liver and stomach and cannot be removed completely through surgery (Figure 1.3). Different surgical procedures involved in pancreatic resection include pancreaticoduodenectomy, which is the partial removal of portions of the pancreas, or total pancreatectomy. Each type is performed depending on which region of the pancreas or the degree to which the pancreas is affected. The Whipple procedure is performed to remove the head of the pancreas, while a distal pancreatectomy is performed to remove the tail of the organ (Chincarini et al., 2018). In many cases, neoadjuvant chemotherapy and/or radiotherapy are used before and after surgery in cases of locally advanced pancreatic cancers or those that are borderline resectable (Botta et al., 2023).

Thanks to advances in surgery and perioperative supportive care, pancreatic cancer surgery now has mortality rates lower than 5% (Cameron et al., 2006). These advances in surgery include minimally invasive surgery such as laparoscopic pancreatic surgery (Cuschieri et al., 1996) and more recently, robotic pancreatectomy (Liu et al., 2017). Vascular reconstruction allows the removal of vascular invasive tumors and the subsequent repair of resected blood vessels, allowing them to retain their function and preserve the integrity of the surrounding tissues and organs (Maley and Yeo, 2017).

Chemotherapy

Since so few pancreatic cancers are resectable when the disease is diagnosed, other treatment options need to be explored. Due to the locally advanced and metastatic nature of these tumors, cytotoxic chemotherapy is still currently the most viable treatment option (Ghosn et al., 2016). Gemcitabine has been the first line of treatment for advanced pancreatic cancer since the end of the last century (Burris et al., 1997). Initially used as a monotherapy, it has been more successful

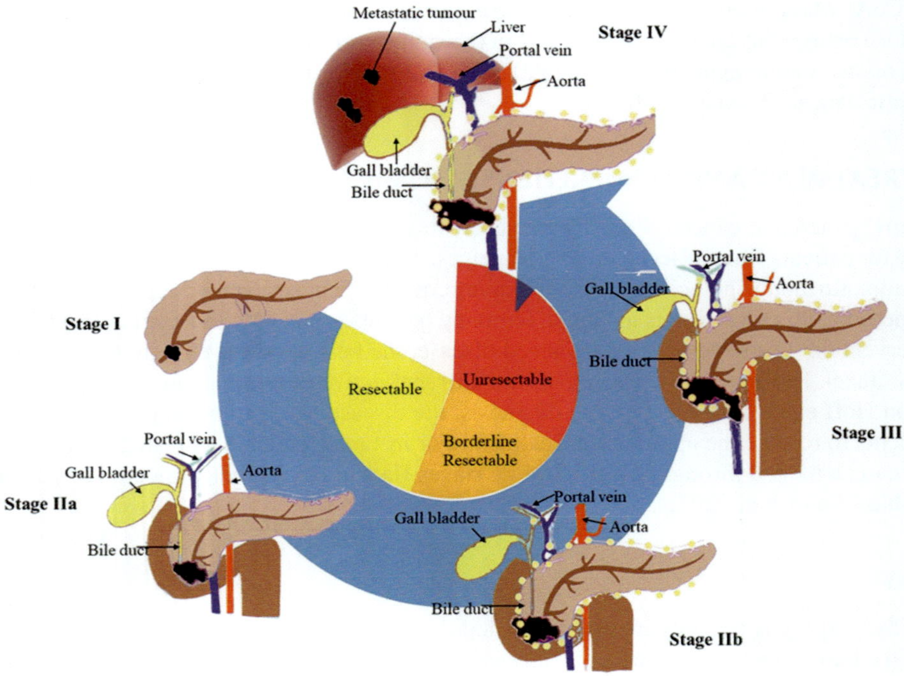

FIGURE 1.3 The eligibility of different stages of pancreatic cancer to be removed through surgical resection. As pancreatic cancer progresses, grows, and spreads, it becomes less likely that it can be completely surgically removed. As such, stage I and IIa pancreatic cancers can be classed as resectable. Stage IIb and stage III cancers are classified as borderline resectable, meaning that there is a chance of successfully treating the cancer through surgery, especially in combination with pre- and postoperative treatments. Finally, most stage III and all stage IV tumors are classified as unresectable.

when used as part of a combination therapy in conjunction with FOLFIRINOX and nab-paclitaxel. FOLFIRINOX itself is a combination therapy consisting of four agents: folic acid (FOL), fluorouracil or 5FU (F), irinotecan (IRIN), and oxaliplatin (OX) (Conroy et al., 2011). These different agents, while not having overlapping functions, are thought to still have synergistic effects (Moore et al., 2007). FOLFIRINOX does, unfortunately, have adverse side effects such as neutropenia, thrombocytopenia, diarrhea, and peripheral neuropathy. Despite these side effects, FOLFIRINOX did result in a significant increase in the quality of life of pancreatic cancer patients (Gourgou-Bourgade et al., 2013). Different regions of the world have adopted different drugs as the standard treatment based on studies performed in those regions. In East Asia, an orally administered prodrug version of 5FU known as S-1 became the standard adjuvant chemotherapy (Uesaka et al., 2016). In Europe, the combination of gemcitabine and S-1 has been studied as a potentially more effective treatment (Neoptolemos et al., 2017).

Treatment of advanced pancreatic cancer generally involves transitioning from one treatment to another, with most studies showing that initial treatment with FOLFIRINOX transitions to gemcitabine/nab-paclitaxel (El Rassy et al., 2017; Zhang et al., 2015). Other second-line therapies include the combination of oxaliplatin and 5-FU in conjunction with folinic acid (OFF). OFF differs from FOLFOX in that it lacks irinotecan and is applied in different ways, with treatment being administered at different stages of the treatment (Gill et al., 2016).

Neoadjuvant therapy when applied to resectable pancreatic cancer usually refers to preoperative therapy, and it aims to reduce the chance of recurrence and increase the effectiveness of any surgical intervention. In addition to this, neoadjuvant therapy when applied to nonresectable pancreatic cancers aims to increase the ability of the tumor to be removed through surgery by decreasing its size or metastatic spread (Heinemann et al., 2013; Papavasiliou et al., 2014). Once again, combination therapy using OLFIRINOX is the standard for neoadjuvant therapy (Tzeng et al., 2014).

Targeted Therapy

Research on the changes in the molecular patterns and TME that drive the development and progression of pancreatic cancer has led to the discovery of targetable pathways, some of which may be population or even patient specific. As such, these therapies may lead to the introduction of personalized therapy for pancreatic cancer patients. Many of the subtypes of PDAC susceptible to targeted therapy include those with specific genetic alterations.

Small molecule inhibitors have been used to target the tyrosine kinase domain of the epidermal growth factor receptor (EGFR), This drug Erlotinib specifically targets mutations in this region of EGFR (Türeci et al., 2019). Mutations in the DNA double-strand break repair genes *brca1* and *brca2* mean that tumors with these mutations are susceptible to drugs that cause this sort of DNA damage. These include platinum salts and topoisomerase inhibitors (Wattenberg et al., 2020; O'Reilly et al., 2020). In addition to this, Olaparib is a small molecule that inhibits PARP and prevents the repair of single-strand breaks, resulting in increased levels of double-strand breaks, which tumors with *brca* mutations struggle to repair. This leads these tumors to then undergo apoptosis (Golan et al., 2019).

Mutations in the *kras* gene can be targeted using the small molecule inhibitor ARS-1620, which binds to the allosteric binding pocket of the *G12C* mutant of KRAS. This keeps these mutant *KRAS* in their inactive GDP-bound form (Janes et al., 2018). Unfortunately, the most common KRAS mutants in pancreatic cancer are *G12D* and *G12V*, and as such require the development of different small molecule inhibitors. One of these is the compound MRTX1133, which specifically targets the *G12D* mutant (Wang et al., 2022). Another option is to target molecules just upstream of the *KRAS* signaling pathway such as SOS1 or SHP2 (Hofmann et al., 2021).

Radiation Therapy

Radiation therapy for the treatment of pancreatic cancer has always been considered an adjuvant therapy, with its use as a first-line therapy being considered unfavorable due to the difficulties in targeting the pancreas without affecting surrounding organs. One of the most important considerations for radiotherapy of locally advanced pancreatic cancer is the ability to escalate the dose of radiation without damaging surrounding tissues and organs. At the same time, radiation therapy plays an important role in providing neoadjuvant therapy for resectable pancreatic cancer and as a means to decrease metastasis and tumor size in borderline resectable pancreatic cancer, allowing it to be recharacterized as resectable (Liu et al., 2022). When it comes to dose escalation, the conventional treatment is 54 Gy in 30 fractions (Hammel et al., 2016). However, there have been advances in radiation techniques that allow for an increase in dosage. These include advances such as stereotactic body radiation therapy (SBRT), which allows for more direct targeting of the affected area with higher doses without affecting surrounding tissue (Katz et al., 2016).

Nearby organs at risk (OARs) can be damaged by increasing doses beyond 33 Gy in five fractions. This can be avoided by splitting the dosage over more fractions, hypofractionation, in a technique known as ablative hypofractionation (Ferini et al., 2021). Another method that can be used to increase dosage includes simultaneous integrated boost (SIB), a technique that applies different doses to different regions (Zhu et al., 2019). Charged particle therapy, such as proton ions or carbon ions, can give high doses of radiation with a sharper division between the targeted area and surrounding areas as there is a steep dose falloff outside the target (Takatori et al., 2014). OARs can be spared through improved beam guidance and targeting using magnetic resonance guided radiation therapy (MRgRT) (Rudra et al., 2019). Dosage escalation is not an option in patients with large tumors or locations too near to OARs or the tumor invading tissues such as the bowel. In these cases, radiosensitizing agents increase the effectiveness of radiotherapy. Some of these compounds include EGFR inhibitors, cell cycle checkpoint inhibitors, DNA-dependent protein kinase inhibitors, ataxia telangiectasia mutated and Rad3-related inhibitors, receptor tyrosine kinase inhibitors, and *KRAS* pathway effector inhibitors (Wolfe and Williams, 2018).

Immunotherapy

The TME also contains factors that suppress immune reactions. These include arginase-I, reactive oxygen species (ROS), and suppressive cytokines (Manji et al., 2017). Further suppression of the immune system occurs due to the recruitment of inhibitory dendritic cells (DCs), Tregs, and cancer-associated macrophages (Manji et al., 2017). As a result, targeting these immunosuppressive factors to reactivate the anticancer immune response or stimulate the anticancer immune response are viable strategies to target pancreatic cancer. Molecules that are targets for immunotherapy include the immune checkpoint inhibitors programmed cell death-1 (PD-1) and cytotoxic T lymphocyte antigen-4 (CTLA-4), both of which can be targeted using antibodies (Graziani et al., 2012). PD-1 inhibitors include nivolumab and pembrolizumab (Eso and Seno, 2020), and the use of these antibodies in conjunction with IL-6 inhibitors increases the number of tumor-infiltrating T lymphocytes (Mace et al., 2018). Pancreatic cancers with a high level of microsatellite instability (MSI) are less sensitive to drugs such as FOLFIRINOX—but are more sensitive to immunotherapies such as the checkpoint protein PD-1 (programmed cell death protein-1) inhibitor, pembrolizumab (Riazy et al., 2015). Selicrelumab is an antibody that acts as a CD40 agonist and can modulate the pancreatic cancer TME, leading to an increased immune response against the tumor (Byrne et al., 2021).

Advancement in Therapeutic Technologies

Research is underway to improve all the treatments discussed above. Advances in surgery mean that more pancreatic cancers that were previously defined as borderline resectable or unresectable can now be redefined as resectable. Some of these advances include improved preoperative imaging, surgical techniques, as well as the use of new neoadjuvant therapies alongside surgery (Kasumova et al., 2018). New combinations of chemotherapeutic agents are being investigated. These include the combination of chemotherapy agents and targeted therapies and are being investigated in various phase III trials. These targeted therapies include STAT3 inhibitors, PARP inhibitors, and EGFR inhibitors. The delivery of chemotherapy drugs is also being improved by packaging them into nanoparticles (Kifle et al., 2021). Circulating tumor DNA (ctDNA) is now being used as an indicator of disease state, and at the same time whole genome and transcriptome sequencing are becoming more prevalent, and new biomarkers for pancreatic cancer are being identified. These can be population and patient specific. This allows for more optimized treatment strategies (Turpin et al., 2020). Vaccines that target cancer cells by containing fragments of telomerase or cytokines are being combined with chemotherapy in phase III clinical trials (Fan et al., 2023).

PALLIATIVE CARE

Due to the high mortality rate of pancreatic cancer, a vital part of managing this disease includes symptom management and palliative care (Ferrell et al., 2021). This includes addressing issues such as abdominal pain, weight loss, and fatigue, as well as the psychological toll the progression and management of the disease can take on the patient, such as depression and social withdrawal. Patients may also develop jaundice as the tumor blocks the bile ducts. Patients with jaundice may need to halt treatment or be hospitalized, which can create more stress and anxiety (Ferrell et al., 2021). Studies have also shown that the quality of life of pancreatic cancer patients is worse than that of individuals suffering from other types of cancer, having worse psychological stress (Bauer et al., 2018). Palliative care requires care provided by a multidisciplinary team to manage these diverse problems. This team should include an oncologist, a nutritionist, a psychologist, and other providers of supportive medicine (Ferrell et al., 2020). Initial treatment, such as surgery and radiation therapy, can improve the situation by delaying disease progression, thereby improving the patient's quality of life. However, recurrence may require further treatment with chemotherapy, which itself may have adverse effects that lower the patient's quality of life, such as nausea and vomiting, fatigue, loss of appetite, diarrhea, and neuropathy. The adverse effects also mean that there are limits to treatment in older patients (Conroy et al., 2011). Apart from jaundice, the obstruction of the pancreatic duct can also cause pancreatic insufficiency, resulting in steatorrhea and protein-calorie malnutrition. The symptoms of these ailments can exacerbate the depression felt by a patient over and above that caused by the initial symptoms of pancreatic cancer progression. These conditions and symptoms need to be managed to minimize their effect on patient survival, which can be compromised through decreased quality of life and feelings of hopelessness (Boulay and Birg, 2016).

It is common for pancreatic cancer patients to take part in early-stage phase I clinical trials once they have exhausted the limited approved chemotherapy treatments that are available. These studies aim to evaluate the safety of new drugs or drug combinations. These studies are intensive, requiring patients to spend extended periods in hospitals as they have blood and biopsy samples taken. This keeps patients away from their homes and loved ones, which may be their main support structures, adding to a decrease in the quality of life (Ferrell et al., 2021)

Surgical intervention to improve the quality of life used to be more common since it was done at the same time as exploratory diagnostic procedures (surgical staging) that identified cancer as being unresectable. These procedures included gastric bypass, hepatico-enteric decompression, and celiac plexus neurolysis for pain relief. This is no longer the case as diagnostic procedures have improved and the need for surgical staging has declined (Beyranvand et al., 2019). Biliary decompression is another palliative surgical intervention that is performed to prevent obstruction of the bile duct and prevent jaundice; however, these procedures are being replaced by using radiology and endoscopy to insert percutaneous or endoluminal stents (Balogun and Atoyebi, 2022). Pain management is very important in palliative care since pain is one of the main reported reasons for a decrease in the quality of life. Opioids are the main method used to manage pain in most pancreatic cancer patients, but this is ineffective in up to a third of patients (Lu et al., 2022). Pain can be treated in these patients through radiation therapy, chemotherapy, and celiac plexus neurolysis. In addition to these treatments, percutaneous neurolytic celiac plexus block can be performed using ethyl alcohol injections (Cornman-Homonoff et al., 2017).

CHALLENGES IN PANCREATIC CANCER MANAGEMENT

Disparities in cancer treatment refer to the unequal distribution of cancer incidence, prevalence, mortality, and survival among different population groups. These disparities can be the result of a population's socioeconomic status, ethnicity, location, and the ability of a group to access healthcare. Rural communities, for instance, generally have lower access to healthcare, as do minorities. At the same time, income and wealth can dictate the ability to access healthcare, with poorer communities having lower access to health care (Zavala et al., 2021). These disparities can be addressed through increased community engagement in these disadvantaged communities and population groups. This engagement results in the active participation of the members of these communities in initiating healthcare initiatives and interventions. Communities will have a better idea of what they need, and by asking their opinion and consulting with them, they can be empowered. This will also ensure that any intervention would be culturally appropriate and accepted by the community (Kale et al., 2023).

Socioeconomic factors can contribute to health disparities since limited finances mean that the community is largely dependent on free or subsidized healthcare as they are also likely to lack health insurance or, in South Africa's case, medical aid. Lower socioeconomic status is also associated with lower access to health care and lower education. As a result, these communities have lower levels of screening and prevention and worse treatment outcomes (Carethers and Doubeni, 2020). Minorities may be the victims of systemic racism and discrimination which prevents them from accessing healthcare. Minorities may also have different cultural beliefs and speak different languages from the majority of the population of a country, and as such the healthcare system may not include sensitivity to these cultural differences or may lack information regarding healthcare and screening programs in a language they can understand (Bhatia et al., 2022). Those living in remote or inaccessible areas only small rural clinics within easy access to these populations. If they need to travel to larger, more well-equipped hospitals with specialized cancer care, it will require transportation, which may cost both time and money (Omotoso et al., 2023)

When it comes to Africa, there are many African-specific barriers to the management of pancreatic cancer. First, there is inadequate funding for African-specific pancreatic cancer research. This means that, in many cases, Africa relies on research done in other parts of the world, which may not reflect the African perspective. African populations may have unique genetic drivers of pancreatic cancer and unique risk factors compared to other regions (Adewole et al., 2014). Second, many communities in Africa lack awareness of pancreatic cancer and lack cancer education in general. This can result in low screening rates and patients only seeking medical attention at advanced stages of the disease. This problem is exacerbated even further by the third factor, which is a lack of facilities capable of performing diagnosis and screening. Fourth, the lack of cancer registries in many African countries makes it difficult to assess the extent of the problem. Fifth, there is a lack of qualified personnel and adequate facilities to treat and care for pancreatic cancer patients across Africa. Finally, the response of many communities to the disease is shaped by myths, misinformation, stigma, and socio-cultural/religious beliefs (World Health Organization, 2022). Technological advances can contribute to resolving these problems. These include the use of social media, broadcast media, and print media to disseminate information and educate the public, increasing awareness (Omotoso et al., 2021).

GLOBAL COLLABORATION IN COMBATING PANCREATIC CANCER

Collaboration between government agencies and non-governmental organizations has developed across multiple countries in Africa and has resulted in the development of measures to increase screening and education programs for various cancers. Various collaborative research projects have been undertaken. These include collaborations between institutes, countries, and even continents. For instance, the EURECCA group was a collaboration between researchers from eight European countries and compared treatments and outcomes. This study showed country and region-specific differences between patients (Groen et al., 2018). Collaborative research projects such as this also demonstrated that it is important to do studies on specific populations as they differ in their molecular makeup and environmental differences. However, all these results should be combined to find universal responses and pathways and identify if results from different regions or centers are comparable (Besselink, 2019). These collaborations also provide the ability to share resources and expertise. These resources include large diverse population types, and these diverse populations allow for research such as providing a large number of diverse subjects for clinical trials. This will make the clinical trials more impactful (Penniston, 2022). Collaborations would also lead to more data sharing and the ability to construct even larger databases with more information both in the number and diversity of samples. This is important for the training of machine-learning algorithms (Obiora et al., 2022). Finally, collaboration will allow for widespread awareness and advocacy for policy changes based on the best available information allowing researchers to have a voice in policy making at an international level (Riley and Su'esu'e, 2023).

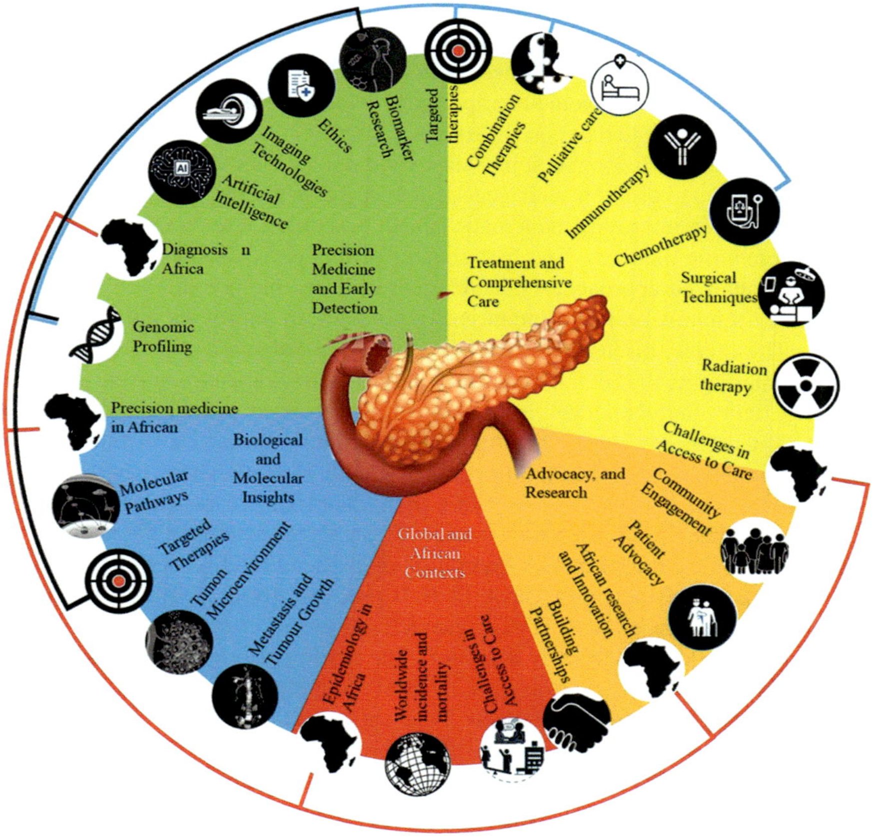

FIGURE 1.4 A summary of the current pancreatic cancer landscape. The four parts of this book discuss pancreatic cancer in an African and global context, the molecular changes driving pancreatic cancer, precision medicine and early detection in pancreatic cancer, treatment and care of pancreatic cancer patients, and finally, new research and patient advocacy. Africa faces unique challenges in all these topics (red line), while the development of precision medicine necessitates the identification of population- and individual-specific molecular patterns (black line). Pancreatic cancer is known to develop resistance to treatments rapidly; as such using specific combinations of treatments and tracking the development of resistance by monitoring molecular changes are important for the successful treatment of the disease (blue line).

CONCLUSION

This book is divided into five parts (Figure 1.4). The first part places pancreatic cancer in a global and African context and focuses on the epidemiological patterns of pancreatic cancer, examining regional disparities, emerging trends, and factors contributing to its prevalence. This will help shed light on the unique challenges faced by African populations, such as limited resources and healthcare infrastructure, as well as challenges around the collection of epidemiological data and cancer surveillance in Africa. This chapter underscores the importance of targeted interventions and tailored strategies for addressing pancreatic cancer in the region. It also discusses the risk factors associated with pancreatic cancer.

The second part provides biological and molecular insights into the development and progression of pancreatic cancer. The first chapter of this part and the third chapter of the book describes the molecular pathways driving pancreatic cancer and explores the key genetic and epigenetic alterations, including mutations in oncogenes. The role of chronic inflammation and its impact on the

TME are examined, highlighting how it fosters a pro-tumorigenic setting. Additionally, the chapter discusses the signaling pathways that are often deregulated in pancreatic cancer. The chapter aims to provide a comprehensive overview of these factors and how this knowledge can be used to pave the way for the development of targeted therapies and personalized treatment strategies.

The third part of this book focuses on the role of early detection and precision medicine in combating pancreatic cancer. Precision medicine has revolutionized cancer treatment by tailoring therapies to individual patients based on their genetic makeup and disease characteristics. While early detection is paramount in improving pancreatic cancer outcomes, it remains challenging due to the lack of reliable screening methods. The first chapter of this part, Chapter 4, focuses on precision medicine and personalized approaches. This chapter explores the potential of precision medicine techniques, such as genomic profiling and targeted therapies, in improving outcomes for pancreatic cancer patients. The chapter discusses the application of personalized approaches in African healthcare settings and highlights the need for equitable access to innovative treatments and technologies. The chapter highlights the challenges in implementing precision medicine in Africa, as well as the opportunities it offers in African healthcare systems, and the ethics of its implementation in this setting. It accomplishes this through the description of case studies of personalized treatment in pancreatic cancer. The second chapter of this part, Chapter 5, covers the strategies used in the diagnosis and early detection of pancreatic cancer. This chapter examines the latest advancements in diagnostic technologies, including the integration of artificial intelligence and machine learning algorithms, to enhance early detection capabilities. It reviews current diagnostic methods and their limitations, as well as advances such as new research in biomarker discovery and application, and the role of imaging technologies in early detection. The chapter also focuses on the implementation of early detection programs in Africa and how, by leveraging these innovative tools, healthcare providers can expedite diagnosis, facilitate timely interventions, and ultimately improve survival rates for pancreatic cancer patients in Africa.

The fourth part of this book discusses the treatment and comprehensive care of pancreatic cancer patients. From surgery and chemotherapy to immunotherapy and targeted therapies, a wide range of treatment options exists for pancreatic cancer patients. At the same time, palliative care plays a crucial role in supporting pancreatic cancer patients by addressing symptoms, managing pain, and improving the quality of life. Despite advancements in diagnosis and treatment, access to quality care remains a significant challenge for pancreatic cancer patients in Africa. The first chapter of this part, Chapter 6, explores both conventional and innovative treatment modalities, highlighting recent advancements in precision medicine, surgery, immunotherapy, as well as chemotherapy and combination therapies in the treatment of pancreatic cancer. This chapter underscores the importance of expanding access to cutting-edge treatments and fostering interdisciplinary collaboration to optimize patient care in African healthcare contexts. It also includes case studies of innovative treatments in African contexts. Chapter 7, the next chapter in this part, explores the principles of palliative care and its integration into pancreatic cancer treatment plans. By emphasizing the importance of holistic care and personalized support, this chapter advocates for the inclusion of palliative care services as an integral component of cancer care delivery in Africa. It details the importance of addressing the psychological and social needs of the patient and includes case studies of palliative care programs in Africa. The following chapter, Chapter 8, examines the barriers to accessing advanced diagnostics, precision treatments, and supportive care services, including issues related to healthcare infrastructure, affordability, insurance coverage, and geographical disparities. The chapter also explores strategies to overcome these obstacles and discusses how stakeholders can work toward creating more equitable healthcare systems that ensure all patients receive timely and comprehensive care. These include the role of telemedicine and mobile health solutions.

The final part of the book, Part 5, covers advocacy, research, and future directions. Pancreatic cancer profoundly impacts patients and their families, often leading to physical, emotional, and financial burdens. By fostering a culture of solidarity and support, communities can play a pivotal role in improving outcomes and quality of life for pancreatic cancer patients across the continent.

Research and innovation are crucial drivers of progress in the fight against pancreatic cancer. By fostering a culture of scientific inquiry and knowledge exchange, Africa can contribute to global efforts to advance pancreatic cancer research and translate discoveries into clinical practice, ultimately improving patient outcomes and survival rates. Collaboration and partnership are essential for addressing the complex challenges of pancreatic cancer in Africa. By leveraging collective expertise, resources, and networks, stakeholders can amplify their impact, accelerate innovation, and advance toward a future where pancreatic cancer is no longer a death sentence but a manageable chronic condition. The first chapter of this part, Chapter 9, amplifies the voices of pancreatic cancer patients in Africa, sharing their experiences, challenges, and resilience in navigating the disease. It highlights the vital role of advocacy organizations, support groups, and community initiatives in raising awareness, reducing stigma, and empowering patients to become advocates for change. It discusses some of the strategies for raising awareness, reducing stigma, and empowering patients and families through education and support. The following chapter, Chapter 10, explores ongoing research initiatives and innovative approaches to pancreatic cancer management in Africa, including efforts to leverage precision medicine, artificial intelligence, and splicing technologies. It gives an overview of current research initiatives and discusses the role of African institutions in global research efforts. It reviews the importance of collaborations with international research bodies and detail the future directions for pancreatic cancer research in Africa. The final chapter in this part, Chapter 11, explores the importance of forging alliances between governments, healthcare organizations, research institutions, industry partners, and advocacy groups to drive progress in pancreatic cancer prevention, diagnosis, treatment, and support services. It discusses strategies for engaging stakeholders, the role of government and policy in supporting cancer care, as well as the creation of a framework for sustainable partnerships. The chapter reviews case studies that demonstrate successful partnerships.

The concluding chapter outlines a roadmap for action to improve pancreatic cancer care in Africa and beyond. It calls for sustained commitment, investment, and advocacy to overcome barriers to access, expand research capacity, and enhance collaboration across sectors. It provides recommendations for policymakers and healthcare providers, as well as discusses strategies for sustained investment in cancer care. The concluding chapter also highlights the importance of patient-centered approaches, while providing a roadmap for the implementation of innovative solutions and a vision for the future of pancreatic cancer care in Africa.

REFERENCES

ABRAHAM, S. C., KLIMSTRA, D. S., WILENTZ, R. E., YEO, C. J., CONLON, K., BRENNAN, M., CAMERON, J. L., WU, T. T. & HRUBAN, R. H. 2002. Solid-pseudopapillary tumors of the pancreas are genetically distinct from pancreatic ductal adenocarcinomas and almost always harbor beta-catenin mutations. *Am J Pathol,* 160, 1361–1369.

ADEWOLE, I., MARTIN, D. N., WILLIAMS, M. J., ADEBAMOWO, C., BHATIA, K., BERLING, C., CASPER, C., ELSHAMY, K., ELZAWAWY, A., LAWLOR, R. T., LEGOOD, R., MBULAITEYE, S. M., ODEDINA, F. T., OLOPADE, O. I., OLOPADE, C. O., PARKIN, D. M., REBBECK, T. R., ROSS, H., SANTINI, L. A., TORODE, J., TRIMBLE, E. L., WILD, C. P., YOUNG, A. M. & KERR, D. J. 2014. Building capacity for sustainable research programmes for cancer in Africa. *Nat Rev Clin Oncol,* 11, 251–259.

AMRI, F., KOULALI, H., JABI, R., ZAZOUR, A., BOUZIANE, M., ISMAILI, Z. & KHARRASSE, G. 2023. Pancreatic cancer: experience from an emerging country in North Africa. *J Cancer Res Clin Oncol,* 149, 14297–14302.

AMUNDADOTTIR, L., KRAFT, P., STOLZENBERG-SOLOMON, R. Z., FUCHS, C. S., PETERSEN, G. M., ARSLAN, A. A., BUENO-DE-MESQUITA, H. B., GROSS, M., HELZLSOUER, K., JACOBS, E. J., LACROIX, A., ZHENG, W., ALBANES, D., BAMLET, W., BERG, C. D., BERRINO, F., BINGHAM, S., BURING, J. E., BRACCI, P. M., CANZIAN, F., CLAVEL-CHAPELON, F., CLIPP, S., COTTERCHIO, M., DE ANDRADE, M., DUELL, E. J., FOX, J. W., JR., GALLINGER, S., GAZIANO, J. M., GIOVANNUCCI, E. L., GOGGINS, M., GONZÁLEZ, C. A., HALLMANS, G.,

HANKINSON, S. E., HASSAN, M., HOLLY, E. A., HUNTER, D. J., HUTCHINSON, A., JACKSON, R., JACOBS, K. B., JENAB, M., KAAKS, R., KLEIN, A. P., KOOPERBERG, C., KURTZ, R. C., LI, D., LYNCH, S. M., MANDELSON, M., MCWILLIAMS, R. R., MENDELSOHN, J. B., MICHAUD, D. S., OLSON, S. H., OVERVAD, K., PATEL, A. V., PEETERS, P. H., RAJKOVIC, A., RIBOLI, E., RISCH, H. A., SHU, X. O., THOMAS, G., TOBIAS, G. S., TRICHOPOULOS, D., VAN DEN EEDEN, S. K., VIRTAMO, J., WACTAWSKI-WENDE, J., WOLPIN, B. M., YU, H., YU, K., ZELENIUCH-JACQUOTTE, A., CHANOCK, S. J., HARTGE, P. & HOOVER, R. N. 2009. Genome-wide association study identifies variants in the ABO locus associated with susceptibility to pancreatic cancer. *Nat Genet,* 41**,** 986–990.

BAILEY, J. M., SWANSON, B. J., HAMADA, T., EGGERS, J. P., SINGH, P. K., CAFFERY, T., OUELLETTE, M. M. & HOLLINGSWORTH, M. A. 2008. Sonic hedgehog promotes desmoplasia in pancreatic cancer. *Clin Cancer Res,* 14**,** 5995–6004.

BAILEY, P., CHANG, D. K., NONES, K., JOHNS, A. L., PATCH, A. M., GINGRAS, M. C., MILLER, D. K., CHRIST, A. N., BRUXNER, T. J., QUINN, M. C., NOURSE, C., MURTAUGH, L. C., HARLIWONG, I., IDRISOGLU, S., MANNING, S., NOURBAKHSH, E., WANI, S., FINK, L., HOLMES, O., CHIN, V., ANDERSON, M. J., KAZAKOFF, S., LEONARD, C., NEWELL, F., WADDELL, N., WOOD, S., XU, Q., WILSON, P. J., CLOONAN, N., KASSAHN, K. S., TAYLOR, D., QUEK, K., ROBERTSON, A., PANTANO, L., MINCARELLI, L., SANCHEZ, L. N., EVERS, L., WU, J., PINESE, M., COWLEY, M. J., JONES, M. D., COLVIN, E. K., NAGRIAL, A. M., HUMPHREY, E. S., CHANTRILL, L. A., MAWSON, A., HUMPHRIS, J., CHOU, A., PAJIC, M., SCARLETT, C. J., PINHO, A. V., GIRY-LATERRIERE, M., ROOMAN, I., SAMRA, J. S., KENCH, J. G., LOVELL, J. A., MERRETT, N. D., TOON, C. W., EPARI, K., NGUYEN, N. Q., BARBOUR, A., ZEPS, N., MORAN-JONES, K., JAMIESON, N. B., GRAHAM, J. S., DUTHIE, F., OIEN, K., HAIR, J., GRÜTZMANN, R., MAITRA, A., IACOBUZIO-DONAHUE, C. A., WOLFGANG, C. L., MORGAN, R. A., LAWLOR, R. T., CORBO, V., BASSI, C., RUSEV, B., CAPELLI, P., SALVIA, R., TORTORA, G., MUKHOPADHYAY, D., PETERSEN, G. M., MUNZY, D. M., FISHER, W. E., KARIM, S. A., ESHLEMAN, J. R., HRUBAN, R. H., PILARSKY, C., MORTON, J. P., SANSOM, O. J., SCARPA, A., MUSGROVE, E. A., BAILEY, U. M., HOFMANN, O., SUTHERLAND, R. L., WHEELER, D. A., GILL, A. J., GIBBS, R. A., PEARSON, J. V., WADDELL, N., et al. 2016. Genomic analyses identify molecular subtypes of pancreatic cancer. *Nature,* 531**,** 47–52.

BALOGUN, O. S. & ATOYEBI, O. A. 2022. Management of malignant obstructive jaundice: defining the relevance of various palliative surgical options in resource-challenged settings: a review article. *J West Afr Coll Surg,* 12**,** 111–119.

BARBEE, C. L., DEMELLO, F. B. & GRAGE, T. B. 1976. Cystadenoma and cystadenocarcinoma of the pancreas. *J Surg Oncol,* 8**,** 1–10.

BAUER, M. R., BRIGHT, E. E., MACDONALD, J. J., CLEARY, E. H., HINES, O. J. & STANTON, A. L. 2018. Quality of life in patients with pancreatic cancer and their caregivers: a systematic review. *Pancreas,* 47**,** 368–375.

BESSELINK, M. 2019. The value of international collaboration in pancreatic cancer research: EURECCA. *Ann Surg Oncol,* 26**,** 705–706.

BEYRANVAND, T., ARYANKHESAL, A. & AGHAEI HASHJIN, A. 2019. Quality improvement in hospitals' surgery-related processes: a systematic review. *Med J Islam Repub Iran,* 33**,** 129.

BHATIA, S., LANDIER, W., PASKETT, E. D., PETERS, K. B., MERRILL, J. K., PHILLIPS, J. & OSAROGIAGBON, R. U. 2022. Rural-urban disparities in cancer outcomes: opportunities for future research. *J Natl Cancer Inst,* 114**,** 940–952.

BIPAT, S., PHOA, S. S., VAN DELDEN, O. M., BOSSUYT, P. M., GOUMA, D. J., LAMÉRIS, J. S. & STOKER, J. 2005. Ultrasonography, computed tomography and magnetic resonance imaging for diagnosis and determining resectability of pancreatic adenocarcinoma: a meta-analysis. *J Comput Assist Tomogr,* 29**,** 438–445.

BLACKFORD, A., PARMIGIANI, G., KENSLER, T. W., WOLFGANG, C., JONES, S., ZHANG, X., PARSONS, D. W., LIN, J. C., LEARY, R. J., ESHLEMAN, J. R., GOGGINS, M., JAFFEE, E. M., IACOBUZIO-DONAHUE, C. A., MAITRA, A., KLEIN, A., CAMERON, J. L., OLINO, K., SCHULICK, R., WINTER, J., VOGELSTEIN, B., VELCULESCU, V. E., KINZLER, K. W. & HRUBAN, R. H. 2009. Genetic mutations associated with cigarette smoking in pancreatic cancer. *Cancer Res,* 69**,** 3681–3688.

BLOOMSTON, M., FRANKEL, W. L., PETROCCA, F., VOLINIA, S., ALDER, H., HAGAN, J. P., LIU, C. G., BHATT, D., TACCIOLI, C. & CROCE, C. M. 2007. MicroRNA expression patterns to differentiate pancreatic adenocarcinoma from normal pancreas and chronic pancreatitis. *Jama,* 297**,** 1901–1908.

BOSETTI, C., LUCENTEFORTE, E., SILVERMAN, D. T., PETERSEN, G., BRACCI, P. M., JI, B. T., NEGRI, E., LI, D., RISCH, H. A., OLSON, S. H., GALLINGER, S., MILLER, A. B., BUENO-DE-MESQUITA, H. B., TALAMINI, R., POLESEL, J., GHADIRIAN, P., BAGHURST, P. A., ZATONSKI, W., FONTHAM, E., BAMLET, W. R., HOLLY, E. A., BERTUCCIO, P., GAO, Y. T., HASSAN, M., YU, H., KURTZ, R. C., COTTERCHIO, M., SU, J., MAISONNEUVE, P., DUELL, E. J., BOFFETTA, P. & LA VECCHIA, C. 2012. Cigarette smoking and pancreatic cancer: an analysis from the International Pancreatic Cancer Case-Control Consortium (Panc4). *Ann Oncol,* 23, 1880–1888.

BOTTA, G. P., HUYNH, T. R., SPIERLING-BAGSIC S. R., AGELIDIS, A., SCHAFFER, R., LIN, R. & SIGAL, D. 2023. Neoadjuvant chemotherapy and radiotherapy outcomes in borderline-resectable and locally-advanced pancreatic cancer patients. *Cancer Med,* 12, 7713–7723.

BOULAY, B. R. & BIRG, A. 2016. Malignant biliary obstruction: from palliation to treatment. *World J Gastrointest Oncol,* 8, 498–508.

BRAY, F., LAVERSANNE, M., SUNG, H., FERLAY, J., SIEGEL, R. L., SOERJOMATARAM, I. & JEMAL, A. 2024. Global cancer statistics 2022: GLOBOCAN estimates of incidence and mortality worldwide for 36 cancers in 185 countries. *CA Cancer J Clin,* 74, 229–263.

BRUNE, K. A., LAU, B., PALMISANO, E., CANTO, M., GOGGINS, M. G., HRUBAN, R. H. & KLEIN, A. P. 2010. Importance of age of onset in pancreatic cancer kindreds. *J Natl Cancer Inst,* 102, 119–126.

BURRIS, H. A., 3RD, MOORE, M. J., ANDERSEN, J., GREEN, M. R., ROTHENBERG, M. L., MODIANO, M. R., CRIPPS, M. C., PORTENOY, R. K., STORNIOLO, A. M., TARASSOFF, P., NELSON, R., DORR, F. A., STEPHENS, C. D. & VON HOFF, D. D. 1997. Improvements in survival and clinical benefit with gemcitabine as first-line therapy for patients with advanced pancreas cancer: a randomized trial. *J Clin Oncol,* 15, 2403–2413.

BYRNE, K. T., BETTS, C. B., MICK, R., SIVAGNANAM, S., BAJOR, D. L., LAHERU, D. A., CHIOREAN, E. G., O'HARA, M. H., LIUDAHL, S. M., NEWCOMB, C., ALANIO, C., FERREIRA, A. P., PARK, B. S., OHTANI, T., HUFFMAN, A. P., VÄYRYNEN, S. A., DIAS COSTA, A., KAISER, J. C., LACROIX, A. M., REDLINGER, C., STERN, M., NOWAK, J. A., WHERRY, E. J., CHEEVER, M. A., WOLPIN, B. M., FURTH, E. E., JAFFEE, E. M., COUSSENS, L. M. & VONDERHEIDE, R. H. 2021. Neoadjuvant selicrelumab, an agonist CD40 antibody, induces changes in the tumor microenvironment in patients with resectable pancreatic cancer. *Clin Cancer Res,* 27, 4574–4586.

CAMERON, J. L., RIALL, T. S., COLEMAN, J. & BELCHER, K. A. 2006. One thousand consecutive pancreaticoduodenectomies. *Ann Surg,* 244, 10–15.

CANCER GENOME ATLAS RESEARCH NETWORK 2017. Integrated genomic characterization of pancreatic ductal adenocarcinoma. *Cancer Cell,* 32, 185–203.e13.

CANTO, M. I., SCHULICK, R. D., KAMEL, I. R., FISHMAN, E. K., TOPAZIAN, M. D. & TAKAHASHI, N. J. G. E. 2010. Screening for familial pancreatic neoplasia: a prospective, multicenter blinded study of EUS, CT, and secretin-MRCP (The NCI-Spore Lustgarten Foundation Cancer of the Pancreas Caps 3 Study). *Gastroint Endoscopy,* 71, AB119.

CARETHERS, J. M. & DOUBENI, C. A. 2020. Causes of socioeconomic disparities in colorectal cancer and intervention framework and strategies. *Gastroenterology,* 158, 354–367.

CARTER, H., SAMAYOA, J., HRUBAN, R. H. & KARCHIN, R. 2010. Prioritization of driver mutations in pancreatic cancer using cancer-specific high-throughput annotation of somatic mutations (CHASM). *Cancer Biol Ther,* 10, 582–587.

CHAN-SENG-YUE, M., KIM, J. C., WILSON, G. W., NG, K., FIGUEROA, E. F., O'KANE, G. M., CONNOR, A. A., DENROCHE, R. E., GRANT, R. C., MCLEOD, J., WILSON, J. M., JANG, G. H., ZHANG, A., DODD, A., LIANG, S. B., BORGIDA, A., CHADWICK, D., KALIMUTHU, S., LUNGU, I., BARTLETT, J. M. S., KRZYZANOWSKI, P. M., SANDHU, V., TIRIAC, H., FROELING, F. E. M., KARASINSKA, J. M., TOPHAM, J. T., RENOUF, D. J., SCHAEFFER, D. F., JONES, S. J. M., MARRA, M. A., LASKIN, J., CHETTY, R., STEIN, L. D., ZOGOPOULOS, G., HAIBE-KAINS, B., CAMPBELL, P. J., TUVESON, D. A., KNOX, J. J., FISCHER, S. E., GALLINGER, S. & NOTTA, F. 2020. Transcription phenotypes of pancreatic cancer are driven by genomic events during tumor evolution. *Nat Genet,* 52, 231–240.

CHEUNG, M. C., YANG, R., BYRNE, M. M., SOLORZANO, C. C., NAKEEB, A. & KONIARIS, L. G. 2010. Are patients of low socioeconomic status receiving suboptimal management for pancreatic adenocarcinoma? *Cancer,* 116, 723–733.

CHINCARINI, M., ZAMBONI, G. A. & POZZI MUCELLI, R. 2018. Major pancreatic resections: normal postoperative findings and complications. *Insights Imaging,* 9, 173–187.

COLLISSON, E. A., BAILEY, P., CHANG, D. K. & BIANKIN, A. V. 2019. Molecular subtypes of pancreatic cancer. *Nat Rev Gastroenterol Hepatol,* 16, 207–220.

COLLISSON, E. A., SADANANDAM, A., OLSON, P., GIBB, W. J., TRUITT, M., GU, S., COOC, J., WEINKLE, J., KIM, G. E., JAKKULA, L., FEILER, H. S., KO, A. H., OLSHEN, A. B., DANENBERG, K. L., TEMPERO, M. A., SPELLMAN, P. T., HANAHAN, D. & GRAY, J. W. 2011. Subtypes of pancreatic ductal adenocarcinoma and their differing responses to therapy. *Nat Med,* 17**,** 500–503.

CONROY, T., DESSEIGNE, F., YCHOU, M., BOUCHÉ, O., GUIMBAUD, R., BÉCOUARN, Y., ADENIS, A., RAOUL, J. L., GOURGOU-BOURGADE, S., DE LA FOUCHARDIÈRE, C., BENNOUNA, J., BACHET, J. B., KHEMISSA-AKOUZ, F., PÉRÉ-VERGÉ, D., DELBALDO, C., ASSENAT, E., CHAUFFERT, B., MICHEL, P., MONTOTO-GRILLOT, C. & DUCREUX, M. 2011. FOLFIRINOX versus gemcitabine for metastatic pancreatic cancer. *N Engl J Med,* 364**,** 1817–1825.

CONROY, T., HAMMEL, P., HEBBAR, M., BEN ABDELGHANI, M., WEI, A. C., RAOUL, J. L., CHONÉ, L., FRANCOIS, E., ARTRU, P., BIAGI, J. J., LECOMTE, T., ASSENAT, E., FAROUX, R., YCHOU, M., VOLET, J., SAUVANET, A., BREYSACHER, G., DI FIORE, F., CRIPPS, C., KAVAN, P., TEXEREAU, P., BOUHIER-LEPORRIER, K., KHEMISSA-AKOUZ, F., LEGOUX, J. L., JUZYNA, B., GOURGOU, S., O'CALLAGHAN, C. J., JOUFFROY-ZELLER, C., RAT, P., MALKA, D., CASTAN, F. & BACHET, J. B. 2018. FOLFIRINOX or gemcitabine as adjuvant therapy for pancreatic cancer. *N Engl J Med,* 379**,** 2395–2406.

CORNMAN-HOMONOFF, J., HOLZWANGER, D. J., LEE, K. S., MADOFF, D. C. & LI, D. 2017. Celiac plexus block and neurolysis in the management of chronic upper abdominal pain. *Semin Intervent Radiol,* 34**,** 376–386.

CUSCHIERI, A., JAKIMOWICZ, J. J. & VAN SPREEUWEL, J. 1996. Laparoscopic distal 70% pancreatectomy and splenectomy for chronic pancreatitis. *Ann Surg,* 223**,** 280–285.

DA SILVA XAVIER, G. 2018. The cells of the Isletsislets of langerhans. *J Clin Med,* 54, 7.

DEL ARCO, H., CHAKIBA-BRUGÈRE, C., SALABERT, L. & BÉCHADE, D. 2019. Adenosquamous carcinoma of the pancreas. *Clin Med Insights Oncol,* 13**,** 1179554919886587.

DIETRICH, C. F., BRADEN, B., HOCKE, M., OTT, M. & IGNEE, A. 2008. Improved characterisation of solitary solid pancreatic tumours using contrast enhanced transabdominal ultrasound. *J Cancer Res Clin Oncol,* 134**,** 635–643.

EL RASSY, E., ASSI, T., EL KARAK, F., GHOSN, M. & KATTAN, J. 2017. Could the combination of Nab-paclitaxel plus gemcitabine salvage metastatic pancreatic adenocarcinoma after folfirinox failure? A single institutional retrospective analysis. *Clin Res Hepatol Gastroenterol,* 41**,** e26–e28.

ELOUBEIDI, M. A., TAMHANE, A., VARADARAJULU, S. & WILCOX, C. M. 2006. Frequency of major complications after EUS-guided FNA of solid pancreatic masses: a prospective evaluation. *Gastrointest Endosc,* 63**,** 622–629.

ESO, Y. & SENO, H. 2020. Current status of treatment with immune checkpoint inhibitors for gastrointestinal, hepatobiliary, and pancreatic cancers. *Therap Adv Gastroenterol,* 13**,** 1756284820948773.

FABREGAS, J. C., RILEY, K. E., BRANT, J. M., GEORGE, T. J., ORAV, E. J. & LAM, M. B. 2022. Association of social determinants of health with late diagnosis and survival of patients with pancreatic cancer. *J Gastrointest Oncol,* 13**,** 1204–1214.

FAN, T., ZHANG, M., YANG, J., ZHU, Z., CAO, W. & DONG, C. 2023. Therapeutic cancer vaccines: advancements, challenges, and prospects. *Signal Transduct Target Therapy,* 8**,** 450.

FERINI, G., MOLINO, L., BOTTALICO, L., DE LUCIA, P. & GAROFALO, F. 2021. A small case series about safety and effectiveness of a hypofractionated electron beam radiotherapy schedule in five fractions for facial non melanoma skin cancer among frail and elderly patients. *Rep Pract Oncol Radiother,* 26**,** 66–72.

FERRELL, B., CHUNG, V., HUGHES, M. T., KOCZYWAS, M., AZAD, N. S., RUEL, N. H., KNIGHT, L., COOPER, R. S. & SMITH, T. J. 2021. A palliative care intervention for patients on phase 1 studies. *J Palliat Med,* 24**,** 846–856.

FERRELL, B. R., CHUNG, V., KOCZYWAS, M. & SMITH, T. J. 2020. Dissemination and implementation of palliative care in oncology. *J Clin Oncol,* 38**,** 995–1001.

FISCHER, C. G. & WOOD, L. D. 2018. From somatic mutation to early detection: insights from molecular characterization of pancreatic cancer precursor lesions. *J Pathol,* 246**,** 395–404.

FOSTER, D. S., JONES, R. E., RANSOM, R. C., LONGAKER, M. T. & NORTON, J. A. 2018. The evolving relationship of wound healing and tumor stroma. *JCI Insight,* e99911, 3.

GANDOLFI, L., TORRESAN, F., SOLMI, L. & PUCCETTI, A. 2003. The role of ultrasound in biliary and pancreatic diseases. *Eur J Ultrasound,* 16**,** 141–59.

GANGI, S., FLETCHER, J. G., NATHAN, M. A., CHRISTENSEN, J. A., HARMSEN, W. S., CROWNHART, B. S. & CHARI, S. T. 2004. Time interval between abnormalities seen on CT and the clinical diagnosis of pancreatic cancer: retrospective review of CT scans obtained before diagnosis. *AJR Am J Roentgenol,* 182**,** 897–903.

GENKINGER, J. M., SPIEGELMAN, D., ANDERSON, K. E., BERGKVIST, L., BERNSTEIN, L., VAN DEN BRANDT, P. A., ENGLISH, D. R., FREUDENHEIM, J. L., FUCHS, C. S., GILES, G. G., GIOVANNUCCI, E., HANKINSON, S. E., HORN-ROSS, P. L., LEITZMANN, M., MÄNNISTÖ, S., MARSHALL, J. R., MCCULLOUGH, M. L., MILLER, A. B., REDING, D. J., ROBIEN, K., ROHAN, T. E., SCHATZKIN, A., STEVENS, V. L., STOLZENBERG-SOLOMON, R. Z., VERHAGE, B. A., WOLK, A., ZIEGLER, R. G. & SMITH-WARNER, S. A. 2009. Alcohol intake and pancreatic cancer risk: a pooled analysis of fourteen cohort studies. *Cancer Epidemiol Biomarkers Prev,* **18,** 765–76.

GHOSN, M., IBRAHIM, T., ASSI, T., EL RASSY, E., KOURIE, H. R. & KATTAN, J. 2016. Dilemma of first line regimens in metastatic pancreatic adenocarcinoma. *World J Gastroenterol,* **22,** 10124–10130.

GILL, S., KO, Y. J., CRIPPS, C., BEAUDOIN, A., DHESY-THIND, S., ZULFIQAR, M., ZALEWSKI, P., DO, T., CANO, P., LAM, W. Y. H., DOWDEN, S., GRASSIN, H., STEWART, J. & MOORE, M. 2016. PANCREOX: a randomized phase III study of fluorouracil/leucovorin with or without oxaliplatin for second-line advanced pancreatic cancer in patients who have received gemcitabine-based chemotherapy. *J Clin Oncol,* **34,** 3914–3920.

GOLAN, T., HAMMEL, P., RENI, M., VAN CUTSEM, E., MACARULLA, T., HALL, M. J., PARK, J. O., HOCHHAUSER, D., ARNOLD, D., OH, D. Y., REINACHER-SCHICK, A., TORTORA, G., ALGÜL, H., O'REILLY, E. M., MCGUINNESS, D., CUI, K. Y., SCHLIENGER, K., LOCKER, G. Y. & KINDLER, H. L. 2019. Maintenance olaparib for germline BRCA-mutated metastatic pancreatic cancer. *N Engl J Med,* **381,** 317–327.

GOURGOU-BOURGADE, S., BASCOUL-MOLLEVI, C., DESSEIGNE, F., YCHOU, M., BOUCHÉ, O., GUIMBAUD, R., BÉCOUARN, Y., ADENIS, A., RAOUL, J. L., BOIGE, V., BÉRILLE, J. & CONROY, T. 2013. Impact of FOLFIRINOX compared with gemcitabine on quality of life in patients with metastatic pancreatic cancer: results from the PRODIGE 4/ACCORD 11 randomized trial. *J Clin Oncol,* **31,** 23–29.

GRAZIANI, G., TENTORI, L. & NAVARRA, P. 2012. Ipilimumab: a novel immunostimulatory monoclonal antibody for the treatment of cancer. *Pharmacol Res,* **65,** 9–22.

GRENACHER, L., KLAUSS, M., DUKIC, L., DELORME, S., KNAEBEL, H. P., DÜX, M., KAUCZOR, H. U., BÜCHLER, M. W., KAUFFMANN, G. W. & RICHTER, G. M. 2004. [Diagnosis and staging of pancreatic carcinoma: MRI versus multislice-CT -- a prospective study]. *Rofo,* **176,** 1624–1633.

GROEN, J. V., SIBINGA MULDER, B. G., VAN EYCKEN, E., VALERIANOVA, Z., BORRAS, J. M., VAN DER GEEST, L. G. M., CAPRETTI, G., SCHLESINGER-RAAB, A., PRIMIC-ZAKELJ, M., RYZHOV, A., VAN DE VELDE, C. J. H., BONSING, B. A., BASTIAANNET, E. & MIEOG, J. S. D. 2018. Differences in treatment and outcome of pancreatic adenocarcinoma stage I and II in the EURECCA Pancreas Consortium. *Ann Surg Oncol,* **25,** 3492–3501.

GROVER, S., STOFFEL, E. M., BUSSONE, L., TSCHOEGL, E. & SYNGAL, S. 2004. Physician assessment of family cancer history and referral for genetic evaluation in colorectal cancer patients. *Clin Gastroenterol Hepatol,* **2,** 813–819.

HAMMEL, P., HUGUET, F., VAN LAETHEM, J. L., GOLDSTEIN, D., GLIMELIUS, B., ARTRU, P., BORBATH, I., BOUCHÉ, O., SHANNON, J., ANDRÉ, T., MINEUR, L., CHIBAUDEL, B., BONNETAIN, F. & LOUVET, C. 2016. Effect of chemoradiotherapy vs chemotherapy on survival in patients with locally advanced pancreatic cancer controlled after 4 months of gemcitabine with or without erlotinib: the LAP07 randomized clinical trial. *Jama,* **315,** 1844–1853.

HARSHA, H. C., KANDASAMY, K., RANGANATHAN, P., RANI, S., RAMABADRAN, S., GOLLAPUDI, S., BALAKRISHNAN, L., DWIVEDI, S. B., TELIKICHERLA, D., SELVAN, L. D., GOEL, R., MATHIVANAN, S., MARIMUTHU, A., KASHYAP, M., VIZZA, R. F., MAYER, R. J., DECAPRIO, J. A., SRIVASTAVA, S., HANASH, S. M., HRUBAN, R. H. & PANDEY, A. 2009. A compendium of potential biomarkers of pancreatic cancer. *PLoS Med,* **6,** e1000046.

HEINEMANN, V., HAAS, M. & BOECK, S. 2013. Neoadjuvant treatment of borderline resectable and non-resectable pancreatic cancer. *Ann Oncol,* **24,** 2484–2492.

HOFMANN, M. H., GMACHL, M., RAMHARTER, J., SAVARESE, F., GERLACH, D., MARSZALEK, J. R., SANDERSON, M. P., KESSLER, D., TRAPANI, F., ARNHOF, H., RUMPEL, K., BOTESTEANU, D. A., ETTMAYER, P., GERSTBERGER, T., KOFINK, C., WUNBERG, T., ZOEPHEL, A., FU, S. C., TEH, J. L., BÖTTCHER, J., POTOTSCHNIG, N., SCHACHINGER, F., SCHIPANY, K., LIEB, S., VELLANO, C. P., O'CONNELL, J. C., MENDES, R. L., MOLL, J., PETRONCZKI, M., HEFFERNAN, T. P., PEARSON, M., MCCONNELL, D. B. & KRAUT, N. 2021. BI-3406, a potent and selective SOS1-KRAS interaction inhibitor, is effective in KRAS-driven cancers through combined MEK inhibition. *Cancer Discov,* **11,** 142–157.

HONG, S. M., HEAPHY, C. M., SHI, C., EO, S. H., CHO, H., MEEKER, A. K., ESHLEMAN, J. R., HRUBAN, R. H. & GOGGINS, M. 2011. Telomeres are shortened in acinar-to-ductal metaplasia lesions associated with pancreatic intraepithelial neoplasia but not in isolated acinar-to-ductal metaplasias. *Mod Pathol,* 24, 256–266.

HOORENS, A., PRENZEL, K., LEMOINE, N. R. & KLÖPPEL, G. 1998. Undifferentiated carcinoma of the pancreas: analysis of intermediate filament profile and Ki-ras mutations provides evidence of a ductal origin. *J Pathol,* 185, 53–60.

HRUBAN, R. H., ADSAY, N. V., ALBORES-SAAVEDRA, J., COMPTON, C., GARRETT, E. S., GOODMAN, S. N., KERN, S. E., KLIMSTRA, D. S., KLÖPPEL, G., LONGNECKER, D. S., LÜTTGES, J. & OFFERHAUS, G. J. 2001. Pancreatic intraepithelial neoplasia: a new nomenclature and classification system for pancreatic duct lesions. *Am J Surg Pathol,* 25, 579–586.

HRUBAN, R. H., MAITRA, A. & GOGGINS, M. 2008. Update on pancreatic intraepithelial neoplasia. *Int J Clin Exp Pathol,* 1, 306–316.

HU, J. X., ZHAO, C. F., CHEN, W. B., LIU, Q. C., LI, Q. W., LIN, Y. Y. & GAO, F. 2021. Pancreatic cancer: a review of epidemiology, trend, and risk factors. *World J Gastroenterol,* 27, 4298–4321.

HUR, C., TRAMONTANO, A. C., DOWLING, E. C., BROOKS, G. A., JEON, A., BRUGGE, W. R., GAZELLE, G. S., KONG, C. Y. & PANDHARIPANDE, P. V. 2016. Early pancreatic ductal adenocarcinoma survival is dependent on size: positive implications for future targeted screening. *Pancreas,* 45, 1062–1066.

IACOBUZIO-DONAHUE, C. A., KLIMSTRA, D. S., ADSAY, N. V., WILENTZ, R. E., ARGANI, P., SOHN, T. A., YEO, C. J., CAMERON, J. L., KERN, S. E. & HRUBAN, R. H. 2000. Dpc-4 protein is expressed in virtually all human intraductal papillary mucinous neoplasms of the pancreas: comparison with conventional ductal adenocarcinomas. *Am J Pathol,* 157, 755–761.

JACOBETZ, M. A., CHAN, D. S., NEESSE, A., BAPIRO, T. E., COOK, N., FRESE, K. K., FEIG, C., NAKAGAWA, T., CALDWELL, M. E., ZECCHINI, H. I., LOLKEMA, M. P., JIANG, P., KULTTI, A., THOMPSON, C. B., MANEVAL, D. C., JODRELL, D. I., FROST, G. I., SHEPARD, H. M., SKEPPER, J. N. & TUVESON, D. A. 2013. Hyaluronan impairs vascular function and drug delivery in a mouse model of pancreatic cancer. *Gut,* 62, 112–120.

JANES, M. R., ZHANG, J., LI, L. S., HANSEN, R., PETERS, U., GUO, X., CHEN, Y., BABBAR, A., FIRDAUS, S. J., DARJANIA, L., FENG, J., CHEN, J. H., LI, S., LI, S., LONG, Y. O., THACH, C., LIU, Y., ZARIEH, A., ELY, T., KUCHARSKI, J. M., KESSLER, L. V., WU, T., YU, K., WANG, Y., YAO, Y., DENG, X., ZARRINKAR, P. P., BREHMER, D., DHANAK, D., LORENZI, M. V., HU-LOWE, D., PATRICELLI, M. P., REN, P. & LIU, Y. 2018. Targeting KRAS mutant cancers with a covalent G12C-specific inhibitor. *Cell,* 172, 578–589.e17.

JOHANSEN, D., STOCKS, T., JONSSON, H., LINDKVIST, B., BJÖRGE, T., CONCIN, H., ALMQUIST, M., HÄGGSTRÖM, C., ENGELAND, A., ULMER, H., HALLMANS, G., SELMER, R., NAGEL, G., TRETLI, S., STATTIN, P. & MANJER, J. 2010. Metabolic factors and the risk of pancreatic cancer: a prospective analysis of almost 580,000 men and women in the Metabolic Syndrome and Cancer Project. *Cancer Epidemiol Biomarkers Prev,* 19, 2307–2317.

JONES, L. E., HUMPHREYS, M. J., CAMPBELL, F., NEOPTOLEMOS, J. P. & BOYD, M. T. 2004. Comprehensive analysis of matrix metalloproteinase and tissue inhibitor expression in pancreatic cancer: increased expression of matrix metalloproteinase-7 predicts poor survival. *Clin Cancer Res,* 10, 2832–2845.

JONES, S., ZHANG, X., PARSONS, D. W., LIN, J. C., LEARY, R. J., ANGENENDT, P., MANKOO, P., CARTER, H., KAMIYAMA, H., JIMENO, A., HONG, S. M., FU, B., LIN, M. T., CALHOUN, E. S., KAMIYAMA, M., WALTER, K., NIKOLSKAYA, T., NIKOLSKY, Y., HARTIGAN, J., SMITH, D. R., HIDALGO, M., LEACH, S. D., KLEIN, A. P., JAFFEE, E. M., GOGGINS, M., MAITRA, A., IACOBUZIO-DONAHUE, C., ESHLEMAN, J. R., KERN, S. E., HRUBAN, R. H., KARCHIN, R., PAPADOPOULOS, N., PARMIGIANI, G., VOGELSTEIN, B., VELCULESCU, V. E. & KINZLER, K. W. 2008. Core signaling pathways in human pancreatic cancers revealed by global genomic analyses. *Science,* 321, 1801–1806.

KALE, S., HIRANI, S., VARDHAN, S., MISHRA, A., GHODE, D. B., PRASAD, R. & WANJARI, M. 2023. Addressing cancer disparities through community engagement: lessons and best practices. *Cureus,* 15, e43445.

KARPIŃSKA, M. & CZAUDERNA, M. 2022. Pancreas-its functions, disorders, and physiological impact on the mammals' organism. *Front Physiol,* 13, 807632.

KASUMOVA, G. G., CONWAY, W. C. & TSENG, J. F. 2018. The role of venous and arterial resection in pancreatic cancer surgery. *Ann Surg Oncol,* 25, 51–58.

KATZ, M. H., SHI, Q., AHMAD, S. A., HERMAN, J. M., MARSH RDE, W., COLLISSON, E., SCHWARTZ, L., FRANKEL, W., MARTIN, R., CONWAY, W., TRUTY, M., KINDLER, H., LOWY, A. M., BEKAII-SAAB, T., PHILIP, P., TALAMONTI, M., CARDIN, D., LOCONTE, N., SHEN, P., HOFFMAN, J. P. & VENOOK, A. P. 2016. Preoperative modified FOLFIRINOX treatment followed by capecitabine-based chemoradiation for borderline resectable pancreatic cancer: Alliance for Clinical Trials in Oncology Trial A021101. *JAMA Surg,* 151, e161137.

KIFLE, Z. D., TADELE, M., ALEMU, E., GEDAMU, T. & AYELE, A. G. 2021. A recent development of new therapeutic agents and novel drug targets for cancer treatment. *SAGE Open Med,* 9, 20503121211067083.

KIM, J. Y. & HONG, S. M. 2018. Precursor lesions of pancreatic cancer. *Oncol Res Treat,* 41, 603–610.

KIM, S. K. & MELTON, D. A. 1998. Pancreas development is promoted by cyclopamine, a hedgehog signaling inhibitor. *Proc Natl Acad Sci U S A,* 95, 13036–41.

KING, J. C., NG, T. T., WHITE, S. C., CORTINA, G., REBER, H. A. & HINES, O. J. 2009. Pancreatic serous cystadenocarcinoma: a case report and review of the literature. *J Gastrointest Surg,* 13, 1864–8.

KLAPMAN, J. & MALAFA, M. P. 2008. Early detection of pancreatic cancer: why, who, and how to screen. *Cancer Control,* 15, 280–287.

KLEIN, A. P., BRUNE, K. A., PETERSEN, G. M., GOGGINS, M., TERSMETTE, A. C., OFFERHAUS, G. J., GRIFFIN, C., CAMERON, J. L., YEO, C. J., KERN, S. & HRUBAN, R. H. 2004. Prospective risk of pancreatic cancer in familial pancreatic cancer kindreds. *Cancer Res,* 64, 2634–2638.

LEE, T., TENG, T. Z. J. & SHELAT, V. G. 2020. Carbohydrate antigen 19-9 - tumor marker: past, present, and future. *World J Gastrointest Surg,* 12, 468–490.

LEE, T. Y., KIM, M. H., PARK, D. H., SEO, D. W., LEE, S. K., KIM, J. S. & LEE, K. T. 2009. Utility of 18F-FDG PET/CT for differentiation of autoimmune pancreatitis with atypical pancreatic imaging findings from pancreatic cancer. *AJR Am J Roentgenol,* 193, 343–348.

LI, A., OMURA, N., HONG, S. M., VINCENT, A., WALTER, K., GRIFFITH, M., BORGES, M. & GOGGINS, M. 2010. Pancreatic cancers epigenetically silence SIP1 and hypomethylate and overexpress miR-200a/200b in association with elevated circulating miR-200a and miR-200b levels. *Cancer Res,* 70, 5226–5237.

LI, M. & ZHAO, Z. W. 2012. Clinical implications of mismatched repair gene promoter methylation in pancreatic cancer. *Med Oncol,* 29, 970–976.

LIU, F. 2001. SMAD4/DPC4 and pancreatic cancer survival. Commentary re: M. Tascilar et al., The SMAD4 protein and prognosis of pancreatic ductal adenocarcinoma. Clin. Cancer Res., 7: 4115-4121, 2001. *Clin Cancer Res,* 7, 3853–3856.

LIU, J., LEE, P., MCGEE, H. M., CHUNG, V., MELSTROM, L., SINGH, G., RAOOF, M., AMINI, A., CHEN, Y. J. & WILLIAMS, T. M. 2022. Advances in radiation oncology for pancreatic cancer: an updated review. *Cancers (Basel),* 5725, 14.

LIU, R., LIU, Q., ZHAO, Z. M., TAN, X. L., GAO, Y. X. & ZHAO, G. D. 2017. Robotic versus laparoscopic distal pancreatectomy: a propensity score-matched study. *J Surg Oncol,* 116, 461–469.

LU, Z., ZHANG, N., GIORDANO, S. H. & ZHAO, H. 2022. Opioid use and associated factors among pancreatic cancer patients diagnosed between 2007 and 2015. *Cancer Med,* 11, 2296–2307.

LUO, G., JIN, K., DENG, S., CHENG, H., FAN, Z., GONG, Y., QIAN, Y., HUANG, Q., NI, Q., LIU, C. & YU, X. 2021. Roles of CA19-9 in pancreatic cancer: biomarker, predictor and promoter. *Biochim Biophys Acta Rev Cancer,* 1875, 188409.

MACE, T. A., SHAKYA, R., PITARRESI, J. R., SWANSON, B., MCQUINN, C. W., LOFTUS, S., NORDQUIST, E., CRUZ-MONSERRATE, Z., YU, L., YOUNG, G., ZHONG, X., ZIMMERS, T. A., OSTROWSKI, M. C., LUDWIG, T., BLOOMSTON, M., BEKAII-SAAB, T. & LESINSKI, G. B. 2018. IL-6 and PD-L1 antibody blockade combination therapy reduces tumour progression in murine models of pancreatic cancer. *Gut,* 67, 320–332.

MAEMURA, K., TAKAO, S., SHINCHI, H., NOMA, H., MATAKI, Y., KURAHARA, H., JINNOUCHI, S. & AIKOU, T. 2006. Role of positron emission tomography in decisions on treatment strategies for pancreatic cancer. *J Hepatobiliary Pancreat Surg,* 13, 435–441.

MALEY, W. R. & YEO, C. J. 2017. Vascular resections during the whipple procedure. *Adv Surg,* 51, 41–63.

MANIRAKIZA, A. V. C., RUBAGUMYA, F. & NGENDAHAYO, L. 2020. Burundi cancer care needs: a call to action. *Oncologist,* 25, 1055–1059.

MANJI, G. A., OLIVE, K. P., SAENGER, Y. M. & OBERSTEIN, P. 2017. Current and emerging therapies in metastatic pancreatic cancer. *Clin Cancer Res,* 23, 1670–1678.

MAPHUMULO, W. T. & BHENGU, B. R. 2019. Challenges of quality improvement in the healthcare of South Africa post-apartheid: a critical review. *Curationis,* 42, e1–e9.

MENG, Q., SHI, S., LIANG, C., LIANG, D., XU, W., JI, S., ZHANG, B., NI, Q., XU, J. & YU, X. 2017. Diagnostic and prognostic value of carcinoembryonic antigen in pancreatic cancer: a systematic review and meta-analysis. *Onco Targets Ther,* 10**,** 4591–4598.

MOORE, M. J., GOLDSTEIN, D., HAMM, J., FIGER, A., HECHT, J. R., GALLINGER, S., AU, H. J., MURAWA, P., WALDE, D., WOLFF, R. A., CAMPOS, D., LIM, R., DING, K., CLARK, G., VOSKOGLOU-NOMIKOS, T., PTASYNSKI, M. & PARULEKAR, W. 2007. Erlotinib plus gemcitabine compared with gemcitabine alone in patients with advanced pancreatic cancer: a phase III trial of the National Cancer Institute of Canada Clinical Trials Group. *J Clin Oncol,* 25**,** 1960–1966.

MOORE, M. J., HAMM, J., DANCEY, J., EISENBERG, P. D., DAGENAIS, M., FIELDS, A., HAGAN, K., GREENBERG, B., COLWELL, B., ZEE, B., TU, D., OTTAWAY, J., HUMPHREY, R. & SEYMOUR, L. 2003. Comparison of gemcitabine versus the matrix metalloproteinase inhibitor BAY 12-9566 in patients with advanced or metastatic adenocarcinoma of the pancreas: a phase III trial of the National Cancer Institute of Canada Clinical Trials Group. *J Clin Oncol,* 21**,** 3296–3302.

NATIONAL CANCER REGISTRY. 2024. *Summary statistics of cancer diagnosed histologically in South Africa in 2022*. Johannesburg: National Institute for Occupational Health (NIOH) Available: https://www.nicd.ac.za/wp-content/uploads/2024/04/NCR_ASR_tables_2022_final.pdf [Accessed August 4, 2024].

NEJADGHADERI, S. A., KOLAHI, A. A., NOORI, M., SULLMAN, M. J. M. & SAFIRI, S. 2023. The burden of pancreatic cancer and its attributable risk factors in the Middle East and North Africa region, 1990-2019. *J Gastroenterol Hepatol,* 38**,** 1535–1545.

NEOPTOLEMOS, J. P., PALMER, D. H., GHANEH, P., PSARELLI, E. E., VALLE, J. W., HALLORAN, C. M., FALUYI, O., O'REILLY, D. A., CUNNINGHAM, D., WADSLEY, J., DARBY, S., MEYER, T., GILLMORE, R., ANTHONEY, A., LIND, P., GLIMELIUS, B., FALK, S., IZBICKI, J. R., MIDDLETON, G. W., CUMMINS, S., ROSS, P. J., WASAN, H., MCDONALD, A., CROSBY, T., MA, Y. T., PATEL, K., SHERRIFF, D., SOOMAL, R., BORG, D., SOTHI, S., HAMMEL, P., HACKERT, T., JACKSON, R. & BÜCHLER, M. W. 2017. Comparison of adjuvant gemcitabine and capecitabine with gemcitabine monotherapy in patients with resected pancreatic cancer (ESPAC-4): a multicentre, open-label, randomised, phase 3 trial. *Lancet,* 389**,** 1011–1024.

NHLEKO, M. L., EDOKA, I. & MUSENGE, E. 2023. Pancreatic cancer mortality in South Africa: a case-control study. *S Afr Med J,* 114**,** 27–32.

O'REILLY, E. M., LEE, J. W., ZALUPSKI, M., CAPANU, M., PARK, J., GOLAN, T., TAHOVER, E., LOWERY, M. A., CHOU, J. F., SAHAI, V., BRENNER, R., KINDLER, H. L., YU, K. H., ZERVOUDAKIS, A., VEMURI, S., STADLER, Z. K., DO, R. K. G., DHANI, N., CHEN, A. P. & KELSEN, D. P. 2020. Randomized, multicenter, phase II trial of gemcitabine and cisplatin with or without veliparib in patients with pancreas adenocarcinoma and a germline BRCA/PALB2 mutation. *J Clin Oncol,* 38**,** 1378–1388.

OBIORA, O. L., OLIVIER, B., SHEAD, D. A. & WITHERS, A. 2022. Data sharing practices of health researchers in Africa: a scoping review protocol. *JBI Evid Synth,* 20**,** 681–688.

OHWADA, S., OGAWA, T., TANAHASHI, Y., NAKAMURA, S., TAKEYOSHI, I., OHYA, T., IKEYA, T., KAWASHIMA, K., KAWASHIMA, Y. & MORISHITA, Y. 1998. Fibrin glue sandwich prevents pancreatic fistula following distal pancreatectomy. *World J Surg,* 22**,** 494–498.

OMOTOSO, O., TEIBO, J. O., ATIBA, F. A., OLADIMEJI, T., PAIMO, O. K., ATAYA, F. S., BATIHA, G. E. & ALEXIOU, A. 2023. Addressing cancer care inequities in sub-Saharan Africa: current challenges and proposed solutions. *Int J Equity Health,* 22**,** 189.

OMOTOSO, O. E., MATARIEK, G., OMOTOSO, E. F., MATAREEK, A., ABDUL-RAFEE, G. M., MALAKAR, S. & CHUTIA, N. 2021. Practice of breast self-examination and knowledge of breast and cervical cancer screening. *SciMed J,* 3**,** 219–229.

OMURA, N. & GOGGINS, M. 2009. Epigenetics and epigenetic alterations in pancreatic cancer. *Int J Clin Exp Pathol,* 2**,** 310–326.

OYAMA, H., TADA, M., TAKAGI, K., TATEISHI, K., HAMADA, T., NAKAI, Y., HAKUTA, R., IJICHI, H., ISHIGAKI, K., KANAI, S., KOGURE, H., MIZUNO, S., SAITO, K., SAITO, T., SATO, T., SUZUKI, T., TAKAHARA, N., MORISHITA, Y., ARITA, J., HASEGAWA, K., TANAKA, M., FUKAYAMA, M. & KOIKE, K. 2020. Long-term risk of malignancy in branch-duct intraductal papillary mucinous neoplasms. *Gastroenterology,* 158**,** 226–237.e5.

OZAWA, F., FRIESS, H., KÜNZLI, B., SHRIKHANDE, S. V., OTANI, T., MAKUUCHI, M. & BÜCHLER, M. W. 2001. Treatment of pancreatic cancer: the role of surgery. *Dig Dis,* 19**,** 47–56.

PAPAVASILIOU, P., HOFFMAN, J. P., COHEN, S. J., MEYER, J. E., WATSON, J. C. & CHUN, Y. S. 2014. Impact of preoperative therapy on patterns of recurrence in pancreatic cancer. *HPB,* 16**,** 34–39.

PENNISTON, K. L. 2022. Academic research collaboration. In: NAKADA, S. Y. & PATEL, S. R. (eds.) *Navigating organized urology: a practical guide* (pp. 69–88). Cham: Springer International Publishing.
PETERSEN, G. M., DE ANDRADE, M., GOGGINS, M., HRUBAN, R. H., BONDY, M., KORCZAK, J. F., GALLINGER, S., LYNCH, H. T., SYNGAL, S., RABE, K. G., SEMINARA, D. & KLEIN, A. P. 2006. Pancreatic cancer genetic epidemiology consortium. *Cancer Epidemiol Biomarkers Prev,* 15, 704–710.
PROVENZANO, P. P., CUEVAS, C., CHANG, A. E., GOEL, V. K., VON HOFF, D. D. & HINGORANI, S. R. 2012. Enzymatic targeting of the stroma ablates physical barriers to treatment of pancreatic ductal adenocarcinoma. *Cancer Cell,* 21, 418–429.
RAIMONDI, S., MAISONNEUVE, P. & LOWENFELS, A. B. 2009. Epidemiology of pancreatic cancer: an overview. *Nat Rev Gastroenterol Hepatol,* 6, 699–708.
REBOURS, V., BOUTRON-RUAULT, M. C., SCHNEE, M., FÉREC, C., MAIRE, F., HAMMEL, P., RUSZNIEWSKI, P. & LÉVY, P. 2008. Risk of pancreatic adenocarcinoma in patients with hereditary pancreatitis: a national exhaustive series. *Am J Gastroenterol,* 103, 111–119.
REZAEI-DALOUEI, H., SEILANIAN-TOOSI, F., NEKOOEI, S., DAVACHI, B., AZARIAN, A. & ABBASI, B. 2018. Comparing the image quality of tissue harmonic and conventional B-mode ultrasound of kidney in over-obese individuals. *Electron Physician,* 10, 7095–7100.
RIAZY, M., KALLOGER, S. E., SHEFFIELD, B. S., PEIXOTO, R. D., LI-CHANG, H. H., SCUDAMORE, C. H., RENOUF, D. J. & SCHAEFFER, D. F. 2015. Mismatch repair status may predict response to adjuvant chemotherapy in resectable pancreatic ductal adenocarcinoma. *Mod Pathol,* 28, 1383–1389.
RIBATTI, D. & PEZZELLA, F. 2021. Overview on the different patterns of tumor vascularization. *Cells,* 639, 10.
RILEY, L. & SU'ESU'E, A. 2023. Community-focused policy advocacy: evaluating Hawai'i's historical trauma legislation. *Hawaii J Health Soc Welf,* 82, 44–50.
ROGGIN, K. K., CHENNAT, J., OTO, A., NOFFSINGER, A., BRIGGS, A. & MATTHEWS, J. B. 2010. Pancreatic cystic neoplasm. *Curr Probl Surg,* 47, 459–510.
ROZENBLUM, E., SCHUTTE, M., GOGGINS, M., HAHN, S. A., PANZER, S., ZAHURAK, M., GOODMAN, S. N., SOHN, T. A., HRUBAN, R. H., YEO, C. J. & KERN, S. E. 1997. Tumor-suppressive pathways in pancreatic carcinoma. *Cancer Res,* 57, 1731–1734.
RUDRA, S., JIANG, N., ROSENBERG, S. A., OLSEN, J. R., ROACH, M. C., WAN, L., PORTELANCE, L., MELLON, E. A., BRUYNZEEL, A., LAGERWAARD, F., BASSETTI, M. F., PARIKH, P. J. & LEE, P. P. 2019. Using adaptive magnetic resonance image-guided radiation therapy for treatment of inoperable pancreatic cancer. *Cancer Med,* 8, 2123–2132.
SAIF, M. W. 2007. Pancreatoblastoma. *Jop,* 8, 55–63.
SCHNEIDERHAN, W., DIAZ, F., FUNDEL, M., ZHOU, S., SIECH, M., HASEL, C., MÖLLER, P., GSCHWEND, J. E., SEUFFERLEIN, T., GRESS, T., ADLER, G. & BACHEM, M. G. 2007. Pancreatic stellate cells are an important source of MMP-2 in human pancreatic cancer and accelerate tumor progression in a murine xenograft model and CAM assay. *J Cell Sci,* 120, 512–519.
SCHÖNLEBEN, F., QIU, W., CIAU, N. T., HO, D. J., LI, X., ALLENDORF, J. D., REMOTTI, H. E. & SU, G. H. 2006. PIK3CA mutations in intraductal papillary mucinous neoplasm/carcinoma of the pancreas. *Clin Cancer Res,* 12, 3851–3855.
SCHUTTE, M., HRUBAN, R. H., GERADTS, J., MAYNARD, R., HILGERS, W., RABINDRAN, S. K., MOSKALUK, C. A., HAHN, S. A., SCHWARTE-WALDHOFF, I., SCHMIEGEL, W., BAYLIN, S. B., KERN, S. E. & HERMAN, J. G. 1997. Abrogation of the Rb/p16 tumor-suppressive pathway in virtually all pancreatic carcinomas. *Cancer Res,* 57, 3126–3130.
SENER, S. F., FREMGEN, A., MENCK, H. R. & WINCHESTER, D. P. 1999. Pancreatic cancer: a report of treatment and survival trends for 100,313 patients diagnosed from 1985-1995, using the National Cancer Database. *J Am Coll Surg,* 189, 1–7.
SHARMA, C., ELTAWIL, K. M., RENFREW, P. D., WALSH, M. J. & MOLINARI, M. 2011. Advances in diagnosis, treatment and palliation of pancreatic carcinoma: 1990-2010. *World J Gastroenterol,* 17, 867–897.
SHI, C., DANIELS, J. A. & HRUBAN, R. H. 2008. Molecular characterization of pancreatic neoplasms. *Adv Anat Pathol,* 15, 185–95.
SHI, C., HRUBAN, R. H. & KLEIN, A. P. 2009. Familial pancreatic cancer. *Arch Pathol Lab Med,* 133, 365–374.
SIEGEL, R. L., MILLER, K. D., FUCHS, H. E. & JEMAL, A. 2021. Cancer statistics, 2021. *CA Cancer J Clin,* 71, 7–33.
STELOW, E. B., SHACO-LEVY, R., BAO, F., GARCIA, J. & KLIMSTRA, D. S. 2010. Pancreatic acinar cell carcinomas with prominent ductal differentiation: Mixed acinar ductal carcinoma and mixed acinar endocrine ductal carcinoma. *Am J Surg Pathol,* 34, 510–518.

STROBEL, O., ROSOW, D. E., RAKHLIN, E. Y., LAUWERS, G. Y., TRAINOR, A. G., ALSINA, J., FERNÁNDEZ-DEL CASTILLO, C., WARSHAW, A. L. & THAYER, S. P. 2010. Pancreatic duct glands are distinct ductal compartments that react to chronic injury and mediate Shh-induced metaplasia. *Gastroenterology,* 138**,** 1166–1177.

SUNG, H., SIEGEL, R. L., ROSENBERG, P. S. & JEMAL, A. 2019. Emerging cancer trends among young adults in the USA: analysis of a population-based cancer registry. *Lancet Public Health,* 4**,** e137–e147.

TAKATORI, K., TERASHIMA, K., YOSHIDA, R., HORAI, A., SATAKE, S., OSE, T., KITAJIMA, N., KINOSHITA, Y., DEMIZU, Y. & FUWA, N. 2014. Upper gastrointestinal complications associated with gemcitabine-concurrent proton radiotherapy for inoperable pancreatic cancer. *J Gastroenterol,* 49**,** 1074–1080.

TERHUNE, P. G., PHIFER, D. M., TOSTESON, T. D. & LONGNECKER, D. S. 1998. K-ras mutation in focal proliferative lesions of human pancreas. *Cancer Epidemiol Biomarkers Prev,* 7**,** 515–521.

TÜRECI, Ö., MITNACHT-KRAUS, R., WÖLL, S., YAMADA, T. & SAHIN, U. 2019. Characterization of zolbetuximab in pancreatic cancer models. *Oncoimmunology,* 8**,** e1523096.

TURPIN, A., EL AMRANI, M., BACHET, J. B., PIETRASZ, D., SCHWARZ, L. & HAMMEL, P. 2020. Adjuvant pancreatic cancer management: towards new perspectives in 2021. *Cancers,* 3866, 12.

TZENG, C. W., TRAN CAO, H. S., LEE, J. E., PISTERS, P. W., VARADHACHARY, G. R., WOLFF, R. A., ABBRUZZESE, J. L., CRANE, C. H., EVANS, D. B., WANG, H., ABBOTT, D. E., VAUTHEY, J. N., ALOIA, T. A., FLEMING, J. B. & KATZ, M. H. 2014. Treatment sequencing for resectable pancreatic cancer: influence of early metastases and surgical complications on multimodality therapy completion and survival. *J Gastrointest Surg,* 18**,** 16–24; discussion 24–25.

UESAKA, K., BOKU, N., FUKUTOMI, A., OKAMURA, Y., KONISHI, M., MATSUMOTO, I., KANEOKA, Y., SHIMIZU, Y., NAKAMORI, S., SAKAMOTO, H., MORINAGA, S., KAINUMA, O., IMAI, K., SATA, N., HISHINUMA, S., OJIMA, H., YAMAGUCHI, R., HIRANO, S., SUDO, T. & OHASHI, Y. 2016. Adjuvant chemotherapy of S-1 versus gemcitabine for resected pancreatic cancer: a phase 3, open-label, randomised, non-inferiority trial (JASPAC 01). *Lancet,* 388**,** 248–257.

VAN HEEK, N. T., MEEKER, A. K., KERN, S. E., YEO, C. J., LILLEMOE, K. D., CAMERON, J. L., OFFERHAUS, G. J., HICKS, J. L., WILENTZ, R. E., GOGGINS, M. G., DE MARZO, A. M., HRUBAN, R. H. & MAITRA, A. 2002. Telomere shortening is nearly universal in pancreatic intraepithelial neoplasia. *Am J Pathol,* 161**,** 1541–1547.

VARGAS, R., NINO-MURCIA, M., TRUEBLOOD, W. & JEFFREY, R. B., JR. 2004. MDCT in Pancreatic adenocarcinoma: prediction of vascular invasion and resectability using a multiphasic technique with curved planar reformations. *AJR Am J Roentgenol,* 182**,** 419–425.

VINCENT, A., DUCOUROUBLE, M. P. & VAN SEUNINGEN, I. 2008. Epigenetic regulation of the human mucin gene MUC4 in epithelial cancer cell lines involves both DNA methylation and histone modifications mediated by DNA methyltransferases and histone deacetylases. *Faseb J,* 22**,** 3035–3045.

VON HOFF, D. D., ERVIN, T., ARENA, F. P., CHIOREAN, E. G., INFANTE, J., MOORE, M., SEAY, T., TJULANDIN, S. A., MA, W. W., SALEH, M. N., HARRIS, M., RENI, M., DOWDEN, S., LAHERU, D., BAHARY, N., RAMANATHAN, R. K., TABERNERO, J., HIDALGO, M., GOLDSTEIN, D., VAN CUTSEM, E., WEI, X., IGLESIAS, J. & RENSCHLER, M. F. 2013. Increased survival in pancreatic cancer with nab-paclitaxel plus gemcitabine. *N Engl J Med,* 369**,** 1691–1703.

VONLAUFEN, A., JOSHI, S., QU, C., PHILLIPS, P. A., XU, Z., PARKER, N. R., TOI, C. S., PIROLA, R. C., WILSON, J. S., GOLDSTEIN, D. & APTE, M. V. 2008. Pancreatic stellate cells: partners in crime with pancreatic cancer cells. *Cancer Res,* 68**,** 2085–2093.

VREELAND, T. J. & KATZ, M. H. G. 2018. Timing of pancreatic resection and patient outcomes: is there a difference? *Surg Clin North Am,* 98**,** 57–71.

WANG, X., ALLEN, S., BLAKE, J. F., BOWCUT, V., BRIERE, D. M., CALINISAN, A., DAHLKE, J. R., FELL, J. B., FISCHER, J. P., GUNN, R. J., HALLIN, J., LAGUER, J., LAWSON, J. D., MEDWID, J., NEWHOUSE, B., NGUYEN, P., O'LEARY, J. M., OLSON, P., PAJK, S., RAHBAEK, L., RODRIGUEZ, M., SMITH, C. R., TANG, T. P., THOMAS, N. C., VANDERPOOL, D., VIGERS, G. P., CHRISTENSEN, J. G. & MARX, M. A. 2022. Identification of MRTX1133, a noncovalent, potent, and selective KRAS(G12D) inhibitor. *J Med Chem,* 65**,** 3123–3133.

WANG, X. Y., YANG, F., JIN, C. & FU, D. L. 2014. Utility of PET/CT in diagnosis, staging, assessment of resectability and metabolic response of pancreatic cancer. *World J Gastroenterol,* 20**,** 15580–15589.

WANG, Y. T., GOU, Y. W., JIN, W. W., XIAO, M. & FANG, H. Y. 2016. Association between alcohol intake and the risk of pancreatic cancer: a dose-response meta-analysis of cohort studies. *BMC Cancer,* 16**,** 212.

WATTENBERG, M. M., ASCH, D., YU, S., O'DWYER, P. J., DOMCHEK, S. M., NATHANSON, K. L., ROSEN, M. A., BEATTY, G. L., SIEGELMAN, E. S. & REISS, K. A. 2020. Platinum response characteristics of patients with pancreatic ductal adenocarcinoma and a germline BRCA1, BRCA2 or PALB2 mutation. *Br J Cancer,* 122**,** 333–339.

WOLFE, A. R. & WILLIAMS, T. M. 2018. Altering the response to radiation: radiosensitizers and targeted therapies in pancreatic ductal adenocarcinoma: preclinical and emerging clinical evidence. *Ann Pancreat Cancer,* 1, 26–49.

WOOD, L. D., CANTO, M. I., JAFFEE, E. M. & SIMEONE, D. M. 2022. Pancreatic cancer: pathogenesis, screening, diagnosis, and treatment. *Gastroenterology,* 163**,** 386–402.e1.

WORLD HEALTH ORGANIzATION. 2022. *Addressing inequities in breast cancer treatment in sub-Saharan Africa: insights from a breast cancer surgeon in Nairobi.* Available: https://www.who.int/news-room/feature-stories/detail/addressing-inequities-in-breast-cancer-treatment-in-sub-saharan-africa-insights-from-a-breast-cancer-surgeon-in-nairobi.

WORLD HEALTH ORGANIzATION. 2024. *Africa to intensify cancer control through cancer registries.* Available: https://www.afro.who.int/news/africa-intensify-cancer-control-through-cancer-registries.

YANG, M. & ZHANG, C. Y. 2021. Diagnostic biomarkers for pancreatic cancer: An update. *World J Gastroenterol,* 27**,** 7862–7865.

ZAVALA, V. A., BRACCI, P. M., CARETHERS, J. M., CARVAJAL-CARMONA, L., COGGINS, N. B., CRUZ-CORREA, M. R., DAVIS, M., DE SMITH, A. J., DUTIL, J., FIGUEIREDO, J. C., FOX, R., GRAVES, K. D., GOMEZ, S. L., LLERA, A., NEUHAUSEN, S. L., NEWMAN, L., NGUYEN, T., PALMER, J. R., PALMER, N. R., PÉREZ-STABLE, E. J., PIAWAH, S., RODRIQUEZ, E. J., SANABRIA-SALAS, M. C., SCHMIT, S. L., SERRANO-GOMEZ, S. J., STERN, M. C., WEITZEL, J., YANG, J. J., ZABALETA, J., ZIV, E. & FEJERMAN, L. 2021. Cancer health disparities in racial/ethnic minorities in the United States. *Br J Cancer,* 124**,** 315–332.

ZHANG, Y., HOCHSTER, H., STEIN, S. & LACY, J. 2015. Gemcitabine plus nab-paclitaxel for advanced pancreatic cancer after first-line FOLFIRINOX: single institution retrospective review of efficacy and toxicity. *Exp Hematol Oncol,* 4**,** 29.

ZHAO, C., CHEN, Z., YE, X., ZHANG, Y., ZHAN, H., ABURANO, T., TIAN, M. & ZHANG, H. 2009. Imaging a pancreatic carcinoma xenograft model with 11C-acetate: a comparison study with 18F-FDG. *Nucl Med Commun,* 30**,** 971–977.

ZHU, X., JU, X., CAO, Y., SHEN, Y., CAO, F., QING, S., FANG, F., JIA, Z. & ZHANG, H. 2019. Patterns of local failure after stereotactic body radiation therapy and sequential chemotherapy as initial treatment for pancreatic cancer: implications of target volume design. *Int J Radiat Oncol Biol Phys,* 104**,** 101–110.

2 Epidemiology and Trends of Pancreatic Cancer in Africa

Zukile Mbita[1], Tebogo Marutha[2], Moshawa Khaba[3], Morgie Govender[4], Abidemi Omonisi[5,6], and Zodwa Dlamini[2]

[1]Department of Biochemistry, Microbiology and Biotechnology, School of Molecular and Life Sciences, University of Limpopo, Polokwane, South Africa.

[2]SAMRC/UP Precision Oncology Research Unit, DSI/NRF SARChI Chair in Precision Oncology (POCP) and Cancer Prevention, Pan African Cancer Research Institute (PACRI), University of Pretoria, Pretoria, South Africa.

[3]Department of Anatomical Pathology, School of Medicine, Sefako Makgatho Health Sciences University, Pretoria, South Africa.

[4]Department of Surgery, Pietermaritzburg Metropolitan Complex and College of Health Sciences, Nelson R Mandela School of Medicine, University of KwaZulu-Natal, Durban, South Africa.

[5]Department of Anatomic Pathology, College of Medicine, Ekiti State University, Ado-Ekiti, Ekiti State, Nigeria.

[6]Department of Laboratory Medicine, College of Health Sciences, University of Liberia, Monrovia, Liberia.

INTRODUCTION

Worldwide, pancreatic cancer (PC) continues to devastate numerous healthcare systems and contributes significantly to the high mortality rates associated with malignancies. According to the World Cancer Research Fund International and the International Agency for Research on Cancer (IARC), PC is the 12th most common cancer, worldwide, it is the 11th most common cancer in men and the 11th most common cancer in women. It is currently the sixth leading cause of cancer-related deaths, globally. It is estimated to contribute 2.5% of all cancer cases worldwide. The highest incidence rates of this cancer are mostly found in North America, Europe, and Australia, while Africa and Asia are associated with the lowest rates (Simoes et al., *2017*). However, due to underreporting in these two continents, PC remains an emergent concern in Africa. The highest rates are reported in North Africa, where incidence rates imitate those found in Europe due to close proximity and the adoption of European lifestyle and dietary habits. Sub-Saharan Africa (SSA) has lower incidence rates; however, due to similar habits adopted from the Western world, the incidence rates are rising. Consequently, South Africa is reported to have the highest increasing incidence rates in Africa.

PC may not be among the most common cancers in Africa and across the world, but the numbers are on the rise, and the statistics may not truly reflect the prevalence of the disease in less developed countries due to a lack of screening, reporting, weak health infrastructure, inadequate human resources, and lack of health insurance, especially in Africa and Asia. Globally, PC is the leading cause of cancer-related deaths, and its worldwide incidence is on the rise and has been reported to have doubled in the last 25 years (Luo et al., 2023; Klein, 2021). These doubling numbers are worrying for the underdeveloped and developing countries as cancer incidences are on the rise. Even

 DOI: 10.1201/9781003544401-3

though PC mortalities were initially reported to be at the highest levels in developed countries (Ilic and Ilic, 2016), incidences of this cancer are on the rise throughout the world and mitigating interventions are of paramount importance.

Thus, this chapter delves into the epidemiological patterns of PC in Africa, examining regional disparities, emerging trends, and factors contributing to its prevalence. By shedding light on the unique challenges faced by African populations, such as limited resources and healthcare infrastructure, this chapter underscores the importance of targeted interventions and tailored strategies for addressing PC in the region.

EPIDEMIOLOGY OF PANCREATIC CANCER

Currently, PC is one of the deadliest cancers worldwide and is associated with the worst prognosis since it is often asymptomatic and difficult to diagnose. There are no fully effective treatment strategies; hence, once diagnosed, death often ensues. On a global scale, PC is identified as the 12th most frequently occurring malignancy and is the sixth most significant contributor to cancer-related deaths (Bray et al., 2018). In the year 2020 alone, the global statistics for PC revealed over 495,000 newly diagnosed cases and approximately 466,000 fatalities (Sung et al., 2021). In the United States, PC is the third leading cause of deaths associated with cancer, with estimates suggesting that there are more than 53,000 new diagnoses and an increase of 41,000 fatalities annually (Siegel et al., 2017; Wong et al., 2017). IARC data indicates that Asia presents with the highest incidence rate, making up 45% of the total world PC cases. The lowest incidence rate was observed in Oceania with a total of 0.95% of cases. While the global burden of PC has been extensively studied, the prevalence within the African continent has received less attention. The World Health Organization (WHO), via IARC, has been at the forefront of gathering and disseminating data on the global burden of cancer. The data revealed that PC prevalence is on the rise in Africa, in both males and females, with countries such as South Africa, Nigeria, Egypt, Algeria, and Morocco among those that recorded the highest (Figure 2.1).

Risk factors for PC are classified as non-modifiable and modifiable. PC predominantly affects older individuals, with cases in those under 30 being exceedingly uncommon. Data indicate that around 90% of newly diagnosed patients are aged 55 and above, with the majority falling within the 70–80 age range (Moshayedi et al., 2022; Pourshams et al., 2019). Males exhibit a greater prevalence of PC than their female counterparts (Moshayedi et al., 2022; Hu et al., 2021). The incidence of PC is not uniform across ethnic groups; it is significantly higher in African Americans than in Caucasians, whereas Asian Americans and Pacific Islanders have the lowest reported incidence rates (Huang et al., 2019; Midha et al., 2016). In the context of modifiable factors, the act of cigarette smoking shows the most pronounced positive association with the risk of cancer onset (Ghadirian et al., 2003).

A comprehensive meta-analysis examining the relationship between smoking and the incidence of PC revealed that the risk is significantly elevated among individuals who consume the highest quantities of cigarettes daily (Wood and Hruban, 2012). Specifically, those who smoke over 35 cigarettes each day exhibit an odds ratio of 3.0 for developing PC when compared to individuals who have never smoked (Wood and Hruban, 2012). Moreover, it has been established that higher body weight or body mass index (BMI) significantly raises the risk of PC by a factor of 1.6 relative to those classified as having a normal weight (Stolzenberg-Solomon et al., 2013; Arslan et al., 2010). Furthermore, family history of PC is recognized as a critical risk factor, with individuals who have one family member diagnosed experiencing a twofold increase in risk compared to the general population (Brune et al., 2010). In cases where multiple family members have been diagnosed, the risk escalates to seven times that of the general population (Brune et al., 2010). Diabetes is also recognized as a significant risk factor for PC, as the emergence of new-onset diabetes correlates with a markedly elevated incidence of PC diagnoses, especially within the initial two years following the diagnosis of diabetes (Gupta et al., 2006). Genetic predisposition through inherited genes represents

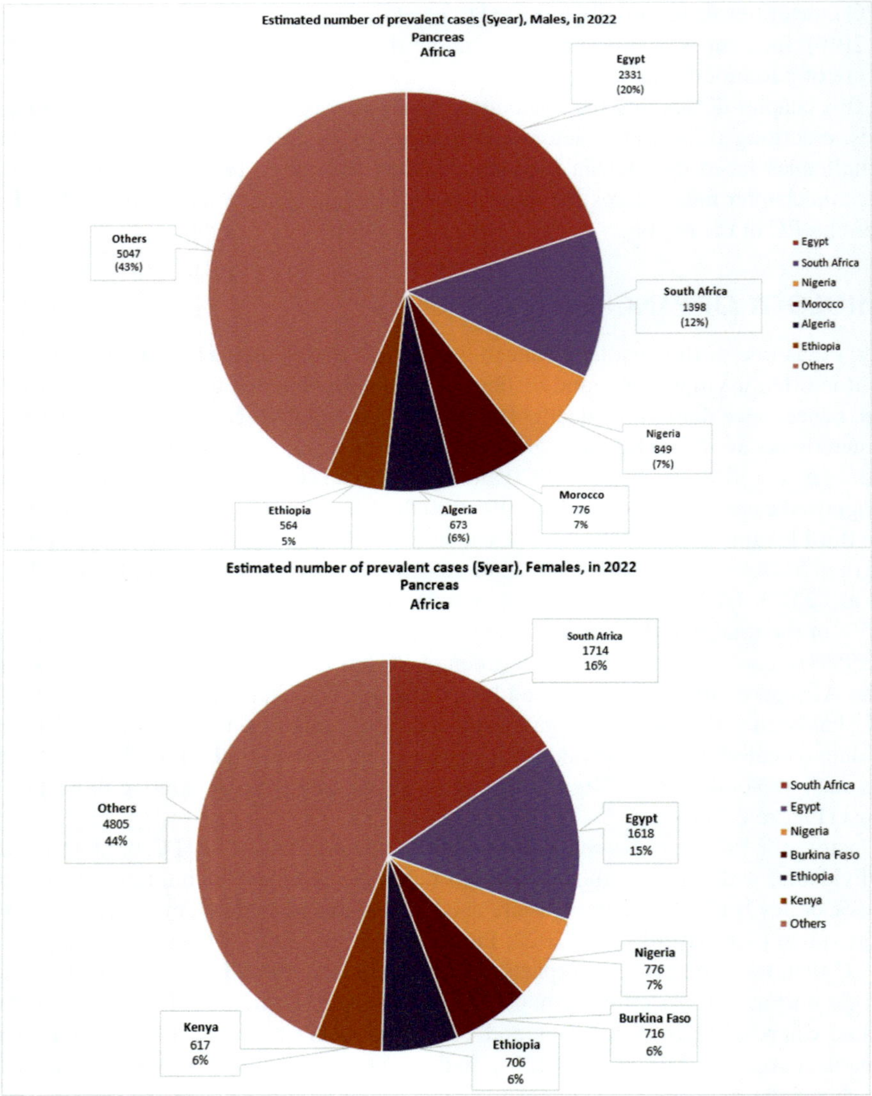

FIGURE 2.1 Estimated number of prevalent cases of pancreatic cancer in Africa (2022). This figure illustrates the estimated number of prevalent cases of pancreatic cancer in both sexes across various African countries by the year 2022. The data, sourced from GLOBOCAN 2022, highlight the geographic distribution and burden of pancreatic cancer on the continent. The prevalence rates are age-standardized to the World Standard Population, providing a comparative view across different regions. The highest prevalence rates are observed in Northern and Southern Africa, while the lowest rates are found in Eastern and Western Africa.

a significant risk factor for PC. Notably, pathogenic variants in the BRCA2 gene have been identified as contributing to the highest percentage of PC cases among patients (Zhen et al., 2015).

Despite advancements in the detection and therapeutic approaches for PC, the five-year survival rate is still only 9% (Gaidhani and Balasubramaniam, 2021; Rawla et al., 2019). However, in the United States, there has been a modest advancement in PC survival rates over recent years, with the overall five-year survival rate now nearing 11% (Siegel et al., 2022; Blackford et al., 2020). Survival rates are higher for patients with localized tumors (37%) compared to those with metastatic disease (3%) (Siegel et al., 2020).

Definition and Classification of Pancreatic Cancer

Pancreatic ductal adenocarcinoma (PDAC) is by far the most common type of pancreatic malignancy, accounting for about 90% of all pancreas neoplasms. It originates from the pancreatic exocrine cells, which are responsible for generating digestive enzymes (Gupta and Yelamanchi, 2021). With a five-year survival rate of less than 10%, this malignancy has an extremely poor prognosis due to its high aggressiveness and tendency to become visible at an advanced stage (Rawla, Sunkara, and Gaduputi, 2019). Treating PDAC is almost impossible because most patients appear at an advanced stage (Luchini et al., 2016). Even under the greatest medical care, the five-year survival rate with PDAC is as low as 2% (McGuigan et al., 2018). The high rates of morbidity and mortality linked to PDAC have prompted researchers worldwide to step up their efforts to find more effective diagnostic and therapeutic approaches. PDAC is most commonly located in the proximal pancreas. Occurrence in the pancreatic body or tail is rarer. The most common location of PDAC at the pancreatic head explains the features of obstruction of the common bile duct, which is most often associated with this histologic type of PC. It can infiltrate surrounding structures (peripancreatic adipose tissue, duodenum, stomach, portal vein, etc.) (Zhang et al., 2018; Haeberle and Esposito, 2019; Ryan et al., 2014).

Squamous cell carcinoma (SCC) is exceedingly uncommon, accounting for about 0.5–1% of all malignant pancreatic tumors (Grant et al., 2023). Pancreatic SCC, unlike the more prevalent PDAC, develops from squamous epithelial cells rather than pancreatic glandular tissue. These cells are normally absent from the pancreas, and their appearance in this situation is frequently explained by squamous metaplasia, a condition in which glandular cells change into squamous cells, presumably because of a history of pancreatic damage or chronic inflammation (Boyd et al., 2020).

Neuroendocrine tumors (NETs), also called islet cell tumors, make up about 5–10% of PCs and are thought to originate from pancreatic hormone-producing cells (Yao et al., 2016). It comes directly from neuroendocrine cells in the pancreas as well as from pancreatic stem cells that can develop into exocrine and endocrine cells (Marchese et al., 2022).

Ampullary cancer is a very rare tumor that arises from the ampulla of Vater, where the pancreatic and bile ducts meet the small intestine (Cloyd and Poultsides, 2018). It is the second most common tumor in the peri-ampullary area, accounting for 6–8% of pancreatic head region tumors and 0.2–0.5% of gastrointestinal malignancies (Rizzo et al., 2021). Roughly 7% of peri-ampullary tumors are ampullary carcinomas. Despite being a subgroup of peri-ampullary tumors, ampullary carcinoma must be distinguished from the larger category of peri-ampullary tumors due to therapeutic and prognostic implications (Regalla et al., 2019).

Intraductal papillary mucinous neoplasm (IPMN) is a precancerous disease that can proceed to invasive PC if left untreated (Nagtegaal et al., 2020). According to Tanaka et al. (2012), such lesions emanate from the pancreatic ducts and generate mucin, which may cause cysts to grow inside the pancreas.

Around 1–2% of all pancreatic neoplasms are acinar cell carcinomas (ACCs), which are uncommon, malignant tumors made up of cells that resemble acinar cells morphologically and provide evidence of the neoplastic cells synthesizing exocrine enzymes. Pancreatic ACC is a high-grade cancer that can originate anywhere in the organ, but it most frequently affects the head. It does not appear to have a propensity for any particular region (Calimano-Ramirez et al., 2022; Ikezawa et al., 2023).

The extremely uncommon pancreatic tumor known as pancreatoblastoma (PB) typically affects infants and young children and is thought to originate from the pancreas' embryonic cells (Liu et al., 2022). It is a rare and difficult disorder to diagnose and treat because it makes up less than 1% of pancreatic tumors. It is distinguished by its unusual histological features, such as the presence of acinar, squamoid, and endocrine components. It can be challenging to distinguish PB from other retroperitoneal cancers due to its varied and vague clinical indications (Moreno et al., 2021; Strobel et al., 2020).

The uncommon exocrine tumor known as pancreatic solid pseudopapillary neoplasm (pSPN) typically affects young women. It makes up 1–3% of all pancreatic tumors and is a low-grade malignant tumor without a distinct line of pancreatic epithelial differentiation. Despite their malignant potential, SPNs often have a good prognosis, owing to their slow growth and the fact that they are frequently detected before metastasis occurs (Milanetto et al., 2021; Dinarvand and Lai, 2017).

A hereditary tendency may be connected to PC, particularly in cases where inherited genetic abnormalities are present. PC risk is greatly increased by mutations in the BRCA2 gene, which are also linked to ovarian and breast malignancies (Goggins, 2022; Yurgelun et al., 2019). Furthermore, another genetic disorder that increases the risk of PC is Lynch syndrome, sometimes referred to as hereditary nonpolyposis colorectal cancer (HNPCC) (Klein, 2020). Moreover, those who have familial atypical multiple mole melanoma (FAMMM) syndrome (a condition marked by numerous melanomas) also have a higher chance of developing PC (Goggins, 2022; Klein, 2020). Other malignancies that can occur in the pancreas, though less common than PDAC, include mesenchymal neoplasms, such as leiomyosarcoma and gastrointestinal stromal tumors (GIST). Additionally, the pancreas can also be the site for secondary tumors (metastatic cancers), which spread from other primary sites such as the lungs, kidneys, or breasts. These malignancies are rare but should be considered in differential diagnoses when evaluating pancreatic masses (Jain et al., 2019). Each type has distinct characteristics, and accurate diagnosis is crucial for effective treatment and management.

RISK FACTORS AND ETIOLOGY

Environmental and Lifestyle Factors

The onset of PC is influenced by both modifiable and non-modifiable risk factors (Zhao and Liu, 2020). Modifiable risk factors include lifestyle factors, dietary patterns, infectious diseases, and socioeconomic factors. Among the lifestyle factors that are associated with PC, smoking is considered a major risk; however, alcohol consumption also plays a significant role. Smoking may contribute to pancreatic carcinogenesis through the effects of tobacco-related carcinogens, which may mutate both proto-oncogenes and tumor suppressor genes. These carcinogens may also induce chronic inflammation and thus increase the development of PC (Duell, 2012). A study by Montes et al. (2020) revealed that the risk of acquiring PC was increased by at least 74% in smokers compared to non-smokers. On the other hand, a study conducted on 2,187 PC patients demonstrated that there was a high risk of the disease in people who consumed more than 30 g of alcohol daily (Genkinger et al., 2009). Moreover, a meta-analysis revealed that low to moderate consumption of alcohol was not associated with PC; thus, indicating that PC has a dose-dependent relationship with alcohol intake (Wang et al., 2016).

Dietary Influences, Infectious Agents, Socioeconomic Determinants, and Genetic Predisposition

A high-fat, high-energy diet aids the absorption of bacterial liposaccharides into the bloodstream (Djuric, 2017), and research indicates that PC patients have an abundance of lipopolysaccharide-producing bacteria in their intestines (Ren et al., 2017). Additionally, increased consumption of red or processed meat, saturated fat, and fried foods is also associated with a high risk of pancreatic carcinogenesis (Stolzenburg-Solomon et al., 2002; Casari and Falasca, 2015). Infectious agents such as hepatitis B (HBV) and hepatitis C (HCV) viruses have been linked to increased risk of PC onset. These viruses are hepatotropic and may cause persistent liver injury (Dandri and Locarnini, 2012). However, a study has identified HBV surface and core antigens, as well as HCV antigens, in pancreatic acinar cells (Montalbano and Neff, 2006). As a result, patients infected with HBV and HCV have been reported to have a high risk of PC (Huang et al., 2013; Hassan et al., 2008). Socioeconomic factors such as education, income, geographic location, and work environment play a crucial role in the prevalence, occurrence, diagnosis, and treatment of PC (Kim et al., 2022; Zhu et al., 2020). Lack of income, education, and limited access to healthcare facilities has been linked to increased risk of PC (Bakens et al., 2017). One important non-modifiable risk factor for PC is genetic predisposition. A study has reported that a history of PC in the family greatly influences the risk of acquiring the disease (Wood and Hruban, 2012). More than 80% of PC is due to sporadic mutations, with a small percentage being attributed to specific gene mutations (Becker et al., 2014; Yeo, 2015). Commonly, point mutations in KRAS, CDKN2A (P16), TP53, SMAD4 (Raimondi et al., 2009; Del Chiaro et al., 2014), BRCA1, BRCA2, STK11, PRSS1, and MMR (Chen et al., 2017; Lanfranca et al., 2019) are involved in PC.

TRENDS IN INCIDENCE AND MORTALITY

Recent Changes in Incidence and Mortality Rates

Historically, the incidence of PC has been reported to be low in developing countries, including Africa, with developed countries showing a high prevalence (Bray et al., 2018). However, challenges such as limited access to healthcare, underreporting, and lack of comprehensive cancer registries have resulted in incorrect incidence rates, especially on the African continent (Arnold et al., 2020). Despite advancements in increasing knowledge and awareness of PC risk factors, the incidence of this disease is expected to rise, with an estimate of 355,317 new cases by 2040 (Prashant et al., 2019). Over the past eight years, there has been an increase in the incidence (Figure 2.2) and mortality rates (Table 2.1) of PC in Africa. In Africa, as reported by WHO via IARC, the highest incidence rates were observed in Burkina Faso and South Africa, with most central African countries having no data reported (Figure 2.2).

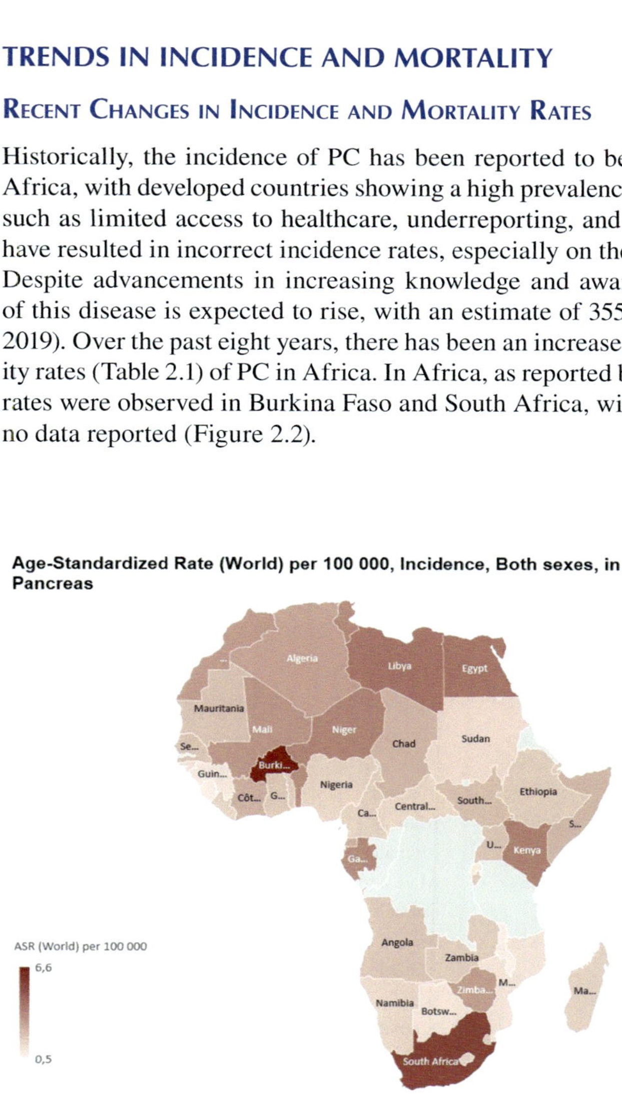

	ASR (World) per 100 000		ASR (World) per 100 000
Burkina Faso	6,6	Togo	1,5
South Africa	5,3	Cameroon	1,4
Libya	3,8	Congo (DRC)	1,4
Mauritius	3,8	Ethiopia	1,4
Egypt	3,8	Djibouti	1,4
Kenya	3,5	Liberia	1,4
Gabon	3,4	Eswatini	1,4
Niger	3,4	Zambia	1,4
Tunisia	3,3	Madagascar	1,3
Benin	3,2	Nigeria	1,3
Mali	2,9	Cape Verde	1,2
Morocco	2,8	Central African Republic	1,2
Zimbabwe	2,7		
Algeria	2,6	Lesotho	1,2
Equatorial Guinea	2,4	Namibia	1,2
		Congo	1,1
Côte d'Ivoire	2,3	Sudan	1,1
Chad	2	Guinea-Bissau	0,94
Somalia	2	Mozambique	0,87
Mauritania	1,8	Guinea	0,81
Senegal	1,8	Comoros	0,76
South Sudan	1,8	Botswana	0,75
Tanzania	1,8	Rwanda	0,73
Ghana	1,7	Sierra Leone	0,65
Angola	1,6	Malawi	0,5
Gambia	1,6	Papua New Guinea	1,66
Uganda	1,6		
Burundi	1,5	Togo	1,5

FIGURE 2.2 Incidence of pancreatic cancer in Africa (2022). The figure depicts the incidence rates of pancreatic cancer across various African countries in 2022. The data, sourced from GLOBOCAN 2022, provide an age-standardized incidence rate per 100,000 population, allowing for comparison across different regions. Northern and Southern Africa showed increased incidence rates, while East and West Africa showed the lowest rates. This figure highlights the geographic disparities in pancreatic cancer incidence and underscores the need for targeted public health interventions and research efforts in the regions with higher incidence rates.

TABLE 2.1
Mortality Rates of Pancreatic Cancer in Africa (2022)

Cancer Type	Country	Age Group (Years)	Sex	Number	ASR (World)
Pancreatic	Burkina Faso	0–85+	Male and female	655	6.1
Pancreatic	France, La Réunion	0–85+	Male and female	90	5.3
Pancreatic	South Africa	0–85+	Male and female	2360	4.7
Pancreatic	Egypt	0–85+	Male and female	3186	3.7
Pancreatic	Kenya	0–85+	Male and female	872	3.5
Pancreatic	Libya	0–85+	Male and female	203	3.5
Pancreatic	Gabon	0–85+	Male and female	48	3.3
Pancreatic	Niger	0–85+	Male and female	333	3.2
Pancreatic	Tunisia	0–85+	Male and female	503	3.2
Pancreatic	Benin	0–85+	Male and female	202	3.1
Pancreatic	Mauritius	0–85+	Male and female	65	2.9
Pancreatic	Morocco	0–85+	Male and female	1181	2.7
Pancreatic	Mali	0–85+	Male and female	236	2.6
Pancreatic	Zimbabwe	0–85+	Male and female	188	2.6
Pancreatic	Algeria	0–85+	Male and female	1136	2.5
Pancreatic	Equatorial Guinea	0–85+	Male and female	16	2.4
Pancreatic	Côte d'Ivoire	0–85+	Male and female	302	2.3
Pancreatic	São Tomé and Príncipe	0–85+	Male and female	3	2.3
Pancreatic	Chad	0–85+	Male and female	146	1.9
Pancreatic	Somalia	0–85+	Male and female	145	1.9
Pancreatic	Mauritania	0–85+	Male and female	48	1.8
Pancreatic	Senegal	0–85+	Male and female	160	1.8
Pancreatic	Ghana	0–85+	Male and female	341	1.7

(*Continued*)

TABLE 2.1 (CONTINUED)
Mortality Rates of Pancreatic Cancer in Africa (2022)

Cancer Type	Country	Age Group (Years)	Sex	Number	ASR (World)
Pancreatic	South Sudan	0–85+	Male and female	103	1.7
Pancreatic	Tanzania	0–85+	Male and female	494	1.7
Pancreatic	Angola	0–85+	Male and female	269	1.6
Pancreatic	Gambia	0–85+	Male and female	18	1.6
Pancreatic	Uganda	0–85+	Male and female	271	1.6
Pancreatic	Burundi	0–85+	Male and female	81	1.5
Pancreatic	Togo	0–85+	Male and female	62	1.5
Pancreatic	Cape Verde	0–85+	Male and female	7	1.4
Pancreatic	Congo (DRC)	0–85+	Male and female	645	1.4
Pancreatic	Djibouti	0–85+	Male and female	11	1.4
Pancreatic	Liberia	0–85+	Male and female	40	1.4
Pancreatic	Eswatini	0–85+	Male and female	9	1.4
Pancreatic	Cameroon	0–85+	Male and female	168	1.3
Pancreatic	Ethiopia	0–85+	Male and female	842	1.3
Pancreatic	Madagascar	0–85+	Male and female	202	1.3
Pancreatic	Nigeria	0–85+	Male and female	1283	1.3
Pancreatic	Zambia	0–85+	Male and female	102	1.3
Pancreatic	Central African Republic	0–85+	Male and female	29	1.2
Pancreatic	Eritrea	0–85+	Male and female	27	1.2
Pancreatic	Lesotho	0–85+	Male and female	20	1.2
Pancreatic	Namibia	0–85+	Male and female	19	1.2
Pancreatic	Congo, Republic of	0–85+	Male and female	37	1.1
Pancreatic	Sudan	0–85+	Male and female	260	0.98

(Continued)

TABLE 2.1 (CONTINUED)
Mortality Rates of Pancreatic Cancer in Africa (2022)

Cancer Type	Country	Age Group (Years)	Sex	Number	ASR (World)
Pancreatic	Guinea-Bissau	0–5+	Male and female	9	0.94
Pancreatic	Mozambique	0–85+	Male and female	130	0.83
Pancreatic	Guinea	0–5+	Male and female	54	0.80
Pancreatic	Comoros	0–85+	Male and female	4	0.76
Pancreatic	Botswana	0–85+	Male and female	14	0.75
Pancreatic	Rwanda	0–5+	Male and female	57	0.70
Pancreatic	Sierra Leone	0–85+	Male and female	27	0.65
Pancreatic	Malawi	0–85+	Male and female	45	0.44
	Total			**17,758**	**2.2**

Information source: Global Cancer Observatory, accessed September 7, 2024.

In 2020, PC was the ninth cause of cancer-related deaths (Sung et al., 2021). The global increase in the prevalence of PC creates challenges in health and economic security; and thus, it is crucial to continue raising awareness on PC risk factors and monitoring population changes in morbidity, mortality, and survival of PC.

Table 2.1 presents the mortality rates of PC across various African countries in 2022. The data, sourced from GLOBOCAN 2022, are age-standardized to the World Standard Population, providing a comparative view of mortality rates per 100,000 population. The highest mortality rates are observed in Northern and Southern Africa, while the lowest rates are found in Eastern and Western Africa.

Age, Gender, Geographic Variations, and Projections for Future Trends in Pancreatic Cancer Rates

Age and gender have a great influence on PC onset. PC is predominantly a disease of older adults, with around 90% of cases occurring in individuals aged 55–80. It is rare in younger people and is very uncommon in those under 30 years old (McMenamin et al., 2017). On a global scale, the incidence of PC is more prevalent in males than in females, with this being highly noticeable in developed countries (McGuigan et al., 2018). High smoking rates and occupational exposure generally put men at higher risk of PC than women. The temporal future trends in PC incidence and mortality from 2022 to 2050, based on data from GLOBOCAN 2022, indicate a significant increase. The incidence is projected to rise by 99.8%, with an estimated 482,093 new cases, while mortality is expected to increase by 104.7%, resulting in approximately 450,072 deaths as reported by IARC. However, these trends show significant international variation with the highest increase in PC incidence being projected in Africa (+85.3%), followed by Latin America and the Caribbean (+64.1%). In contrast, the lowest increase is expected in Europe (+23.9%) as reported by WHO via IARC.

DIAGNOSTIC AND SCREENING CHALLENGES

Availability and Accessibility of Diagnostic Tools

PC is a devastating disease that poses significant diagnostic and screening challenges in Africa. The availability and accessibility of diagnostic tools are major hurdles, as many African countries lack the necessary infrastructure and resources to effectively detect PC in its early stages (Adesina et al., 2013; Farrè and Kibera, 2018). The shortage of pathologists, medical biological scientists in oncology, and inadequate diagnostic infrastructure are contributing factors to the delay in cancer diagnosis across the continent (Farrè and Kibera, 2018). Without access to modern diagnostic technologies, such as imaging and biomarker testing, many patients in Africa face late-stage diagnosis, limiting their treatment options and chances of survival (Jain, 2021; Luu, 2021; Suzuki et al., 2012; Szymoński et al., 2021).

However, the literature suggests that there are significant disparities in the availability of diagnostic and screening services across different regions of Africa, influenced by economic development, awareness, lack of national or regional screening programs, and access to healthcare (Lockhat et al., 2014). Interestingly, while there is a focus on improving diagnostic capabilities for PC, such as the use of molecular spectroscopic methods and immunohistochemistry biomarkers (Luu, 2021; Szymoński et al., 2021), these advanced diagnostic tools may not be widely accessible in many parts of Africa. The continent faces challenges such as lack of awareness, limited access to diagnostic tools, human resources, and inadequate healthcare infrastructure (Lockhat et al., 2014). Moreover, the literature indicates that there is a need for investment in healthcare systems, including the procurement and maintenance of diagnostic imaging equipment, to address the growing cancer burden (Lockhat et al., 2014).

Late-Stage Diagnosis and Its Implications

PC's aggressive nature and poor prognosis are exacerbated by late-stage diagnosis, which is common due to non-specific early symptoms (Hernández-Blanquisett et al., 2024). This issue is not unique to high-income countries; in SSA, despite lower incidence rates, mortality is comparable to high-income countries due to late-stage presentation. The median survival for advanced PC remains around one year, while in Africa it is six months from diagnosis until death, even with new chemotherapy regimens (Park et al., 2024), highlighting the critical need for earlier detection. When PC is detected at an advanced stage, the prognosis is often poor, with limited treatment options available. Patients diagnosed at late stages typically have a very low survival rate, as the disease has already progressed significantly and metastasized to other organs. This lack of early detection not only reduces the chances of successful treatment but also contributes to the high mortality rates associated with PC in Africa. Early detection methods for PC are crucial to improving patient outcomes. Developing more sensitive and specific screening tools could potentially identify the disease at a treatable stage. Increasing awareness among healthcare providers and the general public about the subtle early signs of PC and reducing exposure to the identified and documented predisposing factors may contribute to earlier diagnosis and improved survival rates (Park et al., 2024).

Interestingly, socioeconomic factors, including race, income, and education, have been associated with the stage of PC at diagnosis (Gallegos et al., 2023). This suggests that disparities in healthcare access and awareness contribute to late-stage diagnosis. These factors appear not to be strongly associated with cases seen in Africa. Paradoxically, educated individuals and people with high incomes are also significantly affected by the cancer in Africa. In contrast, treatments like arterial infusion of rAdp53 have shown some promise in late-stage PC, offering symptomatic relief and improved survival in a small cohort (Mo et al., 2017). However, the lack of survival benefit from metformin (Yang and Rustgi, 2016) and the significantly longer survival in patients with locally advanced PC who received local treatment post-neoadjuvant chemotherapy (Güven et al., 2023) underscore the importance of early detection and treatment.

Barriers to Early Detection and Screening Programs

The barriers to early detection and screening programs for PC in Africa are multifaceted. Most countries in Africa do not have a national screening program for cancers, including PC. Furthermore, in countries with national cancer screening programs, PC is not included. The focus has been on breast, prostate, and cervical cancer. While the currently accessible data does not provide a direct analysis of PC screening in Africa, insights can be extrapolated from the challenges faced in the detection and screening of other cancers within the continent. Limited endoscopic capacity, poor knowledge of cancer and screening, healthcare factors, cultural beliefs, and socioeconomic factors are significant barriers to effective cancer screening in Africa (Lee and Holmes, 2023). These challenges are compounded in the case of PC, which is known for its late presentation and poor prognosis, making early detection crucial (Hernández-Blanquisett et al., 2024). Despite the lack of specific focus on PC screening in Africa, the literature on other cancers reveals systemic issues that are likely to affect PC screening. For instance, the lack of infrastructure and trained personnel, as well as the need for systematic changes by governing bodies, are cited as primary barriers to colorectal cancer (CRC) screening (Lee and Holmes, 2023). These systemic issues are likely to be similar for PC, given the need for specialized diagnostic tools and trained healthcare professionals for effective screening (Lee and Holmes, 2023).

Role of Imaging and Biomarker Testing

In recent years, imaging and biomarker testing have emerged as essential tools in the diagnosis of PC, providing clinicians with valuable information for accurate and timely detection of the disease. Imaging techniques, such as computed tomography (CT), magnetic resonance imaging (MRI), and endoscopic ultrasonography (EUS), play a fundamental role in the diagnosis of PC (Thomas, 2018; Lee and Yoon, 2017). These non-invasive methods allow for visualization and assessment of the pancreas, aiding in the identification of tumors and determination of their stage and size. CT scans provide detailed cross-sectional images of the pancreas, allowing for the detection of tumors and the assessment of their relationship to surrounding structures. MRI, on the other hand, offers excellent soft tissue contrast, allowing for the detection and characterization of pancreatic tumors (Thomas, 2018; Lee and Yoon, 2017). EUS, a more advanced technique, combines ultrasound and endoscopy to provide high-resolution images of the pancreas, facilitating the detection of small lesions and the evaluation of lymph node involvement (Chong et al., 2016). Imaging techniques can also guide the collection of tissue samples for further analysis through minimally invasive procedures such as fine-needle aspiration (FNA) or core needle biopsy, which help establish a definitive diagnosis. In addition to imaging, biomarker testing has emerged as a valuable tool in diagnosing PC. Biomarkers, which are measurable substances, can be detected in blood, urine, or other bodily fluids and can aid in the identification of specific diseases or conditions (Kato and Honda, 2020). In the case of PC, several biomarkers have been identified, including carbohydrate antigen 19-9 (CA 19-9) and carcinoembryonic antigen (CEA) (Ma et al., 2009; Zhao et al., 2022). These markers are elevated in the presence of PC and can help in the diagnosis, as well as monitoring the progression and response to treatment. However, it is important to note that biomarker testing alone is not sufficient for diagnosing PC, as these markers can be elevated in other non-malignant conditions, limiting their specificity. Therefore, combining biomarker testing with imaging techniques provides a more comprehensive and accurate diagnosis (Zhao et al., 2022). The availability of these facilities is very limited in Africa. The biomarkers are not available in most countries in Africa (Elmore et al., 2021).

TREATMENT AND MANAGEMENT APPROACHES

Overview of Current Treatment Options

Surgery remains the gold standard for the treatment of PC. However, due to late-stage diagnosis and limited surgical expertise, many cases in Africa are deemed unresectable, where surgery is not a viable option (Kim et al., 2012). In such cases, chemotherapy and radiation therapy play a crucial role in controlling diseases and providing palliative care. Radiotherapy services are also non-existent in about 23 countries in Africa. Most cases are inoperable and are treated with chemotherapy and palliative care services. Chemotherapy drugs such as gemcitabine and fluorouracil are commonly used to target cancer cells, thereby slowing down the progression of the disease. Moreover, emerging therapies, including targeted treatments and immunotherapy, are being explored to enhance outcomes for patients who cannot undergo surgery (Mishra, 2021). As research advances, clinical trials are increasingly focusing on these innovative approaches, aiming to improve survival rates and quality of life for patients facing this challenging diagnosis. Supportive care measures, such as nutritional support and pain management, are essential components of the treatment plan to ensure that patients maintain their strength and comfort throughout their journey (Popescu et al., 2021). These holistic strategies not only address the physical aspects of cancer treatment but also provide emotional and psychological support, fostering resilience and hope in patients and their families. Integrating palliative care early in the treatment process can significantly enhance the overall patient experience, allowing for a more comprehensive approach that prioritizes both medical and emotional well-being (Akinola and Yusuf, 2023).

Accessibility, Affordability of Treatments in Africa and Emerging Treatment Modalities as well as Their Potential Impact

Despite advancements in treatment options, accessibility and affordability remain major barriers to effective PC care in Africa. Many countries on the continent lack the necessary infrastructure, medical personnel, and resources to provide comprehensive cancer treatment (Alatise et al., 2010). Access to specialized cancer centers is limited, and financial constraints prevent many patients from affording the necessary treatments. There is an urgent need for increased investment in healthcare infrastructure, medical training, and affordable medication to improve the availability of treatment options for PC patients in Africa. Public awareness campaigns are essential to educate communities about the symptoms and risks associated with PC, encouraging early detection and timely intervention (Alatise et al., 2010). In recent years, there have been promising advancements in the field of PC treatment. Targeted therapies, immunotherapy, and precision medicine hold immense potential for improving patient outcomes. These innovative approaches aim to precisely target cancer cells, minimizing damage to healthy tissue, and increasing treatment efficiency. Clinical trials are investigating the effectiveness of these emerging treatment modalities in the African context, offering hope for improved survival rates and quality of life for patients in the region. The infrastructure needed for clinical trials is lacking in most countries in Africa. With the exception of South Africa and Egypt, there are limited or non-existent oncology clinical trials, including those for PC, on the continent (Narayanan and Weekes, 2016; Liu et al., 2021). As these trials progress, collaboration between local healthcare providers and international research institutions will be crucial in adapting these therapies to meet the unique needs of African populations. Community engagement and education will play a vital role in ensuring that patients are informed about their options and can access these cutting-edge treatments (Mutebi et al., 2022; Radouani et al., 2020).

Role of Traditional Medicine and Integrative Approaches

Traditional medicine has long been an integral part of African culture and healthcare practices. In the context of PC management, traditional medicine can play a supportive role alongside conventional treatments (James et al., 2022; Mudyiwa et al., 2022). Herbal remedies and plant extracts, such as *Sutherlandia frutescens* and *Phyllanthus niruri*, have shown potential anticancer properties and are being studied for their effectiveness against PC (Triantafillidis et al., 2022). Integrative approaches that combine traditional medicine with standard treatments are gaining attention as they offer a holistic approach to patient care, addressing physical, emotional, and spiritual needs. These approaches not only enhance the overall well-being of patients but also empower them by involving their cultural beliefs and practices in the healing process (Azimi et al., 2015). Moreover, ongoing research is essential to validate these traditional practices and ensure their safety and efficacy, paving the way for more personalized treatment plans that respect and incorporate diverse cultural perspectives. There is also a need to establish an agency that will regulate the practice of traditional medicine in Africa. Otherwise, this could be mixed with traditional African power and incantation. As researchers delve deeper into the mechanisms of action, they are uncovering specific compounds within these plants that may synergize with conventional therapies, potentially leading to improved outcomes and reduced side effects for patients (Gnanaraj et al., 2017; De Cicco et al., 2018).

HEALTHCARE INFRASTRUCTURE AND RESOURCES

State of Healthcare Systems in Africa Regarding Cancer Care

Healthcare systems around the world face various challenges, particularly in low- and middle-income countries where resources are often stretched thin. In Africa, the burden of non-communicable diseases, including cancer, has been steadily increasing, posing a significant threat to public health and economic stability. The existing high burden of communicable diseases such as HIV/AIDS, tuberculosis, etc., still existing in a setting with weak health infrastructure, poverty, corruption, and lack of political support in addressing this concern. What we are now experiencing with the existing high burden of communicable diseases is a superimposed increasing burden of non-communicable diseases, including cancer (Stefan, 2015). As nations strive to improve their healthcare frameworks, the integration of effective cancer care services remains a critical concern. In September 2000, during a United Nations General Assembly, a total of 54 African nations committed themselves to the improvement of healthcare systems in their respective countries by increasing the budget allocated to the health departments from 5% to 15% of the national budget, with only a total of seven nations successfully able to meet this goal by 2006 (Kirigia et al., 2006). International donors played a crucial role in the achievement of this goal by providing financial support. Some African nations went in the opposite direction, reducing the already low 5% allocation, which resulted in citizens having to cover the cost of their healthcare (Kirigia et al., 2006). Despite the fact that Africa represents approximately 15% of the global population, the continent still bears a disease burden of 25% (Bray et al., 2013), with SSA receiving less of the total worldwide health expenditure to date (Jeetoo and Jaunky, 2023).

Healthcare systems, as well as infrastructure, require urgent improvement to address some of the challenges affecting cancer care in Africa (Stefan, 2015). The establishment of comprehensive and reliable cancer registries would aid in the implementation of effective preventative measures, identification of risk factors, as well as understanding the true burden of cancer in Africa (Figure 2.3). Moreover, the burden of cancer in SSA can be mitigated by raising awareness about cancer risk factors and promoting early detection and intervention (Stefan, 2015).

Availability of Specialized Cancer Centers and Expertise

PC, like many other cancer types in Africa, is habitually not diagnosed and treated until the disease presents at an advanced stage due to insufficient access to healthcare facilities that offer healthcare services in the form of effective screening and early treatment (Mutebi et al., 2022). Those who can access some of the available facilities are sometimes diagnosed, but treatment may be unavailable, and if available, better outcomes are not guaranteed. In SSA countries and similar regions, fewer than 30% of patients diagnosed with cancer are likely to receive treatment, in contrast to 90% in high-income countries, as reported by WHO (Omotoso et al., 2023). Limited treatment options, lack of qualified medical experts, high cost of treatment, and limited access to medical care are just a

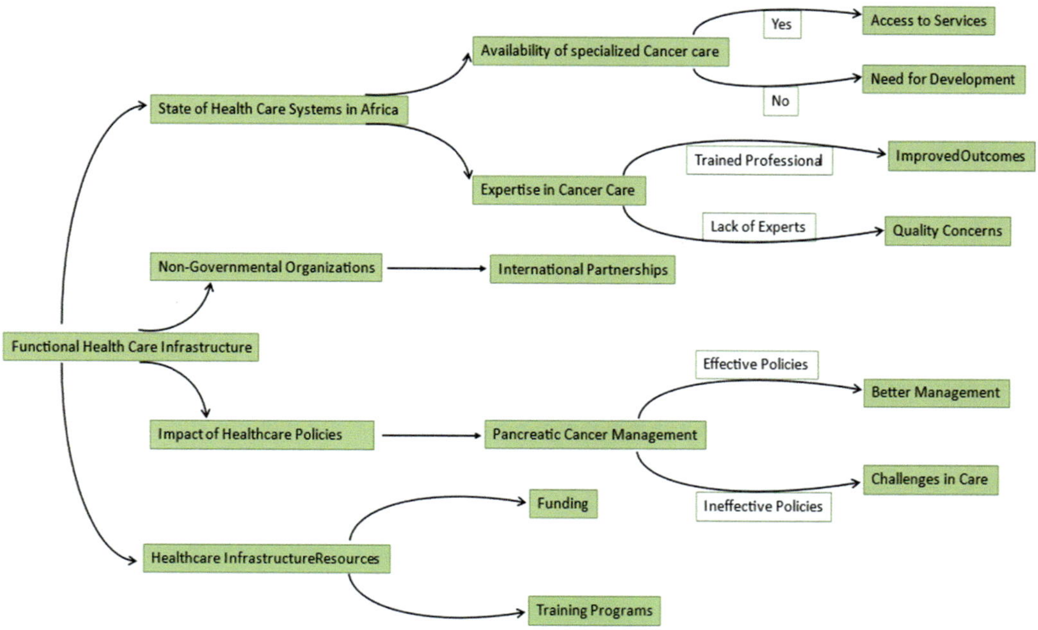

FIGURE 2.3 Comparative analysis of healthcare infrastructure and resources on cancer care outcomes. The tree diagram underscores the critical role of robust healthcare infrastructure and adequate resources in improving cancer care outcomes. Regions with well-developed healthcare systems tend to have higher survival rates, better early detection, and improved quality of life for cancer patients. Conversely, areas with limited resources face significant challenges, resulting in poorer outcomes and higher mortality rates.

few of many factors driving the observed disparity in cancer care in SSA countries, and these need urgent attention (Mutebi et al., 2022).

SSA presents a lack of expertise in surgical and cancer specialists due to the significant shortfall in the number of healthcare workers on the continent with WHO reporting that 18 years ago, about 3% of healthcare workers in the world resided in Africa (Chikezie et al., 2023). The figure has significantly declined to approximately 1.3%, with a further decline of over 6 million projected by 2030 due to factors such as emigration (Naicker et al., 2010). Also, there is a paucity of reliable data concerning the number of oncology experts in SSA, although it is believed to be significantly lower in comparison to other continents, primarily due to the limited access to specialized training in cancer care (Morhason-Bello et al., 2013).

This scarcity of healthcare professionals and oncology specialists exacerbates the existing challenges in dealing with cancer in Africa. It leads to an insufficient capacity for essential aspects of cancer care such as screening, diagnosis, treatment, and post-treatment follow-up, resulting in delayed diagnoses and substandard management of cancer cases (Morhason-Bello et al., 2013). Oncologists in Africa experience a significant workload with a higher patient-to-oncologist ratio because of a shortage of oncology specialists to meet the needs of the large volume of cancer patients seeking care (Vanderpuye et al., 2019). In SSA, countries such as Cameroon present with about three radiologists and four medical oncologists attending to a population of about 26 million people, while in Liberia with a population of 5.6 million, there is only one oncologist (Foxall, 2019; Omotoso et al., 2023). In 2018, Sierra Leone, Benin, Lesotho, Gambia, and South Sudan had no available oncologists (Mathew, 2018). This shortage of oncologists and other healthcare professionals with specialized training further compounds the obstacles faced in delivering effective cancer care across Africa, leading to delays in diagnoses, restricted treatment alternatives, and insufficient post-treatment monitoring and care for individuals with cancer (Mathew, 2018). Because of this expertise shortfall, diagnosis is late with at least 80% of patients in Africa diagnosed with cancer at an advanced stage, consequently leading to an increased mortality ratio, which is higher than that observed in the global north (Kmietowicz, 2007).

Role of Non-governmental Organizations and International Partnerships

Non-governmental organizations (NGOs) as well as international partnerships through such organizations play a vital role in combating cancer in Africa, as these entities often bring together resources, expertise, and advocacy efforts that can significantly enhance cancer care and prevention strategies across SSA and the continent at large (Freireich, 2012). These collaborations can lead to improved access to treatment, increased awareness campaigns, and the development of tailored programs that address the unique challenges faced by different regions. Also, by leveraging local knowledge and fostering community engagement, these initiatives can empower individuals to take charge of their health and promote early detection practices that are vital for successful outcomes (Maree and Wright, 2010).

The African Organisation for Research and Training in Cancer (AORTIC) is one of the prominent NGOs dedicated to the promotion of cancer control as well as palliation in Africa. AORTIC aims to achieve this goal by striving to unite the continent and to make a positive impact throughout SSA by collaborating with health ministries and global cancer organizations (Mohammed et al., 2012). AORTIC has succeeded in putting cancer on the public health agenda in many African countries by highlighting Africa's urgent need for cancer control. AORTIC aims to conduct further research relating to cancers prevalent in Africa, to support training programs in oncology for healthcare workers, to deal with the challenges of creating cancer control and prevention programs, and to raise public awareness of cancer in Africa (Mohammed et al., 2012). The Africa Hepatopancreatobiliary Cancer Consortium (AHPBCC) is also actively involved in building capacity, conducting research, and promoting exchange programs by harnessing the collective expertise and resources in the United States with African investigators to advance the science of hepatopancreatobiliary cancers in Africa (Cooley et al., 2024). AORTIC, as well as other NGOs such as the African Cancer Coalition (ACC), AHPBCC, and the African Palliative Care Association (APC) to name a few, play a pivotal role in promoting cancer control and providing a platform for knowledge exchange and capacity building among healthcare professionals in Africa via the organization of symposia, workshops, meetings, and conferences that support their mission (Mohammed et al., 2012; Cooley et al., 2024).

Impact of Healthcare Policies on Pancreatic Cancer Management

Healthcare policies have a significant impact on the management of PC by shaping the way in which resources are allocated, services are provided, and research is conducted. One of the key factors affecting PC management in Africa is the limited access to specialized healthcare services, as discussed previously (Morhason-Bello et al., 2013). Healthcare policies also influence the availability of essential medication and treatment options for PC in Africa. The high cost of cancer drugs, as well as limited access to advanced treatments such as surgery, chemotherapy, and radiation therapy, can significantly impact the quality of care for patients with PC. The lack of insurance coverage and financial support for cancer patients further exacerbates the challenges in managing PC in Africa (Papageorge et al., 2021).

Examining the role of patient advocacy groups in shaping these policies can provide valuable insights into how community engagement affects legislative priorities and resource allocation. Analysis of the economic implications of these policies on both patients and healthcare systems can also shed some light on the sustainability of PC treatments and the overall burden of the disease (Papageorge et al., 2021).

RESEARCH AND DATA COLLECTION

Current State of Pancreatic Cancer Research in Africa

In Africa, PC is increasingly recognized as a pressing health issue due to several factors including rising life expectancy, evolving dietary patterns, and insufficient awareness regarding the disease (Pourshams et al., 2019). However, research efforts and data collection concerning PC in Africa are notably lagging compared to other global regions. This shortfall impedes a comprehensive understanding of the disease burden and restricts available treatment options for affected patients (Aryannejad et al., 2021). A primary obstacle to advancing research on PC within Africa is the limited funding and inadequate infrastructure dedicated to scientific inquiry. Many African nations face challenges such as insufficient resources and expertise in oncology, which hinder their ability to conduct high-caliber research studies.

There is therefore an urgent need to educate the general public and provide policymakers with high-quality, peer-reviewed, regionally appropriate evidence. This goes hand in hand with increasing the amount of research that is published and read by African oncology professionals so that they are both aware of the latest developments in global cancer research and can bring attention to the region-specific issues surrounding cancer prevention and care in Africa (Foxall, 2019). Publication of research is an essential component of improving the skills of healthcare workers and addressing global inequalities in cancer communication, which inevitably leads to better patient outcomes. African researchers currently only contribute approximately 1% authorship of the world's published cancer research, even though Africa accounts for 15% of the world population (Bray et al., 2013). This inequality is primarily driven by a lack of funding for the training of the already low-to-non-existent medical personnel (oncologists, pathologists, and epidemiologists) in oncology and a lack of experience in applying research methods (Bruijns and Lamanna, 2018).

Another barrier includes the absence of robust data collection systems and comprehensive cancer registries across numerous African countries. The lack of accurate statistics regarding incidence rates, prevalence trends, and patient outcomes presents formidable challenges when attempting to devise effective prevention strategies or therapeutic interventions. In many instances where registries do exist, they may be either incomplete or outdated, thereby obscuring an accurate assessment of pancreas-related cancers within these regions.

Importance of Cancer Registries and Surveillance Systems

Despite ongoing improvements in epidemiological strategies, the health system in Africa faces substantial challenges, one of which is the lack of standardized data registry infrastructure for cancer (Azevedo, 2017; Hamdi et al., 2021). The lack of uniform data registry infrastructure for cancer results in negative downstream effects such as incomplete assessment of the geographic distribution and determinants, exacerbating risk factors, reduced awareness, and hindering patient tracking and referral for preventative care. This includes population-based cancer registries, which are scarce and central to cancer research as they serve as data hubs required for the implementation of preventative measures and national as well as regional planning (WHO). Standard data registry systems for cancer are available in about 25 countries in SSA. The African Cancer Registry Network (AFCRN), with an international office at Oxford University, has been improving the effectiveness of cancer surveillance in SSA since its establishment on March 1, 2021. This network is working very closely with IARC WHO Lyon in France to ensure high-quality data on cancer from SSA. The data from the various population cancer registries in the network are featured in several GLOBOCAN global cancer statistics. However, we must emphasize that some countries, like Central African Republic and the Republic of Congo, do not have an efficient national cancer registry (Omonisi et al., 2020). SSA, which has the most low-to-middle-income countries, has shown a rise in cancer cases, e.g., lung cancer, due to lack of awareness of risk factors associated with the disease, namely exposure to insecticides, asbestos, fumes, and smoking. Meanwhile, in high-income countries, a significant decrease has been observed due to the implementation of increased awareness of the disease in the population (Azevedo, 2017; Dhieb et al., 2019; Hamdi et al., 2021).

African countries that have implemented population-based cancer registries have reliable data on cancer, including PC, in SSA. However, several challenges confront effective cancer registration in some regions. These challenges include a lack of adequate technical staff, funding for the registry, lack of up-to-date census figures, and most regions lack political support and efficiently functional health infrastructure. These challenges mitigate against greater coverage of the population by participating in cancer registries (WHO; Omonisi et al., 2020). In South Africa, PC is recorded as the 24th most common malignancy, with the incidence rate of 0.8 per 100,000 in the National Cancer Registry (NCR). This figure is likely not to be a true representation, given that PC is under-diagnosed and under-reported in SSA (Nhleko et al., 2023). To date there is no national screening program for PC available in all of Africa, nor has any high-risk group been identified. No geographical mapping of the incidence of PC in a South African population group has been reported (Mthunzi and Noel, 2023). Despite these challenges, there are notable research initiatives and institutions in Africa focused on addressing PC. One important effort is the development of national comprehensive cancer control programs, which aim to improve cancer care through the creation of standardized treatment guidelines and the coordination of training, research, and resource allocation (Mutebi et al., 2022). There have been efforts to expand cancer care infrastructure and increase access to diagnostic and treatment services across the continent (Stefan, 2015).

CASE STUDIES AND REGIONAL INSIGHTS

Detailed Analysis of Pancreatic Cancer in Specific African Countries or Regions (Lessons Learned from Regional Variations and Interventions)

PC is a growing concern in many African countries, with rates of the disease on the rise in recent years. Although there are limited data on the prevalence and incidence of PC in specific African countries, several studies have shed light on the burden of the disease in certain regions. Within Northern Africa, age-standardized incidence rates (ASR) range from 1.0 per 100,000 in Sudan to 4.3 per 100,000 in Libya (Bray et al., 2018). In a systematic review conducted by Pearlstein et al. (2021), it was observed that there exists a disparity in PC publication volume, with the majority of scholarship from two countries, Egypt, and Nigeria. In 2012, the incidence and mortality rates of PC in Africa were reported to be about 22.2 per 100,000 and 17 per 100,000, as reported by GLOBOCAN 2012 (Ferlay et al., 2018). In contrast to the rest of the world, these rates were very low, yet within the black populations, evidence showed an increased mortality rate compared to other races (Odedina et al., 2009; Rebbeck et al., 2013). This was further evidenced by the significant variations seen in the presentation as well as patterns of PC between Northern Africa and SSA (Adeloye and Basquill, 2014).

Regional evaluation of PC incidence variation in Egypt's Nile delta by Baum et al. (2020) showed a significant difference by district. The study noted a significantly higher incidence rate in urban districts compared with the rural districts, leading to the hypothesis that such disparities are not due to poor access to health facilities, as approximately 95% of the population in Egypt live close (i.e., within 5 km) to primary healthcare facilities (Baum et al., 2020). Soliman et al. (2006) previously showed that the mortality rate also varied by province in Egypt, with a 2.85 average rate for northern districts in contrast to the southern districts. The differences attributed to the varying levels of environmental exposure between the regions (Soliman et al., 2006). Mthunzi and Noel (2023), in the first study to investigate the geographic distribution of PC in South Africa, showed that regional distribution of incidence provides important details about the disease's epidemiology, similar to the risk factors. It was also discovered that the rates of PC in every district in central South Africa were much higher than those reported nationally by the NCR in the country (Mthunzi and Noel, 2023). In central South Africa, the mean age for PC diagnoses is 62 years, while the regions with the highest

PC rate had a population with a mean age of 25 years (Mthunzi and Noel, 2023). In contrast to international research that claims a male-to-female ratio of 2:1, this study also showed a higher incidence of PC in females than in males (Rawla et al., 2019; da Costa et al., 2020; Mthunzi and Noel, 2023).

Due to inadequate health management information systems, it is still very difficult to accurately quantify the burden of PC in Africa, despite this seemingly disproportionate burden and the ongoing issues (Odedina et al., 2009; Bray et al., 2018). A relative lack of information and data on cancers and many other health issues on the continent has been attributed to poor routine recordkeeping in the majority of African health facilities, according to numerous studies (Omotoso et al., 2023). This has, in turn, had a significant impact on research output, public health interventions, and policy response.

In Africa, the PC burden is still anticipated to rise throughout the continent due to ongoing urbanization, population expansion, and rising life expectancies. The way that the governments of many African countries have responded is still quite concerning. Health programs in Africa urgently need to be reoriented with an eye toward enhancing training and research, screening, diagnosis, as well as treatment. This milestone can only be achieved through strong political will, increased funding, capacity building, the establishment of improved treatment facilities, and the establishment of sustainable palliative care programs in Africa.

FUTURE DIRECTIONS AND RECOMMENDATIONS

Strategies to Improve Early Detection and Diagnosis

The early detection and diagnosis of PC are critical for improving patient outcomes, given the disease's aggressive nature and poor prognosis (Yang et al., 2021). In the context of Africa, where healthcare resources may be limited, strategies to improve early detection are particularly vital. While the reviewed literature does not specifically address the African context for PC, insights can be drawn from the general principles and advances in detection and diagnosis. Innovative technologies such as artificial intelligence (AI) and computer-aided diagnosis systems have shown promise in enhancing the detection of PC through improved imaging analysis (Idoko et al., 2024; Yang et al., 2021). Particle swarm optimization (PSO) has been highlighted as a technique that improves the visibility of CT images, aiding in the accurate detection and diagnosis of pancreatic tumors (Dhruv et al., 2021). The importance of screening high-risk individuals and the potential of liquid biopsies as well as molecular markers are emphasized, although their application in unselected individuals remains limited (Schwingel et al., 2023). The literature does not directly address the situation in Africa, but the strategies that could be adapted and implemented in an African setting include the use of AI and optimized imaging techniques to improve diagnostic accuracy (Dhruv et al., 2021; Yang et al., 2021; Idoko et al., 2024). Focusing on high-risk populations and exploring novel biomarkers and liquid biopsies may offer avenues for early detection (Schwingel et al., 2023). It is crucial to consider the specific healthcare context of African countries and adapt these strategies accordingly, potentially drawing on experiences from other regions with similar healthcare challenges, such as the use of social media technologies in cervical cancer control in SSA (Eilu and Baguma, 2019).

Enhancing Treatment Access and Outcomes

Despite advances in PC treatment, access to care remains a major barrier in many African countries. Strategies to improve treatment access could include the expansion of cancer centers, increasing the availability of essential cancer drugs, and developing national cancer control plans. Improving the integration of palliative care services and supporting patients and caregivers are also crucial to enhance the quality of life for those living with PC (Adesina et al., 2013; Nelson et al., 2016; Farrè and Kibera, 2018). Challenges such as limited healthcare resources, high treatment costs, and inadequate transportation infrastructure have already been noted to be the most common contributors to poor treatment access and outcomes. Therefore, governments and international organizations should address these barriers by fostering collaboration to ensure affordable and accessible healthcare services. This could be achieved through the establishment of specialized PC treatment centers, the training of healthcare professionals in the latest treatment techniques, and the inclusion of PC treatments in national health insurance schemes (Nelson et al., 2016; Farrè and Kibera, 2018).

Strengthening Healthcare Infrastructure, Workforce Research, and International Collaboration

Robust healthcare infrastructure and a well-trained workforce are fundamental to delivering high-quality cancer care to effectively manage PC in Africa. This should involve investing in the construction and equipping of hospitals and clinics, especially in rural and under-served areas (LeJeune et al., 2019). Specialized training programs should also be invested in provision of adequate equipment and supplies, as well as ensuring effective quality assurance measures, which are critical steps to address the current deficits. Training programs should be implemented to increase the number of oncologists, surgeons, nurses, and other healthcare professionals with expertise in cancer research. The decentralization of cancer services and empowering primary care providers to play a greater role in early detection and supportive care can help improve access in rural and underserved areas (Dos Santos Silva et al., 2022). Efforts should be made to improve the coordination and integration of care across the healthcare system to ensure timely and comprehensive treatment delivery (Warner-Cohen et al., 2020).

Advancing PC research in Africa is essential to developing context-specific solutions. Multidisciplinary research teams, improved data collection and sharing, as well as collaborations with international partners can help strengthen the cancer research ecosystem (Mutebi et al., 2022). Prioritizing research on early detection, novel treatment approaches, and implementation science can generate evidence to inform policymaking and improve outcomes. Research is needed to evaluate the prevalence and risk factors of PC in African populations, as well as to identify potential biomarkers for early detection. Clinical trials should be conducted to test the efficacy and safety of new treatments specifically tailored to the African context (Odedina and Rotimi, 2021; Mutebi et al., 2022).

Policy Recommendations for Addressing Pancreatic Cancer Burden in Africa

Addressing the growing PC burden in Africa will require a comprehensive, multi-pronged approach. Integrating PC control into national cancer control plans, allocating sufficient funding, and establishing a robust pan-African specialized cancer registry for PC will greatly facilitate the generation of robust data on the disease (Nelson et al., 2016; Omonisi et al., 2020; dos-Santos-Silva et al., 2022; Mutebi et al., 2022). Governments should prioritize oncology services and allocate sufficient funding for PC prevention, treatment, and research. Policies should be developed to incentivize and facilitate the recruitment and retention of healthcare professionals in areas with a high burden of PC. The implementation of policies to regulate tobacco use, as smoking is a significant risk factor for PC, and to improve nutrition and lifestyle choices should be considered. Promoting public awareness, advocating for increased government commitment, and fostering international partnerships can further support these efforts (Nelson et al., 2016; Mutebi et al., 2022).

CONCLUSION

Summary of Key Findings

The burden of cancer, including PC, is on the rise in Africa. The continent faces significant challenges in addressing this growing health crisis, with limited oncology care and pathology resources (Nelson et al., 2016). While some countries have established single cancer referral centers, the ability to scale up services nationally toward universal health coverage remains a significant obstacle (Park et al., 2021). The lack of education, infrastructure, diagnostic services, and high treatment costs further compound the issue, leading to poor cancer outcomes in many African nations (Nelson et al., 2016).

Regarding PC specifically, the epidemiological data and trends in Africa are limited. However, the available evidence suggests that the incidence and mortality rates for this deadly disease are likely on the rise, mirroring the overall trend of increasing cancer burden in the region (Siegel et al., 2017; Wong et al., 2017; Sung et al., 2021). The lack of comprehensive cancer registries and surveillance systems in many African countries makes it difficult to accurately assess the true magnitude of the PC problem (Nelson et al., 2016; Park et al., 2021).

Emphasis on the Need for Comprehensive Action

The growing burden of PC and cancer in general in Africa underscores the urgent need for a comprehensive and coordinated response. Strategies are required to improve oncology care and pathology resources, which are currently lacking in many parts of the continent (Mutebi et al., 2022). Establishing robust cancer registries and surveillance systems is crucial to better understand the epidemiology and trends of PC in Africa (Papageorge et al., 2021; Chikezie et al., 2023). Several risk factors contribute to the development of PC in Africa. One of the most significant risk factors is tobacco smoking. The prevalence of smoking in Africa has been steadily rising, particularly among men (Ghadirian et al., 2003; Wood and Hruban, 2012). Tobacco control measures, such as increased taxation and strict advertising regulations, should be implemented to reduce the incidence of PC. Obesity and a high-fat diet are other prominent risk factors, indicating the importance of promoting healthy lifestyles and balanced nutrition in the population (Arslan et al., 2010; Stolzenberg-Solomon et al., 2013).

Call to Action for Stakeholders at All Levels

Governments and policymakers should prioritize cancer control and prevention efforts by allocating adequate resources for cancer research, prevention, and treatment. Strengthening healthcare systems is crucial to improving access to cancer care, including early detection programs, treatment facilities, and specialized healthcare professionals. Public awareness campaigns to promote healthy lifestyles and raise awareness about the signs and symptoms of PC can help improve early detection rates. Collaboration among various stakeholders, including governments, healthcare professionals, researchers, and international organizations, is essential for developing and implementing effective strategies to combat PC in Africa. Multidisciplinary approaches that integrate primary prevention, early detection, and treatment should be adopted to address this complex disease. Active involvement and engagement from all levels of society are paramount to successfully tackling the challenges posed by PC in Africa. By working together, we can make significant progress in improving patient outcomes and saving lives.

REFERENCES

ADELOYE, D., BASQUILL, C. 2014. Estimating the prevalence and awareness rates of hypertension in Africa: a systematic analysis. *PLoS One*. 9(8), e104300. 10.1371/journal.pone.0104300

ADESINA, A., CHUMBA, D., NELSON, A.M., OREM, J., ROBERTS, D.J., WABINGA, H., WILSON, M., REBBECK, T.R., 2013. Improvement of pathology in sub-Saharan Africa. *The Lancet Oncology,* 14, e152–e157. https://doi.org/10.1016/S1470-2045(12)70598-3.

AKINOLA, A., YUSUF, N., 2023. Palliative care support for cancer patients: short commentary article. *Indian Can Aware J,* 2, 14–16. https://doi.org/10.25259/ICAJ_10_2022.

ALATISE, O.I., LAWAL, O.O., TOPE, O.O., 2010. Management challenges of pancreatic cancer in a resource scarce setting. *African Journal of Health Sciences,* 17, 36–41.

ARNOLD, M., ABNET, C. C., NEALE, R. E., VIGNAT, J., GIOVANNUCCI, E. L., MCGLYNN, K. A., BRAY, F. 2020. Global burden of 5 major types of gastrointestinal cancer. *Gastroenterology*, 159(1), 335–349.e15.

ARSLAN, A.A., HELZLSOUER, K.J., KOOPERBERG, C., SHU, X.O., STEPLOWSKI, E., BUENO-DE-MESQUITA, H.B., FUCHS, C.S., GROSS, M.D., JACOBS, E.J., LACROIX, A.Z., PETERSEN, G.M., 2010. Anthropometric measures, body mass index, and PC: a pooled analysis from the PC Cohort Consortium (PanScan). *Archives of Internal Medicine*, 170(9), 791–802.

ARYANNEJAD, A., TABARY, M., EBRAHIMI, N., MOHAMMADI, E., FATTAHI, N., ROSHANI, S., MASINAEI, M., NADERIMAGHAM, S., AZADNAJAFABAD, S., JAMSHIDI, K., FATEH, S.M., MOGHIMI, M., KOMPANI, F., REZAEI, N., FARZADFAR, F. 2021Global, regional, and national survey on the burden and quality of care of pancreatic cancer: a systematic analysis for the Global Burden of Disease study 1990-2017*Pancreatology*. 21(8), 1443–1450. 10.1016/j.pan.2021.09.002.

AZEVEDO, M.J. 2017. The state of health system(s) in Africa: challenges and opportunities. *Historical Perspectives on the State of Health and Health Systems in Africa, Volume II*. 3:1–73. 10.1007/978-3-319-32564-4_1.

AZIMI, H., KHAKSHUR, A.A., ABDOLLAHI, M., RAHIMI, R., 2015. Potential new pharmacological agents derived from medicinal plants for the treatment of pancreatic cancer. *Pancreas,* 44, 11–15. https://doi.org/10.1097/mpa.0000000000000175.

BAKENS, M. J., LEMMENS, V. E., DE HINGH, I. H. 2017. Socio-economic status influences the likelihood of undergoing surgical treatment for PC in the Netherlands. *HPB*, 19(5), 443–448.

BAUM, C., SOLIMAN, A.S., BROWN, H. E., SEIFELDIN, I.A., RAMADAN, M., LOTT, B., NGUYEN, A., El-GHAWALBY, A., HABLAS, A. 2020. Regional variation of pancreatic cancer incidence in the nile delta region of Egypt over a twelve-year period. *J Cancer Epidemiol*. 6031708. 10.1155/2020/6031708.

BECKER, A. E., HERNANDEZ, Y. G., FRUCHT, H., LUCAS, A. L. 2014. Pancreatic ductal adenocarcinoma: risk factors, screening, and early detection. 世界胃肠病学杂志：英文版（电子版） , 32, 11182–11198.

BLACKFORD, A.L., CANTO, M.I., KLEIN, A.P., HRUBAN, R.H. AND GOGGINS, M., 2020. Recent trends in the incidence and survival of stage 1A PC: a surveillance, epidemiology, and end results analysis. *JNCI: Journal of the National Cancer Institute*, 112(11), 1162–1169.

BOYD, L., MEIJER, L., IN 'T VELD, S., LE LARGE, T., BEST, M., LELIJ E., V.D., DWARSHUIS, G., SOL, N., MANTINI, G., CAPULA, M., FRAMPTON, A., MORELLI, L., ARKANI, M., POST, E., ZONDERHUIS, B., DAAMS, F., BESSELINK, M., BIJLSMA, M., GIOVANNETTI, E., WURDINGER, T., KAZEMIER, G., 2020. Blood platelet RNA yields a new diagnostic prospective for accurate detection and staging of pancreatic ductal adenocarcinoma. Pancreatology, 52nd meeting of the European Pancreatic Club combined with the International Association of Pancreatology 20, S135. https://doi.org/10.1016/j.pan.2020.07.253

BRAY, F., REN, J.S., MASUYER, E., FERLAY, J. 2013. Global estimates of cancer prevalence for 27 sites in the adult population in 2008. *International Journal of Cancer*. March 1, 132(5), 1133–1145. : 10.1002/ijc.27711

BRAY, F., FERLAY, J., SOERJOMATARAM, I., SIEGEL, R.L., TORRE, L.A. AND JEMAL, A., 2018. Global cancer statistics 2018: GLOBOCAN estimates of incidence and mortality worldwide for 36 cancers in 185 countries. *CA: A Cancer Journal for Clinicians*, 68(6), 394–424.

BRUIJNS, S., LAMANNA, C. 2018. The case for a regional approach to publication impact. *Ecancermedicalscience*. 12, ed78. 10.3332/ecancer.2018.ed78.

BRUNE, K.A., LAU, B., PALMISANO, E., CANTO, M., GOGGINS, M.G., HRUBAN, R.H. AND KLEIN, A.P., 2010. Importance of age of onset in PC kindreds. *Journal of the National Cancer Institute*, 102(2), 119–126.

CALIMANO-RAMIREZ, L. F., DAOUD, T., GOPIREDDY, D. R., MORANI, A. C., WATERS, R., GUMUS, K., KLEKERS, A. R., BHOSALE, P. R., AND VIRARKAR, M. K. 2022. Pancreatic acinar cell carcinoma: a comprehensive review. *World Journal of Gastroenterology*, 28(40), 5827–5844. https://doi.org/10.3748/wjg.v28.i40.5827.

CASARI, I., FALASCA, M. 2015. Diet and PC Prevention. *Cancers*, 7, 2309–2317.

CHEN, F., ROBERTS, N.J., KLEIN, A. P. 2017. Inherited PC. *Chinese Clinical Oncology*, 6(6), 58.

CHIKEZIE, N.C., SHOMUYIWA, D.O., OKOLI, E.A., ONAH, I.M., ADEKOYA, O.O., OWHOR, G.A., ABDULWAHAB, A.A. 2023 Addressing the issue of a depleting health workforce in sub-Saharan Africa. *Lancet,* 401(10389):1649–1650 10.1016/S0140-6736(23)00720-1.

CHONG, C.C.N., TANG, R.S.Y., WONG, J.C.T., CHAN, A.W.H., TEOH, A.Y.B., 2016. Endoscopic ultrasound of pancreatic lesions. *Journal of Visualized Surgery,* 2. https://doi.org/10.21037/jovs.2016.07.10.

CLOYD, J. M., AND POULTSIDES, G. A. 2018. Ampullary cancer: managing the last frontier in periampullary malignancies. *Surgical Oncology Clinics of North America*, 27(2), 341–353.

COOLEY, M., ASOMBANG, A., CHIPAILA, J., NAGALO, B.M., ANTWI, S., ROBERTS, L., 2024. The Africa hepatopancreatobiliary cancer consortium for the study of epidemiology, etiology, pathobiology of liver, biliary tract, and pancreatic cancers in Africa. *HPB,* 26, S503. https://doi.org/10.1016/J.HPB.2024.03.1064.

DA COSTA, W.L., Jr, OLUYOMI, A.O., THRIFT, A.P. 2020. Trends in the incidence of pancreatic adenocarcinoma in All 50 United States examined through an age-period-cohort analysis. *JNCI Cancer Spectr.* 4(4), pkaa033, 10.1093/jncics/pkaa033.

DANDRI, M., LOCARNINI, S. 2012. New insight in the pathobiology of hepatitis B virus infection. *Gut*, 61 (Suppl. 1), i6–i17.

DE CICCO, P., PANZA, E., ARMOGIDA, C., ERCOLANO, G., CIRINO, G., IANARO, A., 2018. Current practices and awareness of anticancer plants in the traditional healthcare system. In: Akhtar, M.S., Swamy, M.K. (eds.) *Anticancer plants: properties and application: volume 1.* Singapore: Springer, pp. 1–36. https://doi.org/10.1007/978-981-10-8548-2_1.

DEL CHIARO, M., SEGERSVARD, R., LOHR, M., VERBEKE, C. 2014. Early detection and prevention of PC: is it really possible today? *World Journal of Gastroenterology*, 20(34), 12118–12131.

DHIEB, D., BELGUITH, I., CAPELLI, L., CHIADINI, E., CANALE, M., BRAVACCINI, S., YANGUI, I., BOUDAWARA, O., JLIDI, R., BOUDAWARA, T., CALISTRI, D., KESKES, L.A., ULIVI, P. 2019. Analysis of genetic alterations in Tunisian patients with lung adenocarcinoma. *Cells.* 8(6), 514. 10.3390/cells8060514.

DHRUV, B., MITTAL, N., MODI, M., 2021. Early and precise detection of pancreatic tumor by hybrid approach with edge detection and artificial intelligence techniques. *EAI Endorsed Transactions on Pervasive Health and Technology,* 7, phat 21(28), e1. http://dx.doi.org/10.4108/eai.31-5-2021.170009 .

DINARVAND, P. AND LAI, J. 2017. Solid pseudopapillary neoplasm of the pancreas: a rare entity with unique features, *Archives of Pathology & Laboratory Medicine*, 141(7), 990–995. https://doi.org/10.5858/arpa.2016-0322-rs.

DJURIC, Z. 2017. Obesity-associated cancer risk: the role of intestinal microbiota in the etiology of the host proinflammatory state. *Translational Research*, 179, 155–167.

DOS-SANTOS-SILVA, I., GUPTA, S., OREM, J., SHULMAN, L.N., 2022. Global disparities in access to cancer care. *Communications Medicine*, 2, 31. https://doi.org/10.1038/s43856-022-00097-5.

DUELL, E. J. 2012. Epidemiology and potential mechanisms of tobacco smoking and heavy alcohol consumption in PC. *Molecular Carcinogenesis*, 51, 40–52.

EILU, E., BAGUMA, R., 2019. Prevention, early detection, and treatment of cervical cancer in Sub-Saharan Africa (SSA): a mobile social media approach. In: Paiva, S. (ed.) *Mobile solutions and their usefulness in everyday life.* Springer International Publishing, Cham, pp. 189–207. https://doi.org/10.1007/978-3-319-93491-4_10.

ELMORE, S.N.C., POLO, A., BOURQUE, J.-M., PYNDA, Y., MERWE, D. VAN DER, GROVER, S., HOPKINS, K., ZUBIZARRETA, E., ABDEL-WAHAB, M., 2021. Radiotherapy resources in Africa: an International Atomic Energy Agency update and analysis of projected needs. *The Lancet. Oncology,* 22, e391. https://doi.org/10.1016/S1470-2045(21)00351-X.

FARRÉ, X., KIBERA, J., 2018. The untapped potential of digital pathology in prostate cancer diagnosis and medical education in sub-Saharan Africa. *African Journal of Urology*, 24, 54–55. https://doi.org/10.1016/j.afju.2017.07.003.

FERLAY, J. E. M., LAM, F., COLOMBET, M., MERY, L., PINEROS, M., ZNAOR, A., SOERJOMATARAM, I. 2018. *Global cancer observatory: cancer tomorrow.* Lyon, France: International Agency for Research on Cancer.

FOXALL, K. 2019. The current state of African oncology research publication: how to increase Africa's research impact. *Ecancermedicalscience,* 13, ed93. 10.3332/ecancer.2019.ed93.

FREIREICH, E.J. 2012 The African organization for research and training in cancer. *Current Oncolology (Toronto Ont).* 19(5), 245. 10.3747/co.19.1102

GAIDHANI, R.H. AND BALASUBRAMANIAM, G., 2021. An epidemiological review of PC with special reference to India. *Indian Journal of Medical Sciences*, 73(1), 99–109.

GALLEGOS, J.M., TAYLOR, A., VARDELL, V., SILBERSTEIN, P.T. 2023. Socioeconomic factors associated with a late-stage pancreatic cancer diagnosis: an analysis of the national cancer database. *Cureus,* 15(3), e35857. https://doi.org/10.7759/cureus.35857.

GENKINGER, J. M., SPIEGELMAN, D., ANDERSON, K. E. 2009. Alcohol intake and PC risk: a pooled analysis of fourteen cohort studies. *Cancer Epidemiology, Biomarkers & Prevention*, 18(3), 765–776.

GHADIRIAN, P., LYNCH, H.T. AND KREWSKI, D., 2003. Epidemiology of PC: an overview. *Cancer Detection and Prevention*, 27(2), 87–93.

GNANARAJ, C., GOTHAI, S., MUNIANDY, K., THAMARAISELVAN, R., ARULSELVAN, P., 2017. Toxicological and pharmacological use of anticancer compounds. In: Akhtar, M.S., SWAMY, M.K. (eds.) *Anticancer clinical trials and nanotechnology: volume 3.* Singapore: Springer, pp. 203–231. https://doi.org/10.1007/978-981-10-8216-0_7.

GOGGINS, M. 2022. Genetic predisposition to PC: An overview. *PC Journal*, 19(2), 89–95.

GRANT, C. R., ARTER, Z. L., TRUC, T., AND LEE, F. C. 2023. Case report: a complete pathologic response in PC with squamous cell differentiation. *Frontiers in Oncology*, 13, 1240405.

GUPTA, N., AND YELAMANCHI, R. 2021. Pancreatic adenocarcinoma: a review of recent paradigms and advances in epidemiology, clinical diagnosis and management. *World Journal of Gastroenterology*, 27(23), 3158–3181. https://doi.org/10.3748/wjg.v27.i23.3158.

GUPTA, S., VITTINGHOFF, E., BERTENTHAL, D., CORLEY, D., SHEN, H., WALTER, L.C. AND MCQUAID, K., 2006. New-onset diabetes and PC. *Clinical Gastroenterology and Hepatology*, 4(11), 1366–1372.

GÜVEN, Z.T., ÖZKAN KARAAHMETOĞLU, S., DOĞAN, M., GÜVEN, Z.T., ÖZKAN KARAAHMETOĞLU, S., DOĞAN, M., 2023. Retrospective evaluation of our patients diagnosed with advanced pancreatic cancer. *Journal of Current Hematology & Oncology Research,* 1, 49–53.

HAEBERLE, L., ESPOSITO, I. 2019. Pathology of pancreatic cancer. *Transl Gastroenterol Hepatology* June 27, 4, 50. doi: 10.21037/tgh.2019.06.02.

HAMDI, Y., ABDELJAOUED-TEJ, I., ZATCHI, A.A., ABDELHAK, S., BOUBAKER, S., BROWN, J.S., BENKAHLA A. 2021. Cancer in Africa: the untold story. *Frontiers in Oncology* 11, 650117. 10.3389/fonc.2021.650117.

HASSAN, M. M., LI, D., EL-DEEB, A. S., WOLFF, R. A., BONDY, M. L., DAVILA, M., ABBRUZZESE, J. L. 2008. Association between hepatitis B virus and PC. *Journal of Clinical Oncology*, 26(28), 4557–4562.

HERNÁNDEZ-BLANQUISETT, A., QUINTERO-CARREÑO, V., MARTÍNEZ-ÁVILA, M.C., PORTO, M., MANZUR-BARBUR, M.C., BUENDÍA, E., 2024. Metastatic pancreatic cancer: where are we? *Oncology Reviews,* 17, 11364. https://doi.org/10.3389/or.2023.11364.

HU, J.X., ZHAO, C.F., CHEN, W.B., LIU, Q.C., LI, Q.W., LIN, Y.Y. AND GAO, F., 2021. PC: A review of epidemiology, trend, and risk factors. *World Journal of Gastroenterology*, 27(27), 4298.

HUANG, B.Z., STRAM, D.O., LE MARCHAND, L., HAIMAN, C.A., WILKENS, L.R., PANDOL, S.J., ZHANG, Z.F., MONROE, K.R. AND SETIAWAN, V.W., 2019. Interethnic differences in PC incidence and risk factors: the Multiethnic Cohort. *Cancer Medicine*, 8(7), 3592–3603.

HUANG, J., MAGNUSSON, M., TÖRNER, A., YE, W., DUBERG, A. S. 2013. Risk of PC among individuals with hepatitis C or hepatitis B virus infection: a nationwide study in Sweden. *British Journal of Cancer*, 109, 2917–2923.

IDOKO, D., ADEGBAJU, M., IJEOMA, N., OKEREKE, E., AGABA, J., IJIGA, A., 2024. Enhancing early detection of pancreatic cancer by integrating AI with advanced imaging techniques. *Magna Scientia Advanced Biology and Pharmacy,* 12, 051–083. https://doi.org/10.30574/msabp.2024.12.2.0044.

IKEZAWA, K., URABE, M., KAI, Y., TAKADA, R., AKITA, H., NAGATA, S., AND OHKAWA, K. 2023. Comprehensive review of pancreatic acinar cell carcinoma: epidemiology, diagnosis, molecular features and treatment. *Japanese Journal of Clinical Oncology*, 54(3), 271–281. https://doi.org/10.1093/jjco/hyad176.

ILIC M, ILIC I. 2016. Epidemiology of PC. *World Journal of Gastroenterology*, 22(44), 9694–9705. https://doi.org/10.3748/wjg.v22.i44.9694.

International Agency for Research on Cancer. 2020. World Health Organization, global cancer observatory.

JAIN, A.G., SALEEM, T., KUMAR, R., KHETPAL, N., ZAFAR, H., RASHID, M.U., ALI, S., MAJEED, U., AHMAD, S., 2019. Chapter 1 – Overview of Pancreatic Cancer Biology. In: Nagaraju, G.P. (ed.) *Breaking tolerance to pancreatic cancer unresponsiveness to chemotherapy, cancer sensitizing agents for chemotherapy.* Academic Press, pp. 1–11. https://doi.org/10.1016/B978-0-12-817661-0.00001-9.

JAIN, S.K., 2021. Challenges in diagnosis of pancreatic cancer. *Asian Journal of Multidimensional Research,* 10. https://doi.org/10.5958/2278-4853.2021.01172.1.

JAMES, P.B., ASIIMWE, J.B., WARDLE, J., MWAKA, A.D., KASILO, O.M.J., 2022. African culture, traditional medicine, and cancer care. *The Lancet Oncology,* 23, 705–706. https://doi.org/10.1016/S1470-2045(22)00157-7.

JEETOO, J., JAUNKY, V.C. 2023. Divergence of healthcare expenditure in sub-Saharan Africa. *Studies in Economics and Econometrics,* 47(1), 75–105. DOI: 10.1080/03796205.2023.2185664.

KATO, S., HONDA, K., 2020. Use of biomarkers and imaging for early detection of pancreatic cancer. *Cancers,* 12, 1965. https://doi.org/10.3390/cancers12071965.

KIM, G., QIN, J., HALL, C. B., IN, H. 2022. Association between socioeconomic and insurance status and delayed diagnosis of gastrointestinal cancers. *Journal of Surgical Research*, 279, 170–186.

KIM, S.C., KIM, Y.H., PARK, K.M., LEE, Y.J., 2012. Pancreatic cancer surgery: the state of the art. *Current Drug Targets,* 13, 764–771. https://doi.org/10.2174/138945012800564185.

KLEIN, A.P. 2020. Genetics, screening, and management of patients with familial PC. *Gastroenterology*, 158(3), 735–746.

KLEIN, A.P. 2021. PC epidemiology: understanding the role of lifestyle and inherited risk factors. *Nature Reviews Gastroenterology & Hepatology*, 18(7), 493–502. https://doi.org/10.1038/s41575-021-00457-x.

KMIETOWICZ, Z. 2007. Tackle cancer in Africa now to prevent catastrophe, say health activists. *BMJ*. 334(7602), 1022–1023. 10.1136/bmj.39215.854016.BE.

KIRIGIA, J.M., PREKER, A., CARRIN, G., MWIKISA, C., DIARRA-NAMA, A,J. 2006. An overview of health financing patterns and the way forward in the WHO African Region. *East African Medical Journal*, 83(9 Suppl), S1-28. doi: 10.4314/eamj.v83i9.9492.

LANFRANCA, P. M., THOMPSON, J. K., BEDNAR, F. 2019. Metabolism and epigenetics of PC stem cells. *Seminars in Cancer Biology*, 57, 19–26.

LEE, J.M., YOON, J.H., 2017. Imaging diagnosis of pancreatic cancer: CT and MRI. In: Kim, S.-W., Yamaue, H. (eds.) *Pancreatic cancer: with special focus on topical issues and surgical techniques*. Berlin and Heidelberg: Springer, pp. 95–114. https://doi.org/10.1007/978-3-662-47181-4_7.

LEE, R., HOLMES, D., 2023 Barriers and recommendations for colorectal cancer screening in Africa. *Global Health Action,* 16, 2181920. https://doi.org/10.1080/16549716.2023.2181920.

LEJEUNE, A., MOHAMMED, F., SEYMOUR, D., HUNTHAUSEN, N., LAKE, D., LEE, M., WEBER, C., HARGAN-CALVOPIÑA, J., MANNER, C., GRAEF, K., DENT, J., 2019. Driven by Africa, for Africa – improving access to cancer medicines and technologies through the African Access Initiative (AAI). In: *IEEE Global Humanitarian Technology Conference (GHTC). Presented at the IEEE Global Humanitarian Technology Conference (GHTC)*, pp. 1–2. https://doi.org/10.1109/GHTC46095.2019.9033082.

LIU, T., ZHAO, T., AND SHI, C. 2022. Pancreatoblastoma in children: clinical Management and literature review. *Translational Oncology*, 18, 101359. https://doi.org/10.1016/j.tranon.2022.101359.

LIU, X., LI, Z., WANG, Y., 2021. Advances in targeted therapy and immunotherapy for pancreatic cancer. *Advanced Biology,* 5, 1900236. https://doi.org/10.1002/adbi.201900236.

LOCKHAT, Z.I., VAN RENSBURG, L.J., DU TOIT, H., 2014. Challenges of cancer imaging in Africa. *Cancer Imaging*, 14, O2. https://doi.org/10.1186/1470-7330-14-S1-O2

LUCHINI, C., CAPELLI, P., AND SCARPA, A. 2016. Pancreatic ductal adenocarcinoma and its variants. *Surgical Pathology Clinics*, 9(4), 547–560.

LUO W, WANG J, CHEN H, YE L, QIU J, LIU Y, WANG R, WENG G, LIU T, SU D, TAO J, DING C, YOU L, ZHANG T. 2023. Epidemiology of PC: new version, new vision. *Chinese Journal of Cancer Research*, 35(5), 438–450. https://doi.org/10.21147/j.issn.1000-9604.2023.05.03.

LUU, T.T., 2021. Review of immunohistochemistry biomarkers in pancreatic cancer diagnosis. *Frontiers in Oncology,* 11, 799025. https://doi.org/10.3389/fonc.2021.799025.

MA, Z., MA, Q., WANG, Z., 2009. An evaluation of the diagnostic value of CA19-9 and CEA levels in patients with pancreatic cancer. *Journal of Nanjing Medical University,* 23, 199–202. https://doi.org/10.1016/S1007-4376(09)60055-1.

MAREE, J.E., WRIGHT, S.C. 2010. How would early detection be possible? An enquiry into cancer related knowledge, understanding and health seeking behaviour of urban Black women in Tshwane, South Africa. *European Journal of Oncology Nursing* Jul, 14(3), 190–196. 10.1016/j.ejon.2009.10.009

MARCHESE, U., GAILLARD, M., PELLAT., A, TZEDAKIS., S, ABOU ALI, E., DOHAN, A., BARAT, M., SOYER, P., FUKS, D., CORIAT, R. 2022. Multimodal management of grade 1 and 2 pancreatic neuroendocrine tumors. *Cancers*, 14(2), 433. https://doi.org/10.3390/cancers14020433.

MARINELLI, V., MAZZI, M. A., RIMONDINI, M., DANZI, O. P., BONAMINI, D., BASSI, C., SALVIA, R., AND DEL PICCOLO, L. 2023. Preoperative anxiety in patients with PC: what contributes to anxiety levels in patients waiting for surgical intervention. *Healthcare*, 11(14). https://doi.org/10.3390/healthcare11142039.

MATHEW, A. 2018. Global survey of clinical oncology workforce. *Journal of Global Oncology*, 4, 1–12. 10.1200/JGO.17.00188.

MCGUIGAN, A., KELLY, P., TURKINGTON, R. C., JONES, C., COLEMAN, H. G., MCCAIN, R. S. 2018. PC: a review of clinical diagnosis, epidemiology, treatment and outcomes. *World Journal of Gastroenterology*, 24(43), 4846–4861.

MCMENAMIN, Ú. C., MCCAIN, S., & KUNZMANN, A. T. 2017. Do smoking and alcohol behaviours influence GI cancer survival? *Best Practice & Research Clinical Gastroenterology*, 31(5), 569–577.

MIDHA, S., CHAWLA, S. AND GARG, P.K., 2016. Modifiable and non-modifiable risk factors for PC: a review. *Cancer Letters*, 381(1), 269–277.

MILANETTO, A.C., GAIS ZÜRCHER, A.L., MACCHI, L., DAVID, A. AND PASQUALI, C., 2021. Pancreatic solid pseudopapillary neoplasm in male patients: systematic review with three new cases. *Updates in Surgery*, 73, 1285–1295.

MISHRA, S., 2021. Anaesthetic implications of chemotherapy and radiotherapy. In: Garg, R., Bhatnagar, S. (eds.) *Textbook of onco-anesthesiology*. Singapore: Springer, pp. 45–68. https://doi.org/10.1007/978-981-16-0006-7_4.

MO, J., LIN, M., HE, B., TAN, K., JIN, C., JIANG, H., PAN, X., LIN, W., 2017. Recombinant human adenovirus-p53 improves the outcome of mid-late stage pancreatic cancer via arterial infusion. *Oncology Letters,* 14, 6829–6832. https://doi.org/10.3892/ol.2017.7058.

MOHAMMED, S.I., WILLIAMS, C.K., NDOM, P., HOLLAND, J.F. 2012. The African organization for research and training in cancer: historical perspective. *Current Oncology* 19(5), 272–276. 10.3747/co.19.1075.

MONTALBANO, M., NEFF, G. W. 2006. Management of recurrent viral hepatitis B and C after liver transplantation. *Current Gastroenterology Reports*, 8, 60–66.

MONTES, E. M., VAN HOOGSTRATEN, L., RUBIO, P. G. 2020. PC risk in relation to lifetime smoking patterns, tobacco type, and dose-response relationships. *Cancer Epidemiology, Biomarkers & Prevention*, 29(5), 1009–1018.

MORENO, C.C., STROSBERG, J., CHOI, J., HAN, S. AND PAN, C. 2021. Pancreatoblastoma in pediatric and adult patients: clinicopathologic and imaging features. *American Journal of Roentgenology*, 216(5), 1121–1130.

MORHASON-BELLO, I.O., ODEDINA, F., REBBECK, T.R., HARFORD, J., DANGOU, J.M., DENNY, L., ADEWOLE, I.F. 2013. Challenges and opportunities in cancer control in Africa: a perspective from the African organisation for research and training in cancer. *The Lancet Oncology,* 14(4), e142–51. 10.1016/S1470-2045(12)70482-5.

MOSHAYEDI, N., ESCOBEDO, A.L., THOMASSIAN, S., OSIPOV, A. AND HENDIFAR, A.E., 2022. Race, sex, age, and geographic disparities in PC incidence. *Journal of Clinical Oncology*, 40, 520–520.

MTHUNZI, R.J., NOEL, C.B. 2023. Geographic distribution of pancreaticobiliary malignancy in central South Africa presenting to the Universitas Academic Hospital Complex. *South African Journal of Surg*ery, 61(3), 7–11. PMID: 37791707.

MUDYIWA, M., SHARMA, M., RAY, S.K., GUNDIDZA, M., MASIMIREMBWA, C., 2022. Opportunities for the discovery of anticancer drugs from traditional herbal medicines in Africa: a review. *afjbs,* 4, 1–14. https://doi.org/10.33472/AFJBS.4.4.2022.1-14.

MUTEBI, M., SCROGGINS, D., SIMONS, V., OTI, N.O., HAMMAD, N., 2020. Engaging patients for clinical trials in Africa: patient-centered approaches. *JCO Global Oncology,* 6, JGO.19.00190. https://doi.org/10.1200/JGO.19.00190.

MUTEBI, M., LEWISON. G., AGGARWAL, A., ALATISE, O.I., BOOTH, C., CIRA, M., GROVER, S., GINSBURG, O., GRALOW, J., GUEYE, S., KITHAKA, B., KINGHAM, T.P., KOCHBATI, L., MOODLEY, J., MOHAMMED, S.I., MUTOMBO, A., NDLOVU, N., NITZIMIRA, C., PARHAM, G.P., WALTER, F., PARKES, J., SHAMLEY, D., HAMMAD, N., SEELEY, J., TORODE, J., SULLIVAN, R., VANDERPUYE, V. 2022. Cancer research across Africa: a comparative bibliometric analysis. *BMJ Glob Health*, 7(11):e009849. doi: 10.1136/bmjgh-2022-009849.

NAGTEGAAL, I. D., ODZE, R. D., KLIMSTRA, D., PARADIS, V., RUGGE, M., SCHIRMACHER, P., WASHINGTON, K. M., CARNEIRO, F., CREE, I. A., AND WHO CLASSIFICATION OF TUMOURS EDITORIAL BOARD. 2020. The 2019 WHO classification of tumours of the digestive system. *Histopathology*, 76(2), 182–188. https://doi.org/10.1111/his.13975.

NAICKER S., EASTWOOD J.B., PLANGE-RHULE J, TUTTR.C. 2010. Shortage of healthcare workers in sub-Saharan Africa: a nephrological perspective. *Clin Nephrol*. Nov; 74 Suppl 1:S129–33. doi: 10.5414/cnp74s129

NARAYANAN, V., WEEKES, C.D., 2016. Molecular therapeutics in pancreas cancer. *World Journal of Gastrointestinal Oncology,* 8, 366–379. https://doi.org/10.4251/wjgo.v8.i4.366.

NELSON, A.M., MILNER, D.A., REBBECK, T.R., ILIYASU, Y., 2016. Oncologic care and pathology resources in Africa: survey and recommendations. *Journal of Clinical Oncology,* 34, 20–26. https://doi.org/10.1200/JCO.2015.61.9767.

NHLEKO, M.L., EDOKA, I., MUSENGE, E. 2023. PC mortality in South Africa: a case-control study. *South African Medical Journal*, 114(1), 27–32. https://doi.org/10.7196/SAMJ.2024.v114i1.1132.

ODEDINA, F.T., AKINREMI, T.O., CHINEGWUNDOH, F., ROBERTS, R., YU, D., REAMS, R.R., FREEDMAN, M.L., RIVERS, B., GREEN, B.L., KUMAR, N. 2009. Prostate cancer disparities in Black men of African descent: a comparative literature review of prostate cancer burden among Black men in the United States, Caribbean, United Kingdom, and West Africa. *Infectious Agents and Cancer*, 4 Suppl 1(Suppl 1), S2. 10.1186/1750-9378-4-S1-S2.

ODEDINA, F.T., ROTIMI, S., 2021. Promoting cancer genomics research in Africa: a roadmap. *Nature Reviews Cancer,* 21, 409–410. https://doi.org/10.1038/s41568-021-00359-9.

OMONISI, A.E., LIU, B., PARKIN, D.M. 2020. Population-Based Cancer Registration in Sub-Saharan Africa: Its Role in Research and Cancer Control. *JCO Global Oncology* 6, 1721–1728. 10.1200/GO.20.00294.

OMOTOSO, O., TEIBO, J.O., ATIBA, F.A,, OLADIMEJI, T., PAIMO, O.K., ATAYA, F.S., BATIHA, G.E., ALEXIOU, A. 2023. Addressing cancer care inequities in sub-Saharan Africa: current challenges and proposed solutions. *International Journal for Equity in Health*, 22(1), 18910.1186/s12939-023-01962-y.

PAPAGEORGE, M.V., EVANS, D.B., TSENG, J.F. 2021. Health care disparities and the future of pancreatic cancer care. *Surgical Oncological Clinics of North America*, 30(4), 759–77110.1016/j.soc.2021.06.012.

PARK, H.J., LEE, K.H., LEE, J.K., LEE, K.T., CHOI, Y.H., PARK, J.K., 2024. A case of five-year survival in metastatic pancreatic cancer. *The Korean Journal of Pancreas and Biliary Tract,* 29, 110–114. https://doi.org/10.15279/kpba.2024.29.3.110.

PARK, P.H., SHYIRAMBERE, C., KATEERA, F., GUPTA, N., RUSANGWA, C., MUKHERJEE, J., COUTINHO, A., LEHMANN, L., BUSWELL, L., SHULMAN, L.N., MUBILIGI, J.M., UWINKINDI, F., HIRSCHHORN, L.R., 2021. Implementing cancer care in Rwanda: capacity building for treatment and scale-up. *Sustainability,* 13, 7216. https://doi.org/10.3390/su13137216.

PEARLSTEIN, E. F., TAZINKENG, N. N., SAWYER, K., CHUKWUDIKE, E.S., BAINS, K., ASANTE, C., IBRAHIM, O.S., PHARIDAH, R., CHISHINGA, N., ASOMBANG, A.W. 2021. S61 Epidemiology of pancreatic cancer in Africa: a systematic review. *The American Journal of Gastroenterology,* Oct 116(), S26–S27 10.14309/01.ajg.0000772224.74784.af

POPESCU, R.A., ROILA, F., ARENDS, J., METRO, G., LUSTBERG, M., 2021. Supportive care: low cost, high value. *American Society of Clinical Oncology Educational Book,* 41, 1–11. https://doi.org/10.1200/EDBK_320041.

POURSHAMS, A., SEPANLOU, S.G., IKUTA, K.S., BISIGNANO, C., SAFIRI, S., ROSHANDEL, G., SHARIF, M., KHATIBIAN, M., FITZMAURICE, C., NIXON, M.R. AND ABBASI, N., 2019. The global, regional, and national burden of PC and its attributable risk factors in 195 countries and territories, 1990–2017: a systematic analysis for the Global Burden of Disease Study 2017. *The Lancet Gastroenterology & Hepatology*, 4(12), 934–947.

PRASHANT, R., TAGORE, S., VINAYA, G. 2019. Epidemiology of PC: global trends, etiology and risk factors. *World Journal of Oncology*, 10(1), 10–27. https://doi.org/10.4021/wjon.v10i1.1166.

RADOUANI, F., ZASS, L., HAMDI, Y., DA ROCHA, J., SALLAM, R., ABDELHAK, S., AHMED, S., AZZOUZI, M., BENAMRI, I., BENKAHLA, A., BOUHAOUALA-ZAHAR, B., CHAOUCH, M., JMEL, H., KEFI, R., KSOURI, A., KUMUTHINI, J., MASILELA, P., MASIMIREMBWA, C., OTHMAN, H., PANJI, S., ROMDHANE, L., SAMTAL, C., SIBIRA, R., GHEDIRA, K., FADLELMOLA, F., KASSIM, S.K., MULDER, N., 2020. A review of clinical pharmacogenetics Studies in African populations. *Personalized Medicine*, 17, 155–170. https://doi.org/10.2217/pme-2019-0110.

RAIMONDI, S., MAISONNEUVE, P., LOWENFELS, A. B. 2009. Epidemiology of PC: an overview. *Nature Reviews Gastroenterology & Hepatology*, 6(12), 699–708.

RAWLA, P., SUNKARA, T. AND GADUPUTI, V. 2019. Epidemiology of PC: global trends, etiology and risk factors. *World Journal of Oncology*, 10(1), 10.

REBBECK, T.R., DEVESA, S.S., CHANG, B.L., BUNKER, C.H., CHENG, I., COONEY, K., EELES, R., FERNANDEZ, P., GIRI, V.N., GUEYE, S.M., HAIMAN, C.A., HENDERSON, B.E., HEYNS, C.F., HU, J.J., INGLES, S.A., ISAACS, W., JALLOH, M., JOHN, E.M., KIBEL, A.S., KIDD, L.R., LAYNE, P., LEACH, R.J., NESLUND-DUDAS, C., OKOBIA, M.N., OSTRANDER, EA, PARK, J.Y., PATRICK, A.L., PHELAN, CM, RAGIN, C., ROBERTS, R.A., RYBICKI, B.A., STANFORD, J.L., STROM, S., THOMPSON, I.M., WITTE, J., XU, J., YEBOAH, E., HSING, A.W., ZEIGLER-JOHNSON, C.M. 2013. Global patterns of prostate cancer incidence, aggressiveness, and mortality in men of African descent. *Prostate Cancer.* 560857. 10.1155/2013/560857

REGALLA, D. K. R., JACOB, R., MANNE, A., AND PALURI, R. K. 2019. Therapeutic options for ampullary carcinomas. A review. *Oncology Reviews*, 13(2), 440. https://doi.org/10.4081/oncol.2019.440.

REN, Z., JIANG, J., XIE, H., LI, A., LU, H., XU, S., ZHOU, L., ZHANG, H., CUI, G., CHEN, X., LIU, Y., WU, L., QIN, N., SUN, R., WANG, W., LI, L., WANG, W., ZHENG, S. 2017. Gut microbial profile analysis by MiSeq sequencing of pancreatic carcinoma patients in China. *Oncotarget*, 8(56), 95176–95191.

RIZZO, A., DADDUZIO, V., LOMBARDI, L., RICCI, A. D., GADALETA-CALDAROLA, G. 2021. Ampullary carcinoma: an overview of a rare entity and discussion of current and future therapeutic challenges. *Current Oncology*, 28(5), 3393–3402. https://doi.org/10.3390/curroncol28050293.

RYAN, D., HONG, T., BARDEESY, N. 2014. Pancreatic adenocarcinoma. *New England Journal of Medicine*, 371(11), 1039–1049. https://doi.org/10.1056/nejmra1404198.

SCHWINGEL, J., DECKER, M., SCHNEIDER, L., STÜRMER, C.J., LUTZ, M.P., 2023. Early detection of pancreatic cancer. *Oncology Research and Treatment,* 46, 259–267. https://doi.org/10.1159/000530790.

SIEGEL, R.L., MILLER, K.D., FUCHS, H.E. AND JEMAL, A., 2022. Cancer statistics, 2022. *CA: A Cancer Journal for Clinicians*, 72(1), 7–33. https://doi.org/10.3322/caac.21654.

SIEGEL, R.L., MILLER, K.D., JEMAL, A. 2017. Cancer statistics, 2017. *CA: A Cancer Journal for Clinicians*, 67(1), 7–30. https://doi.org/10.3322/caac.21387.

SIEGEL, R.L., MILLER, K.D., JEMAL, A. 2020. Cancer statistics, 2020. *CA: A Cancer Journal for Clinicians*, 70(1), 7–30. https://doi.org/10.3322/caac.21590.

SIMOES, P.K., OLSON, S.H., SALDIA, A. AND KURTZ, R.C. 2017. Epidemiology of pancreatic adenocarcinoma. *Chinese Clinical Oncology*, 6(3), 24. https://doi.org/10.21037/cco.2017.06.32.

SOLIMAN A.S., ZHANG Q., SALEh T.,ZARZOUZarzour A., SELIm M., ABDEL-FATTAH M., ABBRUZZESE J. L. 2006. Pancreatic cancer mortality in Egypt: comparison to the United States pancreatic cancer mortality rates. *Cancer Detection and Prevention* 30(5), 473–479. 10.1016/j.cdp.2006.06.006.

STEFAN, D.C. 2015. Cancer care in Africa: an overview of resources. *Journal of Global Oncology*, 1(1), 30–36.10.1200/JGO.2015.000406.

STOLZENBERG-SOLOMON, R. Z., PIETINEN, P., TAYLOR, P. R., VIRTAMO, J., ALBANES, D. 2002. Prospective study of diet and PC in male smokers. *American Journal of Epidemiology*, 155, 783–792.

STOLZENBERG-SOLOMON, R.Z., SCHAIRER, C., MOORE, S., HOLLENBECK, A. AND SILVERMAN, D.T., 2013. Lifetime adiposity and risk of PC in the NIH-AARP Diet and Health Study cohort. *The American Journal of Clinical Nutrition*, 98(4), 1057–1065.

STROBEL, O., HINZ, U., GLUTH, A., HANK, T., HACKERT, T., NEOPTOLEMOS, J.P. AND BÜCHLER, M.W. 2020. Pancreatoblastoma. *The Lancet Oncology*, 21(3), 345–355.

SUNG, H., FERLAY, J., SIEGEL, R.L., LAVERSANNE, M., SOERJOMATARAM, I., JEMAL, A. AND BRAY, F., 2021. Global cancer statistics 2020: GLOBOCAN estimates of incidence and mortality worldwide for 36 cancers in 185 countries. *CA: A Cancer Journal for Clinicians,* 71(3), 209–249. 10.3322/caac.21660.

SUZUKI, R., OHIRA, H., IRISAWA, A., BHUTANI, M.S., 2012. Pancreatic cancer: early detection, diagnosis, and screening. *Clinical Journal of Gastroenterology*, 5, 322–326. https://doi.org/10.1007/s12328-012-0327-0.

SZYMOŃSKI, K., LIPIEC, E., SOFIŃSKA, K., SKIRLIŃSKA-NOSEK, K., MILIAN-CIESIELSKA, K., SZPOR, J., CZAJA, M., SEWERYN, S., WILKOSZ, N., BIRARDA, G., PICCIRILLI, F., VACCARI, L., SZYMOŃSKI, M., 2021. Spectroscopic screening of pancreatic cancer. C*linical Spectroscopy,* 3, 100016. https://doi.org/10.1016/j.clispe.2021.100016.

TANAKA, M., FERNÁNDEZ-DEL CASTILLO, C., ADSAY, V., CHARI, S., FALCONI, M., JANG, J. Y., KIMURA, W., LEVY, P., PITMAN, M. B., SCHMIDT, C. M., SHIMIZU, M., WOLFGANG, C. L., YAMAGUCHI, K., YAMAO, K., AND INTERNATIONAL ASSOCIATION OF PANCREATOLOGY. 2012. International consensus guidelines 2012 for the management of IPMN and MCN of the pancreas. *Pancreatology*, 12(3), 83–197. https://doi.org/10.1016/j.pan.2012.04.004.

THOMAS, C., 2018. Risk factors, biomarker and imaging techniques used for pancreatic cancer screening. *Chinese Clinical Oncology*, 6, 61–61. https://doi.org/10.21037/cco.2017.12.06.

TRIANTAFILLIDIS, J.K., TRIANTAFYLLIDI, E., SIDERIS, M., PITTARAS, T., PAPALOIS, A.E., 2022. Herbals and plants in the treatment of pancreatic cancer: a systematic review of experimental and clinical studies. *Nutrients,* 14, 619. https://doi.org/10.3390/nu14030619.

VANDERPUYE, V., HAMMAD, N., MARTEI, Y., HOPMAN, W.M., FUNDYTUS, A., SULLIVAN, R., SERUGA, B., LOPES, G., SENGAR, M., BRUNDAGE, M.D., BOOTH, C.M. 2019. Cancer care workforce in Africa: perspectives from a global survey. *Infectious Agents and Cancer*, 14, 11. 10.1186/s13027-019-0227-8.

WANG, Y. T., GOU, Y. W., JIN, W. W., XIAO, M., FANG, H. Y. 2016. Association between alcohol intake and the risk of PC: a dose-response meta-analysis of cohort studies. *BMC Cancer*, 16(1), 212.

WARNER-COHEN, J., POLOKOWSKI, A.R., PINTO, D., ZAVADSKY, T., BEYER, K., SAIF, M.W., WEISS, M.J., 2020. Increasing access to multidisciplinary care in pancreatic cancer. *Psycho-Oncology,* 29, 2105–2108. https://doi.org/10.1002/pon.5493.

WONG, M.C., JIANG, J.Y., LIANG, M., FANG, Y., YEUNG, M.S. AND SUNG, J.J., 2017. Global temporal patterns of PC and association with socioeconomic development. *Scientific Reports*, 7(1), 3165.

WOOD, L.D. AND HRUBAN, R.H., 2012. Pathology and molecular genetics of pancreatic neoplasms. *The Cancer Journal,* 18(6), 492–501.

YANG, J., XU, R., WANG, C., QIU, J., REN, B., YOU, L., 2021. Early screening and diagnosis strategies of pancreatic cancer: a comprehensive review. *Cancer Communications,* 41, 1257–1274. https://doi.org/10.1002/cac2.12204.

YANG, Y.-X., RUSTGI, A.K., 2016. Impact of metformin on advanced pancreatic cancer survival: too little, too late? *Clinical Cancer Research,* 22, 1031–1033. https://doi.org/10.1158/1078-0432.CCR-15-2494.

YAO, J. C., PAVEL, M., LOMBARD-BOHAS, C., AND OBERG, K. 2016. A comprehensive review of neuroendocrine tumors of the pancreas. *Endocrine Reviews*, 37(4), 407–455.

YEO, T.P. 2015. Demographics, epidemiology, and inheritance of pancreatic ductal adenocarcinoma. *Seminars in Oncology*, 42(1), 8–18.

YURGELUN, M. B., CHITTENDEN, A. B., MORALES-OYARVIDE, V., RUBINSON, D. A., DUNNE, R. F., KOZAK, M. M., QIAN, Z. R., WELCH, M. W., BRAIS, L. K., DA SILVA, A., BUI, J. L., YUAN, C., LI, T., LI, W., MASUDA, A., GU, M., BULLOCK, A. J., CHANG, D. T., CLANCY, T. E., LINEHAN, D. C., WOLPIN, B. M. 2019. Germline cancer susceptibility gene variants, somatic second hits, and survival outcomes in patients with resected PC. *Genetics in Medicine*, 21(1), 213–223. https://doi.org/10.1038/s41436-018-0009-5.

ZHANG, L., SANAGAPALLI, S. AND STOITA, A. 2018. Challenges in diagnosis of PC. *World Journal of Gastroenterology*, 24(19), 2047–2060. https://doi.org/10.3748/wjg.v24.i19.2047.

ZHAO, B., ZHAO, B., CHEN, F., 2022. Diagnostic value of serum carbohydrate antigen 19-9 in pancreatic cancer: a systematic review and meta-analysis. *European Journal of Gastroenterology & Hepatology,* 34, 891–904. https://doi.org/10.1097/MEG.0000000000002415.

ZHAO, Z., LIU, W. 2020. PC: a review of risk factors, diagnosis, and treatment. *Technology in Cancer Research & Treatment*, 19, 153303382096211.

ZHEN, D.B., RABE, K.G., GALLINGER, S., SYNGAL, S., SCHWARTZ, A.G., GOGGINS, M.G., HRUBAN, R.H., COTE, M.L., MCWILLIAMS, R.R., ROBERTS, N.J. AND CANNON-ALBRIGHT, L.A., 2015. BRCA1, BRCA2, PALB2, and CDKN2A mutations in familial PC: a PACGENE study. *Genetics in Medicine*, 17(7), 569–577.

ZHU, F., WANG, H., ASHAMALLA, H. 2020. Racial and socioeconomic disparities in the treatments and outcomes of PC among different treatment facility types. *Pancreas*, 49(10), 1355–1363.

Part II

Biological and Molecular Insights

3 Molecular Pathways Driving Pancreatic Cancer

Rodney Hull[1], Thulo Molefi[1,2], Richard Khanyile[1,2], Boitumelo Phakathi[3], Fieke Froeling[4,5], and Zodwa Dlamini[1]*

[1]SAMRC/UP Precision Oncology Research Unit, DSI/NRF Starchy Chair in Precision Oncology (POCP) and Cancer Prevention, Pan African Cancer Research Institute (PACRI), University of Pretoria, Pretoria, South Africa.

[2]Department of Medical Oncology, Faculty of Health Sciences, Steve Biko Academic Hospital, University of Pretoria, Pretoria, South Africa.

[3]Department of Surgery, College of Health Sciences, University of KwaZulu Natal, Durban, South Africa.

[4]Department of Oncology, Centre for Cancer Sciences, University of Nottingham Biodiscovery Institute, Academic Unit of Translational Medical Sciences, School of Medicine, University of Nottingham, United Kingdom.

[5]Wolfson Wohl Cancer Research Centre, Pancreatic Cancer Lead Experimental Cancer Centre Glasgow, School of Cancer Sciences and Beatson West of Scotland Cancer Centre, University of Glasgow, United Kingdom.

*Corresponding author - Rodney Hull

INTRODUCTION

While the mortality rates of most other cancers are decreasing, the mortality rates for pancreatic cancers remain high. In addition, incidence rates of cancers like pancreatic ductal adenocarcinoma (PDAC), the most common and most deadly form of pancreatic cancer, are on the rise (R.S., 2021), as such pancreatic cancer is set to become one of the leading causes of cancer-related death (Mathur et al., 2020). This high mortality rate may be due to many factors, one of which is the unique molecular alterations that drive the development and progression of pancreatic cancer. For example, increased expression of chemokine receptors, such as CCR2, CCR5, CCR6, CXCR4, CXCR5, and CX3CR1 (Song et al., 2024), by gene mutations in components of their signalling pathways, such as KRAS and TP53 in the CCL2/CCR2 pathway (Song et al., 2024), promotes cell invasion and metastases resulting in a poor prognosis in a subgroup of pancreatic cancer patients. As a result, these chemokines, which normally function in development and inflammation, now function to promote cell invasion and metastasis (Kim et al., 2012; Lee et al., 2015). The transcriptomes of patients have been used to group them into molecular subtypes. These include a classical subtype, which has a better prognosis, and a squamous/basal-like subtype, with a poor prognosis (Collisson et al., 2011).

The current one-size-fits-all strategy used in chemotherapy is largely ineffective in the treatment of pancreatic cancer (Conroy et al., 2011; Von Hoff et al., 2013; Buanes, 2016). Consequently, much interest has been shown in the development of targeted therapies that rely on targeting specific molecular alterations or specific molecules with a high affinity for particular signalling components. This strategy is one of the central tenets of precision medicine (Min and Lee, 2022). In

DOI: 10.1201/9781003544401-5

pancreatic cancer, this was first proposed over 30 years ago when it was suggested that the presence of mutations in *BRCA1*, *BRCA2*, or *PALB2* can drive therapeutic decision-making. The presence of these mutations indicated that these patients could be effectively treated using platinum salts, such as cisplatin, combined with PARP inhibitors (Principe, 2022). The Pancreas Cancer Olaparib Ongoing (POLO) trial was the first study showing a personalised approach for patients with pancreatic cancer and a germline *BRCA1* or *BRCA2* mutation and platinum-sensitive disease (Golan et al., 2019). Subsequent studies showed similar findings for patients with somatic *BRCA1/2* or *PALB2* alterations (Reiss et al., 2021; Kindler et al., 2022; O'Kane and Lowery, 2022). Whole-genome sequencing has allowed for in-depth characterisation of the genetic drivers of pancreatic cancer. Sequencing has revealed that pancreatic cancers are associated with 119 somatic chromosomal structural variants with 63 confirmed mutations (Waddell et al., 2015; Javadrashid et al., 2021; The Cancer Genome Atlas, 2017; Bailey et al., 2016). These structural variants consist of intrachromosomal deletions, tandem duplications, inversions, amplified inversions, chromosomal rearrangements, and foldback inversions. These changes influence the activity of at least 12 signalling pathways that play various roles. It has been estimated that at least 67–100% of pancreatic cancers have at least one of these mutations (Javadrashid et al., 2021; Jones et al., 2008). The most iconic of these mutations in PDAC are mutations in the *kras* gene, which are thought to drive the development of PDAC and also contribute to how aggressive the cancer is. Other genes commonly found mutated in pancreatic cancers include *TP53*, *CDKN2A/B*, and *SMAD4* (Kanda et al., 2012).

Apart from these mutations, epigenetic changes also play a role in pancreatic cancer, especially alterations in histone acetylation due to changes in HDAC and HAT function, as well as changes in the levels of various non-coding RNAs (Silverman and Shi, 2016). The anticancer immune response plays an important role in the elimination of cancer, with T cells being able to recognise tumour-specific neoantigens and thereby mount an immune response against the cancer (Gnoni et al., 2020; Schumacher and Schreiber, 2015). In pancreatic cancer, the tumour creates a unique microenvironment surrounding it. This tumour microenvironment (TME) suppresses the immune system through various mechanisms, preventing an effective immune response (Gnoni et al., 2020). In addition to the TME, pancreatic cancers exploit immune checkpoints to suppress the immune system (Pardoll, 2012). As such there is interest in the use of immune checkpoint inhibitors to bypass the immune inhibition activity of pancreatic cancer (Li et al., 2021a).

This chapter will examine the genes that are mutated in pancreatic cancer as well as the signalling pathways that are altered, epigenetic changes, and changes to the TME that may play a role in driving pancreatic cancer. It will then go on to examine the use of these alterations as biomarkers for diagnosis, screening, disease management, and as targets for the development of new therapies.

SIGNALLING PATHWAYS

KRAS Pathway

It is estimated that up to 90% of patients with pancreatic cancer have a point mutation in the *kras* gene. In the development of pancreatic cancer, mutations in *kras* occur in low-grade pancreatic intraepithelial neoplasias (PanINs), where they drive the progression of these lesions to full-blown pancreatic cancer (Kanda et al., 2012). Continued KRAS signalling results in the constant proliferation of pancreatic cancer cells, which can avoid or negate all cell death promoting signals (Collins et al., 2012). *KRAS* is a member of the RAS family of the small G protein superfamily. The activity of these proteins is regulated by conformational changes that take place when they bind guanosine triphosphate (GTP), which turns them on, and guanosine diphosphate (GDP), which turns them off. These proteins are activated by guanine nucleotide exchange factors (GEFs) and GTPase-activating proteins (GAPs), which regulate the binding of the guanosine nucleotides (Pantsar, 2020). GEFs serve a catalytic function promoting the exchange of GDP with GTP. GAPs catalyse the GTPase

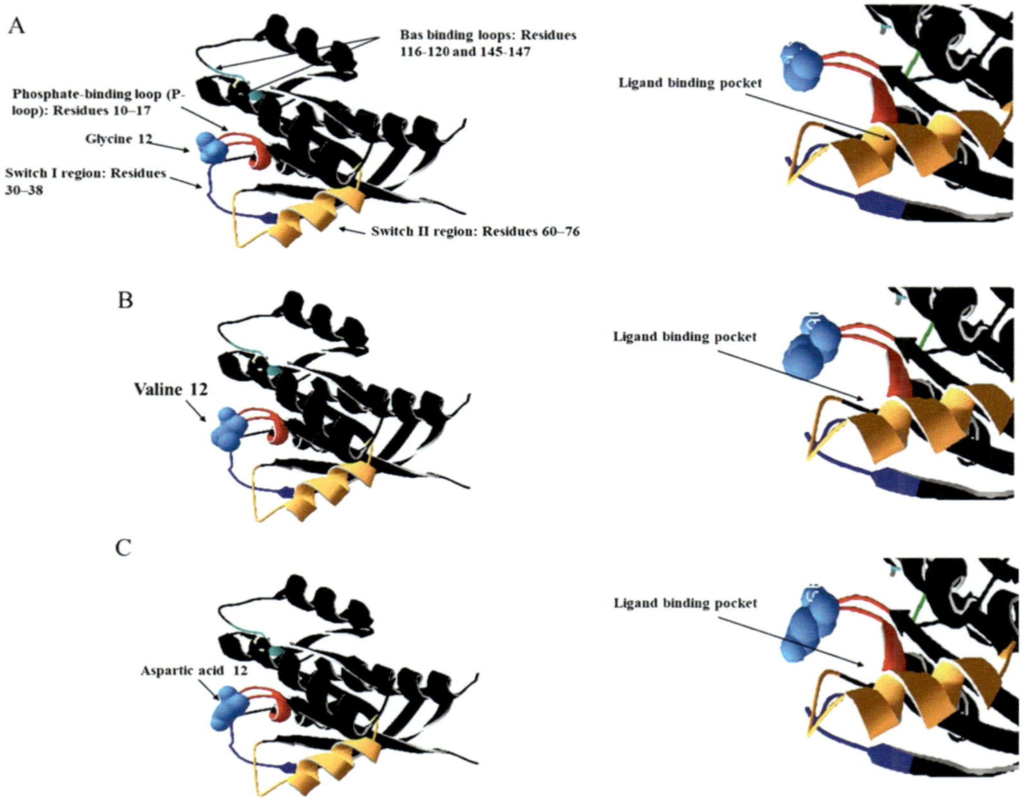

FIGURE 3.1 Effect of common mutation in *KRAS* on the structure of the protein's binding pocket. The two most common mutations in *krasS* that occur in pancreatic cancer are the substitutions that occur at position 12, where glycine is substituted by either valine or aspartic acid. These two amino acids have larger, bulkier residue sidechains than glycine and extend further into the binding pocket of KRAS, with the electrostatic forces they can exert having a larger area of effect on the binding pocket.

(the hydrolysis of GTP into GDP) activity of RAS family members (Downward, 2003). The inactivation of the GTPase activity as a result of mutations leads to constitutively active RAS signalling and the resulting downstream signalling to be upregulated. Mutations in *KRAS* that are known to promote the development of PDAC include the *G12D* or *G12V* mutations (Figure 3.1). Apart from these common KRAS mutants, there are also pancreatic cancer patients who frequently possess other KRAS mutants such as KRAS G12R and *KRAS Q61. KRAS G12R* mutations are more common in stage I pancreatic cancers, with a decreased propensity to spread to lymph nodes and better survival than those patients with other *kras* mutations. Patients with this mutation also did not show increased epithelial-mesenchymal transition but did show increased NF-κB signalling (McIntyre et al., 2024). The rare Q61 mutation is unique in both the position of the mutation and the effect of the mutation. This mutant KRAS has impaired GTPase function, and this results in an overactive *RAS* pathway (Nusrat et al., 2024). Studies of these various *kras* mutations in pancreatic cancers indicate that different *kras* alleles have different roles in the development and progression of pancreatic cancer (Cook et al., 2021).

As the disease progresses, the tumours expressing mutant *kras* also commonly accumulate mutations in various tumour-suppressor genes, such as *cdk2a* and *tp53*. Mutations in these tumour-suppressor genes have been found to occur repeatedly in precancerous pancreatic lesions with *KRAS* mutations (Downward, 2003). KRAS is involved in the activation of downstream signalling pathways such as the Raf/MAPK/extracellular signal-regulated kinase (ERK) pathway; the PI3Ks/(PDK-1)/Akt pathway and the RalGEFs, and phospholipase Cε pathway (Tripathi et al., 2024). In *KRAS*-driven pancreatic cancer, mutations or alterations in expression can also occur in components of these downstream signalling pathways (Eser et al., 2013).

Tumour-Suppressor Genes

The tumour-suppressor P16 is coded for by the *CDKN2A* gene and acts as an inhibitor of cell cycle progression by blocking the activity of *CDK4/6*. *CDK4/6* phosphorylates the retinoblastoma protein (Rb) (Figure 3.2) which, when unphosphorylated, sequesters the E2F transcription factor. Once released, E2F translocates to the nucleus where it initiates the transcription of genes that promote entry into the S phase of the cell cycle. In pancreatic cancer, mutations in the *CDKN2A* gene commonly result in its silencing, and these *CDKN2A* mutations are the most common tumour-suppressor gene mutations in pancreatic cancer. These mutations lead to silencing the expression of P16 through homozygous deletions and loss of its heterozygosity. In addition to this, the expression of *CDKN2A* can be silenced through epigenetic modifications, such as promoter methylation (Li et al., 2011). As previously stated, mutations in *kras* can promote subsequent mutations in *cdkn2a* and the development of pancreatic cancer. This occurs due to the expression of mutant KRAS, leading to a selective pressure that favours the development of *CDKN2A* mutations (Hu et al., 2021).

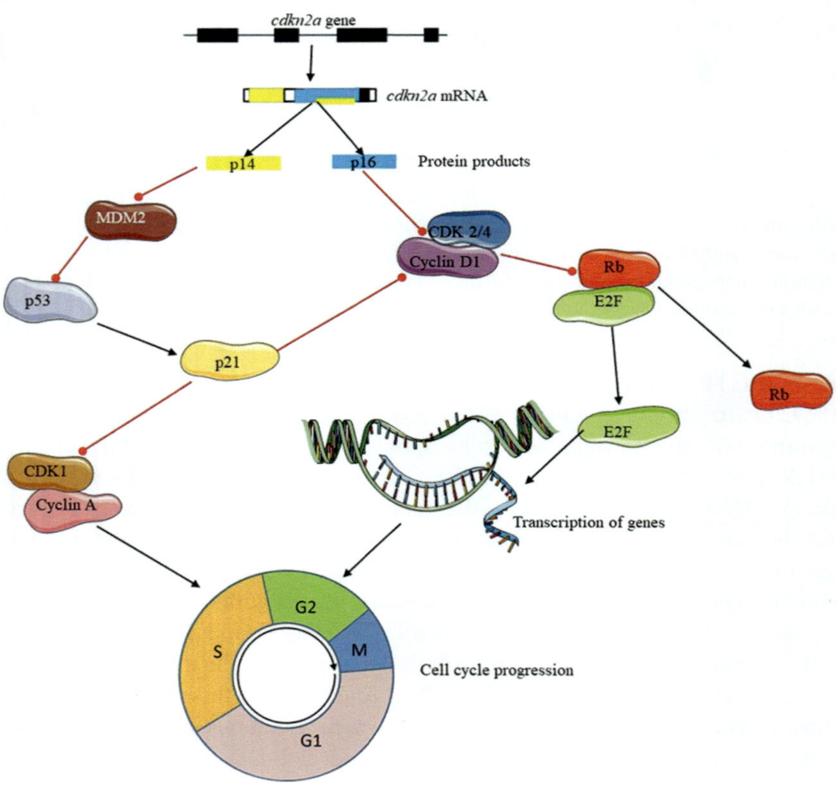

FIGURE 3.2 The role of the *CDKN2A* gene in suppressing the development of pancreatic cancer. The *CDKN2A* gene results in an mRNA transcript which is spliced to give rise to two mRNAs that code for the proteins P14 and P16. Both proteins can act as tumour-suppressor genes by inhibiting the progression of the cell cycle. Mutations in the *CDKN2A* gene result in the silencing of the expression of these two proteins.

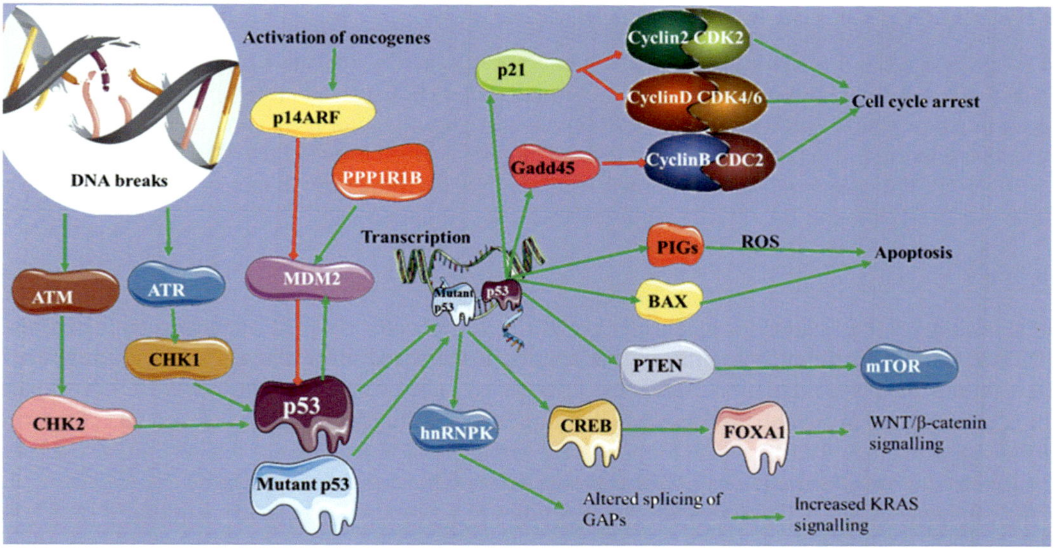

FIGURE 3.3 The role of P53 mutants in pancreatic cancer. The normal function of *P53* is to detect and respond to DNA damage by halting the cell cycle to allow for DNA repair or, in cases where repair is untenable, to initiate apoptosis. Mutant P53 can not only result in a loss of the tumour-suppressor function but also a gain in oncogenic activity, such as increased Wnt signalling. In pancreatic cancer, mutant *P53* can alter the expression of splicing regulators, which result in alterations in KRAS activity due to the expression of different GAP isoforms.

Similar to other types of cancer, the tumour-suppressor *P53* is inactivated in pancreatic cancer. This occurs in 50–75% of PDAC patients. Canonically, *P53* acts as a tumour suppressor by inducing apoptosis or cell cycle arrest in response to DNA damage (Figure 3.3). The inactivation of *tp53* occurs through genetic mutations leading to the loss of the second allele and mutations within the gene which give rise to proteins with no or decreased function. This results in cells that resist the induction of apoptosis, survive, and proliferate in late-stage PanIN, promoting the progression to pancreatic cancer (Hafner et al., 2019). Loss of the expression of Hypoxia Inducible Factor 1 alpha is also known to occur in pancreatic cancer. HIF1A acts as a tumour suppressor by preserving the activity of P53 by preventing its degradation. This is accomplished via the inhibition of the expression of PPP1R1B, which promotes the expression of MDM2. As such, the loss of HIF1A expression in pancreatic cancer is associated with increased invasion and metastasis (Tiwari et al., 2020). Mutations in *tp53* and *kras* have been shown to cooperate in promoting the development and progression of pancreatic cancer. Mutant P53 has been shown to promote the expression of hnRNPK, which regulates the splicing of GTPase-activating proteins (GAPs). These new GAP isoforms lose the ability to associate with the cell membrane and result in increased KRAS activity (Escobar-Hoyos et al., 2020). The transcription factor *FOXA1* promotes metastasis and is upregulated in many prostate cancers. Mutant *P53* and *CREB1* upregulate *FOXA1* transcription. This results in increased WNT/β-catenin signalling and results in increased metastasis (Kim et al., 2021).

Members of the transforming growth factor beta family (TGF-β) regulate cell development and growth. The main transducers of the signal from the TGF-β receptor to the nucleus are the small mothers against decapentaplegic (SMAD) proteins. *SMAD4*, one of the key role players in this pathway, is inactivated in up to 55% of pancreatic cancers. This is the result of mutations in the *SMAD4* gene, combined with the loss of the second allele of the gene. This results in cancer cells in the late PanIN stage proliferating uncontrollably as there is no regulation of this growth pathway (Ezrova et al., 2021)

SIGNALLING OF GROWTH FACTOR RECEPTORS

The expression and activity of many mitogenic growth factors and their ligands are increased in pancreatic cancer. These include epidermal growth factor (EGF) and its receptor (EGFR), fibroblast growth factor (FGF) and its receptor (FGFR), insulin growth factor (IGF) and its receptor (IGFR), vascular endothelial growth factor (VEGF) and platelet-derived growth factor (PDGF) (Javadrashid et al., 2021). Following the binding of their receptor ligands, these growth factors trigger phosphorylation of PI3K/Akt and Ras/MAPK pathway members (Figure 3.4).

EGFR

The binding of the ligands for epidermal growth factor receptor (EGFR), including both EGF and TGF-α, initiates a signalling pathway that results in increased proliferation, survival, migration, and altered cell cycle regulation. EGFR has been found to be overexpressed in up to 90% of all pancreatic cancers. This overexpression was found to be linked to the likelihood of tumour recurrence following treatment. EGFR is also found to be overexpressed in pancreatic cancers that have metastasised to the liver. Inhibition of EGFR is associated with decreased tumour growth (Wagner et al., 2001; Balaz et al., 2001; Troiani et al., 2012). Studies of gene expression patterns in EGFR knockout

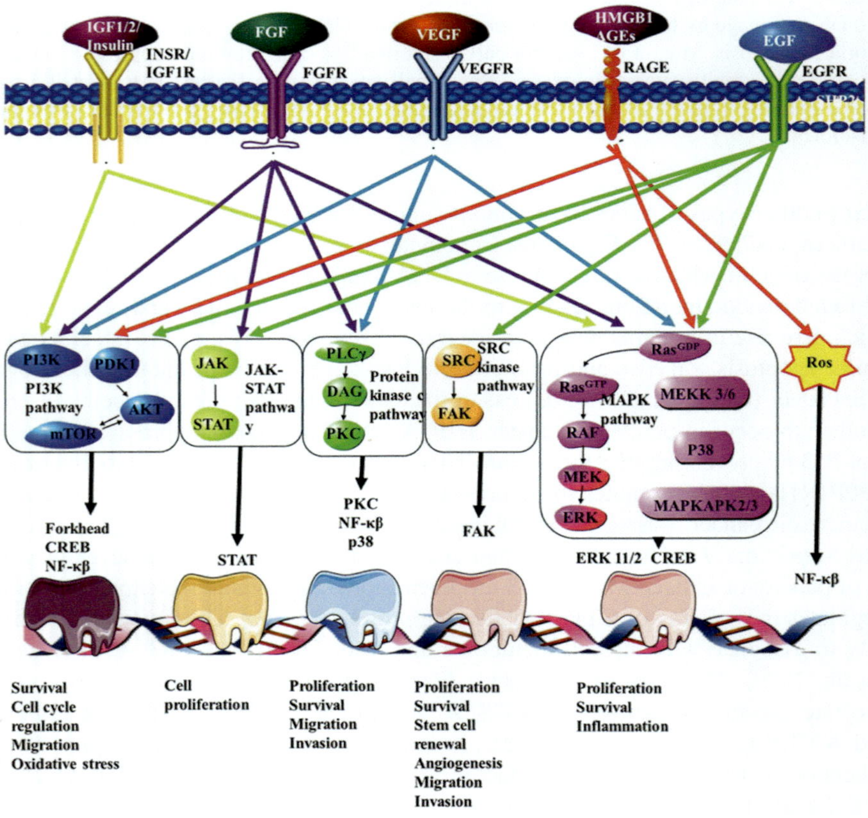

FIGURE 3.4 Growth factor signalling in pancreatic cancer. Various growth factor signalling pathways are activated in pancreatic cancer and show increased signalling pathway activation due to increased expression of the receptors or and their ligands, or alternately mutation that can increase the activity of these receptors and ligands, resulting in increased signalling activity. The end result of these growth factor pathways overlaps with many downstream signalling molecules being activated by more than one pathway.

mice revealed lower expression levels of MEK/ERK signalling in *KRAS* mutants. This decreased tumourigenesis, which implies that single allele *kras* mutations are not sufficient to actively activate MEK/ERK signalling but also require increased EGFR signalling (Daniluk et al., 2012). The increased expression of EGFR has been found to be a requirement for strong KRAS signalling in acinar cells (Feng et al., 2023). Epigenetic changes can affect the expression of EGFR and promote pancreatic cancer. Lysine demethylase 3A (KDM3A) has been found to regulate the response of pancreatic cancers to immunotherapy. KDM3A alters the expression of Krueppel-like factor 5 (KLF5) and *SMAD4*, resulting in altered expression of EGFR (Li et al., 2021b).

IGF

The insulin-like growth factors (IGFs) have been identified as playing an important role in promoting angiogenesis, invasion, and cell survival in multiple different cancers (Moschos and Mantzoros, 2002). This is also true for pancreatic cancer, where increased expression of IGF is associated with poorer outcomes and higher expression of IGF is found in the later stages of the disease (Valsecchi et al., 2012). Experiments performed *in vitro* using pancreatic cancer cells showed that the addition of IGF-1 to these cells resulted in increased growth (Zheng et al., 2010). However, attempts to target IGF signalling using IGF1Rblocking antibodies have not yielded promising results (Rieder et al., 2011). This may be due to the fact that the homologous insulin receptor (IR) is able to bind IGF and initiate the signalling pathways that lead to survival, proliferation, invasion, and alterations in the cell cycle (Figure 3.4) (Cai et al., 2017).

FGFR

Fibroblast growth factor receptor (FGFR) and its ligand fibroblast growth factor (FGF) have been identified as playing a role in the development of pancreatic cancer, where they participate in promoting proliferation, survival, migration, invasion, angiogenesis, and metastasis. Following activation of the FGF pathway by FGF binding to FGFR, FGFR substrate 2 elements are recruited and phosphorylated. In a percentage of pancreatic cancers, it has been found that FGFR-1 and FGFR-2, along with multiple FGFs, are all upregulated. In these cancers, the expression of these proteins is associated with increased proliferation and angiogenesis (Zhang et al., 2014; Kornmann et al., 1998). In preclinical pancreatic cancer models, a tyrosine-kinase inhibitor, shRNA, or dovitinib was used to inhibit FGFR signalling, and it achieved remarkable anticancer effects. Knockout of EGFR1 and EGFR2 or inhibition of FGF/FGFR signalling inhibits the development of pancreatic cancer *in vivo*. In these cases, it has been established that decreased FGFR signalling results in downregulation of SOX2 is downregulated and AKT phosphorylation. This implies that the FGFR/AKT/SOX2 signalling network plays an important role in the development of prostate cancer (Quan et al., 2020).

VEGF

As in most cancers, angiogenesis plays a crucial role in the progression of pancreatic cancer, and the conventional angiogenic pathways rely on the action of growth factors such as the vascular endothelial growth factor protein family (VEGF) (Costache et al., 2015). VEGF binds to VEGFR-1 and VEGFR-2 and stimulates endothelial cell proliferation and survival (Figure 3.4). Of special importance in pancreatic cancer are VEGF-A, VEGF-C, VEGF-D, VEGFR-1, and VEGFR-2, with VEGF-A and VEGFR-2 being the most important (Costache et al., 2015). Samples from PDAC patients show elevated levels of these receptors and their ligands, where their levels correlate with the stage and progression of the cancer. Later stages of PDAC are accompanied by higher microvessel density (Polireddy and Chen, 2016). Inhibition of VEGF or angiogenesis in pancreatic cancer models inhibits the growth, progression, and spread of PDAC (Hotz et al., 2001). One of the dangers of pancreatic cancer is the ease and rapidity with which it spreads. This is due to the molecular signals that promote lymphangiogenesis and lymphovascular invasion in pancreatic cancer. These molecular pathways result from the cooperative signalling between dual-specificity phosphatase-2 (DUSP2) and vascular endothelial growth factor-C (VEGF-C). DUSP2 negatively regulates the MAPK pathways, and its expression is lower in PDAC. This decreased expression of DUSP2 is associated with increased expression and secretion of VEGF-C via extracellular vesicles (EVs). The elevated level of EV-VEGF-C promotes lymphovascular invasion (Wang et al., 2020). VEGF also promotes glycolysis in pancreatic cancer, thereby promoting the metabolic shift favoured by cancer cells. This is the result of VEGF stimulating the expression of Neuropilin 1 (NRP1) and HIF1α (Shi et al., 2016).

RAGE

The exposure of proteins or lipids to sugars can result in them becoming glycated. These molecules are then known as advanced glycation end products. Receptors for these products are transmembrane immunoglobulins known as receptors for advanced glycation end products (RAGE or AGER). RAGE is encoded in the major histocompatibility complex's class III region and functions to promote inflammation and is overexpressed in neurodegenerative diseases and cancers. Knockdown or knockout of RAGE demonstrated that RAGE is involved in driving pancreatic cancer with oncogenic *kras* mutations and promoting drug resistance in pancreatic cancer (Gebhardt et al., 2008). Soluble isoforms of RAGE which lack the transmembrane domain are expressed at different levels and have altered cellular localisation in PDAC (Azizan et al., 2017). RAGE is also implicated in modulating signals between pro-survival and autophagy pathways and is required for hypoxic oncogenic *KRAS*-mediated signalling in pancreatic. RAGE also regulates mitochondrial bioenergetics via inflammation pathways to promote the development of pancreatic cancer (Kang et al., 2014).

EPITHELIAL TO MESENCHYMAL TRANSITION

As previously mentioned, the rapid progression of pancreatic cancer from localised to metastatic disease contributes to the poor prognosis for most patients, since this rapid progression and spread of the disease often occurs before it is diagnosed. The process of epithelial to mesenchymal transition (EMT) plays a key role in this metastasis (Rhim et al., 2012). EMT is characterised by epithelial cells, now cancer cells, ceasing the expression of epithelial cell markers and starting to express mesenchymal cell markers. Epithelial cell markers include proteins that are responsible for the formation of cell junctions and barriers, such as such as E-cadherin, occluding, claudin, and laminin-1. Mesenchymal cell markers include proteins such as N-cadherin, vimentin, and fibronectin (Kalluri and Weinberg, 2009). EMT takes place in three different circumstances within the body. These include type 1 EMT in embryonic development, type 2 EMT in wound healing, and finally, type 3 EMT taking place in cancer metastasis (Nieto, 2009). Type 3 EMT is coordinated through multiple signalling pathways and components of the extracellular matrix (ECM). ECM components involved in EMT include hyaluronic acid, collagen, Wnt, FGF, HGF, Notch, TGF-β family members, TNF-α, and HIF1α. It also requires the activity of multiple transcription factors including Snail 1, Snail 2, Zeb-1, Zeb-2, and bHLH (E12, E-47, and TWIST). Finally, the expression of many proteins is repressed, with the most important of these being the *CDH1* gene which encodes E-cadherin, a protein essential for cell adhesion (Thiery and Sleeman, 2006).

One of the major signalling pathways that induce EMT in pancreatic cancer is the transforming growth factor β (TGF-β) signalling pathway (Figure 3.5). Activation of this pathway is initiated by the binding of TGF-β to the transforming growth factor β type I receptor (TβRI). Once activated, TβRI phosphorylates *SMAD2* and *SMAD3*, leading to the formation of a complex consisting of *SMAD2*, *SMAD3*, and *SMAD4*. This complex is translocated to the nucleus, where it activates the transcription of genes important for EMT. These include the transcription factors Snail, Zeb-1, Slug, and TWIST (Heldin et al., 2012).

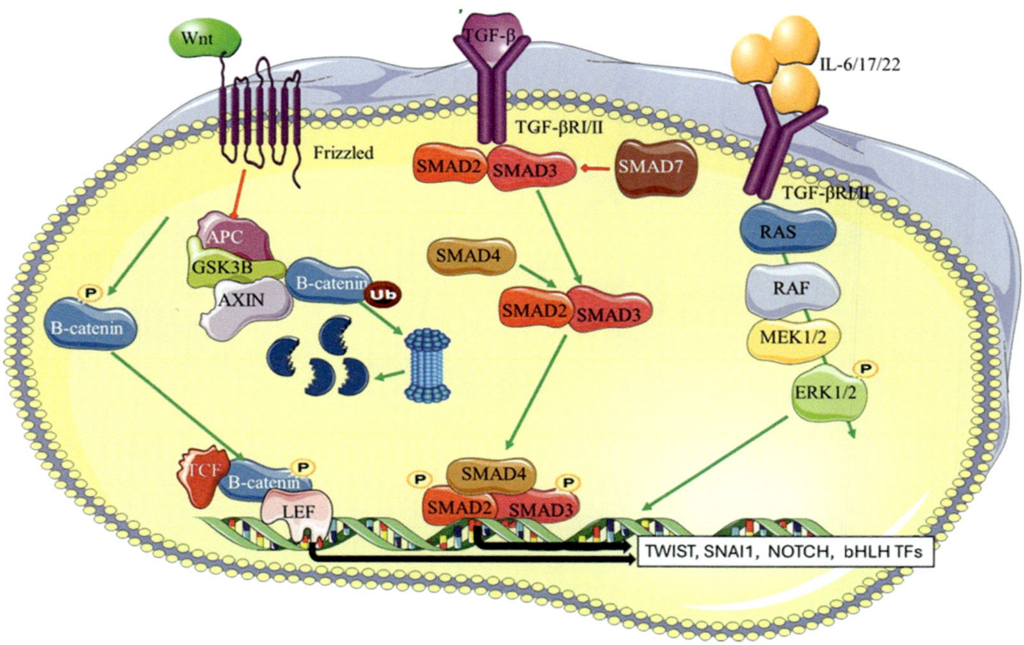

FIGURE 3.5 EMT signalling pathways. Normally, β-catenin levels are kept low via its degradation by a complex of proteins consisting of axin, adenomatous polyposis coli, glycogen synthase kinase-3β, and CK-1. This complex ubiquitinates β-catenin, labelling it for proteasomal degradation. However, the binding of Wnt to frizzled and LRP5/6 receptors leads to complex disbanding via the downstream phosphorylation of axin. β-catenin then forms a complex with TCF/LFE, activating the expression of genes that promote cell growth and proliferation in the nucleus. The TGFβ I and II receptors can be activated through the TGFβ ligand, which activates the canonical *SMAD* signalling pathway or through the binding of interleukins, which activates the MAPK/ERK signalling pathway.

In PDAC, this pathway can function in a SMAD-independent mechanism, involving PI3K, ERK/MAPK, P38, RhoA, and JNK (Dardare et al., 2020). Another important pathway for EMT in pancreatic cancer is the Wnt/β-catenin signalling pathway (Figure 3.5). This pathway promotes cell proliferation by stabilising β-catenin, allowing it to activate gene expression. The pathway also inhibits the degradation of the transcriptional repressor protein SNAI1. SNAI1 normally represses the expression of E-cadherin (Ram Makena et al., 2019). In pancreatic cancer, the Wnt/β-catenin signalling pathway stimulates EMT by upregulating the expression of genes that promote this process. The activation of the Wnt/β-catenin signalling pathway is driven by *kras* activity (Ram Makena et al., 2019). Inhibition of the Wnt/β-catenin signalling pathway has been shown to prevent EMT, accompanied by an increase in the expression of epithelial markers and a decrease in the expression of mesenchymal markers (Zhao et al., 2011).

Signalling Pathways Unconvolved in EMT

The Notch signalling pathway plays a role in cell differentiation and survival (Guruharsha et al., 2012). There are currently four identified Notch receptors and five Notch ligands. These ligands are named Delta-like 1, 2, and 3, as well as Jagged 1 and 2. Activation of Notch signalling happens when the Notch protein binds to a receptor on the adjacent cell (Figure 3.6) (Zhou et al., 2022). The Notch enzyme is cleaved through a cascade of proteolytic cleavages by metalloproteases, tumour necrosis factor alpha-converting enzymes, and γ-secretase (Song et al., 2018). The cleavage of Notch by the

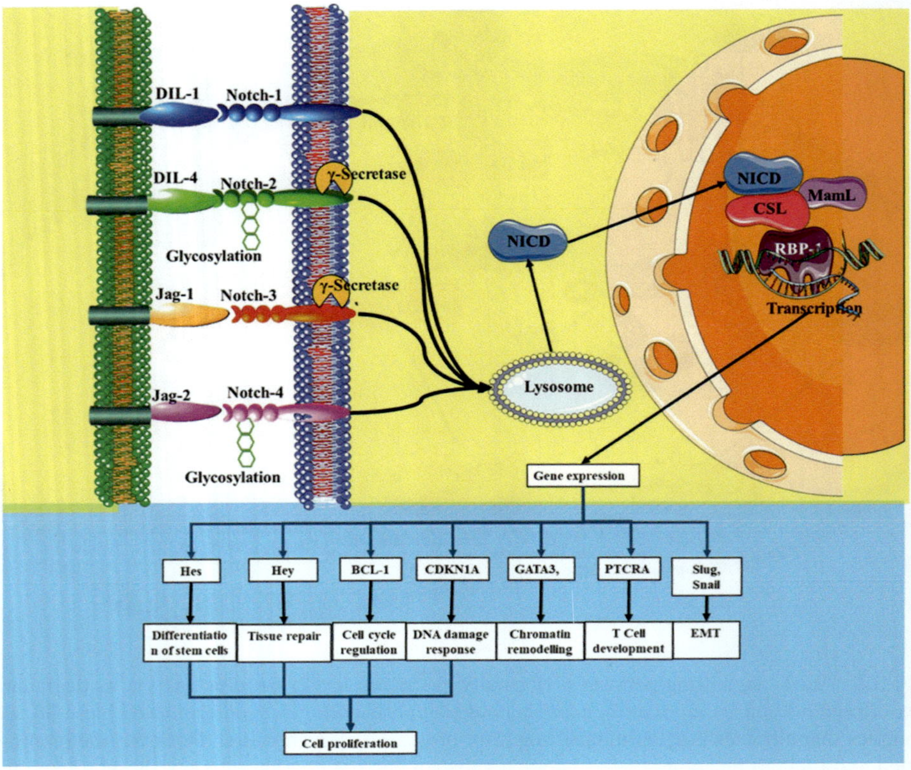

FIGURE 3.6 Notch signalling pathway. The Notch pathway is activated upon the interaction between the DLL and Jagged molecules expressed on the surface of an adjacent cell and the Notch receptors. This results in the γ-secretase complex stimulating the enzymatic cleavage of the Notch receptor to form a smaller protein, NUCD, which translocates to the nucleus where it binds to multiple transcriptional co-activators to initiate the transcription of multiple genes, including the SLU and Snail-1 transcriptional repressors which stimulate EMT.

gamma secretase complex leads to the creation of the Notch intracellular domain fragment (NICD). NICD translocates to the nucleus where it binds to the CBF1, Suppressor of Hairless, and Lag-1 (CSL) transcription factor. The CSL-NICD complex recruits other coactivators of transcription such as MamL, RBP, p300, and others. This activates the transcription of various Notch target genes. These activated genes regulate and activate cell growth, proliferation, angiogenesis, and apoptosis and include Cyclin D1, COX-2, Akt, MMP9, ERK, VEGF, c-Myc, mTOR, NF-κB, TP53, P27, and P21 (Shi et al., 2024). The expression of these genes promotes immune suppression, cell proliferation, chromatin remodelling, DNA repair, regulation of the cell cycle, tissue repair, and differentiation. The Notch pathway also directly promotes the increased expression of Slug and Snail-1 transcription activators which play an important role in epithelial to mesenchymal transition (Güngör et al., 2011).

The Snail family of transcriptional repressors includes Snail (Snail-1), Slug (Snail-2), and Smuc (Snail-3). All these transcription factors share the common characteristic that they possess C2H2 type, zinc finger motifs. They also possess a growth factor independence-1 (GFi-1) domain at the amino terminal of the protein. Also known as a SNAG domain, this domain is required for the transcriptional repressor function of these proteins. The SNAG domain also stabilises the structure of the Snail family members (Yastrebova et al., 2021). EMT is activated through signalling pathways initiated by Slug and Snail during embryonic development, but during cancer development and progression, they are activated abnormally, resulting in the accompanying pathologic EMT

(Lamouille et al., 2014). Snail and Slug expression was found to be increased in 50% and 68% of all PDAC cases, respectively. As such, they are thought to be the main drivers of EMT in a majority of pancreatic cancers. Expression levels of Snail were more likely to be increased in pancreatic cancers that had already invaded the lymph nodes and undergone metastasis to distant sites within the body (Hotz et al., 2007).

Metastasis could be induced in pancreatic cancer cell lines by transfecting them with Snail (Nishioka et al., 2010). Knocking out Snail-enhanced sensitivity to gemcitabine, leading to amplified overall survival in a genetically modified mouse model of PDAC (Zheng et al., 2015). Snail represses the transcription of a variety of genes while stimulating the transcription of others. Some of the targets of Snail whose transcription is repressed include genes involved in maintaining the epithelial phenotype, genes promoting apoptosis, or regulating cell proliferation, and many tumour-suppressor genes. Genes whose transcription is activated by Snail include mesenchymal genes, anti-apoptotic genes, genes that promote growth and proliferation, as well as survival. E-cadherin is the only known direct target of Snail (Figure 3.7) (Peinado et al., 2004).

Zeb-1, a member of the Zeb family of transcription factors, is overexpressed in pancreatic cancers and its role in EMT in cancer is well established. In pancreatic cancer, increased levels of Zeb-1 expression are associated with poor patient outcomes. In addition to this, increased expression of Zeb-1 is accompanied by a decrease in the expression of E-cadherin in pancreatic cancer cell lines (Zhang et al., 2015). Like Snail family members, Zeb transcription factors possess C2H2 zinc finger domains. Zeb-1 possesses four consecutive such domains. It also possesses four of these domains in the centre of the protein and another three consecutive zinc finger domains at the C terminus

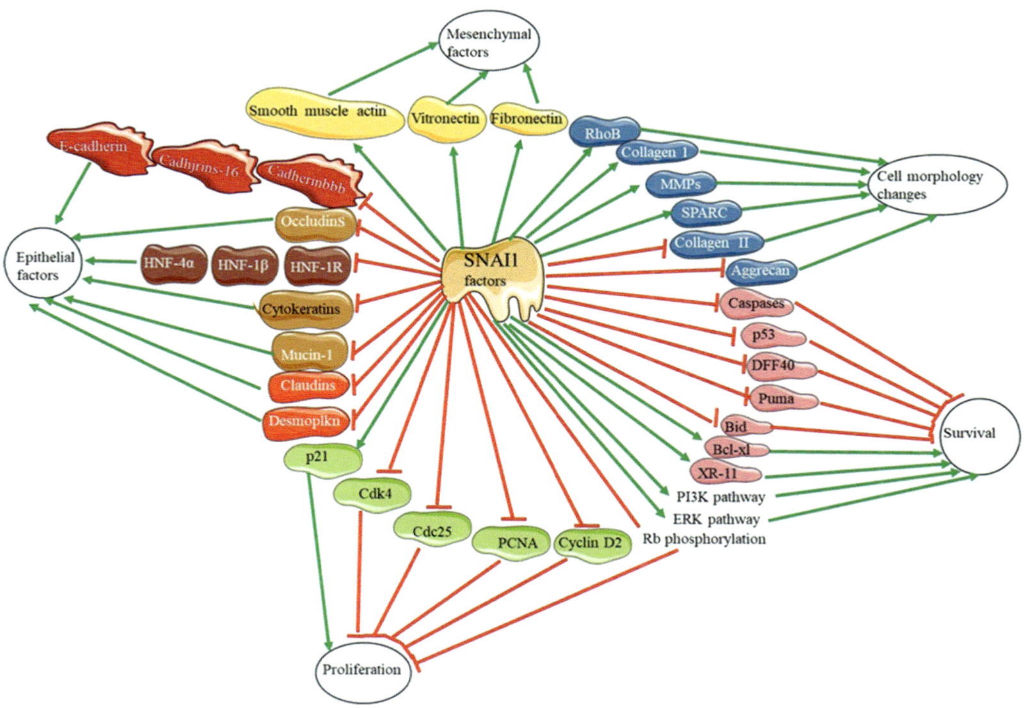

FIGURE 3.7 Various targets of the Snail family of transcriptional repressors. The Snail family of transcriptional repressors repress the expression of proteins known as epithelial markers. These are proteins that maintain the epithelial phenotype of cells while repressing the expression of proteins that characterise a mesenchymal phenotype. They also repress the expression of genes that downregulate cell survival or promote apoptosis, while promoting the expression of proteins that alter cell morphology. Finally, they downregulate proteins which repress cell proliferation or inhibit cell cycle checkpoint expression.

of the proteins. It also possesses a homeobox (HOX) DNA-binding domain (Perez-Oquendo and Gibbons, 2022). Zeb-1 knockdown or silencing has also been shown to decrease levels of migration, metastasis, and tumour development and growth in pancreatic and colon cancer cell lines (Wellner et al., 2009). The transcription of epithelial marker genes is suppressed by Zeb-1, the majority of which are genes involved in cell adhesion and cell polarity (Zhang et al., 2015). It is also involved in the epigenetic silencing of E-cadherin expression by increasing the acetylation of the promoter of the *CDH1* gene by recruiting HDAC-1/2. In addition to this, Zeb-1 also recruits the switch/sucrose non-fermentable chromatin remodelling protein BRG1 to the promoter of the *CDH1* gene. BRG1 can activate or repress gene expression by remodelling chromatin (Aghdassi et al., 2012). ZEB1 inhibited the DNA damage response triggered by ATM, as well as promoting the loss of apical polarity, which promotes EMT as the cell is more uniform. Finally, oestrogen signalling is known to promote EMT through affecting the function of the ERα oestrogen receptor. Various alternative splicing factors can drive either the epithelial or mesenchymal phenotype, and ZEB1 can alter the expression of these splicing factors (Figure 3.8) (Schwab et al., 2024).

The transcription factors E12, E47, TWIST 1, and TWIST 2 are known to play a role in EMT and are all members of the basic helix-loop-helix (bHLH) family of transcription factors. This is one of the largest families of transcription factors, and apart from a DNA-binding domain and the bHLH motif, they have a wide range of structures and functions (Teng and Li, 2014). Both E-47 and E12 repress the expression of E-cadherin, thereby promoting EMT (Perez-Moreno et al., 2001). TWIST 1 and TWIST 2 are the bHLH transcription factors most commonly associated with EMT since they are well-known regulators of pathogenic EMT. TWIST expression is either low or absent from PDAC tissue samples and pancreatic cancer cell lines. TWIST expression is induced due to hypoxic conditions, which can arise during the tissue invasion by pancreatic cancer cells (Zhu et al., 2016). Like E47 and E12, TWIST represses E-cadherin expression. It also induces the expression

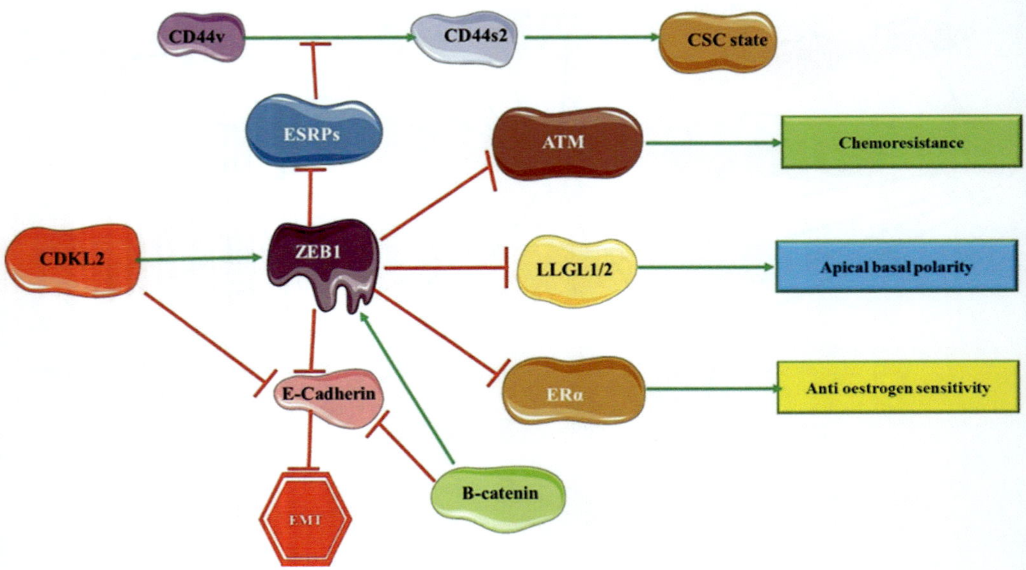

FIGURE 3.8 The role of ZEB1 in stimulating EMT. ZEB1 supresses the expression of E-cadherin, which is an epithelial marker that promotes cell adhesion and inhibits EMT. The Wnt-signalling pathway leads to the stabilisation of the β-catenin protein, which leads to the transcription of ZEB1 and the inhibition of E-cadherin. Cyclin-dependent kinase-like 2 (CDKL2) promotes the expression of ZEB1 by enhancing its transcription. Epithelial splicing regulatory proteins (ESRPs) promote the splicing of genes that reinforce the epithelial cell phenotype. ZEB1 decreases the activity of these splicing proteins, leading to the expression of protein isoforms that promote EMT.

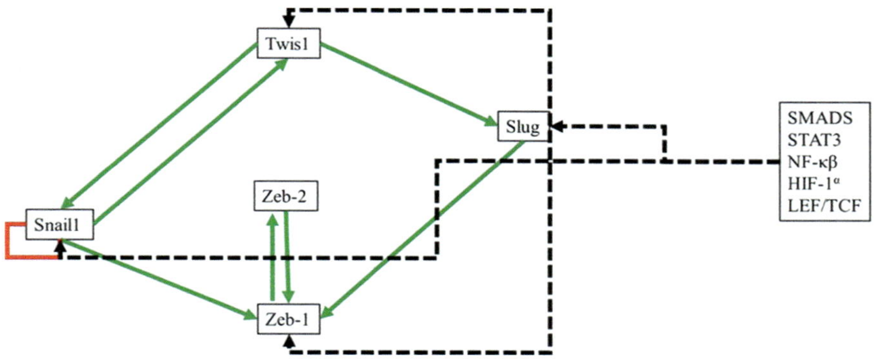

FIGURE 3.9 The interaction of transcription factors that promote EMT. EMT is regulated by multiple transcription factors, and the expression of these factors depends on the activity of other transcription factors. This leads to the regulation of EMT by a complex signalling network.

of N-cadherin. The upregulation of TWIST has been associated with decreased methylation (hypomethylation) of the TWIST promoter (Fu et al., 2011).

These various transcription factors that promote EMT can interact with and promote each other's activity in a complex network (Figure 3.9) (Huang et al., 2022).

THE TUMOUR MICROENVIRONMENT

The tumour microenvironment (TME) is the complex surrounding of a tumour that consists of physical and biochemical characteristics which have been altered due to the presence of cancerous cells and their interactions with stromal cells. The resulting microenvironment surrounding the tumour promotes the growth and spread of the tumour, while also protecting it from the immune response or the induction of apoptotic pathways. The microenvironment can also contribute to the development of treatment resistance, especially resistance to drugs and compounds (Bożyk et al., 2022). The TME of pancreatic cancer is extremely well suited to aid in the tumour's ability to evade an immune response due to its immunosuppressive nature. This immunosuppressive environment is the result of the expression of constituents and pathways that hinder the immune response to the tumour. Cellular constituents of the pancreatic cancer TME include immune cells, pancreatic stellate cells/cancer-associated fibroblasts, and acellular stroma (Karamitopoulou, 2019). The complexity of the TME makes the elucidation of the mechanisms underlying its contribution to disease progression complex and difficult. This is due to the heterogeneous nature of the disease and the presence of various types of immune regulatory cells and cancer-associated fibroblasts within the stroma (Sadeghi Rad et al., 2021).

Desmoplasia is a hallmark of the TME of pancreatic cancer and describes the activation of fibroblasts due to signals from the cancer cells that result in large amounts of extracellular matrix (ECM) being secreted around the tumour, consisting predominantly of fibrillar type I collagen. This dense collagen ECM assists cancer cells in their survival, proliferation, and invasion, as the cells within this ECM secrete higher levels of matrix metalloproteinases, which assist in invasion by degrading the ECM and allowing cell migration. This dense ECM resulting from desmoplasia can also promote angiogenesis and assist in the induction of EMT. The dense ECM may also limit the amount of a drug reaching the tumour cell, thereby promoting drug resistance. This dense physical barrier can also contribute to immune evasion by preventing immune cells from infiltrating the tumour (DeClerck, 2012). Three-dimensional cell culture has demonstrated that collagen density is a major factor influencing T cell activity, with a denser matrix resulting in fewer cellular markers of cytotoxic T cell activity and higher levels of markers for Tregs (Kuczek et al., 2019).

Cancer-associated fibroblasts (CAFs) secrete cytokines and growth factors that are immunosuppressive and act by inhibiting the activity of immune cells such as cytotoxic T cells (Zhang et al., 2024). The population and function of immune cells can also be altered by certain stromal cells. This includes shifts in the T cell population to favour regulatory T cells (Tregs) and myeloid-derived suppressor cells (MDSs), both of which suppress the activity of immune cells towards cancer cells. Finally, desmoplasia can lead to the metabolic reprogramming of tissues due to the hypoxic TME. Hypoxia can lead to the expression of immune checkpoint proteins, increasing immune evasion (Zhang et al., 2024).

There are different forms of the ECM that contribute to the development of the TME in different ways. Two forms of ECM that are important in pancreatic cancer are the basement membrane (BM) and the interstitial matrix (IM). The BM is composed of thin sheets made up of laminin, type IV collagen, and heparan sulphate proteoglycan. BM performs a protective role and also polarises epithelial cell layers, separating them from the tissue layers beneath them. The fibrillar type I collagen within the BM contains mesenchymal cells, such as fibroblasts, which are the main structural component of the IM (Puls et al., 2017). The IM forms the ECM in the spaces between cells and, apart from fibroblasts, consists of polysaccharides and fibrous proteins (Ho et al., 2020). The fibroblasts in the ECM have been found to both promote and inhibit the progression of pancreatic cancer, due largely to their functional heterogeneity. This heterogeneity is due to the existence of multiple classes of fibroblasts, each with their own specific or multiple functions present in the TME (Chen et al., 2021). The ability of fibroblasts to promote pancreatic cancer progression is largely due to ECM remodelling, their contribution to desmoplasia, and immune suppression. These factors have all been discussed above. They also secrete metabolites that promote tumour development, and various tumour-promoting factors such as TGFβ, LIF, CXCL12, IL-6, IGF1, and MAPK (Bryce et al., 2022). Fibroblasts in the TME can inhibit pancreatic cancer development and progression by many of the same mechanisms through which they promote the progression of pancreatic cancer. The increased density of the stroma that the fibroblasts cause can also restrain the development and growth of tumours. This increased density of the TME can also inhibit invasion and cell migration (Bryce et al., 2022). The dense environment created by these fibroblasts has also been associated with the inhibition of cancer-promoting pathways such as the Sonic Hedgehog-SMO signalling pathway (Lee et al., 2014) and the JAK signalling pathway (Biffi et al., 2019).

Irradiation of PDAC tumours can cause the pancreatic cancer cells to penetrate the BM, a process which can be blocked by anti-integrin antibodies. Fibronectin fibres function as a linker between integrin and collagen through the presence of binding sites of both molecules on fibronectin fibres. This migration of cancer cells into the BM can also be blocked through the depletion of fibronectin. This implicates fibronectin as playing a key role in the creation of the TME of pancreatic cancers. Fibronectin is known to bind to proteins that can bind to TGF. This recruitment of TGF results in the activation of the TGF pathway, stimulating further growth and proliferation (Weniger et al., 2018). Hyaluronan (HA), another major component of the TME in pancreatic cancer, also promotes the progression of the disease. This has been demonstrated in studies where HA was found to accumulate at abnormally elevated levels in the stroma of the TME of malignant cancers. Under normal non-malignant conditions, the level of HA remains constant as it is continuously degraded as new HA is synthesised (Sato et al., 2016). This is due to the overexpression of both HA and its receptors which occurs in pancreatic cancer cells, and high levels of HA are associated with a poorer prognosis in pancreatic cancer patients. Elevated levels of HA were found to promote tumour growth in PDAC models. High levels of HA were also found to promote various hallmarks of cancer, such as increased angiogenesis, metastatic processes, as well as resistance to chemotherapy drugs (Sato et al., 2016). Despite promising early clinical results in the phase II trials of therapeutic targeting of high HA levels with pegvorhyaluronidase alfa (PEGPH20) combined with gemcitabine/nab-paclitaxel in pancreatic cancer (Hingorani et al., 2018), its subsequent randomised phase III trial unfortunately was negative (Van Cutsem et al., 2020).

The formation of the TME is accompanied by the formation of connective tissue between the various cells that constitute the microenvironment. These include cancer cells, epithelial cells, stromal cells, and newly transitioned mesenchymal cells. This new connective tissue can serve as a framework for cell migration and angiogenesis, thereby promoting the progression of pancreatic cancer (Zhao et al., 2023).

THE NERVOUS SYSTEM AND PANCREATIC CANCER

Perineural invasion (PNI) refers to the process where cancer cells invade the space surrounding a nerve cell or nerve tissue. For over a century, it has been known that the process of PNI facilitates the spread and metastasis of various cancers (Demir et al., 2010). The activation of nerve-dependent signalling pathways in the development of pancreatic cancer is similar to the activation of these pathways observed in the normal development of the pancreas. The role of the nervous system in the development and patterning of the pancreas can be studied through examining the development of similar, easier-to-access glands such as the submandibular gland. Initial steps in this process are known as the nerve recruitment phase and involve the secretion of neurotrophins by epithelial, mesenchymal, and stromal cells, which recruit the peripheral nerves (Zahalka and Frenette, 2020). Neurotrophins that play a role in this process include nerve growth factor (NGF), brain-derived neurotrophic factor (BDNF), neurotrophin 3 (NT-3), and neurotrophins 4/5 (NT-4/5) (Skaper, 2018). In addition to these, factors that have neurotrophic activity, such as ciliary neurotrophic factor (CNTF) and glial cell line-derived neurotrophic factor (GDNF), also play a role (Ip and Yancopoulos, 1996). These molecules bind to one of three tropomyosin-related kinases (TrkA, TrkB, and TrkC) and the p75 neurotrophin receptor ($p75^{NTR}$), activating downstream signalling pathways that promote gene expression and exogenesis (Skaper, 2018). This results in parasympathetic nerves initiating the next stage of ductal epithelial tubulogenesis and glandular or acinar epithelium growth. This development is accompanied by the development of the vasculature, leading to a rapid phase of glandular growth (Borden et al., 2013).

Cancer cells and the cells in the TME secrete high levels of neurotrophins, leading to sensory nerve (Cervantes-Villagrana et al., 2020), resulting in nerve recruitment and increased growth with the generation of neovasculature (Zahalka et al., 2017). Stimulation of the parasympathetic nervous system stimulates the growth of cancer cells, providing them with migratory cues. Newly formed as well as existing nerves also provide a pathway for these migrating cancer cells to move along. At the same time, the release mediators of cell growth by peripheral nerves and macrophages stimulates the invasion of nerve tissue by cancer cells (Tibensky and Mravec, 2021). This situation is exacerbated in pancreatic cancer due to the abundant nerve supply to the pancreas, which results in a higher frequency and intensity of PNI in the pancreas. Thus, increased innervation may be one of the factors contributing to the rapid spread and high mortality rate observed in pancreatic cancer (Zahalka et al., 2017; Biankin et al., 2012). Pancreatic intraepithelial neoplasia (PanIN) tissue has increased nerve density levels with a higher number of intrapancreatic nerves. This increases as the cancer progresses to PDAC, a situation known as pancreatic cancer-associated neural remodelling (PANR) (Hruban et al., 2008). Unsurprisingly PDACs show increased levels of expression of both neurotrophins and their receptors (Weitz et al., 2023).

APPLICATIONS FOR THE MANAGEMENT OF PANCREATIC CANCER

This has been made possible through the development of high-throughput technologies such as next-generation sequencing, which can provide transcriptome or genomic information for patients that can then be analysed to detect molecular patterns associated with pancreatic cancer, or to identify molecules that may be targets for therapy (Wood et al., 2022). In this way, the most appropriate therapy may be selected. The same is true for proteomic analysis using technologies such as mass spectrometry. This holds the promise for the development of personalised medicine, where the unique molecular characteristics of a patient can be targeted for treatment that is tailored to them, or even screening tests tailored to specific populations or ethnicities. The use of these technologies requires the rapid and accurate analysis of substantial amounts of data. The solution to this problem can be provided using artificial intelligence and machine learning to detect these patterns in large datasets (Brittain et al., 2017).

Diagnosis

Currently, there are no reliable markers that can be used for the routine diagnosis of pancreatic cancer. However, there are promising examples of markers whose expression changes in pancreatic cancer are reliable enough that there is hope for developing reliable molecular screening and diagnosis methods (Singhi et al., 2019). One of the best-known serological biomarkers for pancreatic cancer is the tetrasaccharide, CA19-9. This marker will be discussed in detail as it is a good representative of the problems associated with the use of biomarkers in pancreatic cancer. Other potential markers are listed in Table 3.1. It is important to note that all the markers listed in Table 3.1 are still under investigation, and all have concerns regarding their sensitivity and specificity. This molecule is expressed on the surface of cancer cells. As early as 1979, it was proposed as a marker for the detection of colorectal cancer using an antibody raised against it named NS19-9 (Luo et al., 2021). CA19-9 has previously been used to monitor both the progression of pancreatic cancer as well as its response to treatment (Azizian et al., 2020). However, it is only upregulated in approximately 80% of all pancreatic cancer patients. Apart from pancreatic cancer and colorectal cancer, it can also be used to detect gastric and biliary tract cancer (Ballehaninna and Chamberlain, 2012). It is also upregulated in benign conditions such as chronic pancreatitis, biliary obstruction, and cholangitis (Zhao et al., 2021). This lack of specificity is the main barrier to its use as a reliable marker for pancreatic cancer. Another disadvantage of CA19-9 is that it is absent from at least 10% of the population. This is because it is a Lewis blood group antigen and will only be expressed in patients with the Le (α−β+) or Le (α+β−) blood groups. However, at least 10% of the population belongs to the Le (α−β−) blood group and lacks CA19-9 since this phenotype does not express 1,4-fucosyltransferase, the enzyme required to produce CA19-9 (Kannagi, 2007). Studies conducted in Japanese, Korean, and Taiwanese populations evaluated serum levels of CA19-9 in the diagnosis of prostate cancer. These studies aimed to establish the positive predictive value (PPV) for the presence of CA19-9. These studies indicated that the PPV for pancreatic cancer using CA19-9 was low (Satake et al., 1994; Kim et al., 2004; Chang et al., 2006). The use of a single marker to diagnose pancreatic cancer is not advisable, and a range of different markers should be used. For instance, CA19-9 is used with a variety of other biomarkers. Some of these other markers include sialylated tumour-related antigen (Duffy et al., 2010). Duke pancreatic monoclonal antigen type 2 is the precursor molecule of CA19-9 and is elevated in pancreatic cancer patients, regardless of their Lewis phenotype, and as such has potential advantages over CA19-9 (Shimizu et al., 2023).

Another option for diagnostic biomarkers is the use of circulating tumour cells (CTCs), circulating tumour DNA (ctDNA), noncoding RNAs (ncRNAs), and extracellular vesicles (EVs) or exosomes. These markers can all be assessed using liquid biopsies, providing the added advantages of being cost-effective, non-invasive, rapid, and convenient (Wang et al., 2023).

Treatment

The development of technologies such as next-generation sequencing, bioinformatics, and artificial intelligence has led to a revolution in the development of new cancer therapies. However, only a minority of pancreatic cancer patients have benefitted from these developments. The identification of driver mutations and molecular characterisation of pancreatic cancers not only allows for

TABLE 3.1
Biomarkers under Investigation for Probable Diagnostic or Prognostic Roles in Pancreatic Cancer

Marker	Role	Reference
CA125	Mucin-like glycoprotein associated with ovarian cancer and CRC. It predicts the resectability of pancreatic cancer	Napoli et al. (2023)
CEA	Foetal glycoprotein used as biomarkers for CRC, ovarian, cervical, lung, and breast cancer. Inconsistent results about its use in pancreatic cancer	Kurlinkus et al. (2021)
CA242	High expression in pancreatic cancer patients	Dou et al. (2019)
Osteonectin	Promotes invasion and metastasis in pancreatic cancer	Rossi et al. (2016)
Osteopontin	ECM protein upregulated in pancreatic cancer with a sensitivity and specificity of 80–97%	Rossi et al. (2016)
Laminin γ2	ECM glycoprotein inversely related to pancreatic cancer patient survival	Mitsunaga et al. (2010)
Soluble CD40 ligand	Diagnostic and prognostic marker for pancreatic cancer	Chung and Lim (2014)
Cofilin-1	Chemotaxis, cell migration, and tumour cell invasion, elevated in prostate cancer	Satoh et al. (2017)
Leucine-rich a2-glycoprotein-1	Inflammatory protein able to distinguish prostate cancer from benign conditions. Not effective for early diagnosis	Furukawa et al. (2015)
DKK1	Inhibitor of Wnt/β-catenin signalling and is over-expressed in a number of solid organ malignancies	Osada et al. (2008)
Thrombospondin-1, 2	Glycoprotein that mediates cell-to-cell and cell-to-matrix interactions	Wang et al. (2016)
Anterior gradient homolog 2 protein	Expressed in embryogenesis and multiple solid organ malignancies	Chen et al. (2013)
Regenerating protein family1β	Proteins present in pancreatic acinar cells, implicated in the regeneration of pancreatic islets, implicated in PC	Li et al. (2016)
Syncollin	Significantly elevated in the serum of pancreatic patients	Kalus et al. (2002)
Lysyl oxidase-like 2	Promote EMT. Elevated in pancreatic cancer where over-expression is associated with poorer survival	Makawita et al. (2013)
PARK7/DJ-1	Oncogene over-expressed in [pancreatic cancers	Cao et al. (2015)
Transthyretin	Elevated in patients with endocrine tumours	Ehmann et al. (2007)
Peroxiredoxin-1	Elevated in pancreatic cancer patients	Cai et al. (2015)
Tissue factor pathway inhibitor	Proteinase inhibitor that controls coagulation implicated in malignancy.	Lwaleed and Bass (2006)
TIMP-1	Inhibits MMPs, regulates cell proliferation and apoptosis. Sensitivity of 47.1% in detecting pancreatic cancer	Lee et al. (2003)
Insulin-like growth factor binding protein	Inhibits IGF-1 activity, with low levels predicting an increased risk of prostate cancer	Yoneyama et al. (2016)
Osteoprotegerin	TNF receptor superfamily member specificity of 78% in detecting prostate cancer	Holen et al. (2005)

(Continued)

TABLE 3.1 (CONTINUED)

Biomarkers under Investigation for Probable Diagnostic or Prognostic Roles in Pancreatic Cancer

Marker	Role	Reference
Kisspeptin	Elevated in pancreatic cancer	Loosen et al. (2019)
Galectin-3	β-galactoside-binding protein family member, is over-expressed in pancreatic cancer	Xie et al. (2012)
Mucins	Promotes metastasis chemoresistance and tumourigenicity in pancreatic cancer	Liu et al. (2016)
Heat shock protein 27	Molecular chaperone, promotes metastasis elevated in pancreatic cancer and benign pancreatic diseases	Liao et al. (2009)
Cacodylate haptoglobin	Expression correlates with disease stage	Kamada et al. (2013)
Aminopeptidase N	Membrane bound metallo-proteinase that plays a role in tumour progression, proliferation, invasion, and angiogenesis	Bhagwat et al. (2001)
M2-pyruvate kinase	Plays a role in cancer metabolism. Elevated in nearly all pancreatic cancers but lacks specificity	Vander Heiden et al. (2009)
Vascular endothelial growth factor	Predicts metastasis and poor prognosis, but there is little information on the usefulness of serum VEGF as a diagnostic biomarker	Abdel-Razik et al. (2016)
Fibroblast growth factor 10/ keratinocyte growth factor-2	Regulates pancreatic cancer EMT and is overexpressed in pancreatic cancer	Lanki et al. (2023)
Macrophage inhibitory cytokine-1/growth differentiation factor-15	TGF-β cytokine superfamily member, elevated in many pathologies	Yang et al. (2018)
Oncostatin M	IL-6 cytokine family member, promotes EMT, elevated in pancreatic cancer	Lanki et al. (2023)
Macrophage colony-stimulating factor	Elevated in prostate cancer patients	Yako et al. (2016)
CXCL11/interferon inducible T cell alpha chemokine	Over-expressed in prostate cancer patients	Miekus et al. (2010)
Stem cell factor	Involved in cell proliferation, differentiation, and cell survival and is implicated in pancreatic cancer	Yasuda et al. (2006)
CEA-related cell adhesion molecules	CEACAM immunoglobulin supergene family member. Relatively specific for pancreatic cancer but few supporting studies	Pandey et al. (2019)
ICAM-1	Functions in cell–cell and cell–ECM adhesion. A promising marker but cannot distinguish between early and late-stage disease	Silverman and Shi (2016)

precise targeting of specific molecules and pathways for treatment but also allows for classifying the cancer into specific subtypes, which can assist in designing a treatment strategy (Kolbeinsson et al., 2023). Molecular subtyping allows for better delineation of pancreatic cancers that were previously assumed to be homogeneous into heterogenous types and facilitates the development of personalised or precision oncology. This molecular characterisation can be based on genomic, transcriptomic, exonic, or epigenomic alterations (Froeling et al., 2021).

It is common for PDAC patients to have somatic/germline mutations in DNA damage repair (DDR) pathways that result in the loss of DNA damage repair function. For instance, the loss of genes involved in homologous recombination can result in the loss of DNA integrity, including mutations in genes such as *ATM*, *BAP11*, *BARD1*, *BLM*, *BRIP1*, *CHEK2*, *AM175A*, *FANCA*, *FANCC*, *NBN*, *RAD50*, *RAD51*, *RT11* and most importantly *BRCA1* and B*RCA2*. The resulting HR efficiency is known as "BRCAness" and can result in accumulated DNA damage leading to tumourigenesis. Initial studies using older technologies detected homologous repair disorders (HRD) in 25% of pancreatic cancer patients; however, the use of whole genome/whole exome sequencing led to the detection of HRD in a greater percentage of patients, up to 44% (Casolino et al., 2021). Detection of these mutations allows the patient to be classified into a particular subtype of pancreatic cancer. This subtype of pancreatic cancer with DNA damage repair deficiencies is more sensitive to platinum-based therapy in combination with PARP inhibitors (Dalmasso et al., 2022).

The frequency of *kras* mutations and alterations in *KRAS*-related pathways makes it an attractive target for therapeutic interventions. As discussed earlier (Figure 3.1) the common mutations in pancreatic cancer occur at codon 12. Specific drugs have been designed to target these mutations. The drug Sotorasib was designed to target the *G12C KRAS* mutant, which is common in lung cancer. However, in pancreatic cancer, this mutation only accounts for 1.5% of *KRAS* mutants, meaning this drug is of limited value in treating pancreatic cancer (Parums, 2022). An RNAi-based therapy was designed to target the *G12D KRAS* mutant. This mutant is found in 42% of pancreatic cancers and is therefore a more useful strategy for treating pancreatic cancer. This was evaluated in a phase I/II trial on small biodegradable tumour implants. This study showed that the RNAi of these mutants was a promising treatment as it resulted in longer overall survival times (Golan et al., 2015). Another treatment targeting the *G12D KRAS* mutant used genetically engineered T cells that expressed T cell receptors (TCRs) that targeted this mutant *KRAS* (Leidner et al., 2022). Other mutations that can be targeted are shown in Table 3.2.

TABLE 3.2

Treatments Targeting Specific Mutations in Pancreatic Cancer

Drug	Target	Side Effects	References
Dabrafenib	BRAFV600E mutation	Chills, fever, night sweats, tiredness, skin rash, nausea, vomiting, diarrhoea, abdominal pain, swelling of hands or feet, joint aches, decreased appetite, muscle aches, and headache	Subbiah et al. (2023)
Trametinib	BRAFV600E mutation		
Larotrectinib Entrectinib	*ntrrk* gene	Dizziness, fatigue, nausea, vomiting, constipation, weight gain, and diarrhoea	Hong et al. (2020)
Selpercatinib	*ret* gene	Abnormal liver function test values, low white blood cell counts, dry mouth, diarrhoea, constipation, high blood pressure, tiredness, and skin rash	Bhamidipati et al. (2023)
Adagrasib	*KRAS G12C*	Nausea, decreased appetite, vomiting, diarrhoea, muscle aches, and changes in liver and kidney test values,	Bannoura et al. (2022)
Erlotinib	EGFR	Acne-like rash on the face and neck, diarrhoea, loss of appetite, and fatigue	Moore et al. (2007)
Olaparib	PARP inhibitor	Nausea, vomiting, diarrhoea or constipation, fatigue, feeling dizzy, loss of appetite, taste changes, low red blood cell counts (anaemia), low white blood cell counts (with an increased risk of infection), belly pain, and muscle and joint pain	Golan et al. (2019)

Limitations

Despite the promise offered by targeting the precise molecular alterations in individual pancreatic cancer patients or populations, there are also definite drawbacks to the use of these methods as well as barriers to their implementation. One of the major problems is that many studies find that these interventions targeting a specific molecular change only have small effects on disease outcomes (Salazar et al., 2023). This is also due to the lack of a dominant molecular phenotype with large heterogeneity in pancreatic cancer (apart from KRAS, P53, CDKN2A, and SMAD4). As a result, the prevalence of each molecular alteration that could be therapeutically targeted is very low (<5%). Another major concern is the lack of specificity and universal applicability. This is because pancreatic cancers are diverse and heterogeneous, meaning that not all pancreatic cancer patients or population groups possess all the molecular changes targeted for diagnosis or treatment (Robatel and Schenk, 2022). In terms of diagnosis, many of the molecular changes that can be targeted for diagnosis are not specific to pancreatic cancer and occur in many other pancreatic pathologies (Turner et al., 2023). Resistance to targeted therapies is also a developing problem (Robatel and Schenk, 2022). Resistance is partly due to the unique TME in pancreatic cancer, which can alter the ability of the drug to reach its target. Alternatively, cancer cells can adapt to use new signalling pathways to bypass the pathways and molecules that are targeted. This is true of drugs such as MAKP inhibitors that are used to target *kras* mutant pancreatic cancer (Kinsey et al., 2019). Like all treatments and therapies, there is always the potential for adverse side effects of these drugs. These can range from mild side effects such as fatigue, nausea, and diarrhoea to more severe side effects like liver problems and heart issues (Table 3.2). The cost of developing these therapies is high due to genotyping and phenotyping of pancreatic cancers. This is exacerbated by the fact that genomic and proteomic sequencing needs to be carried out for each patient since it is vital to thoroughly detect the population of patients who have a specific mutation and phenotype (Turpin et al., 2022).

CONCLUSION

Research has shown that the complexity of pancreatic cancer is partly due to various molecular pathways and gene expression patterns that are altered in pancreatic cancer. It has also been shown that these changes drive the development and progression of the disease. These molecular changes can be found in the genome, proteome, transcriptome, and epigenome and are identified and studied using high-throughput technologies. Once identified, these molecular alterations are now being applied to the diagnosis and treatment of pancreatic cancer, a process which is being streamlined and optimised by systems biology approaches and artificial intelligence. These computational methods can organise and classify large amounts of sequencing data to enable the association of molecular patterns with various subtypes, stages, and outcomes of pancreatic cancer. Another change that promotes the progression of pancreatic cancer is changes that give rise to the complex TME.

Pancreatic cancer has a unique microenvironment that is immunosuppressive and protects the tumour from drugs and other treatments. These changes can also be targeted for therapeutic intervention to make immunotherapy to be more effective and to allow other drugs to access the tumour. These molecular changes in pancreatic cancer can be targeted by gene therapy to address individual genes or entire pathways. Other specific therapeutic targets include disrupting angiogenesis, cell proliferation, and survival by targeting various growth factor pathways. Small molecules can be used as drugs to target the mutated proteins by binding with higher affinity to these proteins. Synthetic oligonucleotides can be used to target miRNAs. Finally, since nerves that are surrounded by the TME influence tumour progression, the interaction between cancer cells and neurones can be targeted to inhibit cancer progression. The multitude of therapeutic avenues described above, made possible through the characterisation of the molecular changes in pancreatic cancer, allows for the development of precision medicine with the aid of novel computational methods, including artificial intelligence and machine learning. Despite the promise, there are issues since pancreatic cancer, by its nature develops resistance to targeted therapies. In some cases, pancreatic cancer has mutations in the same genes as other cancers, but the actual mutations are different, resulting in different changes in the amino acid sequence, requiring the development of unique small molecules that bind to these proteins. However, the promise these therapies hold means that solutions to these problems are being actively sought. Current solutions include the use of combination therapies and novel drug delivery methods to bypass the TME. Ultimately, solving these problems will be a collaborative effort, aided by advances in both artificial intelligence and high-throughput technologies such as next generation sequencing and mass spectrometry. Advances in these technologies will make them cheaper and easier to access, allowing for the generation of data from more diverse populations and a greater number of individuals.

REFERENCES

ABDEL-RAZIK, A., ELMAHDY, Y., HANAFY, E. E., ELHELALY, R., ELZEHERY, R., A, M. T. & ELDARS, W. 2016. Insulin-like growth factor-1 and vascular endothelial growth factor in malignant and benign biliary obstructions. *Am J Med Sci,* 351, 259–264.

AGHDASSI, A., SENDLER, M., GUENTHER, A., MAYERLE, J., BEHN, C. O., HEIDECKE, C. D., FRIESS, H., BÜCHLER, M., EVERT, M., LERCH, M. M. & WEISS, F. U. 2012. Recruitment of histone deacetylases HDAC1 and HDAC2 by the transcriptional repressor ZEB1 downregulates E-cadherin expression in pancreatic cancer. *Gut,* 61, 439–448.

AZIZAN, N., SUTER, M. A., LIU, Y. & LOGSDON, C. D. 2017. RAGE maintains high levels of NFκB and oncogenic Kras activity in pancreatic cancer. *Biochem Biophys Res Commun,* 493, 592–597.

AZIZIAN, A., RÜHLMANN, F., KRAUSE, T., BERNHARDT, M., JO, P., KÖNIG, A., KLEIß, M., LEHA, A., GHADIMI, M. & GAEDCKE, J. 2020. CA19-9 for detecting recurrence of pancreatic cancer. *Sci Rep,* 10, 1332.

BAILEY, P., CHANG, D. K., NONES, K., JOHNS, A. L., PATCH, A. M., GINGRAS, M. C., MILLER, D. K., CHRIST, A. N., BRUXNER, T. J., QUINN, M. C., NOURSE, C., MURTAUGH, L. C., HARLIWONG, I., IDRISOGLU, S., MANNING, S., NOURBAKHSH, E., WANI, S., FINK, L., HOLMES, O., CHIN, V., ANDERSON, M. J., KAZAKOFF, S., LEONARD, C., NEWELL, F., WADDELL, N., WOOD, S., XU, Q., WILSON, P. J., CLOONAN, N., KASSAHN, K. S., TAYLOR, D., QUEK, K., ROBERTSON, A., PANTANO, L., MINCARELLI, L., SANCHEZ, L. N., EVERS, L., WU, J., PINESE, M., COWLEY, M. J., JONES, M. D., COLVIN, E. K., NAGRIAL, A. M., HUMPHREY, E. S., CHANTRILL, L. A., MAWSON, A., HUMPHRIS, J., CHOU, A., PAJIC, M., SCARLETT, C. J., PINHO, A. V., GIRY-LATERRIERE, M., ROOMAN, I., SAMRA, J. S., KENCH, J. G., LOVELL, J. A., MERRETT, N. D., TOON, C. W., EPARI, K., NGUYEN, N. Q., BARBOUR, A., ZEPS, N., MORAN-JONES, K., JAMIESON, N. B., GRAHAM, J. S., DUTHIE, F., OIEN, K., HAIR, J., GRÜTZMANN, R., MAITRA, A., IACOBUZIO-DONAHUE, C. A., WOLFGANG, C. L., MORGAN, R. A., LAWLOR, R. T., CORBO, V., BASSI, C., RUSEV, B., CAPELLI, P., SALVIA, R., TORTORA, G., MUKHOPADHYAY, D., PETERSEN, G. M., MUNZY, D. M., FISHER, W. E., KARIM, S. A., ESHLEMAN, J. R., HRUBAN, R. H., PILARSKY, C., MORTON, J. P., SANSOM, O. J., SCARPA, A., MUSGROVE, E. A., BAILEY, U. M., HOFMANN, O., SUTHERLAND, R. L., WHEELER, D. A., GILL, A. J., GIBBS, R. A., PEARSON, J. V., WADDELL, N., et al. 2016. Genomic analyses identify molecular subtypes of pancreatic cancer. *Nature,* 531, 47–52.

BALAZ, P., FRIESS, H. & BÜCHLER, M. W. 2001. Growth factors in pancreatic health and disease. *Pancreatology,* 1, 343–355.

BALLEHANINNA, U. K. & CHAMBERLAIN, R. S. 2012. The clinical utility of serum CA 19-9 in the diagnosis, prognosis and management of pancreatic adenocarcinoma: an evidence based appraisal. *J Gastrointest Oncol,* 3, 105–119.

BANNOURA, S. F., KHAN, H. Y. & AZMI, A. S. 2022. KRAS G12D targeted therapies for pancreatic cancer: has the fortress been conquered? *Front Oncol,* 12, 1013902.

BHAGWAT, S. V., LAHDENRANTA, J., GIORDANO, R., ARAP, W., PASQUALINI, R. & SHAPIRO, L. H. 2001. CD13/APN is activated by angiogenic signals and is essential for capillary tube formation. *Blood,* 97, 652–659.

BHAMIDIPATI, D., YEDURURI, S., HUSE, J., CHINAPUVVULA, S. V., WU, J. & SUBBIAH, V. 2023. Exceptional responses to selpercatinib in RET fusion-driven metastatic pancreatic cancer. *JCO Precis Oncol,* 7, e2300252.

BIANKIN, A. V., WADDELL, N., KASSAHN, K. S., GINGRAS, M. C., MUTHUSWAMY, L. B., JOHNS, A. L., MILLER, D. K., WILSON, P. J., PATCH, A. M., WU, J., CHANG, D. K., COWLEY, M. J., GARDINER, B. B., SONG, S., HARLIWONG, I., IDRISOGLU, S., NOURSE, C., NOURBAKHSH, E., MANNING, S., WANI, S., GONGORA, M., PAJIC, M., SCARLETT, C. J., GILL, A. J., PINHO, A. V., ROOMAN, I., ANDERSON, M., HOLMES, O., LEONARD, C., TAYLOR, D., WOOD, S., XU, Q., NONES, K., FINK, J. L., CHRIST, A., BRUXNER, T., CLOONAN, N., KOLLE, G., NEWELL, F., PINESE, M., MEAD, R. S., HUMPHRIS, J. L., KAPLAN, W., JONES, M. D., COLVIN, E. K., NAGRIAL, A. M., HUMPHREY, E. S., CHOU, A., CHIN, V. T., CHANTRILL, L. A., MAWSON, A., SAMRA, J. S., KENCH, J. G., LOVELL, J. A., DALY, R. J., MERRETT, N. D., TOON, C., EPARI, K., NGUYEN, N. Q., BARBOUR, A., ZEPS, N., KAKKAR, N., ZHAO, F., WU, Y. Q., WANG, M., MUZNY, D. M., FISHER, W. E., BRUNICARDI, F. C., HODGES, S. E., REID, J. G., DRUMMOND, J., CHANG, K., HAN, Y., LEWIS, L. R., DINH, H., BUHAY, C. J., BECK, T., TIMMS, L., SAM, M., BEGLEY, K., BROWN, A., PAI, D., PANCHAL, A., BUCHNER, N., DE BORJA, R., DENROCHE, R. E., YUNG, C. K., SERRA, S., ONETTO, N., MUKHOPADHYAY, D., TSAO, M. S., SHAW, P. A., PETERSEN, G. M., GALLINGER, S., HRUBAN, R. H., MAITRA, A., IACOBUZIO-DONAHUE, C. A., SCHULICK, R. D., WOLFGANG, C. L., MORGAN, R. A., et al. 2012. Pancreatic cancer genomes reveal aberrations in axon guidance pathway genes. *Nature,* **491,** 399–405.

BIFFI, G., ONI, T. E., SPIELMAN, B., HAO, Y., ELYADA, E., PARK, Y., PREALL, J. & TUVESON, D. A. 2019. IL1-induced JAK/STAT signaling is antagonized by TGFβ to shape CAF heterogeneity in pancreatic ductal adenocarcinoma. *Cancer Discov,* **9,** 282–301.

BORDEN, P., HOUTZ, J., LEACH, S. D. & KURUVILLA, R. 2013. Sympathetic innervation during development is necessary for pancreatic islet architecture and functional maturation. *Cell Rep,* **4,** 287–301.

BOŻYK, A., WOJAS-KRAWCZYK, K., KRAWCZYK, P. & MILANOWSKI, J. 2022. Tumor microenvironment-a short review of cellular and interaction diversity. *Biology (Basel),* 929, 11.

BRITTAIN, H. K., SCOTT, R. & THOMAS, E. 2017. The rise of the genome and personalised medicine. *Clin Med (Lond),* 17, 545–551.

BRYCE, A. S., DREYER, S. B., FROELING, F. E. M. & CHANG, D. K. 2022. Exploring the biology of cancer-associated fibroblasts in pancreatic cancer. *Cancers (Basel),* 5302, 14.

BUANES, T. A. 2016. Updated therapeutic outcome for patients with periampullary and pancreatic cancer related to recent translational research. *World J Gastroenterol,* 22, 10502–10511.

CAI, C. Y., ZHAI, L. L., WU, Y. & TANG, Z. G. 2015. Expression and clinical value of peroxiredoxin-1 in patients with pancreatic cancer. *Eur J Surg Oncol,* 41**,** 228–35.

CAI, W., SAKAGUCHI, M., KLEINRIDDERS, A., GONZALEZ-DEL PINO, G., DREYFUSS, J. M., O'NEILL, B. T., RAMIREZ, A. K., PAN, H., WINNAY, J. N., BOUCHER, J., ECK, M. J. & KAHN, C. R. 2017. Domain-dependent effects of insulin and IGF-1 receptors on signalling and gene expression. *Nat Commun,* 8**,** 14892.

CAO, J., LOU, S., YING, M. & YANG, B. 2015. DJ-1 as a human oncogene and potential therapeutic target. *Biochem Pharmacol,* 93**,** 241–250.

CASOLINO, R., PAIELLA, S., AZZOLINA, D., BEER, P. A., CORBO, V., LORENZONI, G., GREGORI, D., GOLAN, T., BRACONI, C., FROELING, F. E. M., MILELLA, M., SCARPA, A., PEA, A., MALLEO, G., SALVIA, R., BASSI, C., CHANG, D. K. & BIANKIN, A. V. 2021. Homologous recombination deficiency in pancreatic cancer: a systematic review and prevalence meta-analysis. *J Clin Oncol,* 39**,** 2617–2631.

CERVANTES-VILLAGRANA, R. D., ALBORES-GARCÍA, D., CERVANTES-VILLAGRANA, A. R. & GARCÍA-ACEVEZ, S. J. 2020. Tumor-induced neurogenesis and immune evasion as targets of innovative anti-cancer therapies. *Signal Transduct Target Ther,* 5**,** 99.

CHANG, C. Y., HUANG, S. P., CHIU, H. M., LEE, Y. C., CHEN, M. F. & LIN, J. T. 2006. Low efficacy of serum levels of CA 19-9 in prediction of malignant diseases in asymptomatic population in Taiwan. *Hepatogastroenterology,* 53**,** 1–4.

CHEN, Y., MCANDREWS, K. M. & KALLURI, R. 2021. Clinical and therapeutic relevance of cancer-associated fibroblasts. *Nat Rev Clin Oncol,* 18**,** 792–804.

CHEN, Y.-T., HO, C.-L., CHEN, P.-K., CHEN, Y.-L. & CHANG, C.-F. 2013. Anterior gradient 2: a novel sensitive tumor marker for metastatic oral cancer. *Cancer Letters,* 339**,** 270–278.

CHUNG, H. W. & LIM, J.-B. 2014. Clinical significance of elevated serum soluble CD40 ligand levels as a diagnostic and prognostic tumor marker for pancreatic ductal adenocarcinoma. *J Transl Med,* 12**,** 102.

COLLINS, M. A., BEDNAR, F., ZHANG, Y., BRISSET, J. C., GALBÁN, S., GALBÁN, C. J., RAKSHIT, S., FLANNAGAN, K. S., ADSAY, N. V. & PASCA DI MAGLIANO, M. 2012. Oncogenic Kras is required for both the initiation and maintenance of pancreatic cancer in mice. *J Clin Invest,* 122**,** 639–653.

COLLISSON, E. A., SADANANDAM, A., OLSON, P., GIBB, W. J., TRUITT, M., GU, S., COOC, J., WEINKLE, J., KIM, G. E., JAKKULA, L., FEILER, H. S., KO, A. H., OLSHEN, A. B., DANENBERG, K. L., TEMPERO, M. A., SPELLMAN, P. T., HANAHAN, D. & GRAY, J. W. 2011. Subtypes of pancreatic ductal adenocarcinoma and their differing responses to therapy. *Nat Med,* 17, 500–503.

CONROY, T., DESSEIGNE, F., YCHOU, M., BOUCHÉ, O., GUIMBAUD, R., BÉCOUARN, Y., ADENIS, A., RAOUL, J. L., GOURGOU-BOURGADE, S., DE LA FOUCHARDIÈRE, C., BENNOUNA, J., BACHET, J. B., KHEMISSA-AKOUZ, F., PÉRÉ-VERGÉ, D., DELBALDO, C., ASSENAT, E., CHAUFFERT, B., MICHEL, P., MONTOTO-GRILLOT, C. & DUCREUX, M. 2011. FOLFIRINOX versus gemcitabine for metastatic pancreatic cancer. *N Engl J Med,* 364, 1817–1825.

COOK, J. H., MELLONI, G. E. M., GULHAN, D. C., PARK, P. J. & HAIGIS, K. M. 2021. The origins and genetic interactions of KRAS mutations are allele- and tissue-specific. *Nat Commun,* 12, 1808.

COSTACHE, M. I., IOANA, M., IORDACHE, S., ENE, D., COSTACHE, C. A. & SĂFTOIU, A. 2015. VEGF expression in pancreatic cancer and other malignancies: a review of the literature. *Rom J Intern Med,* 53, 199–208.

DALMASSO, B., PUCCINI, A., CATALANO, F., BOREA, R., IAIA, M. L., BRUNO, W., FORNARINI, G., SCIALLERO, S., REBUZZI, S. E. & GHIORZO, P. 2022. Beyond BRCA: the emerging significance of DNA damage response and personalized treatment in pancreatic and prostate cancer patients. *Int J Mol Sci,* 4709, 23.

DANILUK, J., LIU, Y., DENG, D., CHU, J., HUANG, H., GAISER, S., CRUZ-MONSERRATE, Z., WANG, H., JI, B. & LOGSDON, C. D. 2012. An NF-κB pathway-mediated positive feedback loop amplifies Ras activity to pathological levels in mice. *J Clin Invest,* 122, 1519–1528.

DARDARE, J., WITZ, A., MERLIN, J. L., GILSON, P. & HARLÉ, A. 2020. SMAD4 and the TGFβ pathway in patients with pancreatic ductal adenocarcinoma. *Int J Mol Sci,* 3534, 21.

DECLERCK, Y. A. 2012. Desmoplasia: a response or a niche? *Cancer Discovery,* 2, 772–774.

DEMIR, I. E., CEYHAN, G. O., LIEBL, F., D'HAESE, J. G., MAAK, M. & FRIESS, H. 2010. Neural invasion in pancreatic cancer: the past, present and future. *Cancers (Basel),* 2, 1513–1527.

DOU, H., SUN, G. & ZHANG, L. 2019. Chapter Twelve – CA242 as a biomarker for pancreatic cancer and other diseases. In: ZHANG, L. (ed.) *Progress in molecular biology and translational science*162 (pp. 229–239). Academic Press.

DOWNWARD, J. 2003. Targeting RAS signalling pathways in cancer therapy. *Nat Rev Cancer,* 3, 11–22.

DUFFY, M. J., STURGEON, C., LAMERZ, R., HAGLUND, C., HOLUBEC, V. L., KLAPDOR, R., NICOLINI, A., TOPOLCAN, O. & HEINEMANN, V. 2010. Tumor markers in pancreatic cancer: a European Group on Tumor Markers (EGTM) status report. *Annals of Oncology,* 21, 441–447.

EHMANN, M., FELIX, K., HARTMANN, D., SCHNÖLZER, M., NEES, M., VORDERWÜLBECKE, S., BOGUMIL, R., BÜCHLER, M. W. & FRIESS, H. 2007. Identification of potential markers for the detection of pancreatic cancer through comparative serum protein expression profiling. *Pancreas,* 34, 205–214.

ESCOBAR-HOYOS, L. F., PENSON, A., KANNAN, R., CHO, H., PAN, C. H., SINGH, R. K., APKEN, L. H., HOBBS, G. A., LUO, R., LECOMTE, N., BABU, S., PAN, F. C., ALONSO-CURBELO, D., MORRIS, J. P. T., ASKAN, G., GRBOVIC-HUEZO, O., OGRODOWSKI, P., BERMEO, J., SAGLIMBENI, J., CRUZ, C. D., HO, Y. J., LAWRENCE, S. A., MELCHOR, J. P., GODA, G. A., BAI, K., PASTORE, A., HOGG, S. J., RAGHAVAN, S., BAILEY, P., CHANG, D. K., BIANKIN, A., SHROYER, K. R., WOLPIN, B. M., AGUIRRE, A. J., VENTURA, A., TAYLOR, B., DER, C. J., DOMINGUEZ, D., KÜMMEL, D., OECKINGHAUS, A., LOWE, S. W., BRADLEY, R. K., ABDEL-WAHAB, O. & LEACH, S. D. 2020. Altered RNA splicing by mutant p53 activates oncogenic RAS signaling in pancreatic cancer. *Cancer Cell,* 38, 198–211.e8.

ESER, S., REIFF, N., MESSER, M., SEIDLER, B., GOTTSCHALK, K., DOBLER, M., HIEBER, M., ARBEITER, A., KLEIN, S., KONG, B., MICHALSKI, C. W., SCHLITTER, A. M., ESPOSITO, I., KIND, A. J., RAD, L., SCHNIEKE, A. E., BACCARINI, M., ALESSI, D. R., RAD, R., SCHMID, R. M., SCHNEIDER, G. & SAUR, D. 2013. Selective requirement of PI3K/PDK1 signaling for Kras oncogene-driven pancreatic cell plasticity and cancer. *Cancer Cell,* 23, 406–20.

EZROVA, Z., NAHACKA, Z., STURSA, J., WERNER, L., VLCAK, E., KRALOVA VIZIOVA, P., BERRIDGE, M. V., SEDLACEK, R., ZOBALOVA, R., ROHLENA, J., BOUKALOVA, S. & NEUZIL, J. 2021. SMAD4 loss limits the vulnerability of pancreatic cancer cells to complex I inhibition via promotion of mitophagy. *Oncogene,* 40, 2539–2552.

FENG, J., HU, Z., XIA, X., LIU, X., LIAN, Z., WANG, H., WANG, L., WANG, C., ZHANG, X. & PANG, X. 2023. Feedback activation of EGFR/wild-type RAS signaling axis limits KRASG12D inhibitor efficacy in KRASG12D-mutated colorectal cancer. *Oncogene,* 42, 1620–1633.

FROELING, F. E. M., CASOLINO, R., PEA, A., BIANKIN, A. V. & CHANG, D. K. 2021. Molecular subtyping and precision medicine for pancreatic cancer. *J Clin Med,* 149, 10.

FU, J., QIN, L., HE, T., QIN, J., HONG, J., WONG, J., LIAO, L. & XU, J. 2011. The TWIST/Mi2/NuRD protein complex and its essential role in cancer metastasis. *Cell Res,* **21,** 275–289.

FURUKAWA, K., KAWAMOTO, K., EGUCHI, H., TANEMURA, M., TANIDA, T., TOMIMARU, Y., AKITA, H., HAMA, N., WADA, H., KOBAYASHI, S., NONAKA, Y., TAKAMATSU, S., SHINZAKI, S., KUMADA, T., SATOMURA, S., ITO, T., SERADA, S., NAKA, T., MORI, M., DOKI, Y., MIYOSHI, E. & NAGANO, H. 2015. Clinicopathological significance of leucine-rich α2-glycoprotein-1 in sera of patients with pancreatic cancer. *Pancreas,* **44,** 93–98.

GEBHARDT, C., RIEHL, A., DURCHDEWALD, M., NÉMETH, J., FÜRSTENBERGER, G., MÜLLER-DECKER, K., ENK, A., ARNOLD, B., BIERHAUS, A., NAWROTH, P. P., HESS, J. & ANGEL, P. 2008. RAGE signaling sustains inflammation and promotes tumor development. *J Exp Med,* **205,** 275–285.

GNONI, A., BRUNETTI, O., LONGO, V., CALABRESE, A., ARGENTIERO, A. L., CALBI, R., SOLIMANDO ANTONIO, G. & LICCHETTA, A. 2020. Immune system and bone microenvironment: rationale for targeted cancer therapies. *Oncotarget,* 11, 480–487.

GOLAN, T., HAMMEL, P., RENI, M., VAN CUTSEM, E., MACARULLA, T., HALL, M. J., PARK, J. O., HOCHHAUSER, D., ARNOLD, D., OH, D. Y., REINACHER-SCHICK, A., TORTORA, G., ALGÜL, H., O'REILLY, E. M., MCGUINNESS, D., CUI, K. Y., SCHLIENGER, K., LOCKER, G. Y. & KINDLER, H. L. 2019. Maintenance olaparib for germline BRCA-mutated metastatic pancreatic cancer. *N Engl J Med,* 381**,** 317–327.

GOLAN, T., KHVALEVSKY, E. Z., HUBERT, A., GABAI, R. M., HEN, N., SEGAL, A., DOMB, A., HARARI, G., DAVID, E. B., RASKIN, S., GOLDES, Y., GOLDIN, E., ELIAKIM, R., LAHAV, M., KOPLEMAN, Y., DANCOUR, A., SHEMI, A. & GALUN, E. 2015. RNAi therapy targeting KRAS in combination with chemotherapy for locally advanced pancreatic cancer patients. *Oncotarget,* **6,** 24560–24570.

GÜNGÖR, C., ZANDER, H., EFFENBERGER, K. E., VASHIST, Y. K., KALININA, T., IZBICKI, J. R., YEKEBAS, E. & BOCKHORN, M. 2011. Notch signaling activated by replication stress-induced expression of midkine drives epithelial-mesenchymal transition and chemoresistance in pancreatic cancer. *Cancer Res,* 71**,** 5009–5019.

GURUHARSHA, K. G., KANKEL, M. W. & ARTAVANIS-TSAKONAS, S. 2012. The Notch signalling system: recent insights into the complexity of a conserved pathway. *Nat Rev Genet,* 13**,** 654–666.

HAFNER, A., BULYK, M. L., JAMBHEKAR, A. & LAHAV, G. 2019. The multiple mechanisms that regulate p53 activity and cell fate. *Nat Rev Mol Cell Biol,* 20**,** 199–210.

HELDIN, C. H., VANLANDEWIJCK, M. & MOUSTAKAS, A. 2012. Regulation of EMT by TGFβ in cancer. *FEBS Lett,* 586**,** 1959–1970.

HINGORANI, S. R., ZHENG, L., BULLOCK, A. J., SEERY, T. E., HARRIS, W. P., SIGAL, D. S., BRAITEH, F., RITCH, P. S., ZALUPSKI, M. M., BAHARY, N., OBERSTEIN, P. E., WANG-GILLAM, A., WU, W., CHONDROS, D., JIANG, P., KHELIFA, S., PU, J., ALDRICH, C. & HENDIFAR, A. E. 2018. HALO 202: Randomized phase II study of PEGPH20 plus nab-paclitaxel/gemcitabine versus nab-paclitaxel/gemcitabine in patients with untreated, metastatic pancreatic ductal adenocarcinoma. *J Clin Oncol,* 36**,** 359–366.

HO, W. J., JAFFEE, E. M. & ZHENG, L. 2020. The tumour microenvironment in pancreatic cancer – clinical challenges and opportunities. *Nat Rev Clin Oncol,* 17**,** 527–540.

HOLEN, I., CROSS, S. S., NEVILLE-WEBBE, H. L., CROSS, N. A., BALASUBRAMANIAN, S. P., CROUCHER, P. I., EVANS, C. A., LIPPITT, J. M., COLEMAN, R. E. & EATON, C. L. 2005. Osteoprotegerin (OPG) expression by breast cancer cells in vitro and breast tumours in vivo – a role in tumour cell survival? *Breast Cancer Res Treat,* 92**,** 207–215.

HONG, D. S., DUBOIS, S. G., KUMMAR, S., FARAGO, A. F., ALBERT, C. M., ROHRBERG, K. S., VAN TILBURG, C. M., NAGASUBRAMANIAN, R., BERLIN, J. D., FEDERMAN, N., MASCARENHAS, L., GEOERGER, B., DOWLATI, A., PAPPO, A. S., BIELACK, S., DOZ, F., MCDERMOTT, R., PATEL, J. D., SCHILDER, R. J., TAHARA, M., PFISTER, S. M., WITT, O., LADANYI, M., RUDZINSKI, E. R., NANDA, S., CHILDS, B. H., LAETSCH, T. W., HYMAN, D. M. & DRILON, A. 2020. Larotrectinib in patients with TRK fusion-positive solid tumours: a pooled analysis of three phase 1/2 clinical trials. *Lancet Oncol,* 21**,** 531–540.

HOTZ, B., ARNDT, M., DULLAT, S., BHARGAVA, S., BUHR, H. J. & HOTZ, H. G. 2007. Epithelial to mesenchymal transition: expression of the regulators snail, slug, and twist in pancreatic cancer. *Clin Cancer Res,* 13**,** 4769–47676.

HOTZ, H. G., REBER, H. A., HOTZ, B., SANGHAVI, P. C., YU, T., FOITZIK, T., BUHR, H. J. & HINES, O. J. 2001. Angiogenesis inhibitor TNP-470 reduces human pancreatic cancer growth. *J Gastrointest Surg,* 5, 131–138.

HRUBAN, R. H., MAITRA, A. & GOGGINS, M. 2008. Update on pancreatic intraepithelial neoplasia. *Int J Clin Exp Pathol,* 1, 306–316.

HU, H.-F., YE, Z., QIN, Y., XU, X.-W., YU, X.-J., ZHUO, Q.-F. & JI, S.-R. 2021. Mutations in key driver genes of pancreatic cancer: molecularly targeted therapies and other clinical implications. *Acta Pharmacol Sin,* 42, 1725–1741.

HUANG, Y., HONG, W. & WEI, X. 2022. The molecular mechanisms and therapeutic strategies of EMT in tumor progression and metastasis. *J Hematol Oncol,* 15, 129.

IP, N. Y. & YANCOPOULOS, G. D. 1996. The neurotrophins and CNTF: two families of collaborative neurotrophic factors. *Annu Rev Neurosci,* 19, 491–515.

JAVADRASHID, D., BAGHBANZADEH, A., DERAKHSHANI, A., LEONE, P., SILVESTRIS, N., RACANELLI, V., SOLIMANDO, A. G. & BARADARAN, B. 2021. Pancreatic cancer signaling pathways, genetic alterations, and tumor microenvironment: the barriers affecting the method of Treatmenttreatment. *Biomedicines,* 373, 9.

JONES, S., ZHANG, X., PARSONS, D. W., LIN, J. C., LEARY, R. J., ANGENENDT, P., MANKOO, P., CARTER, H., KAMIYAMA, H., JIMENO, A., HONG, S. M., FU, B., LIN, M. T., CALHOUN, E. S., KAMIYAMA, M., WALTER, K., NIKOLSKAYA, T., NIKOLSKY, Y., HARTIGAN, J., SMITH, D. R., HIDALGO, M., LEACH, S. D., KLEIN, A. P., JAFFEE, E. M., GOGGINS, M., MAITRA, A., IACOBUZIO-DONAHUE, C., ESHLEMAN, J. R., KERN, S. E., HRUBAN, R. H., KARCHIN, R., PAPADOPOULOS, N., PARMIGIANI, G., VOGELSTEIN, B., VELCULESCU, V. E. & KINZLER, K. W. 2008. Core signaling pathways in human pancreatic cancers revealed by global genomic analyses. *Science,* 321, 1801–1806.

KALLURI, R. & WEINBERG, R. A. 2009. The basics of epithelial-mesenchymal transition. *J Clin Invest,* 119, 1420–1428.

KALUS, I., HODEL, A., KOCH, A., KLEENE, R., EDWARDSON, J. M. & SCHRADER, M. 2002. Interaction of syncollin with GP-2, the major membrane protein of pancreatic zymogen granules, and association with lipid microdomains. *Biochem J,* 362, 433–342.

KAMADA, Y., KINOSHITA, N., TSUCHIYA, Y., KOBAYASHI, K., FUJII, H., TERAO, N., KAMIHAGI, K., KOYAMA, N., YAMADA, S., DAIGO, Y., NAKAMURA, Y., TANIGUCHI, N. & MIYOSHI, E. 2013. Reevaluation of a lectin antibody ELISA kit for measuring fucosylated haptoglobin in various conditions. *Clin Chim Acta,* 417, 48–53.

KANDA, M., MATTHAEI, H., WU, J., HONG, S. M., YU, J., BORGES, M., HRUBAN, R. H., MAITRA, A., KINZLER, K., VOGELSTEIN, B. & GOGGINS, M. 2012. Presence of somatic mutations in most early-stage pancreatic intraepithelial neoplasia. *Gastroenterology,* 142, 730–733.e9.

KANG, R., TANG, D., SCHAPIRO, N. E., LOUX, T., LIVESEY, K. M., BILLIAR, T. R., WANG, H., VAN HOUTEN, B., LOTZE, M. T. & ZEH, H. J. 2014. The HMGB1/RAGE inflammatory pathway promotes pancreatic tumor growth by regulating mitochondrial bioenergetics. *Oncogene,* 33, 567–577.

KANNAGI, R. 2007. Carbohydrate antigen sialyl Lewis a – its pathophysiological significance and induction mechanism in cancer progression. *Chang Gung Med J,* 30, 189–209.

KARAMITOPOULOU, E. 2019. Tumour microenvironment of pancreatic cancer: immune landscape is dictated by molecular and histopathological features. *British J Cancer,* 121, 5–14.

KIM, J., YIP, M. L., SHEN, X., LI, H., HSIN, L. Y., LABARGE, S., HEINRICH, E. L., LEE, W., LU, J. & VAIDEHI, N. 2012. Identification of anti-malarial compounds as novel antagonists to chemokine receptor CXCR4 in pancreatic cancer cells. *PLoS One,* 7, e31004.

KIM, J.-E., LEE, K. T., LEE, J. K., PAIK, S. W., RHEE, J. C. & CHOI, K. W. 2004. Clinical usefulness of carbohydrate antigen 19-9 as a screening test for pancreatic cancer in an asymptomatic population. *J Gastroenterol Hepatol*, 19, 182–186.

KIM, M. P., LI, X., DENG, J., ZHANG, Y., DAI, B., ALLTON, K. L., HUGHES, T. G., SIANGCO, C., AUGUSTINE, J. J., KANG, Y., MCDANIEL, J. M., XIONG, S., KOAY, E. J., MCALLISTER, F., BRISTOW, C. A., HEFFERNAN, T. P., MAITRA, A., LIU, B., BARTON, M. C., WASYLISHEN, A. R., FLEMING, J. B. & LOZANO, G. 2021. Oncogenic KRAS recruits an expansive transcriptional network through mutant p53 to drive pancreatic cancer metastasis. *Cancer Discov,* 11, 2094–2111.

KINDLER, H. L., HAMMEL, P., RENI, M., VAN CUTSEM, E., MACARULLA, T., HALL, M. J., PARK, J. O., HOCHHAUSER, D., ARNOLD, D., OH, D. Y., REINACHER-SCHICK, A., TORTORA, G., ALGÜL, H., O'REILLY, E. M., BORDIA, S., MCGUINNESS, D., CUI, K., LOCKER, G. Y. &

GOLAN, T. 2022. Overall survival results from the POLO trial: a phase III study of active maintenance olaparib versus placebo for germline BRCA-mutated metastatic pancreatic cancer. *J Clin Oncol,* 40, 3929–3939.

KINSEY, C. G., CAMOLOTTO, S. A., BOESPFLUG, A. M., GUILLEN, K. P., FOTH, M., TRUONG, A., SCHUMAN, S. S., SHEA, J. E., SEIPP, M. T., YAP, J. T., BURRELL, L. D., LUM, D. H., WHISENANT, J. R., GILCREASE, G. W., CAVALIERI, C. C., REHBEIN, K. M., CUTLER, S. L., AFFOLTER, K. E., WELM, A. L., WELM, B. E., SCAIFE, C. L., SNYDER, E. L. & MCMAHON, M. 2019. Protective autophagy elicited by RAF→MEK→ERK inhibition suggests a treatment strategy for RAS-driven cancers. *Nat Med,* 25, 620–627.

KOLBEINSSON, H. M., CHANDANA, S., WRIGHT, G. P. & CHUNG, M. 2023. Pancreatic cancer: a review of current treatment and novel therapies. *J Invest Surg,* 36, 2129884.

KORNMANN, M., BEGER, H. G. & KORC, M. 1998. Role of fibroblast growth factors and their receptors in pancreatic cancer and chronic pancreatitis. *Pancreas,* 17, 169–175.

KUCZEK, D. E., LARSEN, A. M. H., THORSETH, M. L., CARRETTA, M., KALVISA, A., SIERSBÆK, M. S., SIMÕES, A. M. C., ROSLIND, A., ENGELHOLM, L. H., NOESSNER, E., DONIA, M., SVANE, I. M., STRATEN, P. T., GRØNTVED, L. & MADSEN, D. H. 2019. Collagen density regulates the activity of tumor-infiltrating T cells. *J Immunother Cancer,* 7, 68.

KURLINKUS, B., GER, M., KAUPINIS, A., JASIUNAS, E., VALIUS, M. & SILEIKIS, A. 2021. CEACAM6's role as a chemoresistance and prognostic biomarker for pancreatic cancer: a comparison of CEACAM6's diagnostic and prognostic capabilities with Thosethose of CA19-9 and CEA. *Life (Basel),* 542, 11.

LAMOUILLE, S., XU, J. & DERYNCK, R. 2014. Molecular mechanisms of epithelial–mesenchymal transition. *Nat Rev Mol Cell Biol,* 15, 178–196.

LANKI, M., MUSTONEN, H., SALMI, M., JALKANEN, S., HAGLUND, C. & SEPPÄNEN, H. 2023. Serum cytokine profiles in patients with pancreatic cancer and chronic pancreatitis. *Pancreatology,* 23, 657–662.

LEE, J. J., PERERA, R. M., WANG, H., WU, D. C., LIU, X. S., HAN, S., FITAMANT, J., JONES, P. D., GHANTA, K. S., KAWANO, S., NAGLE, J. M., DESHPANDE, V., BOUCHER, Y., KATO, T., CHEN, J. K., WILLMANN, J. K., BARDEESY, N. & BEACHY, P. A. 2014. Stromal response to Hedgehog signaling restrains pancreatic cancer progression. *Proc Natl Acad Sci U S A,* 111, E3091–E3100.

LEE, S., HEINRICH, E. L., LI, L., LU, J., CHOI, A. H., LEVY, R. A., WAGNER, J. E., YIP, M. L., VAIDEHI, N. & KIM, J. 2015. CCR9-mediated signaling through β-catenin and identification of a novel CCR9 antagonist. *Mol Oncol,* 9, 1599–1611.

LEE, S.-J., YOO, H. J., BAE, Y. S., KIM, H.-J. & LEE, S.-T. 2003. TIMP-1 inhibits apoptosis in breast carcinoma cells via a pathway involving pertussis toxin-sensitive G protein and c-Src. *Biochem Biophys Res Commun,* 312, 1196–1201.

LEIDNER, R., SANJUAN SILVA, N., HUANG, H., SPROTT, D., ZHENG, C., SHIH, Y. P., LEUNG, A., PAYNE, R., SUTCLIFFE, K., CRAMER, J., ROSENBERG, S. A., FOX, B. A., URBA, W. J. & TRAN, E. 2022. Neoantigen T-cell receptor gene therapy in pancreatic cancer. *N Engl J Med,* 386, 2112–2119.

LI, H.-B., YANG, Z.-H. & GUO, Q.-Q. 2021a. Immune checkpoint inhibition for pancreatic ductal adenocarcinoma: limitations and prospects: a systematic review. *Cell Commun Signal,* 19, 117.

LI, J., POI, M. J. & TSAI, M. D. 2011. Regulatory mechanisms of tumor suppressor P16(INK4A) and their relevance to cancer. *Biochemistry,* 50, 5566–5582.

LI, J., YUAN, S., NORGARD, R. J., YAN, F., SUN, Y. H., KIM, I. K., MERRELL, A. J., SELA, Y., JIANG, Y., BHANU, N. V., GARCIA, B. A., VONDERHEIDE, R. H., BLANCO, A. & STANGER, B. Z. 2021b. Epigenetic and transcriptional control of the epidermal growth factor receptor regulates the tumor immune microenvironment in pancreatic cancer. *Cancer Discov,* 11, 736–753.

LI, Q., WANG, H., ZOGOPOULOS, G., SHAO, Q., DONG, K., LV, F., NWILATI, K., GUI, X. Y., CUGGIA, A., LIU, J. L. & GAO, Z. H. 2016. Reg proteins promote acinar-to-ductal metaplasia and act as novel diagnostic and prognostic markers in pancreatic ductal adenocarcinoma. *Oncotarget,* 7, 77838–77853.

LIAO, W.-C., WU, M.-S., WANG, H.-P., TIEN, Y.-W. & LIN, J.-T. 2009. Serum heat shock protein 27 is increased in chronic pancreatitis and pancreatic carcinoma. *Pancreas,* 38, 422–426.

LIU, L., XU, H. X., WANG, W. Q., WU, C. T., XIANG, J. F., LIU, C., LONG, J., XU, J., FU DE, L., NI, Q. X., HOUCHEN, C. W., POSTIER, R. G., LI, M. & YU, X. J. 2016. Serum CA125 is a novel predictive marker for pancreatic cancer metastasis and correlates with the metastasis-associated burden. *Oncotarget,* 7, 5943–5956.

LOOSEN, S. H., LUEDDE, M., LURJE, G., SPEHLMANN, M., PAFFENHOLZ, P., ULMER, T. F., TACKE, F., VUCUR, M., TRAUTWEIN, C., NEUMANN, U. P., LUEDDE, T. & RODERBURG, C. 2019. Serum levels of kisspeptin are elevated in patients with pancreatic cancer. *Dis Markers*, 2019, 5603474.

LUO, G., JIN, K., DENG, S., CHENG, H., FAN, Z., GONG, Y., QIAN, Y., HUANG, Q., NI, Q., LIU, C. & YU, X. 2021. Roles of CA19-9 in pancreatic cancer: Biomarker, predictor and promoter. *Biochim Biophys Acta Rev Cancer,* 1875, 188409.

LWALEED, B. A. & BASS, P. S. 2006. Tissue factor pathway inhibitor: structure, biology and involvement in disease. *J Pathol,* 208, 327–39.

MAKAWITA, S., DIMITROMANOLAKIS, A., SOOSAIPILLAI, A., SOLEAS, I., CHAN, A., GALLINGER, S., HAUN, R. S., BLASUTIG, I. M. & DIAMANDIS, E. P. 2013. Validation of four candidate pancreatic cancer serological biomarkers that improve the performance of CA19.9. *BMC Cancer,* 13, 404.

MATHUR, P., SATHISHKUMAR, K., CHATURVEDI, M., DAS, P., SUDARSHAN, K. L., SANTHAPPAN, S., NALLASAMY, V., JOHN, A., NARASIMHAN, S. & ROSELIND, F. S. 2020. Cancer statistics, 2020: report from National Cancer Registry Programme, India. *JCO Glob Oncol,* 6, 1063–1075.

MCINTYRE, C. A., GRIMONT, A., PARK, J., MENG, Y., SISSO, W. J., SEIER, K., JANG, G. H., WALCH, H., AVESON, V. G., FALVO, D. J., FALL, W. B., CHAN, C. W., WENGER, A., ECKER, B. L., PULVIRENTI, A., GELFER, R., ZAFRA, M. P., SCHULTZ, N., PARK, W., O'REILLY, E. M., HOULIHAN, S. L., ALONSO, A., HISSONG, E., CHURCH, G. M., MASON, C. E., SIOLAS, D., NOTTA, F., GONEN, M., DOW, L. E., JARNAGIN, W. R. & CHANDWANI, R. 2024. Distinct clinical outcomes and biological features of specific *KRAS* mutants in human pancreatic cancer. *Cancer Cell,* 42, 1614–1629.e5.

MIEKUS, K., JAROCHA, D., TRZYNA, E. & MAJKA, M. 2010. Role of I-TAC-binding receptors CXCR3 and CXCR7 in proliferation, activation of intracellular signaling pathways and migration of various tumor cell lines. *Folia Histochem Cytobiol,* 48, 104–111.

MIN, H.-Y. & LEE, H.-Y. 2022. Molecular targeted therapy for anticancer treatment. *Exp Mol Med,* 54, 1670–1694.

MITSUNAGA, S., FUJII, S., ISHII, G., KINOSHITA, T., HASEBE, T., AOYAGI, K., SASAKI, H. & OCHIAI, A. 2010. Nerve invasion distance is dependent on laminin gamma2 in tumors of pancreatic cancer. *Int J Cancer,* 127, 805–819.

MOORE, M. J., GOLDSTEIN, D., HAMM, J., FIGER, A., HECHT, J. R., GALLINGER, S., AU, H. J., MURAWA, P., WALDE, D., WOLFF, R. A., CAMPOS, D., LIM, R., DING, K., CLARK, G., VOSKOGLOU-NOMIKOS, T., PTASYNSKI, M. & PARULEKAR, W. 2007. Erlotinib plus gemcitabine compared with gemcitabine alone in patients with advanced pancreatic cancer: a phase III trial of the National Cancer Institute of Canada Clinical Trials Group. *J Clin Oncol,* 25, 1960–1966.

MOSCHOS, S. J. & MANTZOROS, C. S. 2002. The role of the IGF system in cancer: from basic to clinical studies and clinical applications. *Oncology,* 63, 317–332.

NAPOLI, N., KAUFFMANN, E. F., GINESINI, M., LAMI, L., LOMBARDO, C., VISTOLI, F., CAMPANI, D. & BOGGI, U. 2023. Ca 125 is an independent prognostic marker in resected pancreatic cancer of the head of the pancreas. *Updates Surg,* 75, 1481–1496.

NIETO, M. A. 2009. Epithelial-Mesenchymal Transitions in development and disease: old views and new perspectives. *Int J Dev Biol,* 53, 1541–1547.

NISHIOKA, R., ITOH, S., GUI, T., GAI, Z., OIKAWA, K., KAWAI, M., TANI, M., YAMAUE, H. & MURAGAKI, Y. 2010. SNAIL induces epithelial-to-mesenchymal transition in a human pancreatic cancer cell line (BxPC3) and promotes distant metastasis and invasiveness in vivo. *Exp Mol Pathol,* 89, 149–157.

NUSRAT, F., KHANNA, A., JAIN, A., JIANG, W., LAVU, H., YEO, C. J., BOWNE, W. & NEVLER, A. 2024. The clinical implications of KRAS mutations and variant allele frequencies in pancreatic ductal adenocarcinoma. *J Clin Med,* 13, 2103.

O'KANE, G. M. & LOWERY, M. A. 2022. Moving the needle on precision medicine in pancreatic cancer. *J Clin Oncol,* 40, 2693–2705.

OSADA, H., TOMIDA, S., YATABE, Y., TATEMATSU, Y., TAKEUCHI, T., MURAKAMI, H., KONDO, Y., SEKIDO, Y. & TAKAHASHI, T. 2008. Roles of achaete-scute homologue 1 in DKK1 and E-cadherin repression and neuroendocrine differentiation in lung cancer. *Cancer Res,* 68, 1647–1655.

PANDEY, R., ZHOU, M., ISLAM, S., CHEN, B., BARKER, N. K., LANGLAIS, P., SRIVASTAVA, A., LUO, M., COOKE, L. S., WETERINGS, E. & MAHADEVAN, D. 2019. Carcinoembryonic antigen cell adhesion molecule 6 (CEACAM6) in Pancreatic Ductal Adenocarcinoma (PDA): an integrative analysis of a novel therapeutic target. *Sci Rep,* 9, 18347.

PANTSAR, T. 2020. The current understanding of KRAS protein structure and dynamics. *Comput Struct Biotechnol J,* 18**,** 189–198.

PARDOLL, D. M. 2012. The blockade of immune checkpoints in cancer immunotherapy. *Nat Rev Cancer,* 12**,** 252–264.

PARUMS, D. V. 2022. Editorial: recent approval of sotorasib as the first targeted therapy for KRAS G12C-mutated advanced non-small cell lung cancer (NSCLC). *Med Sci Monit,* 28**,** e938746.

PEINADO, H., BALLESTAR, E., ESTELLER, M. & CANO, A. 2004. Snail mediates E-cadherin repression by the recruitment of the Sin3A/histone deacetylase 1 (HDAC1)/HDAC2 complex. *Mol Cell Biol,* 24**,** 306–319.

PEREZ-MORENO, M. A., LOCASCIO, A., RODRIGO, I., DHONDT, G., PORTILLO, F., NIETO, M. A. & CANO, A. 2001. A new role for E12/E47 in the repression of E-cadherin expression and epithelial-mesenchymal transitions. *J Biol Chem,* 276**,** 27424–27431.

PEREZ-OQUENDO, M. & GIBBONS, D. L. 2022. Regulation of ZEB1 function and molecular associations in tumor progression and Metastasismetastasis. *Cancers (Basel),* 1864, 14.

POLIREDDY, K. & CHEN, Q. 2016. Cancer of the pancreas: molecular pathways and current advancement in treatment. *J Cancer,* 7**,** 1497–514.

PRINCIPE, D. R. 2022. Precision medicine for BRCA/PALB2-mutated pancreatic cancer and emerging strategies to improve therapeutic Responsesresponses to PARP inhibition. *Cancers (Basel),* 897, 14.

PULS, T. J., TAN, X., WHITTINGTON, C. F. & VOYTIK-HARBIN, S. L. 2017. 3D collagen fibrillar microstructure guides pancreatic cancer cell phenotype and serves as a critical design parameter for phenotypic models of EMT. *PLoS One,* 12**,** e0188870.

QUAN, M. Y., GUO, Q., LIU, J., YANG, R., BAI, J., WANG, W., CAI, Y., HAN, R., LV, Y. Q., DING, L., BILLADEAU, D. D., LOU, Z., BELLUSCI, S., LI, X. & ZHANG, J. S. 2020. An FGFR/AKT/SOX2 signaling axis controls pancreatic cancer stemness. *Front Cell Dev Biol,* 8**,** 287.

R.S., G. 2021. *WHO classification. PathologyOutlines.com website.* Available: https://www.pathologyoutlines.com/topic/pancreaswho.html.

RAM MAKENA, M., GATLA, H., VERLEKAR, D., SUKHAVASI, S., M, K. P. & K, C. P. 2019. Wnt/β-catenin signaling: the culprit in pancreatic carcinogenesis and therapeutic resistance. *Int J Mol Sci,* 20(17), 4242. doi: 10.3390/ijms20174242. PMID: 31480221; PMCID: PMC6747343.

REISS, K. A., MICK, R., O'HARA, M. H., TEITELBAUM, U., KARASIC, T. B., SCHNEIDER, C., COWDEN, S., SOUTHWELL, T., ROMEO, J., IZGUR, N., HANNAN, Z. M., TONDON, R., NATHANSON, K., VONDERHEIDE, R. H., WATTENBERG, M. M., BEATTY, G. & DOMCHEK, S. M. 2021. Phase II study of maintenance rucaparib in patients with platinum-sensitive advanced pancreatic cancer and a pathogenic germline or somatic variant in BRCA1, BRCA2, or PALB2. *J Clin Oncol,* 39**,** 2497–2505.

RHIM, A. D., MIREK, E. T., AIELLO, N. M., MAITRA, A., BAILEY, J. M., MCALLISTER, F., REICHERT, M., BEATTY, G. L., RUSTGI, A. K., VONDERHEIDE, R. H., LEACH, S. D. & STANGER, B. Z. 2012. EMT and dissemination precede pancreatic tumor formation. *Cell,* 148**,** 349–361.

RIEDER, S., MICHALSKI, C. W., FRIESS, H. & KLEEFF, J. 2011. Insulin-like growth factor signaling as a therapeutic target in pancreatic cancer. *Anticancer Agents Med Chem,* 11**,** 427–433.

ROBATEL, S. & SCHENK, M. 2022. Current limitations and novel perspectives in pancreatic cancer treatment. *Cancers*, 14**,** 985.

ROSSI, M. K., GNANAMONY, M. & GONDI, C. S. 2016. The 'SPARC' of life: analysis of the role of osteonectin/SPARC in pancreatic cancer (Review). *Int J Oncol,* 48**,** 1765–71.

SADEGHI RAD, H., MONKMAN, J., WARKIANI, M. E., LADWA, R., O'BYRNE, K., REZAEI, N. & KULASINGHE, A. 2021. Understanding the tumor microenvironment for effective immunotherapy. *Med Res Rev,* 41**,** 1474–1498.

SALAZAR, J., BRACCHIGLIONE, J., SAVALL-ESTEVE, O., ANTEQUERA, A., BOTTARO-PARRA, D., GUTIÉRREZ-VALENCIA, M., MARTÍNEZ-PERALTA, S., PERICAY, C., TIBAU, A., BONFILL, X., ACOSTA-DIGHERO, R., AULADELL-RISPAU, A., CANTERO-FORTIZ, Y., HERNANDEZ, E., IRASSAR, J., MEADE, A.-G., MEINARDI, P., MERCHÁN-GALVIS, A., MEZA, N., QUINTANA, M. J., REQUEIJO, C., RODRIGUEZ-GRIJALVA, G., SALAS-GAMA, K., SANTERO, M., SAVALL-ESTEVE, O., SELVA, A., SOLÀ, I., URRÚTIA, G. & APPROPRIATENESS OF SYSTEMIC ONCOLOGICAL TREATMENTS FOR ADVANCED CANCER RESEARCH, G. 2023. Treatment with anticancer drugs for advanced pancreatic cancer: a systematic review. *BMC Cancer,* 23**,** 748.

SATAKE, K., TAKEUCHI, T., HOMMA, T. & OZAKI, H. 1994. CA19-9 as a screening and diagnostic tool in symptomatic patients: the Japanese experience. 9**,** 703–706.

SATO, N., KOHI, S., HIRATA, K. & GOGGINS, M. 2016. Role of hyaluronan in pancreatic cancer biology and therapy: once again in the spotlight. *Cancer Sci,* 107, 569–75.

SATOH, M., TAKANO, S., SOGAWA, K., NODA, K., YOSHITOMI, H., ISHIBASHI, M., MOGUSHI, K., TAKIZAWA, H., OTSUKA, M., SHIMIZU, H., MIYAZAKI, M. & NOMURA, F. 2017. Immune-complex level of cofilin-1 in sera is associated with cancer progression and poor prognosis in pancreatic cancer. *Cancer Sci,* 108**,** 795–803.

SCHUMACHER, T. N. & SCHREIBER, R. D. 2015. Neoantigens in cancer immunotherapy. *Science,* 348**,** 69–74.

SCHWAB, A., RAO, Z., ZHANG, J., GOLLOWITZER, A., SIEBENKÄS, K., BINDEL, N., D'AVANZO, E., VAN ROEY, R., HAJJAJ, Y., ÖZEL, E., ARMSTARK, I., BEREUTER, L., SU, F., GRANDER, J., BONYADI RAD, E., GROENEWOUD, A., ENGEL, F. B., BELL, G. W., HENRY, W. S., ANGELI, J. P. F., STEMMLER, M. P., BRABLETZ, S., KOEBERLE, A. & BRABLETZ, T. 2024. Zeb1 mediates EMT/plasticity-associated ferroptosis sensitivity in cancer cells by regulating lipogenic enzyme expression and phospholipid composition. *Nat Cell Biol* September; 26(9), 1470–1481. doi: 10.1038/s41556-024-01464-1. Epub 2024 Jul 15. PMID: 39009641; PMCID: PMC11392809..

SHI, Q., XUE, C., ZENG, Y., YUAN, X., CHU, Q., JIANG, S., WANG, J., ZHANG, Y., ZHU, D. & LI, L. 2024. Notch signaling pathway in cancer: from mechanistic insights to targeted therapies. *Signal Transduct Target Therapy,* 9**,** 128.

SHI, S., XU, J., ZHANG, B., JI, S., XU, W., LIU, J., JIN, K., LIANG, D., LIANG, C., LIU, L., LIU, C., QIN, Y. & YU, X. 2016. VEGF promotes glycolysis in pancreatic cancer via HIF1α up-regulation. *Curr Mol Med,* 16**,** 394–403.

SHIMIZU, Y., SUGIURA, T., ASHIDA, R., OHGI, K., YAMADA, M., OTSUKA, S. & UESAKA, K. 2023. Prognostic role of preoperative duke pancreatic monoclonal antigen type 2 levels in patients with pancreatic cancer. *Ann Surg Oncol,* 30**,** 5792–5800.

SILVERMAN, B. R. & SHI, J. 2016. Alterations of epigenetic regulators in pancreatic cancer and their clinical Implicationsimplications. *Int J Mol Sci,* 2138, 17.

SINGHI, A. D., KOAY, E. J., CHARI, S. T. & MAITRA, A. 2019. Early detection of pancreatic cancer: opportunities and challenges. *Gastroenterology,* 156**,** 2024–2040.

SKAPER, S. D. 2018. Neurotrophic factors: an overview. *Methods Mol Biol,* 1727**,** 1–17.

SONG, H. Y., WANG, Y., LAN, H. & ZHANG, Y. X. 2018. Expression of Notch receptors and their ligands in pancreatic ductal adenocarcinoma. *Exp Ther Med,* 16**,** 53–60.

SONG, N., CUI, K., ZENG, L., LI, M., FAN, Y., SHI, P., WANG, Z., SU, W. & WANG, H. 2024. Advance in the role of chemokines/chemokine receptors in carcinogenesis: Focus on pancreatic cancer. *Eur J Pharmacol,* 967**,** 176357.

SUBBIAH, V., KREITMAN, R. J., WAINBERG, Z. A., GAZZAH, A., LASSEN, U., STEIN, A., WEN, P. Y., DIETRICH, S., DE JONGE, M. J. A., BLAY, J. Y., ITALIANO, A., YONEMORI, K., CHO, D. C., DE VOS, F., MOREAU, P., FERNANDEZ, E. E., SCHELLENS, J. H. M., ZIELINSKI, C. C., REDHU, S., BORAN, A., PASSOS, V. Q., ILANKUMARAN, P. & BANG, Y. J. 2023. Dabrafenib plus trametinib in BRAFV600E-mutated rare cancers: the phase 2 ROAR trial. *Nat Med,* 29**,** 1103–1112.

TENG, Y. & LI, X. 2014. The roles of HLH transcription factors in epithelial mesenchymal transition and multiple molecular mechanisms. *Clin Exp Metastasis,* 31**,** 367–377.

THE CANCER GENOME ATLAS 2017. Integrated genomic characterization of pancreatic ductal adenocarcinoma. *Cancer Cell,* 32**,** 185–203.e13.

THIERY, J. P. & SLEEMAN, J. P. 2006. Complex networks orchestrate epithelial-mesenchymal transitions. *Nat Rev Mol Cell Biol,* 7**,** 131–142.

TIBENSKY, M. & MRAVEC, B. 2021. Role of the parasympathetic nervous system in cancer initiation and progression. *Clin Transl Oncol,* 23**,** 669–681.

TIWARI, A., TASHIRO, K., DIXIT, A., SONI, A., VOGEL, K., HALL, B., SHAFQAT, I., SLAUGHTER, J., PARAM, N., LE, A., SAUNDERS, E., PAITHANE, U., GARCIA, G., CAMPOS, A. R., ZETTERVALL, J., CARLSON, M., STARR, T. K., MARAHRENS, Y., DESHPANDE, A. J., COMMISSO, C., PROVENZANO, P. P. & BAGCHI, A. 2020. Loss of HIF1A from pancreatic cancer cells increases expression of PPP1R1B and degradation of p53 to promote invasion and metastasis. *Gastroenterology,* 159**,** 1882–1897.e5.

TRIPATHI, P. K., MITTAL, K. R., JAIN, N., SHARMA, N. & JAIN, C. K. 2024. KRAS pathways: a potential gateway for cancer therapeutics and diagnostics. *Recent Pat Anticancer Drug Discov,* 19**,** 268–279.

TROIANI, T., MARTINELLI, E., CAPASSO, A., MORGILLO, F., ORDITURA, M., DE VITA, F. & CIARDIELLO, F. 2012. Targeting EGFR in pancreatic cancer treatment. *Curr Drug Targets,* 13**,** 802–10.

TURNER, S. A., ABOU SHAAR, R. & YANG, Z. 2023. The basics of commonly used molecular techniques for diagnosis, and application of molecular testing in cytology. *Diagn Cytopathol,* 51**,** 83–94.

TURPIN, A., NEUZILLET, C., COLLE, E., DUSETTI, N., NICOLLE, R., CROS, J., DE MESTIER, L., BACHET, J. B. & HAMMEL, P. 2022. Therapeutic advances in metastatic pancreatic cancer: a focus on targeted therapies. *Ther Adv Med Oncol,* 14, 17588359221118019.

VALSECCHI, M. E., MCDONALD, M., BRODY, J. R., HYSLOP, T., FREYDIN, B., YEO, C. J., SOLOMIDES, C., PEIPER, S. C. & WITKIEWICZ, A. K. 2012. Epidermal growth factor receptor and insulinlike growth factor 1 receptor expression predict poor survival in pancreatic ductal adenocarcinoma. *Cancer,* 118, 3484–93.

VAN CUTSEM, E., TEMPERO, M. A., SIGAL, D., OH, D. Y., FAZIO, N., MACARULLA, T., HITRE, E., HAMMEL, P., HENDIFAR, A. E., BATES, S. E., LI, C. P., HINGORANI, S. R., DE LA FOUCHARDIERE, C., KASI, A., HEINEMANN, V., MARAVEYAS, A., BAHARY, N., LAYOS, L., SAHAI, V., ZHENG, L., LACY, J., PARK, J. O., PORTALES, F., OBERSTEIN, P., WU, W., CHONDROS, D. & BULLOCK, A. J. 2020. Randomized phase III trial of pegvorhyaluronidase alfa with nab-paclitaxel plus gemcitabine for patients with hyaluronan-high metastatic pancreatic adenocarcinoma. *J Clin Oncol,* 38, 3185–3194.

VANDER HEIDEN, M. G., CANTLEY, L. C. & THOMPSON, C. B. 2009. Understanding the Warburg effect: the metabolic requirements of cell proliferation. *Science,* 324, 1029–33.

VON HOFF, D. D., ERVIN, T., ARENA, F. P., CHIOREAN, E. G., INFANTE, J., MOORE, M., SEAY, T., TJULANDIN, S. A., MA, W. W., SALEH, M. N., HARRIS, M., RENI, M., DOWDEN, S., LAHERU, D., BAHARY, N., RAMANATHAN, R. K., TABERNERO, J., HIDALGO, M., GOLDSTEIN, D., VAN CUTSEM, E., WEI, X., IGLESIAS, J. & RENSCHLER, M. F. 2013. Increased survival in pancreatic cancer with nab-paclitaxel plus gemcitabine. *N Engl J Med,* 369, 1691–703.

WADDELL, N., PAJIC, M., PATCH, A. M., CHANG, D. K., KASSAHN, K. S., BAILEY, P., JOHNS, A. L., MILLER, D., NONES, K., QUEK, K., QUINN, M. C., ROBERTSON, A. J., FADLULLAH, M. Z., BRUXNER, T. J., CHRIST, A. N., HARLIWONG, I., IDRISOGLU, S., MANNING, S., NOURSE, C., NOURBAKHSH, E., WANI, S., WILSON, P. J., MARKHAM, E., CLOONAN, N., ANDERSON, M. J., FINK, J. L., HOLMES, O., KAZAKOFF, S. H., LEONARD, C., NEWELL, F., POUDEL, B., SONG, S., TAYLOR, D., WADDELL, N., WOOD, S., XU, Q., WU, J., PINESE, M., COWLEY, M. J., LEE, H. C., JONES, M. D., NAGRIAL, A. M., HUMPHRIS, J., CHANTRILL, L. A., CHIN, V., STEINMANN, A. M., MAWSON, A., HUMPHREY, E. S., COLVIN, E. K., CHOU, A., SCARLETT, C. J., PINHO, A. V., GIRY-LATERRIERE, M., ROOMAN, I., SAMRA, J. S., KENCH, J. G., PETTITT, J. A., MERRETT, N. D., TOON, C., EPARI, K., NGUYEN, N. Q., BARBOUR, A., ZEPS, N., JAMIESON, N. B., GRAHAM, J. S., NICLOU, S. P., BJERKVIG, R., GRÜTZMANN, R., AUST, D., HRUBAN, R. H., MAITRA, A., IACOBUZIO-DONAHUE, C. A., WOLFGANG, C. L., MORGAN, R. A., LAWLOR, R. T., CORBO, V., BASSI, C., FALCONI, M., ZAMBONI, G., TORTORA, G., TEMPERO, M. A., GILL, A. J., ESHLEMAN, J. R., PILARSKY, C., SCARPA, A., MUSGROVE, E. A., PEARSON, J. V., BIANKIN, A. V. & GRIMMOND, S. M. 2015. Whole genomes redefine the mutational landscape of pancreatic cancer. *Nature,* 518, 495–501.

WAGNER, M., GRETEN, F. R., WEBER, C. K., KOSCHNICK, S., MATTFELDT, T., DEPPERT, W., KERN, H., ADLER, G. & SCHMID, R. M. 2001. A murine tumor progression model for pancreatic cancer recapitulating the genetic alterations of the human disease. *Genes Dev,* 15, 286–93.

WANG, C. A., CHANG, I. H., HOU, P. C., TAI, Y. J., LI, W. N., HSU, P. L., WU, S. R., CHIU, W. T., LI, C. F., SHAN, Y. S. & TSAI, S. J. 2020. DUSP2 regulates extracellular vesicle-VEGF-C secretion and pancreatic cancer early dissemination. *J Extracell Vesicles,* 9, 1746529.

WANG, K., WANG, X., PAN, Q. & ZHAO, B. 2023. Liquid biopsy techniques and pancreatic cancer: diagnosis, monitoring, and evaluation. *Mol Cancer,* 22, 167.

WANG, X., ZHANG, L., LI, H., SUN, W., ZHANG, H. & LAI, M. 2016. THBS2 is a potential prognostic biomarker in colorectal cancer. *Sci Rep,* 6, 33366.

WEITZ, J., GARG, B., MARTSINKOVSKIY, A., PATEL, S., TIRIAC, H. & LOWY, A. M. 2023. Pancreatic ductal adenocarcinoma induces neural injury that promotes a transcriptomic and functional repair signature by peripheral neuroglia. *Oncogene,* 42, 2536–2546.

WELLNER, U., SCHUBERT, J., BURK, U. C., SCHMALHOFER, O., ZHU, F., SONNTAG, A., WALDVOGEL, B., VANNIER, C., DARLING, D., ZUR HAUSEN, A., BRUNTON, V. G., MORTON, J., SANSOM, O., SCHÜLER, J., STEMMLER, M. P., HERZBERGER, C., HOPT, U., KECK, T., BRABLETZ, S. & BRABLETZ, T. 2009. The EMT-activator ZEB1 promotes tumorigenicity by repressing stemness-inhibiting microRNAs. *Nat Cell Biol,* 11, 1487–1495.

WENIGER, M., HONSELMANN, K. C. & LISS, A. S. 2018. The extracellular matrix and pancreatic cancer: a complex relationship. *Cancers (Basel),* 316, 10.

WOOD, L. D., CANTO, M. I., JAFFEE, E. M. & SIMEONE, D. M. 2022. Pancreatic cancer: pathogenesis, screening, diagnosis, and treatment. *Gastroenterology,* 163, 386–402.e1.

XIE, L., NI, W.-K., CHEN, X.-D., XIAO, M.-B., CHEN, B.-Y., HE, S., LU, C.-H., LI, X.-Y., JIANG, F. & NI, R.-Z. 2012. The expressions and clinical significances of tissue and serum galectin-3 in pancreatic carcinoma. *J Cancer Res Clin Oncol,* 138, 1035–1043.

YAKO, Y. Y., KRUGER, D., SMITH, M. & BRAND, M. 2016. Cytokines as biomarkers of pancreatic ductal adenocarcinoma: a systematic review. *PLOS ONE,* 11, e0154016.

YANG, Y., YAN, S., TIAN, H. & BAO, Y. 2018. Macrophage inhibitory cytokine-1 versus carbohydrate antigen 19-9 as a biomarker for diagnosis of pancreatic cancer: a PRISMA-compliant meta-analysis of diagnostic accuracy studies. *Medicine (Baltimore),* 97, e9994.

YASTREBOVA, M. A., KHAMIDULLINA, A. I., TATARSKIY, V. V. & SCHERBAKOV, A. M. 2021. Snail-family proteins: role in carcinogenesis and prospects for antitumor therapy. *Acta Naturae,* 13, 76–90.

YASUDA, A., SAWAI, H., TAKAHASHI, H., OCHI, N., MATSUO, Y., FUNAHASHI, H., SATO, M., OKADA, Y., TAKEYAMA, H. & MANABE, T. 2006. The stem cell factor/c-kit receptor pathway enhances proliferation and invasion of pancreatic cancer cells. *Mol Cancer,* 5, 46.

YONEYAMA, T., OHTSUKI, S., HONDA, K., KOBAYASHI, M., IWASAKI, M., UCHIDA, Y., OKUSAKA, T., NAKAMORI, S., SHIMAHARA, M., UENO, T., TSUCHIDA, A., SATA, N., IOKA, T., YASUNAMI, Y., KOSUGE, T., KANEDA, T., KATO, T., YAGIHARA, K., FUJITA, S., HUANG, W., YAMADA, T., TACHIKAWA, M. & TERASAKI, T. 2016. Identification of IGFBP2 and IGFBP3 as compensatory biomarkers for CA19-9 in early-stage pancreatic cancer using a combination of antibody-based and LC-MS/MS-based proteomics. *PLOS ONE,* 11, e0161009.

ZAHALKA, A. H., ARNAL-ESTAPÉ, A., MARYANOVICH, M., NAKAHARA, F., CRUZ, C. D., FINLEY, L. W. S. & FRENETTE, P. S. 2017. Adrenergic nerves activate an angio-metabolic switch in prostate cancer. *Science,* 358, 321–326.

ZAHALKA, A. H. & FRENETTE, P. S. 2020. Nerves in cancer. *Nat Rev Cancer,* 20, 143–157.

ZHANG, H., HYLANDER, B. L., LEVEA, C., REPASKY, E. A., STRAUBINGER, R. M., ADJEI, A. A. & MA, W. W. 2014. Enhanced FGFR signalling predisposes pancreatic cancer to the effect of a potent FGFR inhibitor in preclinical models. *Br J Cancer,* 110, 320–9.

ZHANG, H., LI, S., WANG, D., LIU, S., XIAO, T., GU, W., YANG, H., WANG, H., YANG, M. & CHEN, P. 2024. Metabolic reprogramming and immune evasion: the interplay in the tumor microenvironment. *Biomark Res,* 12, 96.

ZHANG, P., SUN, Y. & MA, L. 2015. ZEB1: at the crossroads of epithelial-mesenchymal transition, metastasis and therapy resistance. *Cell Cycle,* 14, 481–7.

ZHAO, B., CHENG, Q., CAO, H., ZHOU, X., LI, T., DONG, L. & WANG, W. 2021. Dynamic change of serum CA19-9 levels in benign and malignant patients with obstructive jaundice after biliary drainage and new correction formulas. *BMC Cancer,* 21, 517.

ZHAO, J. H., LUO, Y., JIANG, Y. G., HE, D. L. & WU, C. T. 2011. Knockdown of β-Catenin through shRNA cause a reversal of EMT and metastatic phenotypes induced by HIF-1α. *Cancer Invest,* 29, 377–82.

ZHAO, Y., SHEN, M., WU, L., YANG, H., YAO, Y., YANG, Q., DU, J., LIU, L., LI, Y. & BAI, Y. 2023. Stromal cells in the tumor microenvironment: accomplices of tumor progression? *Cell Death Dis,* 14, 587.

ZHENG, D., GOLUBOVSKAYA, V., KURENOVA, E., WOOD, C., MASSOLL, N. A., OSTROV, D., CANCE, W. G. & HOCHWALD, S. N. 2010. A novel strategy to inhibit FAK and IGF-1R decreases growth of pancreatic cancer xenografts. *Mol Carcinog,* 49, 200–9.

ZHENG, X., CARSTENS, J. L., KIM, J., SCHEIBLE, M., KAYE, J., SUGIMOTO, H., WU, C. C., LEBLEU, V. S. & KALLURI, R. 2015. Epithelial-to-mesenchymal transition is dispensable for metastasis but induces chemoresistance in pancreatic cancer. *Nature,* 527, 525–530.

ZHOU, B., LIN, W., LONG, Y., YANG, Y., ZHANG, H., WU, K. & CHU, Q. 2022. Notch signaling pathway: architecture, disease, and therapeutics. *Signal Transduct Target Therapy,* 7, 95.

ZHU, Q. Q., MA, C., WANG, Q., SONG, Y. & LV, T. 2016. The role of TWIST1 in epithelial-mesenchymal transition and cancers. *Tumour Biol,* 37, 185–97.

Part III

Precision Medicine and Early Detection

4 Precision Medicine and Personalized Approaches

Rahaba Marima[1], Egnesious Sambo[1,2], Shonisani Tshidino[3], Francina Lebogang Bopape[4], Thabo Rampai[5], Geoffrey Liu[6], Thabiso Victor Miya[1], Ntombikayise Xelwa[1], Demetra Demetriou[1], and Zodwa Dlamini[1]*

[1]SAMRC Precision Oncology Research Unit (PORU), DSI/NRF SARChI Chair in Precision Oncology and Cancer Prevention (POCP), Pan African Cancer Research Institute (PACRI), University of Pretoria, Pretoria, South Africa.

[2]Department of Medical Oncology, Faculty of Health Sciences, Steve Biko Academic Hospital, University of Pretoria, Pretoria, South Africa.

[3]Department of Biochemistry and Microbiology, Faculty of Science, Engineering and Agriculture, University of Venda, Limpopo, South Africa.

[4]Agricultural Research Council, Plant Health and Protection, Pretoria, South Africa.

[5]Department of Surgery, Faculty of Health Sciences, Steve Biko Academic Hospital, University of Pretoria, Pretoria, South Africa.

[6]Princess Margaret Cancer Centre, Dalla Lana School of Public Health, Departments of Medicine, Pharmacology and Toxicology, Epidemiology, Institute of Medical Science, University of Toronto, Ontario, Canada.*Corresponding author: Rahaba Marima

INTRODUCTION

Pancreatic cancer precision medicine and targeted therapy efforts continue to advance. Even though there may be challenges with these personalized approaches, standard therapies still yield poor pancreatic cancer patient outcomes. Genomic mutations in pancreatic cancer are primary oncogenic drivers. These alterations have been reported to vary among different populations. This is aggravated by the lack of genomic data from African populations, making it more challenging to treat underserved populations (Shaya et al., 2023).

Common genomic mutations in pancreatic cancer were reported to be Kirsten rat sarcoma (KRAS), tumor protein p53 (TP53), suppressor of mothers against decapentaplegic 4 (SMAD4), and cyclin-dependent kinase inhibitor 2A (CDKN2A). It has been observed that products of these mutated genes are difficult to treat with targeted therapies (Zarkavelis et al., 2019). Clinical studies for targeted therapies in pancreatic cancer are restricted by inadequate specificity for genetic changes across various populations. Various studies have shown that targeted therapy matching genomic alterations can improve cancer patient outcomes (Massard et al., 2017; Tsimberidou et al., 2020, 2019, 2017). Specific to pancreatic cancer, multiple genomic alterations may also require a combinatorial targeted therapeutic approach, targeting different genomic drivers (Sicklick et al., 2019). This combinatorial precision approach requires larger population cohorts to be studied.

DOI: 10.1201/9781003544401-7

Exocrine and neuroendocrine pancreatic cancers are the two main types, with exocrine accounting for more than 90% of all cases. Adenocarcinoma is the most common subtype. Although both are challenging to diagnose and treat, neuroendocrine pancreatic cancer is the rarer type and more challenging to manage (Cortiana et al., 2024; Magi et al., 2023). Disparities in pancreatic cancer care render targeted therapies inadequate, highlighting the need for more personalized treatment approaches. This chapter discusses the potential use of precision medicine approaches, such as genomic profiling and targeted therapies, with the overall goal of improving pancreatic cancer patient outcomes even in underserved African populations.

MOLECULAR PROFILING IN PANCREATIC CANCER

Tumor genomic profiling, or tumor profiling, is the personalization of cancer treatment based on an individual's genetic sequences. In this personalized approach, patients' genomic information is used to decide the specific or appropriate chemotherapy and radiation for pancreatic cancer patients (Liu et al., 2023).

Pancreatic cancer, particularly pancreatic ductal adenocarcinoma (PDAC), has one of the worst prognoses. This is due to its aggressive progression nature with a high mortality rate (Birnbaum et al., 2011; Frank et al., 2015; Vitorakis et al., 2024). It has been ranked as the fourth leading cause of cancer-associated mortalities worldwide, and it has a poor five-year survival rate of less than 10% (Ramírez-Maldonado et al., 2024). This pancreatic cancer is mainly resistant to standard systemic therapy due to its high rates of genomic mutation, translocation, insertion, and deletion, and its location (surrounded by other key organs) renders it difficult to treat surgically and with radiation.

Early detection of pancreatic cancer is rare, with a maximum survival of five years (Ansari et al., 2019; Birnbaum et al., 2011; Frank et al., 2015; Yamai et al., 2023). Pancreatic cancer is reported to be one of the most aggressive forms of human cancer, primarily due to late-stage diagnosis in the majority of patients, as well as the absence of early detection strategies and effective treatment (Ansari et al., 2019; Frank et al., 2015). Since there is a lack of proper molecular techniques for the diagnosis and prognosis of pancreatic cancer, the selection of proper treatment methods is also adversely affected. The current treatment uses gemcitabine-based systemic therapy with radiation therapy for pancreatic cancer (Ansari et al., 2019).

The genetic profile of pancreatic tumors may assist the medical team in determining the probability of tumor response to chemotherapy, radiation, and surgery, and can help inform novel drug discovery strategies. Information from genomic profiling may reduce the multiple drug resistance of pancreatic cancer as treatment might be boosted and intensified. Current cancer treatment strategies predominantly focus on targeting gene mutations, while genomic studies at the molecular level provide insights into the underlying mechanisms driving pancreatic cancer progression (Ansari et al., 2019; Yamai et al., 2023). This may lead to the development of novel diagnostic and prognostic biomarkers. Patients diagnosed with tumors with high mutation rates of SMAD4/KRAS/CDKN2A/TP53 usually have homologous recombination deficiency (HRD) related to genetic mutation. In contrast, others have targeted therapy based on comprehensive genomic profile (CGP) test results. The HRD group has been reported to have a longer survival rate (Yamai et al., 2023). Hence, CGP with next-generation sequencing enables the simultaneous detection of genetic alterations across hundreds of genes and multiple molecular biomarkers, which has prognostic implications.

The combined utilization of proteomics and genomics techniques has facilitated the comprehension of pancreatic cancer and the development of cancer biomarkers through a multifaceted approach. When several markers such as osteopontin (OPN), carbohydrate antigen 19-9 (CA 19-9), hepatocyte growth factor (HGF), carcinoembryonic antigen (CEA), and gene mutations in KRAS were used, about 99.5% specificity was achieved for pancreatic cancer diagnosis, and control measures improved (Ansari et al., 2019). The novel molecular and genetic biomarkers for pancreatic cancer are essential to improve early and accurate diagnosis. Furthermore, germline testing can determine whether an individual has inherited pancreatic cancer genetic mutations in genes such as

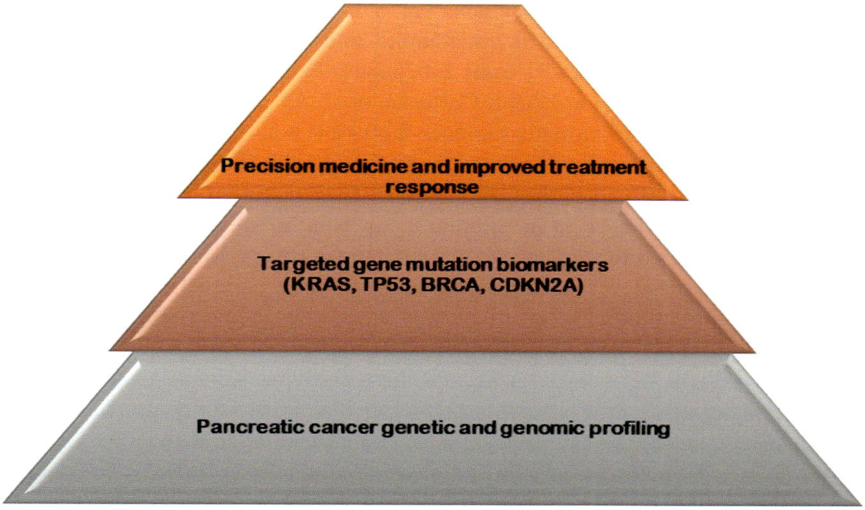

FIGURE 4.1 Genetic and genomic profiling in pancreatic cancer. The discovery and profiling of novel gene targets may improve the success of precision medicine.

BRCA-related genes (*BRCA1* and *BRCA2*) and Lynch syndrome-related genes (*MLH1*, *SH2*, *SH6*, *PMS2*, and *EPCAM*). Therefore, genomic sequencing has great potential to improve diagnosis and guide proper treatment options. The rapid improvement and accuracy obtained through mass spectrophotometry-based proteomics combined with RNA sequencing and whole genome sequencing (WGS) hold the potential to improve pancreatic cancer diagnosis and treatment (Mota Reyes et al., 2021; Sun et al., 2024). Genomic profiling research contributes to accuracy in medicine by defining cancer types and subtypes based on their genetics. It provides cancer patients with a more precise diagnosis and improves the chances of personalized treatment options.

Recently, liquid biopsies have been utilized to analyze different body fluids. These include saliva, pancreatic juice, and urine to investigate the early detection of PDAC. KRAS mutations were identified in biliary cytobrush and pancreatic juice, with KRAS2 mutation levels being significantly elevated in PDAC patients in contrast to those with chronic pancreatitis (CP). Telomerase activity and genetic mutations, including TP53, CDKN2A/p16, and SMAD4/DPC4, could facilitate the diagnosis of PDAC in CP. The analysis of saliva biomarkers presents a non-invasive and readily accessible method for investigating RNA and miRNA alterations in various genes, including *KRAS*, *MBD3L2*, *DPMI*, and *ACRV1*. This approach could potentially facilitate the differentiation of patients with pancreatic cancer from those with chronic pancreatic conditions and healthy individuals. Notably, the efficacy of liquid biopsy techniques to detect PDAC biomarkers differs depending on the tumor stage and biomarker type. For instance, resectable pancreatic cancer patients demonstrate a lower frequency of detectable circulating tumor DNA (ctDNA) in comparison to those with unresectable disease (Ramírez-Maldonado et al., 2024). Figure 4.1 illustrates the potential use of genomic profiling in pancreatic cancer care.

NOVEL PRECLINICAL STRATEGIES FOR PRECISION MEDICINE IN PANCREATIC CANCER

The primary focus of current clinical efforts is on the therapeutic significance of molecular subgroups. However, given the complexity of pancreatic cancer, the current use of mainly germline genetic and tumor genomic alterations is insufficient in precision medicine. The epigenomic, proteomic, and metabolomic modifications significantly impact phenotypic changes in pancreatic cancer and are crucial in PDAC patient classification (Daemen et al., 2015; Law et al., 2020; Lomberk et al., 2018). Studies found proliferative, progenitor-like, metabolic, and inflammatory PDAC subtypes with an upregulated concordance to the transcriptome subclasses. This was in proteome-based analysis of PDAC liver metastases (Bailey et al., 2016; Collisson et al., 2011; Law et al., 2020; Moffitt et al., 2015). The designation of an inflammatory cancer subtype is noteworthy because it highlights how crucial the tumor microenvironment (TME) is in influencing PDAC features. A good discussion of the significant and intricate effects of the inflammatory TME on tumor biology and therapeutic efficacy has been previously reviewed in the literature (Christenson et al., 2020; Storz and Crawford, 2020). Using progression-free survival as the main outcome, this group was able to stratify patients using a signature made up of several leukocyte subpopulations and stromal composition, which demonstrated the potential predictive utility of the immune microenvironment in PDAC (Mahajan et al., 2018; Wartenberg et al., 2018). Likewise, three unique immunological subtypes of PDAC were found in a study by Wartenberg et al. (2018) and these subtypes were connected to outcomes for the patient (Wartenberg et al., 2018). Finding predictive biomarkers for responders to immunotherapy could be a significant future accomplishment. Numerous preclinical investigations are being conducted to find new biomarkers for cancer subtyping, diagnosis, clinical outcome, or treatment response prediction. Yet, knowledge transfer to clinical practice is still difficult (Regel et al., 2020). The preclinical efforts to address epigenetic and metabolic mechanisms for precision medicine in pancreatic cancer are explored below.

Utilizing Metabolic Profiles to Create Targeted Therapies

Most metabolomic cancer research aims to identify cancer-specific biomarkers that can be used to diagnose, forecast, or treat a patient (Armitage and Barbas, 2014). The National Institutes of Health's Biomarkers Definitions Working Group defines a biomarker as a trait that reflects normal pathological processes, biological processes, or responses to environmental influences or interventions, including treatments. An excellent biomarker should meet the following criteria: (i) availability in minimally invasive sources (e.g., urine, blood, regularly available routine-collected tissue biopsies), (ii) high sensitivity and specificity, (iii) association with disease progression and rapid response to specific treatments, (iv) enhanced knowledge of the mechanism of the disease, and (v) utility in research (Bennett and Devarajan, 2011). Compared to other "omics," metabolomics has some advantages. In reality, it is simpler to translate metabolomic research into clinical practice since metabolite alterations are typically visible in biofluids that are easily accessible, such as blood and urine (Davis et al., 2011). Compared to other biological samples, biofluids often require minimal sample preparation, making them the easiest samples to work with (Regel et al., 2020). The metabolome is extremely dynamic, responsive to a variety of host and environmental conditions, and reflects ongoing fluxes of both metabolic and signaling processes. Because of these special characteristics, metabolomics can detect a wide range of minute alterations. Therefore, metabolomics has the potential to assess numerous intricate pathways and their effects at the same time (Monteiro et al., 2013). Additionally, metabolomics studies are less costly than transcriptomic and proteomic methods (Armitage and Barbas, 2014; Fiehn et al., 2006; Griffin, 2004; Monteiro et al., 2013). These biomarkers can then be utilized to create customized prognostic, diagnostic, and therapeutic methods. However, some metabolomic biomarkers have limitations to clinical utility, especially if the clinical associations are influenced by many external forces. Indeed, the strength of metabolomic biomarkers may be their utility to track the course of the disease, the effectiveness of treatment, the likelihood of side effects from medication, and the possibility of relapse (Koen et al., 2016).

While focusing on cancer metabolism is a potentially effective treatment approach, proper diagnostic identification of tumor subtypes with particular metabolic needs is essential for clinical success. PDAC was effectively divided into three very different metabolic subtypes using extensive metabolite profiling: lipogenic, slow proliferating, and glycolytic subtypes (Daemen et al., 2015).

Reduced proliferative capacity characterized one group, whereas different metabolite levels linked to lipogenesis, redox pathways, and glycolysis were seen in the glycolytic and lipogenic subtypes. While the glycolytic subtype was correlated with a mesenchymal subtype, the lipogenic subtype was linked to an epithelial phenotype, indicating potential functional relevance in the course of the disease (Daemen et al., 2015, 2018). The discovery of unique PDAC metabolic subgroups may help choose novel treatment targets and patients for experimental metabolic inhibitors (Daemen et al., 2015; Jain et al., 2012). CA 19-9 is the only serum marker for PDAC patients that is frequently used and clinically beneficial in therapeutic monitoring and early detection of recurrent disease after therapy (Hidalgo, 2010). A high CA 19-9 level indicates advanced PDAC and a dismal prognosis (Kaur et al., 2012; Locker et al., 2006). However, only 65% of patients with resectable PDAC had elevated CA 19-9 (Goggins, 2005; Kaur et al., 2012). Pancreatitis, cirrhosis, and cholestasis can all induce this elevation (Goonetilleke and Siriwardena, 2007).

Furthermore, even in more advanced stages of the illness, people who do not have Lewis antigen a or b (about 10% of patients with PDAC) are incapable of synthesizing CA 19-9. While blood CA 19-9 level measurement is helpful for individuals with confirmed pancreatic cancer, its usage as a screening tool produced unsatisfactory outcomes. Thus, it is not advised as a PDAC screening marker (Hidalgo, 2010; Kaur et al., 2012; Kunovsky et al., 2018). Mayerle and colleagues discovered a metabolite-based biomarker signature that can far more accurately separate PDAC from CP than CA 19-9 alone can (Mayerle et al., 2018). They conducted a comprehensive investigation of blood metabolites, such as lipids, in 914 patients. This was to identify potential metabolites that differentiate PDAC from CP. Interestingly, in plasma and serum samples from the training set, 29 metabolites—out of 477 metabolites across ten ontology classes—were significantly changed between CP and PDAC. Ultimately, with an AUC of 0.96, a nine-metabolite signature combined with the CA 19-9 model effectively distinguished PDAC from CP, and it could be successfully verified in a separate cohort (Mayerle et al., 2018). The goal of the study was to precisely rule out pancreatic cancer in patients with CP, with a focus on maximizing the negative predictive value (NPV) (Costello, 2018; Mayerle et al., 2018). Phua and colleagues developed metabotyping of PDAC-resected tissue in an adjuvant situation as a means of predicting gemcitabine therapy response through another metabolomic technique. This proof-of-principle study demonstrated that different tissue metabolomes, which include different nucleobase intermediates, amino acids, fatty acids, and inorganic acids, might be utilized to illustrate how gemcitabine works as a medication (Phua et al., 2018). A delicately chosen metabolome biomarkers panel that would enable early detection of PDAC, simpler staging determination, and, ideally, prognostic or outcome indications, is likely what the near future holds (Regel et al., 2020).

Epigenetic Treatments for Defined Subtypes

Different patterns of histone modification and DNA methylation at regulatory genomic regions dictate the epigenetic landscape, which in turn determines molecular profiles particular to cell types. For instance, trimethylation of histone 3 on lysine 27 (H3K27me3), DNA hypermethylation at promoter regions, and/or mono-ubiquitination of histone 2A at lysine 119 at different gene regulatory elements are all linked to the silencing of genes. On the other hand, gene activation is linked to the enrichment of trimethylation of lysine 4 and/or histone acetylation on histone 3 (H3K4me3) (Schuettengruber et al., 2011). Without altering the underlying DNA sequence, epigenetic mechanisms can translate internal or external stimuli into gene expression profiles that affect cellular phenotypes. The various subtypes of pancreatic cancer cannot be entirely explained by genetic variations; hence, the epigenetic status of PDAC tumor tissues may be important for predicting clinical fate and therapeutic response. Furthermore, altering the epigenetic landscape and the reversible nature of epigenetic alterations hold significant promise for the therapy of pancreatic cancer. Benitz et al. (2019) demonstrated that the development of pancreatic cancer requires a thorough epigenetic remodeling process together with the epigenetic silencing of differentiation genes (Benitz et al.,

2019). Findings from this study suggest that the probability of a malignant transformation could be predicted by a well-defined epigenetic gene signature. Additionally, this study has shown that pancreatic cancer cells are reprogrammed to have a less aggressive tumor phenotype when the epigenetic remodeler Ring1b is depleted or therapeutically inhibited (Benitz et al., 2019). Despite the outstanding results of various epidrugs (drugs that target the epigenome) in preclinical research, their efficacy in solid cancer clinical trials is fairly low (Hessmann et al., 2017).

Epigenetic cancer markers may have a role in PDAC patients' benefits by pre-classifying them as candidates for epigenetic therapy. Epigenetic profiles in pancreatic cancer have been studied in a growing body of pre-clinical research, and the results show a relationship with the phenotypes and traits of the disease. In addition to the upregulated expression of an epigenetic remodeler, numerous oncogenic pathways exhibit epigenetic regulation, indicating a significant involvement of epigenetic mechanisms in the genesis of PDAC (Benitz et al., 2019; Lomberk et al., 2018). Additionally, two unique patterns were found via a cluster analysis based on chromatin states. Moreover, these patterns mostly resembled the classical and basal transcriptional subtypes that Moffitt et al. (2015) had previously reported (Lomberk et al., 2018; Moffitt et al., 2015). Histone acetylation levels in distant regulatory genomic regions, in particular, are very informative and influence phenotypic differences between high- and low-grade PDAC (Diaferia et al., 2016). However, the process of targeting histone acetylation patterns using inhibitors of either histone acetyltransferases (HAT) or histone deacetylases (HDAC) is non-selective. This could lead to either a global fall in histone acetylation levels or an increase in them. The basal-like subtype exhibits a reduction of histone acetylation at differentiation genes, in contrast to the classical subtype (Lomberk et al., 2018). Thus, basal-like PDAC cells treated with HDAC inhibitors may acquire histone acetylation at differentiation genes and undergo reprogramming to a more restrained phenotype. Therefore, it may be crucial to characterize the histone acetylation patterns on classifier genes to forecast the tumor cells' ability to undergo epigenetic reprogramming. Crucially, applying epigenetic treatment necessitates the prior identification of eligible individuals, just like any other targeted therapy. Further clinical studies are unlikely to succeed without the stratification of patients with PDAC based on transcriptome or epigenetic signatures, even though epigenetic therapy may indicate a high possibility for a subgroup of cancer patients (Regel et al., 2020).

PRECISION MEDICINE AND CASE STUDIES IN PANCREATIC CANCER

Clinicians can better treat patients with pancreatic cancer, improve their prognosis, and improve their quality of life by recognizing and comprehending the many advantages and possibilities of precision medicine in this difficult disease. Clinicians can maximize available therapies and anticipate therapy responses by identifying patient subgroups most likely to benefit from particular interventions. As a result, this personalized strategy increases survival rates while reducing needless side effects and improving patient tolerability of treatment (Pishvaian and Brody, 2022). Furthermore, integrating customized therapy with sophisticated diagnostic procedures enables a more accurate approach to cancer care. Healthcare professionals can better manage patients' conditions by identifying actionable mutations and customizing medicines based on these findings through the use of molecular profiling and genetic testing. This personalized strategy is especially important for pancreatic cancer, as the tumor's heterogeneity frequently necessitates customized treatment regimens. The possibility of unequal advantages emphasizes the significance of ongoing investigation and advancement in the management of pancreatic cancer. Finding patients who will gain the most from new treatments and diagnostic methods will become crucial when these approaches are developed. In addition to enhancing clinical results, this will help allocate healthcare resources more effectively and guarantee that the most effective medicines are available to people who need them the most (Pishvaian and Brody, 2022).

The ultimate objective is to transition to a more accurate, patient-centered care approach that improves the overall quality of life for pancreatic cancer patients in addition to extending their longevity. To provide complete and efficient care, a multifaceted strategy is adopted, combining the knowledge of geneticists, surgeons, oncologists, and other medical specialists. We can get closer to developing cancer medicines that are consistently more effective and customized as long as we continue to comprehend the unique advantages that particular treatments can offer to specific patient populations (Cortiana et al., 2024).

Continuous Improvements in the Treatment of Pancreatic Cancer via Precision Medicine and Biomarker-Based Clinical Trials

Personalized approaches to treatment outcomes are the goal of biomarker-based clinical studies and therapeutic regimens, which thus represent an innovative advance in pancreatic cancer care. Utilizing biomarkers, these novel studies determine patient subgroups most likely to benefit from particular treatment interventions, such as immunotherapies, combination regimens, and targeted therapies. Treatment plans can be customized to each patient's unique molecular profile by incorporating molecular profiling into clinical trial designs, including umbrella trials, adaptive trials, and basket trial designs (Le et al., 2015; Von Hoff et al., 2010).

Various clinical trials are currently being conducted to investigate the effectiveness of biomarker-driven treatments in pancreatic cancer (Cortiana et al., 2024). For example, patients enrolled in the precision medicine cancer treatment clinical trial Molecular Analysis for Therapy Choice (NCI-MATCH) are assigned to targeted therapy based on specified mutations found in their tumors, not the type of cancer they have (O'Dwyer et al., 2023). The Targeted Agent and Profiling Utilization Registry (TAPUR) study is another noteworthy example. It evaluates the safety and effectiveness of FDA-approved targeted therapies in patients with advanced cancer with unique genetic mutations (Mangat et al., 2018). Furthermore, using complete genetic profiling to inform therapy choices for patients with advanced pancreatic cancer is the main goal of the Comprehensive Molecular Characterization of Advanced Pancreatic Ductal Adenocarcinoma for Better Treatment Selection (COMPASS) trial (Aung et al., 2018). The objective of this study is to find functional mutations and categorize patients into subsets that could profit from enrollment in clinical trials or certain targeted treatments (Aung et al., 2018). Another such trial is called Integrative Molecular Profiling of Pancreatic Cancer Therapy (IMPaCT). It aims to match patients with targeted medication according to the molecular features of their tumors. In return, this enables more individualized treatment plans (Stockley et al., 2016).

Numerous studies show that immunotherapy in combination with targeted therapy is a viable strategy, notably for metastatic pancreatic cancer treatment. In the KEYNOTE-158 Phase II study, pembrolizumab was administered to 22 patients with microsatellite instability-high (MSI-H) or mismatch repair-deficient (dMMR) cancers. The patients' outcomes included an 18.2% response rate (RR), 4 months of overall survival (OS), and 2.1 months of progression-free survival (PFS) (Marabelle et al., 2020). A total of 38 patients with KRAS G12C mutations receiving sotorasib treatment were involved in the CODEBREAK-100 Phase Ib/II trial; the results included a 21% RR, 6.9 months OS, and 4 months PFS (Skoulidis et al., 2021). Adagrasib was administered to 21 patients with KRAS G12C mutations in the KRYSTAL-1 Phase II trial; the patients achieved a 33.3% RR, 8 months OS, and 5.4 months PFS (Cortiana et al., 2024). In their 2022 Phase II research study, Schram et al. (2022) treated ten patients with NRG1 fusions containing zenocutuzumab, which resulted in a 40% RR (Schram et al., 2022). In a phase I research study, Sacher et al. (2023) found that 22 patients with KRAS G12C mutations receiving diravasib had a 36% RR (Sacher et al., 2023). In their 2024 Phase I/II trial, Li et al. (2024) showed a 33% RR in treating 24 patients with KRAS G12C mutations with olomorasib (Li et al., 2024). Lastly, 28 patients treated with glecirasib for KRAS G12C mutations in a 2024 Phase I/II research study by Hollebecque and colleagues showed a 46.4% RR and 5.5 months PFS (Kim et al., 2024).

Additionally, the National Cancer Institute (NCI) funds numerous research initiatives to use various strategies to combat pancreatic cancer more successfully. These initiatives encompass a wide range of issues, from basic science researching the underlying causes of cancer to clinical research aimed at using these discoveries to improve patient outcomes (Cortiana et al., 2024). To enhance the understanding of pancreatic cancer's origins and natural history, the Pancreatic Cancer Cohort Consortium, for example, comprises over a dozen prospective epidemiologic cohort investigations. This collaboration encompasses the PanScan genome-wide association study (GWAS). The primary goal of the Pancreatic Cancer Detection Consortium (PCDC) is to develop and evaluate biomarkers for detecting pancreatic cancer early and to identify people at high risk. The PDAC Stromal Reprogramming Consortium (PSRC) is an interdisciplinary network of researchers that connects fundamental biological studies with pre-clinical and translational research. Their objective is to pinpoint and assess TME components that influence PDAC development and therapeutic response. The Pancreatic Specialized Programs of Research Excellence (Pancreatic SPOREs) support innovative approaches for the early detection, diagnosis, prevention, and treatment of pancreatic cancer, designed to expedite the translation of fundamental scientific discoveries into clinical settings. With over 90% of pancreatic tumors linked to mutations in the *RAS* gene, the RAS Initiative seeks to comprehend these abnormalities to create novel, potent treatments for RAS-related cancers (Cortiana et al., 2024).

Patient stratification is aided in the selection of the best therapy medicines by predictive biomarkers such as protein markers (like CA 19-9), gene expression profiles, and genetic alterations (like KRAS, BRCA1/2) (Hyman et al., 2015; Schwaederle et al., 2015; Von Hoff et al., 2010). As for PDAC treatment, addressing KRAS mutations is a major achievement. Around 90% of PDAC cases include KRAS mutations, which activate oncogenic signaling pathways that encourage tumor development and survival. The treatment landscape for this aggressive disease may change as a result of recent therapeutic approaches that directly block mutant KRAS, which have demonstrated encouraging preclinical outcomes. For those with KRAS-mutant PDAC, the introduction of direct KRAS inhibitors has given them fresh hope. Examples of these are adagrasib (MRTX849) and sotorasib (AMG 510), which have shown effectiveness in early clinical trials (Hong et al., 2020).

The findings demonstrate how precision medicine has the power to dramatically enhance results. However, several challenges, including assay standardization, patient heterogeneity, and regulatory considerations, stand in the way of the effective integration of biomarker-based techniques into clinical studies. To overcome these obstacles, strong biomarker validation procedures and efficient clinical practice integration need to be established by researchers, physicians, and regulatory bodies working together. Biomarker-based clinical trials have the potential to improve patient outcomes, increase treatment efficacy, and progress the area of pancreatic cancer treatments despite these obstacles (Cortiana et al., 2024). While the MAP pathway seems to be less important, recent research has demonstrated the critical importance of the ERK pathway in KRAS-mutant cell lines. This demonstrates that ERK pathway activation is essential for the growth of the tumor and its survival (Klomp et al., 2024). These findings imply that ERK pathway targeting may result in more potent treatment approaches. To fully utilize precision medicine in this difficult illness, these strategies must be developed and improved upon indefinitely (Table 4.1) (Cortiana et al., 2024).

PANCREATIC CANCER IN AFRICAN POPULATIONS

According to Global Cancer Observatory (GLOBOCAN) data in 2022, pancreatic cancer incidence rates and mortalities continue to rise rapidly worldwide. Pancreatic cancer accounts for almost as many mortalities (467,005) as incidences (511,000). This malignancy of the pancreas is responsible for approximately 5% of all cancer deaths worldwide (Bray et al., 2024; Sung et al., 2021) and is projected to become the second most fatal cancer around 2030 (Rahib et al., 2014). For decades, the morbidity of pancreatic cancer in the African population has been 30–70% higher compared to other races (Scarton et al., 2018). Furthermore, on the African continent, the reported incidence is increasing more rapidly among young females compared to their male counterparts (Samaan et al., 2023). This could be attributed to alterations in trends of certain identified modifiable and non-modifiable risk factors associated with racial and gender disparities (Samaan et al., 2023), which

TABLE 4.1
An Overview of the Current Clinical Trials Using Biomarkers and Precision Medicine in the Treatment of Pancreatic Cancer

Clinical Trial	Description	References
NCI-MATCH (ID: NCT02465060)	Directs therapeutic intervention toward tumor mutations rather than the classification of malignancy.	O'Dwyer et al. (2023)
TAPUR (ID: NCT02693535)	Assesses the safety and efficacy of FDA-approved targeted therapies in advanced cancer patients with specified genetic mutations.	Mangat et al. (2018)
COMPASS	Employs molecular profiling to treat advanced pancreatic cancer.	Aung et al. (2018)
IMPaCT (ID: NCT03337087)	Matches patients with treatments according to the molecular features of their tumors.	Stockley et al. (2016)
KEYNOTE-158 (Phase II) (ID: NCT02628067)	Pembrolizumab in MSI-H/dMMR tumors: 18.2% RR, 2.1 months PFS	Marabelle et al. (2020)
KRYSTAL-1 (Phase II) (ID: NCT03785249)	Adagrasib in KRAS G12C mutations: 33.3% RR, 8 months OS, 5.4 months PFS	Cortiana et al. (2024)
CODEBREAK-100 (Phase Ib/II) (ID: NCT03600883)	Sotorasib in KRAS G12C mutations: 21% RR, 6.9 months OS, 4 months PFS.	Skoulidis et al. (2021)
Schram's 2021 Phase II (ID: NCT02912949)	Zenocutuzumab in NRG1 fusions: 40% RR.	Schram et al. (2022)
Sacher's 2023 Phase I (ID: NCT04449874)	Diravasib in KRAS G12C mutations: 36% RR.	Sacher et al. (2023)
Li's 2024 Phase I/II (ID: NCT05009329)	Olomorasib in KRAS G12C mutations: 33% RR.	Li et al. (2024)
Hollebecque's 2024 Phase I/II (ID: NTC02912949)	Glecirasib in KRAS G12C mutations: 46.4% RR, 5.5 months PFS.I	Kim et al. (2024)
Pancreatic Cancer Cohort Consortium	Examines the causes and progression of pancreatic cancer, including PanScan studies.	Cortiana et al. (2024)
PCDC	Creates and evaluates biomarkers for early detection.	Cortiana et al. (2024)
RAS Initiative (ID: NCT03600883)	Develops treatments for pancreatic cancer caused by RAS mutations.	Hong et al. (2020)
Pancreatic SPOREs (ID:NCT03122106)	Transforms basic research into clinical applications.	Cortiana et al. (2024)
PSRC	Examines the tumor microenvironment's impact on PDAC progression and treatment efficacy.	Blackford et al. (2024)

may predict increased odds of developing pancreatic cancer (Midha et al., 2016). Modifiable or environmental risk factors include obesity and increased body mass index (BMI), heavy alcohol consumption and smoking, dietary factors (for example, zinc deficiency), and/or infections (Li and Abbruzzese, 2010; Risch et al., 2010; Wang et al., 2016; Zheng and Lee, 2009). Non-modifiable risk factors cannot be controlled and may include genetically predisposed factors, ethnicity, gender, diabetes mellitus (DM), and/or CP (Lowenfels and Maisonneuve, 2006; Midha et al., 2016). Diabetes, obesity, and heavy alcohol intake have been identified as becoming increasingly prevalent among the African population and other racial minorities compared to their white counterparts (Ben et al., 2011; Benoit et al., 2019), reflecting the surging pancreatic cancer morbidities and mortality rates. Compared to other ethnic groups, Black people with pancreatic cancer have a poor overall prognosis at each stage of the disease. They experience higher incidence rates, lower utilization of current treatment modalities, and increased mortality rates (Nipp et al., 2018; Sun et al., 2014).

Pancreatic Cancer Statistics in Africa and High HDI Countries

Pancreatic cancer incidence rates are approximately four times higher in countries with a higher Human Development Index (HDI) compared to those with a lower HDI. The highest number of cases are observed in North America, Australia/New Zealand, and Europe. However, on a global scale, the highest rates are recorded in Armenia, Western Asia, for males, and Uruguay, South America, for females (Bray et al., 2024). The HDI is used by the United Nations to appraise a country's socioeconomic status (SES) (Dasic et al., 2020). The HDI includes three basic indicators: life expectancy at birth, level of education, and gross national income (Dasic et al., 2020). Therefore, socioeconomic stance is one of the components of HDI. Four HDI classes or tiers exist: low (HDI < 0.55), which indicates that a country has a relatively low standard of living due to decreased life expectancy, poor education, high multidimensional poverty index, lack of access to healthcare, and underdeveloped SES and tends to have an HDI score below 0.55, in the "low human development" category. Medium ($0.55 \leq$ HDI < 0.70) HDI countries are characterized by the HDI score ranging from 0.55 to 0.69 and mostly continue to grapple with issues of access to proper healthcare resources, low education system, and high incomes. A high ($0.70 \leq$ HDI < 0.80) HDI essentially means that the country possesses a significantly high level of human development indicated by a cut-off point between 0.70 and 0.79. Countries with high HDI typically offer a high standard of living, with access to decent healthcare, a strong education system, and opportunities to earn high incomes. Very high ($0.80 \leq$ HDI < 1.00) HDI regions are signified by a score of 0.80 or higher, and these nations typically possess higher levels of education, income, and life expectancy (Grimm et al., 2008; Ringdén et al., 2019).

A positive correlation exists between pancreatic cancer and the HDI in both males and females. This is due to lifestyle factors associated with Westernization, particularly obesity and/or smoking, which are recognized as risk factors (Randi, Franceschi, and La Vecchia, 2006; Raimondi, Maisonneuve, and Lowenfels, 2009). Utilizing GLOBOCAN data from 184 countries, age-standardized incidence rates (ASRs) were analyzed across four HDI levels: low, medium, high, and very high (Bray et al., 2018), demonstrating a strong positive correlation between the overall incidence of cancer and the level of HDI (Bray et al., 2012, 2024; Sung et al., 2021). The age-standardized incidence/mortality rate (ASR) is a weighted average of the age-specific incidence/mortality rates per 100,000 people, where the weights are the percentage of people in the corresponding age groups of the World Health Organization (WHO) standard population (Bray et al., 2024). A comparative analysis of ASRs between regions with very high and low HDI demonstrated a positive correlation, with pancreatic cancer incidence being 2–14 times higher in males and 2–11 times higher in females in regions with very high HDI (Fidler, Soerjomataram, and Bray, 2016). Figure 4.2 demonstrates the reported ASRs for incidence and mortalities between the top ten very high HDI Western countries and the African continent.

According to GLOBOCAN reports in 2022, there were no African countries under the very high HDI category based on their socioeconomic stance, as depicted by Figure 4.2. The top ten countries with a very high HDI included Uruguay, Hungary, Japan, France, Austria, Czechia, Sweden, Latvia, Malta, and Finland. These countries have age-specific incidence rates (ASIRs) of 11.4, 10.4, 9.8, 9.4, 9.4, 9.3, 9.3, 9.3, 9.0, and 8.9, respectively. Their age-specific mortality rates (ASMRs) are 9.4, 9.6, 8.8, 8.1, 8.6, 7.5, 8.4, 8.3, and 8.4. For high HDI countries in Africa, only nine leading countries were reported, namely (in descending order) South Africa, Mauritius, Algeria, Egypt, Tunisia, Libya, Gabon, Botswana, and Seychelles. These nations have ASIRs of 5.3, 3.8, 2.6, 3.8, 3.3, 3.8, 3.4, 0.75, and 6, respectively, along with the corresponding ASMRs of 4.7, 2.9, 2.5, 3.7, 3.2, 3.5, 3.3, 0.75, and 5.3. For medium HDI countries in Africa, the ASIRs reported were 2.8, 3.5, 2.7, 2.4, 1.2, 1.2, 1.4, 1.7, 2.3, and 1.6. The ASMRs for these countries were 2.7, 3.5, 2.6, 2.4, 1.2, 1.2, 1.4, 1.7, 2.3, and 1.6, respectively. The top ten countries identified here are Morocco, Kenya, Zimbabwe, Equatorial Guinea, Cape Verde, Namibia, Eswatini, Ghana, São Tomé and Príncipe, and Angola.

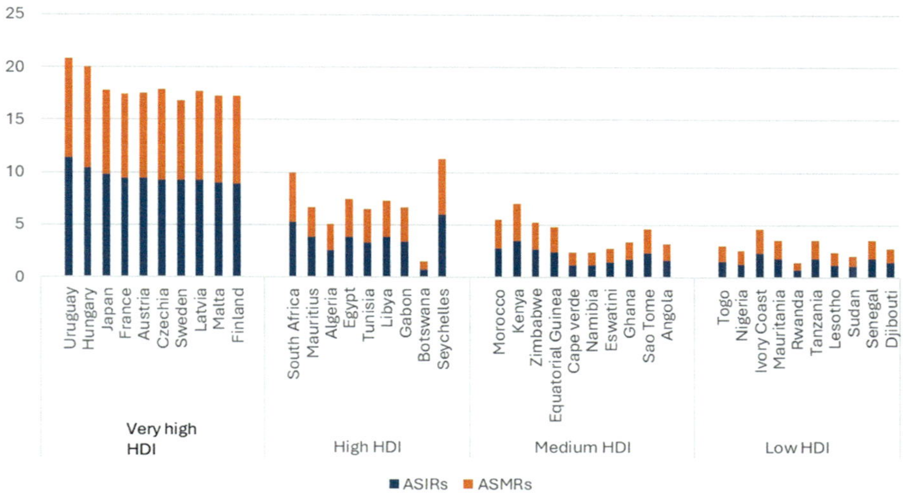

FIGURE 4.2 The visual overlay of ASRs for incidence and mortality interlinked with pancreatic cancer statistics between the top ten very high HDI nations (Western countries) and top ten African countries categorized under high HDI, medium HDI, and low HDI tiers or classes, respectively. Rates were approximately four times greater in higher versus lower HDI countries, with ASRs for incidence and mortality highest in Western Europe, North America, and Australia/New Zealand (very high HDI countries) as depicted by the diagram. The ASIRs and ASMRs analysis outlined the second-highest category (designated by the high HDI tier) being linked with the top ten upper-middle-income African countries. Furthermore, the consecutive third and last categories (identified by medium HDI and low HDI labels, respectively) corresponded with Africa's top ten middle-income countries (MICs) and lower-middle-income countries (LMICs), respectively. This highlights the varying levels of human development across the continent.

Lastly, the ten leading low HDI countries in Africa included Togo, Nigeria, Ivory Coast, Mauritania, Rwanda, Tanzania, Lesotho, Sudan, Senegal, and Djibouti. These low HDI countries were found to possess ASIRs of 1.5, 1.3, 2.3, 1.8, 0.73, 1.8, 1.2, 1.1, 1.8, and 1.4, respectively, with corresponding ASMRs of 1.5, 1.3, 2.3, 1.8, 0.70, 1.8, 1.2, 0.98, 1.8, and 1.4 (Fidler, Soerjomataram, and Bray, 2016; Dasic et al., 2020; Bray et al., 2024). Individuals living in developed countries have higher ASIRs of pancreatic cancer (more than three-fold) and slightly higher ASMRs compared to those living in developing countries (Bray et al., 2024). The relationship between HDI and the incidence and mortality of pancreatic cancer is strongly positively correlated. Thus, a significant directly proportional trend was observed between the HDI of each country and the incidence and mortality rates of pancreatic cancer.

Moreover, the Metabolic Syndrome and Cancer Project with 577,315 participants was conducted to investigate the components of metabolic syndrome associated with pancreatic cancer risk. Throughout a 12-year follow-up period, 315 women and 547 men were diagnosed with pancreatic cancer. In women, obesity was observed to significantly increase the risk of pancreatic cancer (adjusted relative risk, 1.54; 95% CI, 1.04–2.29), whereas this association was less pronounced in men (Johansen et al., 2010). In sub-Saharan Africa (SSA), obesity and excess weight were predominantly observed among women of higher SES (Hashan et al., 2021; Seidu et al., 2021). Furthermore, there is an increasing trend in obesity and overweight reported in SSA among females who can procreate (Amugsi et al., 2017; Moise et al., 2019). These results suggest that obesity may be a contributory factor to the increasing pancreatic cancer incidence and mortality rates in the African population, with a more pronounced effect observed among females compared to males (Samaan et al., 2023).

The prevalence of diabetes increases in parallel with the rising rates of obesity (Klein et al., 2022). The burden of diabetes is relatively high among Africans (Scarton et al., 2018). Currently, the African population living with type 2 diabetes mellitus (T2DM) is estimated to be around 25 million and is projected to be approximately 33 million and 55 million by 2030 and 2045, respectively (Mapa-Tassou et al., 2019). Furthermore, the African population is estimated to be approximately 528 million, and this figure is expected to reach 696 million and 1.05 billion by 2030 and 2045 (Mapa-Tassou et al., 2019). For T2DM, which accounts for about 90% of the total, the increasing trend can be attributed to aging, a fast increase in urbanization, high-energy dietary intake, and obesogenic environments (Saeedi et al., 2019). Immunologic, abnormal metabolic, and hormonal characteristics of T2DM may promote cancer development. Additionally, insulin resistance and hyperinsulinemia may upregulate the production of insulin-like growth factor-1 (IGF-1), which may result in enhanced cell proliferation and promote pancreatic cancer development (Ulanet et al., 2010). A meta-analysis of 36 cohorts found that diabetes was linked with almost a twofold increase in pancreatic cancer (relative risk (RR) = 1.94; 95% CI, 1.66–2.27) in contrast to the general population (Ben et al., 2011). Diabetes, like obesity, may be a risk factor for pancreatic cancer in the African population (Motala et al., 2022).

Cigarette smoking is a recognized etiological factor in cancer development, potentially contributing to disparities across various cancer types (Yuan et al., 2017). A prospective study based on the United States cohort discovered that >25% of cancer mortalities (including pancreas, lung, bladder, esophagus, kidney, pharynx, and oral cavity) are a result of cigarette smoking (Ho and Elo, 2012). A meta-analysis encompassing 12 prospective cohorts and one case-control study proposed that current cigarette smokers exhibit an 80% higher risk of developing pancreatic cancer with earlier onset compared to individuals who have never smoked (Lynch et al., 2009). As tobacco products become increasingly accessible and marketing efforts intensify in Africa, the prevalence of smoking has been observed to increase. This trend is anticipated to persist, with a substantial rise in smoking rates projected in the future (Vellios et al., 2018). Given the projected population growth and increasing life expectancy, the anticipated rise in the number of smokers coupled with an aging population is expected to place Africa at elevated risk for a future burden of smoking-related diseases. Thus, improved tobacco regulatory measures are necessary, particularly in this population (Peer, 2018).

ETHICAL CONSIDERATIONS AND PATIENT RIGHTS

Socioeconomic factors contribute to pancreatic cancer health disparities. However, the specific involvement of these factors in the higher morbidity and mortality rates of pancreatic cancer in the African population is still to be determined (Scarton et al., 2018). SES affects the individual's ability to access physical and social environments, proper healthcare services, behaviors, and treatment interventions. Individuals who reside in more underserved communities have lower income status, and receive less education demonstrate increased incidence and mortality rates with poor prognosis for different cancer types (Singh et al., 2011). Modifiable cancer risk factors, such as diet, obesity, smoking, occupational and environmental exposures, physical inactivity, and cancer screening, are influenced by individual-level and area-level socioeconomic factors (Kogevinas et al., 1997). Evaluating the socioeconomic patterns in cancer incidence and mortality is essential for quantifying health disparities between the most and least advantaged social groups. This evaluation facilitates the identification of areas and populations with increased susceptibility to cancer diagnosis and mortality, thereby benefiting targeted social and medical interventions (Powers et al., 2021). Moreover, patients with government-sponsored health insurance had a lower likelihood of receiving multimodality treatment for pancreatic cancer compared to those with private insurance (Dimou et al., 2016). Consequently, healthcare policies should aim to enhance accessibility to early cancer screening programs for underserved regions and disadvantaged populations, thereby facilitating timely treatment interventions (Kale et al., n.d.).

FIGURE 4.3 Pancreatic cancer precision medicine will improve patient outcomes across the globe when minorities in cancer research are included in medical and translational advancements.

CHALLENGES WITH PANCREATIC CANCER PERSONALIZED CARE

Pancreatic cancer is highly heterogeneous and primarily driven by genetic mutations. Although some of these mutations have been discovered in European populations, they significantly vary among populations. Thus, targeted therapies may not be effective, especially in African populations, due to a lack of genomic data from African populations. Furthermore, there is a lack of clinical trials in African populations, which is a major cause of pancreatic cancer racial disparities. Late diagnosis is a significant clinical problem across all races and populations. The effectiveness of liquid biopsies in detecting pancreatic cancer biomarkers depends on the biomarker type and tumor stage. Inconsistencies in biomarker standardization also exist. Novel pancreatic cancer precision medicine preclinical approaches are still challenging to implement, even with the omicsapproaches.

CONCLUSIONS AND PERSPECTIVES

Genetic factors play a major role in pancreatic cancer. In addition, risk factors including external factors such as the environment, diet, lifestyle, and SES also play a significant role. Pancreatic cancer has been reported to be more aggressive in African populations. The precise reasons driving these disparities are yet to be established. Nonetheless, the lack of genomics data from African populations and the lack of inclusion in clinical trials are contributing factors. Intra- and inter-tumor variability in pancreatic cancer further drives these disparities. It is evident that targeted therapies are inadequate to treat pancreatic cancer; thus, more advanced and personalized multimodal approaches are necessary, Figure 4.3.

REFERENCES

AMUGSI, D.A., DIMBUENE, Z.T., MBERU, B., MUTHURI, S., EZEH, A.C., 2017. Prevalence and time trends in overweight and obesity among urban women: an analysis of demographic and health surveys data from 24 African countries, 1991–2014. *BMJ Open* 7, e017344. https://doi.org/10.1136/bmjopen-2017-017344

ANSARI, D., TORÉN, W., ZHOU, Q., HU, D., ANDERSSON, R., 2019. Proteomic and genomic profiling of pancreatic cancer. *Cell Biology and Toxicology* 35, 333–343. https://doi.org/10.1007/s10565-019-09465-9

ARMITAGE, E.G., BARBAS, C., 2014. Metabolomics in cancer biomarker discovery: Current trends and future perspectives. *Journal of Pharmaceutical and Biomedical Analysis, Review Papers on Pharmaceutical and Biomedical Analysis* 87, 1–11. https://doi.org/10.1016/j.jpba.2013.08.041

AUNG, K.L., FISCHER, S.E., DENROCHE, R.E., JANG, G.H., DODD, A., CREIGHTON, S., SOUTHWOOD, B., LIANG, S.B., CHADWICK, D., ZHANG, A., O'KANE, G.M., ALBABA, H., MOURA, S., GRANT, R.C., MILLER, J.K., MBABAALI, F., PASTERNACK, D., LUNGU, I.M., BARTLETT, J.M.S.,...KNOX, J.J. 2018. Genomics-driven precision medicine for advanced pancreatic cancer: early results from the COMPASS trial. *Clinical Cancer Research* 24(6), 1344–1354. https://doi.org/10.1158/1078-0432.Ccr-17-2994

BAILEY, P., CHANG, D.K., NONES, K., JOHNS, A.L., PATCH, A.-M., GINGRAS, M.-C., MILLER, D.K., CHRIST, A.N., BRUXNER, T.J.C., QUINN, M.C., NOURSE, C., MURTAUGH, L.C., HARLIWONG, I., IDRISOGLU, S., MANNING, S., NOURBAKHSH, E., WANI, S., FINK, L., HOLMES, O., CHIN, V., ANDERSON, M.J., KAZAKOFF, S., LEONARD, C., NEWELL, F., WADDELL, NICK, WOOD, S., XU, Q., WILSON, P.J., CLOONAN, N., KASSAHN, K.S., TAYLOR, D., QUEK, K., ROBERTSON, A., PANTANO, L., MINCARELLI, L., SANCHEZ, L.N., EVERS, L., WU, J., PINESE, M., COWLEY, M.J., JONES, M.D., COLVIN, E.K., NAGRIAL, A.M., HUMPHREY, E.S., CHANTRILL, L.A., MAWSON, A., HUMPHRIS, J., CHOU, A., PAJIC, M., SCARLETT, C.J., PINHO, A.V., GIRY-LATERRIERE, M., ROOMAN, I., SAMRA, J.S., KENCH, J.G., LOVELL, J.A., MERRETT, N.D., TOON, C.W., EPARI, K., NGUYEN, N.Q., BARBOUR, A., ZEPS, N., MORAN-JONES, K., JAMIESON, N.B., GRAHAM, J.S., DUTHIE, F., OIEN, K., HAIR, J., GRÜTZMANN, R., MAITRA, A., IACOBUZIO-DONAHUE, C.A., WOLFGANG, C.L., MORGAN, R.A., LAWLOR, R.T., CORBO, V., BASSI, C., RUSEV, B., CAPELLI, P., SALVIA, R., TORTORA, G., MUKHOPADHYAY, D., PETERSEN, G.M., MUNZY, D.M., FISHER, W.E., KARIM, S.A., ESHLEMAN, J.R., HRUBAN, R.H., PILARSKY, C., MORTON, J.P., SANSOM, O.J., SCARPA, A., MUSGROVE, E.A., BAILEY, U.-M.H., HOFMANN, O., SUTHERLAND, R.L., WHEELER, D.A., GILL, A.J., GIBBS, R.A., PEARSON, J.V., WADDELL, NICOLA, BIANKIN, A.V., GRIMMOND, S.M., 2016. Genomic analyses identify molecular subtypes of pancreatic cancer. *Nature* 531, 47–52. https://doi.org/10.1038/nature16965

BEN, Q., XU, M., NING, X., LIU, J., HONG, S., HUANG, W., ZHANG, H., LI, Z., 2011. Diabetes mellitus and risk of pancreatic cancer: a meta-analysis of cohort studies. *European Journal of Cancer* 47, 1928–1937. https://doi.org/10.1016/j.ejca.2011.03.003

BENITZ, S., STRAUB, T., MAHAJAN, U.M., MUTTER, J., CZEMMEL, S., UNRUH, T., WINGERATH, B., DEUBLER, S., FAHR, L., CHENG, T., NAHNSEN, S., BRUNS, P., KONG, B., RAULEFS, S., CEYHAN, G.O., MAYERLE, J., STEIGER, K., ESPOSITO, I., KLEEFF, J., MICHALSKI, C.W., REGEL, I., 2019. Ring1b-dependent epigenetic remodelling is an essential prerequisite for pancreatic carcinogenesis. *Gut* 68, 2007–2018. https://doi.org/10.1136/gutjnl-2018-317208

BENNETT, M.R., DEVARAJAN, P., 2011. Chapter 1 – characteristics of an ideal biomarker of kidney diseases. In: Edelstein, C.L. (ed.) *Biomarkers of kidney disease*. San Diego, CA: Academic Press, pp. 1–24. https://doi.org/10.1016/B978-0-12-375672-5.10001-5

BENOIT, S.R., HORA, I., ALBRIGHT, A.L., GREGG, E.W., 2019. New directions in incidence and prevalence of diagnosed diabetes in the USA. *BMJ Open Diabetes Research and Care* 7, e000657. https://doi.org/10.1136/bmjdrc-2019-000657

BIRNBAUM, D.J., ADÉLAÏDE, J., MAMESSIER, E., FINETTI, P., LAGARDE, A., MONGES, G., VIRET, F., GONÇALVÈS, A., TURRINI, O., DELPERO, J.-R., IOVANNA, J., GIOVANNINI, M., BIRNBAUM, D., CHAFFANET, M., 2011. Genome profiling of pancreatic adenocarcinoma. *Genes, Chromosomes and Cancer* 50, 456–465. https://doi.org/10.1002/gcc.20870.

BLACKFORD, A. L., CANTO, M. I., DBOUK, M., HRUBAN, R. H., KATONA, B. W., CHAK, A., BRAND, R. E., SYNGAL, S., FARRELL, J., KASTRINOS, F., STOFFEL, E. M., RUSTGI, A., KLEIN, A. P., KAMEL, I., FISHMAN, E., HE, J., BURKHART, R., SHIN, E. J., LENNON, A. M., & GOGGINS, M. (2024). Pancreatic cancer surveillance and survival of high-risk individuals. *Jama Oncology, 10*(8), 1087–1096. https://doi.org/10.1001/jamaoncol.2024.1930

BRAY, F., FERLAY, J., SOERJOMATARAM, I., SIEGEL, R.L., TORRE, L.A., JEMAL, A., 2018. Global cancer statistics 2018: GLOBOCAN estimates of incidence and mortality worldwide for 36 cancers in 185 countries. *CA: A Cancer Journal for Clinicians* 68, 394–424. https://doi.org/10.3322/caac.21492.

BRAY, F., JEMAL, A., GREY, N., FERLAY, J., FORMAN, D., 2012. Global cancer transitions according to the Human Development Index (2008–2030): a population-based study. *The Lancet Oncology* 13, 790–801. https://doi.org/10.1016/S1470-2045(12)70211-5.

BRAY, F., LAVERSANNE, M., SUNG, H., FERLAY, J., SIEGEL, R.L., SOERJOMATARAM, I., JEMAL, A., 2024. Global cancer statistics 2022: GLOBOCAN estimates of incidence and mortality worldwide for 36 cancers in 185 countries. *CA: A Cancer Journal for Clinicians* 74, 229–263. https://doi.org/10.3322/caac.21834

CHRISTENSON, E.S., JAFFEE, E., AZAD, N.S., 2020. Current and emerging therapies for patients with advanced pancreatic ductal adenocarcinoma: a bright future. *The Lancet Oncology* 21, e135–e145. https://doi.org/10.1016/S1470-2045(19)30795-8

COLLISSON, E.A., SADANANDAM, A., OLSON, P., GIBB, W.J., TRUITT, M., GU, S., COOC, J., WEINKLE, J., KIM, G.E., JAKKULA, L., FEILER, H.S., KO, A.H., OLSHEN, A.B., DANENBERG, K.L., TEMPERO, M.A., SPELLMAN, P.T., HANAHAN, D., GRAY, J.W., 2011. Subtypes of pancreatic ductal adenocarcinoma and their differing responses to therapy. *Nature Medicine* 17, 500–503. https://doi.org/10.1038/nm.2344

CORTIANA, V., ABBAS, R.H., CHORYA, H., GAMBILL, J., MAHENDRU, D., PARK, C.H., LEYFMAN, Y., 2024. Personalized medicine in pancreatic cancer: the promise of biomarkers and molecular targeting with Dr. Michael J. Pishvaian. *Cancers* 16, 2329. https://doi.org/10.3390/cancers16132329

COSTELLO, E., 2018. A metabolomics-based biomarker signature discriminates pancreatic cancer from chronic pancreatitis. *Gut* 67, 2–3. https://doi.org/10.1136/gutjnl-2016-313665

DAEMEN, A., LIU, B., SONG, K., KWONG, M., GAO, M., HONG, R., NANNINI, M., PETERSON, D., LIEDERER, B.M., CRUZ, C. DE LA, SANGARAJU, D., JAOCHICO, A., ZHAO, X., SANDOVAL, W., HUNSAKER, T., FIRESTEIN, R., LATHAM, S., SAMPATH, D., EVANGELISTA, M., HATZIVASSILIOU, G., 2018. Pan-cancer metabolic signature predicts co-dependency on glutaminase and de novo glutathione synthesis linked to a high-mesenchymal cell state. *Cell Metabolism* 28, 383–399.e9. https://doi.org/10.1016/j.cmet.2018.06.003

DAEMEN, A., PETERSON, D., SAHU, N., MCCORD, R., DU, X., LIU, B., KOWANETZ, K., HONG, R., MOFFAT, J., GAO, M., BOUDREAU, A., MROUE, R., CORSON, L., O'BRIEN, T., QING, J., SAMPATH, D., MERCHANT, M., YAUCH, R., MANNING, G., SETTLEMAN, J., HATZIVASSILIOU, G., EVANGELISTA, M., 2015. Metabolite profiling stratifies pancreatic ductal adenocarcinomas into subtypes with distinct sensitivities to metabolic inhibitors. *Proceedings of the National Academy of Sciences* 112, E4410–E4417. https://doi.org/10.1073/pnas.1501605112

DASIC, B., DEVIC, Z., DENIC, N., ZLATKOVIC, D., ILIC, I.D., CAO, Y., JERMSITTIPARSERT, K., LE, H.V., 2020. Human development index in a context of human development: review on the western Balkans countries. *Brain and Behavior* 10, e01755. https://doi.org/10.1002/brb3.1755

DAVIS, V.W., BATHE, O.F., SCHILLER, D.E., SLUPSKY, C.M., SAWYER, M.B., 2011. Metabolomics and surgical oncology: Potential role for small molecule biomarkers. *Journal of Surgical Oncology* 103, 451–459. https://doi.org/10.1002/jso.21831

DIAFERIA, G.R., BALESTRIERI, C., PROSPERINI, E., NICOLI, P., SPAGGIARI, P., ZERBI, A., NATOLI, G., 2016. Dissection of transcriptional and cis-regulatory control of differentiation in human pancreatic cancer. *The EMBO Journal* 35, 595–617. https://doi.org/10.15252/embj.201592404

DIMOU, F., SINESHAW, H., PARMAR, A.D., TAMIRISA, N.P., JEMAL, A., RIALL, T.S., 2016. Trends in receipt and timing of multimodality therapy in early-stage pancreatic cancer. *Journal of Gastrointestinal Surgery* 20, 93–103. https://doi.org/10.1007/s11605-015-2952-7

FIDLER, M.M., SOERJOMATARAM, I., BRAY, F., 2016. A global view on cancer incidence and national levels of the human development index. *International Journal of Cancer* 139, 2436–2446. https://doi.org/10.1002/ijc.30382

FIEHN, O., KRISTAL, B., OMMEN, B.V., SUMNER, L.W., SANSONE, S.-A., TAYLOR, C., HARDY, N., KADDURAH-DAOUK, R., 2006. Establishing reporting standards for metabolomic and metabonomic studies: a call for participation. *OMICS: A Journal of Integrative Biology* 10, 158–163. https://doi.org/10.1089/omi.2006.10.158

FRANK, T.S., SUN, X., ZHANG, Y., YANG, J., FISHER, W.E., GINGRAS, M.-C., LI, M., 2015. Genomic profiling guides the choice of molecular targeted therapy of pancreatic cancer. *Cancer Letters* 363, 1–6. https://doi.org/10.1016/j.canlet.2015.04.009

GOGGINS, M., 2005. Molecular markers of early pancreatic cancer. *JCO* 23, 4524–4531. https://doi.org/10.1200/JCO.2005.19.711

GOONETILLEKE, K.S., SIRIWARDENA, A.K., 2007. Systematic review of carbohydrate antigen (CA 19-9) as a biochemical marker in the diagnosis of pancreatic cancer. *European Journal of Surgical Oncology (EJSO)* 33, 266–270. https://doi.org/10.1016/j.ejso.2006.10.004

GRIFFIN, J.L., 2004. Metabolic profiles to define the genome: can we hear the phenotypes? *Philosophical Transactions of the Royal Society of London. Series B: Biological Sciences* 359, 857–871. https://doi.org/10.1098/rstb.2003.1411

GRIMM, M., HARTTGEN, K., KLASEN, S., MISSELHORN, M., 2008. A human development index by income groups. *World Development* 36, 2527–2546. https://doi.org/10.1016/j.worlddev.2007.12.001

HASHAN, M.R., HAIDER, S.S., PIAL, R.H., HOSSAIN, M.A., E-ELAHEE, M.-, GUPTA, R.D., 2021. Association between television viewing frequency and overweight/obesity among reproductive age women: cross-sectional evidence from South Africa Demographic and Health Survey 2016. *Obesity Medicine* 25, 100362. https://doi.org/10.1016/j.obmed.2021.100362

HESSMANN, E., JOHNSEN, S.A., SIVEKE, J.T., ELLENRIEDER, V., 2017. Epigenetic treatment of pancreatic cancer: is there a therapeutic perspective on the horizon? *Gut* 66, 168–179. https://doi.org/10.1136/gutjnl-2016-312539

HIDALGO, M., 2010. Pancreatic cancer. *New England Journal of Medicine* 362, 1605–1617. https://doi.org/10.1056/NEJMra0901557

HO, J.Y., ELO, I.T., 2012. The contribution of smoking to Black-White differences in U.S. mortality. *Demography* 50, 545–568. https://doi.org/10.1007/s13524-012-0159-z

HONG, D.S., FAKIH, M.G., STRICKLER, J.H., DESAI, J., DURM, G.A., SHAPIRO, G.I., FALCHOOK, G.S., PRICE, T.J., SACHER, A., DENLINGER, C.S., BANG, Y.-J., DY, G.K., KRAUSS, J.C., KUBOKI, Y., KUO, J.C., COVELER, A.L., PARK, K., KIM, T.W., BARLESI, F., MUNSTER, P.N., RAMALINGAM, S.S., BURNS, T.F., MERIC-BERNSTAM, F., HENARY, H., NGANG, J., NGARMCHAMNANRITH, G., KIM, J., HOUK, B.E., CANON, J., LIPFORD, J.R., FRIBERG, G., LITO, P., GOVINDAN, R., LI, B.T., 2020. KRASG12C inhibition with sotorasib in advanced solid tumors. *New England Journal of Medicine* 383, 1207–1217. https://doi.org/10.1056/NEJMoa1917239

HYMAN, D.M., PUZANOV, I., SUBBIAH, V., FARIS, J.E., CHAU, I., BLAY, J.-Y., WOLF, J., RAJE, N.S., DIAMOND, E.L., HOLLEBECQUE, A., GERVAIS, R., ELEZ-FERNANDEZ, M.E., ITALIANO, A., HOFHEINZ, R.-D., HIDALGO, M., CHAN, E., SCHULER, M., LASSERRE, S.F., MAKRUTZKI, M., SIRZEN, F., VERONESE, M.L., TABERNERO, J., BASELGA, J., 2015. Vemurafenib in multiple nonmelanoma cancers with BRAF V600 mutations. *New England Journal of Medicine* 373, 726–736. https://doi.org/10.1056/NEJMoa1502309

JAIN, M., NILSSON, R., SHARMA, S., MADHUSUDHAN, N., KITAMI, T., SOUZA, A.L., KAFRI, R., KIRSCHNER, M.W., CLISH, C.B., MOOTHA, V.K., 2012. Metabolite profiling identifies a key role for glycine in rapid cancer cell proliferation. *Science* 336, 1040–1044. https://doi.org/10.1126/science.1218595

JOHANSEN, D., STOCKS, T., JONSSON, H., LINDKVIST, B., BJÖRGE, T., CONCIN, H., ALMQUIST, M., HÄGGSTRÖM, C., ENGELAND, A., ULMER, H., HALLMANS, G., SELMER, R., NAGEL, G., TRETLI, S., STATTIN, P., MANJER, J., 2010. Metabolic factors and the risk of pancreatic cancer: a prospective analysis of almost 580,000 men and women in the metabolic syndrome and cancer project. *Cancer Epidemiology, Biomarkers & Prevention* 19, 2307–2317. https://doi.org/10.1158/1055-9965.EPI-10-0234

KAUR, S., BAINE, M.J., JAIN, M., SASSON, A.R., BATRA, S.K., 2012. Early diagnosis of pancreatic cancer: challenges and new developments. *Biomarkers in Medicine* 6, 597–612. https://doi.org/10.2217/bmm.12.69

KIM D. W., SCHRAM, A. M., HOLLEBECQUE, A., NISHINO, K., MACARULLA, T., RHA, S. Y., DURUISSEAUX, M., LIU, S. V., AL HALLAK, M. N., UMEMOTO, K., WESSELER, C., CLEARY, J. M., SPRINGFELD, C., NEUZILLET, C., JOE, A., JAUHARI, S., FORD, J., & GOTO, K. (2024). The phase I/II eNRGy trial: Zenocutuzumab in patients with cancers harboring NRG1 gene fusions. *Future Oncology* 20(16), 1057–1067. https://doi.org/10.2217/fon-2023-0824

KLEIN, S., GASTALDELLI, A., YKI-JÄRVINEN, H., SCHERER, P.E., 2022. Why does obesity cause diabetes? *Cell Metabolism* 34, 11–20. https://doi.org/10.1016/j.cmet.2021.12.012

KLOMP, J.E., DIEHL, J.N., KLOMP, J.A., EDWARDS, A.C., YANG, R., MORALES, A.J., TAYLOR, K.E., DRIZYTE-MILLER, K., BRYANT, K.L., SCHAEFER, A., JOHNSON, J.L., HUNTSMAN, E.M., YARON, T.M., PIEROBON, M., BALDELLI, E., PREVATTE, A.W., BARKER, N.K., HERRING, L.E., PETRICOIN, E.F., GRAVES, L.M., CANTLEY, L.C., COX, A.D., DER, C.J., STALNECKER, C.A., 2024. Determining the ERK-regulated phosphoproteome driving KRAS-mutant cancer. *Science* 384, eadk0850. https://doi.org/10.1126/science.adk0850

KOEN, N., DU PREEZ, I., LOOTS, D.T., 2016. Chapter three – metabolomics and personalized medicine. In: Donev, R. (ed.) *Advances in protein chemistry and structural biology, personalized medicine*. Academic Press, pp. 53–78. https://doi.org/10.1016/bs.apcsb.2015.09.003

KUNOVSKY, L., TESARIKOVA, P., KALA, Z., KROUPA, R., KYSELA, P., DOLINA, J., TRNA, J., 2018. The use of biomarkers in early diagnostics of pancreatic cancer. *Canadian Journal of Gastroenterology and Hepatology* 2018, 5389820. https://doi.org/10.1155/2018/5389820

LAW, H.C.-H., LAGUNDŽIN, D., CLEMENT, E.J., QIAO, F., WAGNER, Z.S., KRIEGER, K.L., COSTANZO-GARVEY, D., CAFFREY, T.C., GREM, J.L., DIMAIO, D.J., GRANDGENETT, P.M., COOK, L.M., FISHER, K.W., YU, F., HOLLINGSWORTH, M.A., WOODS, N.T., 2020. The proteomic landscape of pancreatic ductal adenocarcinoma liver metastases identifies molecular subtypes and associations with clinical response. *Clinical Cancer Research* 26, 1065–1076. https://doi.org/10.1158/1078-0432.CCR-19-1496

LE, D.T., URAM, J.N., WANG, H., BARTLETT, B.R., KEMBERLING, H., EYRING, A.D., SKORA, A.D., LUBER, B.S., AZAD, N.S., LAHERU, D., BIEDRZYCKI, B., DONEHOWER, R.C., ZAHEER, A., FISHER, G.A., CROCENZI, T.S., LEE, J.J., DUFFY, S.M., GOLDBERG, R.M., CHAPELLE, A. DE LA, KOSHIJI, M., BHAIJEE, F., HUEBNER, T., HRUBAN, R.H., WOOD, L.D., CUKA, N., PARDOLL, D.M., PAPADOPOULOS, N., KINZLER, K.W., ZHOU, S., CORNISH, T.C., TAUBE,

J.M., ANDERS, R.A., ESHLEMAN, J.R., VOGELSTEIN, B., DIAZ, L.A., 2015. PD-1 blockade in tumors with mismatch-repair deficiency. *New England Journal of Medicine* 372, 2509–2520. https://doi.org/10.1056/NEJMoa1500596

LI, D., ABBRUZZESE, J.L., 2010. New strategies in pancreatic cancer: emerging epidemiologic and therapeutic concepts. *Clinical Cancer Research* 16, 4313–4318. https://doi.org/10.1158/1078-0432.CCR-09-1942

LI, J., SHEN, L., GU, Y., CALLES, A., WU, L., BA, Y., LI, Z., BAI, C., YAO, Y., HUBERT, A., PEREZ, J.M., ZUGAZAGOITIA, J., DING, Y., LIU, S., RAO, Z., HE, Z., 2024. Preliminary activity and safety results of KRAS G12C inhibitor glecirasib (JAB-21822) in patients with pancreatic cancer and other solid tumors. *Journal of Clinical Oncology* 42, 604–604. https://doi.org/10.1200/JCO.2024.42.3_suppl.604

LIU, J., MROCZEK, M., MACH, A., STĘPIEŃ, M., APLAS, A., PRONOBIS-SZCZYLIK, B., BUKOWSKI, S., MIELCZAREK, M., GAJEWSKA, E., TOPOLSKI, P., KRÓL, Z.J., SZYDA, J., DOBOSZ, P., 2023. Genetics, genomics and emerging molecular therapies of pancreatic cancer. *Cancers* 15, 779. https://doi.org/10.3390/cancers15030779

LOCKER, G., HAMILTON, S., HARRIS, J., JESSUP, J., KEMENY, N., MACDONALD, J., SOMERFIELD, M., HAYES, D., BAST, R., 2006. ASCO 2006 update of recommendations for the use of tumor markers in gastrointestinal cancer. *Journal of Clinical Oncology* 24, 5313–5327. https://doi.org/10.1200/JCO.2006.08.2644

LOMBERK, G., BLUM, Y., NICOLLE, R., NAIR, A., GAONKAR, K.S., MARISA, L., MATHISON, A., SUN, Z., YAN, H., ELAROUCI, N., ARMENOULT, L., AYADI, M., ORDOG, T., LEE, J.-H., OLIVER, G., KLEE, E., MOUTARDIER, V., GAYET, O., BIAN, B., DUCONSEIL, P., GILABERT, M., BIGONNET, M., GARCIA, S., TURRINI, O., DELPERO, J.-R., GIOVANNINI, M., GRANDVAL, P., GASMI, M., SECQ, V., DE REYNIÈS, A., DUSETTI, N., IOVANNA, J., URRUTIA, R., 2018. Distinct epigenetic landscapes underlie the pathobiology of pancreatic cancer subtypes. *Nature Communications* 9, 1978. https://doi.org/10.1038/s41467-018-04383-6

LOWENFELS, A.B., MAISONNEUVE, P., 2006. Epidemiology and risk factors for pancreatic cancer. *Best Practice & Research Clinical Gastroenterology, Pancreatic Cancer* 20, 197–209. https://doi.org/10.1016/j.bpg.2005.10.001

LYNCH, S.M., VRIELING, A., LUBIN, J.H., KRAFT, P., MENDELSOHN, J.B., HARTGE, P., CANZIAN, F., STEPLOWSKI, E., ARSLAN, A.A., GROSS, M., HELZLSOUER, K., JACOBS, E.J., LACROIX, A., PETERSEN, G., ZHENG, W., ALBANES, D., AMUNDADOTTIR, L., BINGHAM, S.A., BOFFETTA, P., BOUTRON-RUAULT, M.-C., CHANOCK, S.J., CLIPP, S., HOOVER, R.N., JACOBS, K., JOHNSON, K.C., KOOPERBERG, C., LUO, J., MESSINA, C., PALLI, D., PATEL, A.V., RIBOLI, E., SHU, X.-O., RODRIGUEZ SUAREZ, L., THOMAS, G., TJØNNELAND, A., TOBIAS, G.S., TONG, E., TRICHOPOULOS, D., VIRTAMO, J., YE, W., YU, K., ZELENIUCH-JACQUETTE, A., BUENO-DE-MESQUITA, H.B., STOLZENBERG-SOLOMON, R.Z., 2009. Cigarette smoking and pancreatic cancer: a pooled analysis from the Pancreatic Cancer Cohort Consortium. *American Journal of Epidemiology* 170, 403–413. https://doi.org/10.1093/aje/kwp134

MAGI, L., MARASCO, M., RINZIVILLO, M., FAGGIANO, A., PANZUTO, F., 2023. Management of functional pancreatic neuroendocrine neoplasms. *Current Treatment Options in Oncology* 24, 725–741. https://doi.org/10.1007/s11864-023-01085-0

MAHAJAN, U.M., LANGHOFF, E., GONI, E., COSTELLO, E., GREENHALF, W., HALLORAN, C., ORMANNS, S., KRUGER, S., BOECK, S., RIBBACK, S., BEYER, G., DOMBROSWKI, F., WEISS, F.-U., NEOPTOLEMOS, J.P., WERNER, J., D'HAESE, J.G., BAZHIN, A., PETERHANSL, J., PICHLMEIER, S., BÜCHLER, M.W., KLEEFF, J., GANEH, P., SENDLER, M., PALMER, D.H., KOHLMANN, T., RAD, R., REGEL, I., LERCH, M.M., MAYERLE, J., 2018. Immune cell and stromal signature associated with progression-free survival of patients with resected pancreatic ductal adenocarcinoma. *Gastroenterology* 155, 1625–1639.e2. https://doi.org/10.1053/j.gastro.2018.08.009

MANGAT, P. K., HALABI, S., BRUINOOGE, S. S., GARRETT-MAYER, E., ALVA, A., JANEWAY, K. A., STELLA, P. J., VOEST, E., YOST, K. J., PERLMUTTER, J., PINTO, N., KIM, E. S., & SCHILSKY, R. L. 2018. Rationale and design of the Targeted Agent and Profiling Utilization Registry (TAPUR) Study. *JCO Precision Oncology* 2018. https://doi.org/10.1200/PO.18.00122

MAPA-TASSOU, C., KATTE, J.-C., MBA MAADJHOU, C., MBANYA, J.C., 2019. Economic impact of diabetes in Africa. *Current Diabetes Reports* 19, 5. https://doi.org/10.1007/s11892-019-1124-7

MARABELLE, A., LE, D.T., ASCIERTO, P.A., DI GIACOMO, A.M., DE JESUS-ACOSTA, A., DELORD, J.-P., GEVA, R., GOTTFRIED, M., PENEL, N., HANSEN, A.R., PIHA-PAUL, S.A., DOI, T., GAO, B., CHUNG, H.C., LOPEZ-MARTIN, J., BANG, Y.-J., FROMMER, R.S., SHAH, M., GHORI, R.,

JOE, A.K., PRUITT, S.K., DIAZ JR, L.A., 2020. Efficacy of pembrolizumab in patients with non-colorectal high microsatellite instability/mismatch repair–deficient cancer: results from the Phase II KEYNOTE-158 Study. *Journal of Clinical Oncology* 38, 1–10. https://doi.org/10.1200/JCO.19.02105

MASSARD, C., MICHIELS, S., FERTÉ, C., LE DELEY, M.-C., LACROIX, L., HOLLEBECQUE, A., VERLINGUE, L., ILEANA, E., ROSELLINI, S., AMMARI, S., NGO-CAMUS, M., BAHLEDA, R., GAZZAH, A., VARGA, A., POSTEL-VINAY, S., LORIOT, Y., EVEN, C., BREUSKIN, I., AUGER, N., JOB, B., DE BAERE, T., DESCHAMPS, F., VIELH, P., SCOAZEC, J.-Y., LAZAR, V., RICHON, C., RIBRAG, V., DEUTSCH, E., ANGEVIN, E., VASSAL, G., EGGERMONT, A., ANDRÉ, F., SORIA, J.-C., 2017. High-throughput genomics and clinical outcome in hard-to-treat advanced cancers: results of the MOSCATO 01 Trial. *Cancer Discovery* 7, 586–595. https://doi.org/10.1158/2159-8290.CD-16-1396

MAYERLE, J., KALTHOFF, H., RESZKA, R., KAMLAGE, B., PETER, E., SCHNIEWIND, B., MALDONADO, S.G., PILARSKY, C., HEIDECKE, C.-D., SCHATZ, P., DISTLER, M., SCHEIBER, J.A., MAHAJAN, U.M., WEISS, F.U., GRÜTZMANN, R., LERCH, M.M., 2018. Metabolic biomarker signature to differentiate pancreatic ductal adenocarcinoma from chronic pancreatitis. *Gut* 67, 128–137. https://doi.org/10.1136/gutjnl-2016-312432

MIDHA, S., CHAWLA, S., GARG, P.K., 2016. Modifiable and non-modifiable risk factors for pancreatic cancer: a review. *Cancer Letters* 381, 269–277. https://doi.org/10.1016/j.canlet.2016.07.022

MOFFITT, R.A., MARAYATI, R., FLATE, E.L., VOLMAR, K.E., LOEZA, S.G.H., HOADLEY, K.A., RASHID, N.U., WILLIAMS, L.A., EATON, S.C., CHUNG, A.H., SMYLA, J.K., ANDERSON, J.M., KIM, H.J., BENTREM, D.J., TALAMONTI, M.S., IACOBUZIO-DONAHUE, C.A., HOLLINGSWORTH, M.A., YEH, J.J., 2015. Virtual microdissection identifies distinct tumor- and stroma-specific subtypes of pancreatic ductal adenocarcinoma. *Nature Genetics* 47, 1168–1178. https://doi.org/10.1038/ng.3398

MOISE, I.K., KANGMENNAANG, J., HALWIINDI, H., GRIGSBY-TOUSSAINT, D.S., FULLER, D.O., 2019. Increase in obesity among women of reproductive age in Zambia, 2002–2014. *Journal of Women's Health* 28, 1679–1687. https://doi.org/10.1089/jwh.2018.7577

MONTEIRO, M.S., CARVALHO, M., BASTOS, M.L., GUEDES DE PINHO, P., 2013. Metabolomics analysis for biomarker discovery: advances and challenges. *Current Medicinal Chemistry* 20, 257–271.

MOTA REYES, C., DOĞRUÖZ, A., ISTVANFFY, R., FRIESS, H., CEYHAN, G.O., DEMIR, I.E., 2021. Molecular profiling in pancreatic cancer: current role and its impact on primary surgery. *Visceral Medicine* 38, 37–41. https://doi.org/10.1159/000519755

MOTALA, A.A., MBANYA, J.C., RAMAIYA, K., PIRIE, F.J., EKORU, K., 2022. Type 2 diabetes mellitus in sub-Saharan Africa: challenges and opportunities. *Nature Reviews Endocrinology* 18, 219–229. https://doi.org/10.1038/s41574-021-00613-y

NIPP, R., TRAMONTANO, A.C., KONG, C.Y., PANDHARIPANDE, P., DOWLING, E.C., SCHRAG, D., HUR, C., 2018. Disparities in cancer outcomes across age, sex, and race/ethnicity among patients with pancreatic cancer. *Cancer Medicine* 7, 525–535. https://doi.org/10.1002/cam4.1277

O'DWYER, P.J., GRAY, R.J., FLAHERTY, K.T., CHEN, A.P., LI, S., WANG, V., MCSHANE, L.M., PATTON, D.R., TRICOLI, J.V., WILLIAMS, P.M., IAFRATE, A.J., SKLAR, J., MITCHELL, E.P., TAKEBE, N., SIMS, D.J., COFFEY, B., FU, T., ROUTBORT, M., RUBINSTEIN, L.V., ... DOROSHOW, J.H. 2023. The NCI-MATCH trial: lessons for precision oncology. *Nature Medicine* 29(6), 1349–1357. https://doi.org/10.1038/s41591-023-02379-4

PEER, N. 2018. Current strategies are inadequate to curb the rise of tobacco use in Africa. *South African Medical Journal,* June 26, 108(7), 551–556 10.7196/SAMJ.2018.v108i7.12978

PHUA, L.C., GOH, S., TAI, D.W.M., LEOW, W.Q., ALKAFF, S.M.F., CHAN, C.Y., KAM, J.H., LIM, T.K.H., CHAN, E.C.Y., 2018. Metabolomic prediction of treatment outcome in pancreatic ductal adenocarcinoma patients receiving gemcitabine. *Cancer Chemotherapy and Pharmacology* 81, 277–289. https://doi.org/10.1007/s00280-017-3475-6

PISHVAIAN, M.J., BRODY, J.R., 2022. Molecular profiling and precision medicine for pancreatic cancer. In: Bhutani, M.S., Katz, M.H.G., Maitra, A., Herman, J.M., Wolff, R.A. (eds.) *Pancreatic cancer: a multidisciplinary approach.* Cham: Springer International Publishing, pp. 255–267. https://doi.org/10.1007/978-3-031-05724-3_21

POWERS, B.D., FULP, W., DHAHRI, A., DEPERALTA, D.K., OGAMI, T., ROTHERMEL, L., PERMUTH, J.B., VADAPARAMPIL, S.T., KIM, J.-K., PIMIENTO, J., HODUL, P.J., MALAFA, M.P., ANAYA, D.A., FLEMING, J.B., 2021. The impact of socioeconomic deprivation on clinical outcomes for pancreatic adenocarcinoma at a high-volume cancer center: a retrospective cohort analysis. *Annals of Surgery* 274, e564. https://doi.org/10.1097/SLA.0000000000003706

RAHIB, L., SMITH, B.D., AIZENBERG, R., ROSENZWEIG, A.B., FLESHMAN, J.M., MATRISIAN, L.M., 2014. Projecting cancer incidence and deaths to 2030: the unexpected burden of thyroid, liver, and pancreas cancers in the United States. *Cancer Research* 74, 2913–2921. https://doi.org/10.1158/0008-5472.CAN-14-0155

RAIMONDI, S., MAISONNEUVE, P., LOWENFELS, A.B., 2009. Epidemiology of pancreatic cancer: an overview. *Nature Reviews Gastroenterology & Hepatology* 6, 699–708. https://doi.org/10.1038/nrgastro.2009.177

RAMÍREZ-MALDONADO, E., LÓPEZ GORDO, S., MAJOR BRANCO, R.P., PAVEL, M.-C., ESTALELLA, L., LLÀCER-MILLÁN, E., GUERRERO, M.A., LÓPEZ-GORDO, E., MEMBA, R., JORBA, R., 2024. Clinical application of liquid biopsy in pancreatic cancer: a narrative review. *International Journal of Molecular Sciences* 25, 1640. https://doi.org/10.3390/ijms25031640

RANDI, G., FRANCESCHI, S., LA VECCHIA, C., 2006. Gallbladder cancer worldwide: geographical distribution and risk factors. *International Journal of Cancer* 118, 1591–1602. https://doi.org/10.1002/ijc.21683

REGEL, I., MAYERLE, J., UJJWAL MUKUND, M., 2020. Current strategies and future perspectives for precision medicine in pancreatic cancer. *Cancers* 12, 1024. https://doi.org/10.3390/cancers12041024

RINGDÉN, O., BOUMENDIL, A., LABOPIN, M., CANAANI, J., BEELEN, D., EHNINGER, G., NIEDERWIESER, D., FINKE, J., STELLJES, M., GERBITZ, A., GANSER, A., KRÖGER, N., KANTZ, L., BRECHT, A., SAVANI, B., SADEGHI, B., MOHTY, M., NAGLER, A., 2019. Outcome of allogeneic hematopoietic stem cell transplantation in patients age >69 years with acute myelogenous leukemia: on behalf of the acute leukemia working party of the European Society for Blood and Marrow Transplantation. *Biology of Blood and Marrow Transplantation* 25, 1975–1983. HTTPS://DOI.ORG/10.1016/J.BBMT.2019.05.037

RISCH, H.A., YU, H., LU, L., KIDD, M.S., 2010. ABO blood group, helicobacter pylori seropositivity, and risk of pancreatic cancer: a case–control study. *JNCI: Journal of the National Cancer Institute* 102, 502–505. https://doi.org/10.1093/jnci/djq007

SACHER, A., LORUSSO, P., PATEL, M.R., MILLER, W.H., GARRALDA, E., FORSTER, M.D., SANTORO, A., FALCON, A., KIM, T.W., PAZ-ARES, L., BOWYER, S., MIGUEL, M. DE, HAN, S.-W., KREBS, M.G., LEE, J.-S., CHENG, M.L., ARBOUR, K., MASSARELLI, E., CHOI, Y., SHI, Z., MANDLEKAR, S., LIN, M.T., ROYER-JOO, S., CHANG, J., DHARIA, N.V., SCHUTZMAN, J.L., DESAI, J., 2023. Single-agent divarasib (GDC-6036) in solid tumors with a KRAS G12C mutation. *New England Journal of Medicine* 389, 710–721. https://doi.org/10.1056/NEJMoa2303810

SAEEDI, P., PETERSOHN, I., SALPEA, P., MALANDA, B., KARURANGA, S., UNWIN, N., COLAGIURI, S., GUARIGUATA, L., MOTALA, A.A., OGURTSOVA, K., SHAW, J.E., BRIGHT, D., WILLIAMS, R., 2019. Global and regional diabetes prevalence estimates for 2019 and projections for 2030 and 2045: results from the International Diabetes Federation Diabetes Atlas, 9th edition. *Diabetes Research and Clinical Practice* 157, 107843. Https://doi.org/10.1016/j.diabres.2019.107843

SAMAAN, J.S., ABBOUD, Y., OH, J., JIANG, Y., WATSON, R., PARK, K., LIU, Q., ATKINS, K., HENDIFAR, A., GONG, J., OSIPOV, A., LI, D., NISSEN, N.N., PANDOL, S.J., LO, S.K., GADDAM, S., 2023. Pancreatic cancer incidence trends by race, ethnicity, age and sex in the United States: a population-based study, 2000–2018. *Cancers* 15, 870. https://doi.org/10.3390/cancers15030870

SCARTON, L., YOON, S., OH, S., AGYARE, E., TREVINO, J., HAN, B., LEE, E., SETIAWAN, V.W., PERMUTH, J.B., SCHMITTGEN, T.D., ODEDINA, F.G., WILKIE, D.J., 2018. Pancreatic cancer related health disparities: a commentary. *Cancers* 10, 235. https://doi.org/10.3390/cancers10070235

SCHRAM, A.M., ODINTSOV, I., ESPINOSA-COTTON, M., KHODOS, I., SISSO, W.J., MATTAR, M.S., LUI, A.J.W., VOJNIC, M., SHAMEEM, S.H., CHAUHAN, T., TORRISI, J., FORD, J., O'CONNOR, M.N., GEUIJEN, C.A.W., SCHACKMANN, R.C.J., LAMMERTS VAN BUEREN, J.J., WASSERMAN, E., DE STANCHINA, E., O'REILLY, E.M., LADANYI, M., DRILON, A., SOMWAR, R., 2022. Zenocutuzumab, a HER2xHER3 bispecific antibody, is effective therapy for tumors driven by NRG1 gene rearrangements. *Cancer Discovery* 12, 1233–1247. https://doi.org/10.1158/2159-8290.CD-21-1119

SCHUETTENGRUBER, B., MARTINEZ, A.-M., IOVINO, N., CAVALLI, G., 2011. Trithorax group proteins: switching genes on and keeping them active. *Nature Reviews Molecular Cell Biology* 12, 799–814. https://doi.org/10.1038/nrm3230

SCHWAEDERLE, M., ZHAO, M., LEE, J.J., EGGERMONT, A.M., SCHILSKY, R.L., MENDELSOHN, J., LAZAR, V., KURZROCK, R., 2015. Impact of precision medicine in diverse cancers: a meta-analysis of phase II clinical trials. *Journal of Clinical Oncology* 33, 3817–3825. https://doi.org/10.1200/JCO.2015.61.5997

SEIDU, A.-A., AHINKORAH, B.O., AGBAGLO, E., NYAABA, A.A., 2021. Overweight and obesity among women of reproductive age in Mali: what are the determinants? *International Health* 13, 428–435. https://doi.org/10.1093/inthealth/ihaa094

SHAYA, J., KATO, S., ADASHEK, J.J., PATEL, H., FANTA, P.T., BOTTA, G.P., SICKLICK, J.K., KURZROCK, R., 2023. Personalized matched targeted therapy in advanced pancreatic cancer: a pilot cohort analysis. *npj Genomic Medicine* 8, 1–8. https://doi.org/10.1038/s41525-022-00346-5

SICKLICK, J.K., KATO, S., OKAMURA, R., SCHWAEDERLE, M., HAHN, M.E., WILLIAMS, C.B., DE, P., KRIE, A., PICCIONI, D.E., MILLER, V.A., ROSS, J.S., BENSON, A., WEBSTER, J., STEPHENS, P.J., LEE, J.J., FANTA, P.T., LIPPMAN, S.M., LEYLAND-JONES, B., KURZROCK, R., 2019. Molecular profiling of cancer patients enables personalized combination therapy: the I-PREDICT study. *Nature Medicine* 25, 744–750. https://doi.org/10.1038/s41591-019-0407-5

SINGH, G.K., WILLIAMS, S.D., SIAHPUSH, M., MULHOLLEN, A., 2011. Socioeconomic, rural-urban, and racial inequalities in US cancer mortality: part I—all cancers and lung cancer and part II—colorectal, prostate, breast, and cervical cancers. *Journal of Cancer Epidemiology* 2011, 107497. https://doi.org/10.1155/2011/107497

SKOULIDIS, F., LI, B.T., DY, G.K., PRICE, T.J., FALCHOOK, G.S., WOLF, J., ITALIANO, A., SCHULER, M., BORGHAEI, H., BARLESI, F., KATO, T., CURIONI-FONTECEDRO, A., SACHER, A., SPIRA, A., RAMALINGAM, S.S., TAKAHASHI, T., BESSE, B., ANDERSON, A., ANG, A., TRAN, Q., MATHER, O., HENARY, H., NGARMCHAMNANRITH, G., FRIBERG, G., VELCHETI, V., GOVINDAN, R., 2021. Sotorasib for lung cancers with KRAS p.G12C mutation. *New England Journal of Medicine* 384, 2371–2381. https://doi.org/10.1056/NEJMoa2103695

STOCKLEY, T.L., OZA, A.M., BERMAN, H.K., LEIGHL, N.B., KNOX, J.J., SHEPHERD, F.A., CHEN, E.X., KRZYZANOWSKA, M.K., DHANI, N., JOSHUA, A.M., TSAO, M.S., SERRA, S., CLARKE, B., ROEHRL, M.H., ZHANG, T., SUKHAI, M.A., CALIFARETTI, N., TRINKAUS, M., SHAW, P., . . . BEDARD, P. L. 2016. Molecular profiling of advanced solid tumors and patient outcomes with genotype-matched clinical trials: the Princess Margaret IMPACT/COMPACT trial. *Genome Medicine* 8(1), 109. https://doi.org/10.1186/s13073-016-0364-2

STORZ, P., CRAWFORD, H.C., 2020. Carcinogenesis of pancreatic ductal adenocarcinoma. *Gastroenterology* 158, 2072–2081. https://doi.org/10.1053/j.gastro.2020.02.059

SUN, HUANHUAN, MA, H., HONG, G., SUN, HONGLIU, WANG, J., 2014. Survival improvement in patients with pancreatic cancer by decade: a period analysis of the SEER database, 1981–2010. *Scientific Reports* 4, 6747. https://doi.org/10.1038/srep06747

SUN, X., WANG, S., WONG, C.C.L., 2024. Mass spectrometry–based proteomics technology in pancreatic cancer research. *Journal of Pancreatology* 7, 145. https://doi.org/10.1097/JP9.0000000000000152

SUNG, H., FERLAY, J., SIEGEL, R.L., LAVERSANNE, M., SOERJOMATARAM, I., JEMAL, A., BRAY, F., 2021. Global cancer statistics 2020: GLOBOCAN estimates of incidence and mortality worldwide for 36 cancers in 185 countries. *CA: A Cancer Journal for Clinicians* 71, 209–249. https://doi.org/10.3322/caac.21660

TSIMBERIDOU, A.M., FOUNTZILAS, E., NIKANJAM, M., KURZROCK, R., 2020. Review of precision cancer medicine: evolution of the treatment paradigm. *Cancer Treatment Reviews* 86, 102019. https://doi.org/10.1016/j.ctrv.2020.102019

TSIMBERIDOU, A.-M., HONG, D.S., WHELER, J.J., FALCHOOK, G.S., JANKU, F., NAING, A., FU, S., PIHA-PAUL, S., CARTWRIGHT, C., BROADDUS, R.R., NOGUERAS GONZALEZ, G.M., HWU, P., KURZROCK, R., 2019. Long-term overall survival and prognostic score predicting survival: the IMPACT study in precision medicine. *Journal of Hematology & Oncology* 12, 145. https://doi.org/10.1186/s13045-019-0835-1

TSIMBERIDOU, A.-M., HONG, D.S., YE, Y., CARTWRIGHT, C., WHELER, J.J., FALCHOOK, G.S., NAING, A., FU, S., PIHA-PAUL, S., JANKU, F., MERIC-BERNSTAM, F., HWU, P., KEE, B., KIES, M.S., BROADDUS, R., MENDELSOHN, J., HESS, K.R., KURZROCK, R., 2017. Initiative for Molecular Profiling and Advanced Cancer Therapy (IMPACT): an MD Anderson Precision Medicine Study. *JCO Precision Oncology,* 1–18. https://doi.org/10.1200/PO.17.00002

ULANET, D.B., LUDWIG, D.L., KAHN, C.R., HANAHAN, D., 2010. Insulin receptor functionally enhances multistage tumor progression and conveys intrinsic resistance to IGF-1R targeted therapy. *Proceedings of the National Academy of Sciences* 107, 10791–10798. https://doi.org/10.1073/pnas.0914076107

VELLIOS, N., ROSS, H., PERUCIC, A.-M., 2018. Trends in cigarette demand and supply in Africa. *PLoS ONE* 13, e0202467. https://doi.org/10.1371/journal.pone.0202467

VITORAKIS, N., GARGALIONIS, A.N., PAPAVASSILIOU, K.A., ADAMOPOULOS, C., PAPAVASSILIOU, A.G., 2024. Precision targeting strategies in pancreatic cancer: the role of tumor microenvironment. *Cancers* 16, 2876. https://doi.org/10.3390/cancers16162876

VON HOFF, D.D., STEPHENSON, J.J., ROSEN, P., LOESCH, D.M., BORAD, M.J., ANTHONY, S., JAMESON, G., BROWN, S., CANTAFIO, N., RICHARDS, D.A., FITCH, T.R., WASSERMAN, E., FERNANDEZ, C., GREEN, S., SUTHERLAND, W., BITTNER, M., ALARCON, A., MALLERY, D., PENNY, R., 2010. Pilot study using molecular profiling of patients' tumors to find potential targets and select treatments for their refractory cancers. *Journal of Clinical Oncology* 28, 4877–4883. https://doi.org/10.1200/JCO.2009.26.5983

WANG, Y.-T., GOU, Y.-W., JIN, W.-W., XIAO, M., FANG, H.-Y., 2016. Association between alcohol intake and the risk of pancreatic cancer: a dose–response meta-analysis of cohort studies. *BMC Cancer* 16, 212. https://doi.org/10.1186/s12885-016-2241-1

WARTENBERG, M., CIBIN, S., ZLOBEC, I., VASSELLA, E., EPPENBERGER-CASTORI, S., TERRACCIANO, L., EICHMANN, M.D., WORNI, M., GLOOR, B., PERREN, A., KARAMITOPOULOU, E., 2018. Integrated genomic and immunophenotypic classification of pancreatic cancer reveals three distinct subtypes with prognostic/predictive significance. *Clinical Cancer Research* 24, 4444–4454. https://doi.org/10.1158/1078-0432.CCR-17-3401

YAMAI, T., IKEZAWA, K., SUGIMOTO, N., URABE, M., KAI, Y., TAKADA, R., NAKABORI, T., UEHARA, H., KAWAMURA, T., KUNIMASA, K., YAMAMOTO, S., WAKAMATSU, T., HAYASHI, T., KUKITA, Y., FUJISAWA, F., INOUE, T., YAMAGUCHI, Y., YAMASAKI, T., HONMA, K., OHKAWA, K., 2023. Utility of comprehensive genomic profiling tests for patients with incurable pancreatic cancer in clinical practice. *Cancers* 15, 970. https://doi.org/10.3390/cancers15030970

YUAN, C., MORALES-OYARVIDE, V., BABIC, A., CLISH, C.B., KRAFT, P., BAO, Y., QIAN, Z.R., RUBINSON, D.A., NG, K., GIOVANNUCCI, E.L., OGINO, S., STAMPFER, M.J., GAZIANO, J.M., SESSO, H.D., COCHRANE, B.B., MANSON, J.E., FUCHS, C.S., WOLPIN, B.M., 2017. Cigarette smoking and pancreatic cancer survival. *Journal of Clinical Oncology* 35, 1822–1828. https://doi.org/10.1200/JCO.2016.71.2026

ZARKAVELIS, G., KOTOULA, V., KOLLIOU, G.-A., PAPADOPOULOU, K., TIKAS, I., KARAVASILIS, V., SAMANTAS, E., DERVENIS, C., EFSTRATIOU, I., NICOLAOU, I., APESSOU, D., KAFIRI, G., KOLETSA, T., BOMPOLAKI, I., RALLIS, G., BATISTATOU, A., GLANTZOUNIS, G., PECTASIDES, D., FOUNTZILAS, G., PENTHEROUDAKIS, G., 2019. Genetic mapping of pancreatic cancer by targeted next-generation sequencing in a cohort of patients managed with nab-paclitaxel-based chemotherapy or agents targeting the EGFR axis: a retrospective analysis of the Hellenic Cooperative Oncology Group (HeCOG). *ESMO Open* 4, e000525. https://doi.org/10.1136/esmoopen-2019-000525

ZHENG, W., LEE, S.-A., 2009. Well-done meat intake, heterocyclic amine exposure, and cancer risk. *Nutrition and Cancer* 61, 437–446. https://doi.org/10.1080/01635580802710741

5 Diagnosis and Early Detection Strategies of Pancreatic Cancer

Sikhumbuzo Z. Mbatha[1,2*], *Onyemaechi O. Azu*[3], *Innocent A. Edagha*[4], *Laura Grech*[5], *Botle Damane*[1], *and Zodwa Dlamini*[2]

[1]Department of Surgery, Steve Biko Academic Hospital, Faculty of Health Sciences, University of Pretoria, Pretoria, South Africa.

[2]SAMRC Precision Oncology Research Unit (PORU), DSI/NRF SARChI Chair in Precision Oncology and Cancer Prevention (POCP), Pan African Cancer Research Institute (PACRI), University of Pretoria, Pretoria, South Africa.

[3]Department of Medical Biosciences, University of the Western Cape, Cape Town, South Africa.

[4]Department of Human Anatomy, School of Medicine, Moi University, Eldoret, Kenya.

[5]Department of Applied Biomedical Science, Faculty of Health Sciences Centre for Molecular Medicine and Biobanking, Room 331, Biomedical Science Building, University of Malta, Malta.

*Corresponding author - Sikhumbuzo Z. Mbatha

INTRODUCTION

Cancer of the pancreas can originate from the exocrine or endocrine pancreas and is made up of multiple histological subtypes. Cancers arising from the exocrine portion account for more than 95% of all pancreatic cancers, with pancreatic ductal adenocarcinoma (PDAC) responsible for around 85–90% of all pancreatic malignancies. As a result, pancreatic cancer has virtually become synonymous with PDAC in discussions and publications (Luo et al., 2019; Pokrzywa et al., 2019). The are other rare subtypes of pancreatic cancer derived from the ductal system, including adenosquamous carcinoma, osteoclastic giant cell carcinoma, colloid carcinoma, medullary carcinoma, and non-ductal variants such as pancreatic neuroendocrine tumors, acinar cell carcinomas, pancreatoblastomas, and solid pseudopapillary neoplasm (Mostafa et al., 2017). Table 5.1 provides a list of histological subtypes of pancreatic tumors. The majority of pancreatic cancers develop from small intraductal precursor lesions called pancreatic intraepithelial neoplasms (PanIN), which are further subdivided into low-grade lesions with minimal dysplasia and high-grade lesions with high-grade dysplasia. The PanINs are typically microscopic lesions that lack detectable clinical or radiological signs. Other precursor lesions are radiologically detectable and include intraductal papillary mucinous neoplasms (IPMNs) and mucinous cystic neoplasms (MCNs) (Kenner et al., 2021). Pancreatic cancer is a lethal malignancy associated with dismal outcomes, with a yearly death rate that closely resembles the annual incidence rate; in North America, it ranks as the fourth most frequent cause of cancer-related mortality (Al-Sukhni et al., 2012). Pancreatic cancer is a disease seen more often in advanced age and is associated with high mortality, with the risk of death increasing dramatically with age. For example, in North America, the mortality risk from pancreatic cancer increases from less than two deaths per 100,000 person-years among people of 35–39 years of age to more than 90

DOI: 10.1201/9781003544401-8

TABLE 5.1
Histological Types of Pancreatic Tumors

Histological Type	Subtypes
Pancreatic ductal adenocarcinoma	
	Adenosquamous carcinoma
	Colloid carcinoma
	Medullary carcinoma
	Hepatoid carcinoma
	Signet ring cell carcinoma
	Undifferentiated carcinoma
Adenocarcinoma precursor lesions	
	Pancreatic intraepithelial neoplasia (PanIN I–III)
	Intraductal papillary mucinous neoplasm (IPMN); low–high grade)
	Mucinous cystic neoplasm (MCN)
	Intraductal tubule-papillary neoplasm
Adenomas	
	Pyloric cyst adenoma
	Acinar cell cystadenoma
	Serous cyst adenoma
Neuroendocrine tumors	
	Non-functional Neuroendocrine tumors (Grades I–III)
	Neuroendocrine carcinoma
	Insulinoma, gastrinoma, VIPoma, somastatinoma, glucagonoma

deaths per 100,000 person-years in the age group above 80 years. Between 1990 and 2020, there was more than twofold increase in the number of new cases of pancreatic cancer. This increase was mainly attributed to an aging global population and advancements in pancreatic cancer diagnostic technologies, especially in high-income nations. Sadly, the number of reported new cases of pancreatic cancer is expected to continue increasing globally across all regions in the coming decades as a result of improvements in access to healthcare, healthcare delivery, and subsequent increases in the human lifespan (Klein, 2021).

Pancreatic cancer is a catastrophic malignancy with an extremely poor prognosis. In 2022, the projected number of new cases of pancreatic cancer for both sexes in the United States was 62,210, with an estimated death rate of 49,830 cases (i.e., 80% of the incidence rate for the same year) (Siegel et al., 2022). Despite advances in treatment of PDAC, it still has an incredibly poor prognosis, with five-year overall survival remaining below 10%, and it is estimated that by 2030 it will rank as the second biggest cause of cancer-related mortalities that are caused by cancers (Haeberle and Esposito, 2019). A study by He et al. examined results from patients diagnosed with periampullary tumors who underwent pancreaticoduodenectomy with the aim of cure in a high-volume facility during a three-decade period, from 1980 to 2011. During that period, 2,564 patients were treated and operated on, of whom 66% had pancreatic cancer. They found that pancreatic cancer typically presented with more advanced disease and that, despite advancements in surgical technique and oncological therapy, there was no significant alteration in the long-term survival rate among pancreatic cancer patients over the said 30-year period (He et al., 2014). The primary difficulties identified as causes of the poor outcomes include the absence of symptoms at the early stages of the disease, leading to late clinical presentations, unfavorable aggressive tumor biology associated with early dissemination and metastasis, and resistance to chemotherapy and radiotherapy (Kamisawa et al., 2016).

Meanwhile, Blackford and colleagues analyzed the Surveillance, Epidemiology and End Results (SEER) database, looking at trends in early-stage pancreatic cancer between 2004 and 2016. They identified a significant improvement in the five-year overall survival of stage IA PDAC, which increased from 44.7% in 2004 to 83.7% in 2012, confirming that the available treatment modalities together with early detection and diagnosis can lead to significantly improved overall survival (Blackford et al., 2020). If the tumors can be identified and surgically removed when they are still very small, or if premalignant precursor lesions such as pancreatic intraepithelial neoplasia, IPMNs, and mucinous cystic neoplasms are diagnosed and managed accordingly, the cure rate of pancreatic adenocarcinoma can be significantly improved (Sipos et al., 2009; Al-Sukhni et al., 2012). The poor prognosis of pancreatic cancer is primarily due to its delayed presentation, often at an unresectable stage (Gangi et al., 2004). Argawal et al. demonstrated that smaller tumors have a low rate of locoregional invasion and distant spread; they found that tumors >3 cm were associated with a higher prevalence of vascular invasion, higher lymph node involvement, and a median survival of 7.6 months, while tumors ≤2 cm had median survival of 17.2 months (Agarwal et al., 2008). Therefore, since surgical excision is the only treatment with the potential to cure pancreatic cancer, early diagnosis while the disease is still surgically resectable is paramount if we are to improve overall survival. (Bilici, 2014).

Table 5.1 shows that PDAC is the most common type of pancreatic cancer and is the subject of the vast majority of discussions about pancreatic cancer, having virtually become synonymous with pancreatic cancer. Nonetheless, there are other histological types, and this table illustrates the several histological types of pancreatic cancer. Some of these variations may be categorized as subtypes of PDAC. Additionally, PDAC precursor lesions such as pancreatic intraepithelial neoplasm, MCNs, and IPMNs have been included (Haeberle and Esposito, 2019; Mostafa et al., 2017).

CURRENT DIAGNOSTIC METHODS

Clinical Presentation

The earliest indication of a potential pancreatic cancer diagnosis may be suggestive clinical symptoms and signs, with or without identifiable risk factors. This information will guide the selection of appropriate diagnostic tests and management. Even though pancreatic cancer is recognized for its relatively rapid spread, the illness tends to begin cryptically, with vague symptoms that frequently appear several months before a patient is actually diagnosed (Holly et al., 2004). The presenting symptoms and signs are variable, may be nonspecific, and are influenced by the location of the tumor. Sudden atypical onset of diabetes mellitus in the elderly may raise suspicion of pancreatic cancer. About 67–75% of PDACs arise in the pancreas head and present with abdominal pain together with symptoms of obstructive jaundice such as pruritus, jaundice, dark urine, and pale stools, whereas cancers arising in the body and tail typically exhibit nonspecific constitutional symptoms such as unexplained weight loss, nausea and vomiting, abdominal pain, anorexia, constipation, and fatigue (De La Cruz et al., 2014; Tummala et al., 2011). Early recognition and evaluation of pancreatic cancer symptoms and signs will significantly increase early detection and, consequently, overall survival (Holly et al., 2004).

New Onset Diabetes in Pancreatic Cancer

Although long-standing diabetes is sometimes observed in pancreatic cancer, its utility as a potential marker is modest due to a large global population living with diabetes compared to the relatively low incidence of pancreatic cancer. However, new-onset diabetes (NOD) in patients over the age of 50 years may be a manifestation of cancer. The likelihood of pancreatic cancer becomes even more pronounced if the NOD develops at age ≥70. Within the first three years of being diagnosed with diabetes, the chance of developing pancreatic cancer in older adults is eight times higher than in the general population with a comparable profile (Chari et al., 2005). It is notable that, NOD diagnosed by physicians is typically associated it with late-stage disease at the time of the cancer diagnosis. However, similar to type 2 diabetes, where glycemic alterations occur years before the diagnosis, changes in blood sugar levels in NOD have been shown to occur more than 12 months prior to the diagnosis of pancreatic cancer. Therefore, detecting glycemic alterations rather than relying on physician-based diagnosis may improve the early detection of pancreatic cancer (Sharma et al., 2018). While type 2 diabetes is typically associated with progressive weight gain, diabetes induced by pancreatic cancer often leads to gradual weight loss, which can become evident months before the cancer diagnosis. Understanding the mechanism behind this phenomenon could be crucial in the early detection of pancreatic cancer (Sah et al., 2013). Sharma et al. created a model that can be utilized to evaluate pancreatic cancer risk in NOD. They called this model Enriching New-Onset Diabetes for Pancreatic Cancer (END-PAC). The model analyzed scores for three variables: weight change, blood glucose change, and age at diabetes onset that were found to be more convincingly related to pancreatic cancer in the discovery cohort. According to their model, an END-PAC score of ≥3 correctly detected the cancer with 78% sensitivity, 82% specificity, and was associated with a 4.4-fold increase in pancreatic cancer prevalence when compared to using NOD alone. Furthermore, from the 48% of the validation cohort that had END-PAC ≤ 0, no patient developed pancreatic cancer throughout the course of the three-year study (Sharma et al., 2018).

Diagnostic Imaging Technologies

Imaging examinations that are employed in the diagnosis of pancreatic cancer and pancreatic disease include non-invasive modalities such as abdominal ultrasound, computed tomography (CT) scan of the abdomen, magnetic resonance imaging (MRI), magnetic resonance cholangiopancreatography (MRCP), and invasive studies such as endoscopic ultrasound (EUS) and endoscopic retrograde cholangiopancreatogram (ERCP).

Transabdominal Ultrasound

Ultrasound (US) is a non-invasive, cheap diagnostic tool that is usually the first choice for patients presenting with abdominal discomfort or jaundice. Features of the head of pancreatic cancer seen on ultrasound include a hypoechoic mass, pancreatic ductal dilatation, and bile duct dilatation. It is plagued by variable sensitivity due to operator dependence and difficulty visualizing the pancreas due to bowel gas and patients' body habitus. It generally has variable sensitivity ranging between 50% and 90%, and as such is not a reliable diagnostic tool for pancreatic cancer (Lee and Lee, 2014). Nonetheless, specialized centers develop and employ strategies that can significantly improve ultrasonic visualization of pancreatic lesions to up to 97.5% sensitivity for pancreas head masses. These strategies include body positioning, filling the stomach with liquids, evaluating the pancreas from multiple angles, using high-frequency probes, and using contrast-enhanced ultrasound (Ashida et al., 2018).

Contrast-Enhanced Ultrasound

Contrast-enhanced ultrasound (CEUS) has proved invaluable in enhancing the diagnostic accuracy of pancreatic neoplasia. To evaluate the pancreas using CEUS, intravenous microbubble contrast agents are administered to expose the vascularity of tumors and improve the characterization of solid and cystic masses. One benefit of CEUS over fixed scanning (i.e., CT or MRI) is its capacity to enable real-time dynamic assessment of the perfusion status and microvascularity of pancreatic lesions (D'Onofrio et al., 2010). Because extensive fibrosis and desmoplasia are present in over 90% of cases, pancreatic adenocarcinoma typically shows up as a hypo-vascular, hypo-enhancing mass in all dynamic phases when contrasted with normal parenchymal tissue. Neuroendocrine tumors, on the other hand, manifest as hyper-enhancing tumors in the arterial phase (Sidhu et al., 2018).

In their meta-analysis, Lin et al. evaluated 18 studies on the diagnostic accuracy of CEUS for pancreatic adenocarcinoma, reporting a pooled sensitivity of 0.90 (95% CI, 0.89–0.92), specificity of 0.88. In 12 of these studies, which assessed CEUS's ability to differentiate between neoplastic lesions and non-neoplastic masses, the average sensitivity and specificity were 0.95 (0.94–0.96) and 0.83 (0.77–0.87) respectively. In the studies that they selected, PDAC diagnosis was confirmed via histology or cytology evaluation (Lin et al., 2016).

D'Onofrio et al. conducted a systematic review looking at the diagnostic utility of CEUS in pancreatic adenocarcinoma and its ability to differentiate pancreatic neoplastic from non-neoplastic tumors. They included 27 studies in the final analysis. For PDAC diagnostic accuracy, they detected pooled estimates with a confidence interval of 95%. The sensitivity was 0.89, specificity was 0.84, and the diagnostic odds ratio was 61.12 in 14 studies. For distinguishing between neoplastic and non-neoplastic lesions, the pooled sensitivity and specificity were 0.95 and 0.72 from 13 studies, respectively. Therefore, the use of CEUS could prove to be highly effective in identifying and differentiating adenocarcinoma from other pancreatic disease entities, enabling accurate and timely treatment of the disease. (D'Onofrio et al., 2014).

A recent meta-analysis compared CEUS and contrast-enhanced CT (CECT) scans in their ability to distinguish between pancreatic lesions and diagnose pancreatic carcinoma. The meta-analysis employed stringent criteria for article inclusion and ultimately included seven articles with 588 patients. In evaluating the diagnostic performance in PDAC, the authors found that CEUS exhibited an overall sensitivity of 0.91, specificity of 0.83, and area under the curve (AUC) of 0.94, while CECT demonstrated a specificity of 0.87, sensitivity of 0.88, and AUC of 0.93. The study indicated that CEUS tended to have higher sensitivity but lower specificity compared to CECT; however, both methods yielded similar findings in terms of area under the curve, diagnostic odds ratio, and post-test probability, suggesting comparable diagnostic performance. Furthermore, CEUS showed superior diagnostic performance for tumors smaller than 20 mm (Yan et al., 2022).

Computed Tomography

Tri-phasic pancreatic protocol computed tomography (CT) is the most useful initial diagnostic modality in patients with clinical symptoms of pancreatic cancer. Furthermore, it is the most effective staging modality currently in use, with a resectability prediction accuracy of approximately 80% when staging the disease. As a result, when there is a high suspicion of cancer, histological confirmation with biopsy before surgical resection is not necessary because of the excellent diagnostic and staging abilities of the imaging technique. Moreover, attempting to obtain a pancreatic biopsy prior to surgical management could potentially delay effective and timely treatment, while resection will confer the desired therapeutic benefit (Vincent et al., 2011). Whereas a contrast-enhanced multi-slice multidetector CT scan (MDCT) provides the most comprehensive first-line diagnostic and staging evaluation of pancreatic cancer, non-contrast CT scans have poor diagnostic accuracy and should ideally not be used except in individuals with severe kidney failure or allergy to contrast media (Tummala et al., 2011). Pancreatic adenocarcinoma typically arises from the pancreatic duct epithelium and therefore invariably obstructs the main pancreatic duct early, inducing the secondary signs of PDAC, including pancreatic duct dilatation and distal pancreatic atrophy with an associated dilated biliary system in tumors involving the head. These secondary signs, together with hypoattenuating masses, constitute the major diagnostic radiological findings in CT scans that are indicative of pancreatic cancer. Additional major findings focus on the interaction between the surrounding structures and the pancreatic lesion, particularly the adjacent vascular structures (Tamada et al., 2016). Figures 5.1 and 5.2 depict axial multi-detector CT images of patients with pancreatic adenocarcinoma.

Following a stringent pancreatic protocol using the CT scanning method, Catalano and colleagues analyzed 46 patients who had a clinical diagnosis of pancreatic cancer when they presented to their institution. Their objective was to evaluate multi-slice CT's (MSCT) ability to accurately diagnose and stage pancreatic cancer. They reported that radiologists correctly diagnosed 44 patients

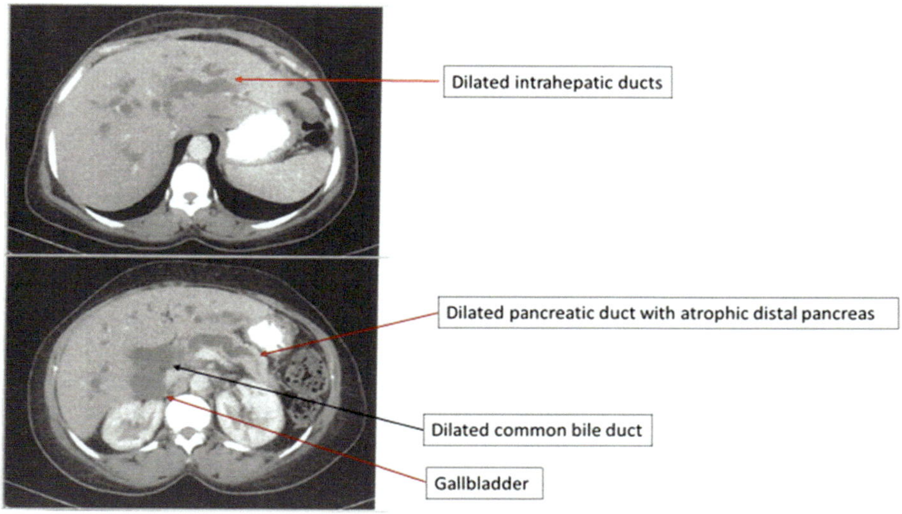

FIGURE 5.1 Axial CT images of 41-year-old female with head-of-pancreas adenocarcinoma who presented with typical clinical features of obstructive jaundice. The selected images are from a contrast-enhanced pancreas protocol scan, illustrating the typical radiological features of pancreatic head tumors, including the double-duct sign (characterized by a dilated common bile duct and a dilated pancreatic duct), dilated intrahepatic ducts, and an atrophic distal pancreas.

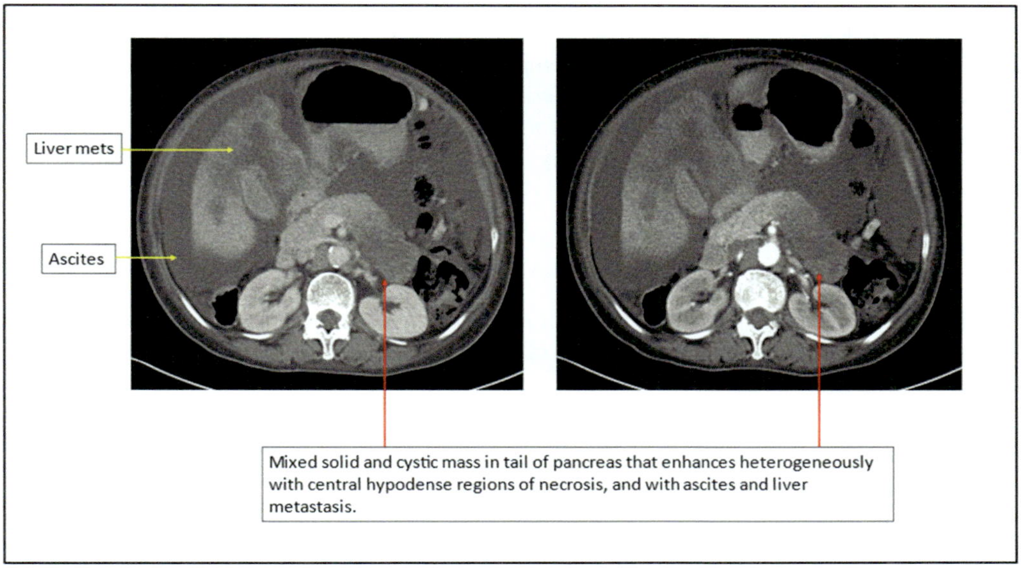

FIGURE 5.2 Porto-venous and arterial phase CT scan images of a 61-year-old female, known diabetic, and hypertensive, who presented with three-month history of worsening constitutional symptoms and jaundice. The CT scan revealed a tail-of-pancreas tumor with extensive liver metastatic diseases and gross ascites.

with a sensitivity of 97%, specificity of 80%, accuracy of 96%, positive predictive value of 97%, and negative predictive value of 80% after examining the unenhanced, arterial, and porta-venous phase images of the MSCT scans. Furthermore, 26 out of the 27 individuals with unresectable disease were correctly identified using the scans while establishing the disease stage. After acquiring the scans, the correct diagnoses were subsequently histologically verified through laparotomy, laparoscopy, or percutaneous biopsy (Catalano et al., 2003).

Although research indicates that CT has undeniable value in the diagnosis of symptomatic patients, it has also proved helpful in asymptomatic individuals with early disease. Park et al. retrospectively examined the pre-diagnostic CT scans carried out on patients treated for pancreatic cancer at their facility over a 12-year period. They found that out of the 736 individuals with pancreatic cancer, 27 had undergone CT scans of the abdomen (n = 25) and chest (n = 2) during the 12 months prior to getting the cancer diagnosis. These pre-diagnostic scans were performed for other purposes that were not linked to pancreatic cancer. They discovered that 51.9% (n = 14) of the 27 individuals had exhibited lesions on the pre-diagnostic CT scan that were consistent with the radiological characteristics of pancreatic cancer. The research's conclusion indicated that a comprehensive examination of the pancreas in every abdomen CT scan will boost the likelihood of incidentally discovering PDACs, thereby improving early cancer identification. (Park et al., 2023).

Although multidetector CT is the primary modality for PDAC evaluation, it has some limitations, including decreased accuracy in identifying small (<2 cm) and cystic lesions, difficulty distinguishing residual PDAC, recurrent tumor, and chemo-radiation-induced tumor fibrosis after treatment, decreased accuracy in evaluating chronic pancreatitis and ductal injuries, and repeated radiation exposure in young patients requiring regular evaluations during follow-up (Almeida et al., 2018). Therefore, in an effort to improve the CT characterization of pancreatic conditions, certain advances have been made and new techniques developed. These include (a) low-voltage acquisition CTs, which is associated with a 25% decrease in radiation dose while increasing contrast enhancement and visibility of PDAC; (b) dual energy CT scans, which take images using two different energy levels, generate monochromatic images at low voltage to improve contrast enhancement, lesion detection, PDAC visibility, and employ iodine maps and virtual unenhanced images to enhance pancreas differentiation and PDAC conspicuousness; (c) quantitative perfusion CT scans, which can more effectively distinguish mass-forming chronic pancreatitis from PDAC; and (d) other applications currently used in research settings (pancreatic volumetry, texture analysis, and radiogenomics) (Agostini et al., 2020).

Magnetic Resonance Studies

Magnetic Resonance Imaging

Magnetic resonance imaging (MRI) allows for non-invasive and reliable analysis of solid and cystic pancreatic lesions, along with the evaluation of pancreatic ducts, pancreatic parenchyma, adjacent soft tissues, and peritumoral vessel architecture (Ansari et al., 2015). It is particularly useful for patients who cannot undergo CECT scans due to contrast allergies or renal failure (Vincent et al., 2011). Furthermore, MRI can be helpful in the evaluation of occult or indeterminate abnormalities picked up on a CT scan (Jha et al., 2018).

Because MRI allows for better differentiation and high contrast resolution between soft tissues more accurately than CT, it provides a superior modality for evaluating small pancreatic tumors that do not alter the normal pancreas contours. Pancreatic adenocarcinoma shows up as a hypointense mass compared to normal pancreatic tissue on T1-weighted fat-suppressed images, which become more conspicuous during the arterial phase with progressive enhancement on delayed sequences. Additionally, MRI may be more sensitive in identifying metastatic lesions in distant organs (Miller et al., 2006).

In another study on pancreatic cancer, MRI demonstrated a moderate advantage over a CT scan. MRI demonstrated 89.1% diagnostic accuracy with 89.5% sensitivity and 63.4% specificity, while CT exhibited a diagnostic accuracy of 83.3% with 81.4% sensitivity (Costache et al., 2017).

Diffusion-Weighted Imaging MRI

A novel technique, diffusion-weighted imaging (DWI), has been integrated into conventional MRI techniques for application in various malignant tumors, yielding encouraging results. It exhibited remarkable diagnostic performance in PDAC detection, achieving an accuracy of 96%, with a 96% sensitivity, and 99% specificity, respectively (Messina et al., 2020). DWI explores the movement of water molecules within the body. In the external environment, water molecules are in free diffusion with constant random Brownian motion, whereas in biological systems, the diffusion is restricted due to interactions with cellular membranes and macromolecules. Movement of water molecules in biological tissues takes place both intracellularly and extracellularly, and also within the vasculature. Diffusion is typically inversely correlated with the tissue's cellular composition and the structural integrity of cell membranes (Koh and Collins, 2007). DWI offers quantitative as well as qualitative data about the diffusion properties, adding functional information to the primarily anatomical information acquired by traditional MRI series (Baliyan et al., 2016).

Kurita et al. imaged 37 patients with main duct strictures without solid tumors using DW-MRI/MRI, CT, and EUS; 20 of these patients were diagnosed with malignancies, whereas the remaining 17 were not. According to their research, 71.4% of patients with carcinoma in situ had high signal intensity in DWI sequences, thereby emphasizing the potential value of DWI in the early identification of PDAC (Kurita et al., 2021).

Magnetic Resonance Cholangiopancreatography

Magnetic resonance cholangiopancreatography (MRCP) is commonly incorporated into abdominal MRI techniques to examine ductal obstruction, dilatation, and shape (Ansari et al., 2015). MRCP performs comparably with MRI in terms of early detection and characterization, but both outperform CT in assessing the luminal characteristics of the pancreas and pancreatic lesions. This includes evaluating ductal communication of the cystic lesions and illustrating ductal features such as septations and mural nodules (Farr et al., 2022).

Adamek et al. examined the diagnostic accuracy of MRCP and ERCP for the identification of pancreatic cancer. The study included 124 patients (76 men and 48 women, mean age 55.1 years, range 19–80 years), who underwent complete ERCP and MRCP. Thirty-seven patients were proven to have PDAC either via histopathology or 12-month follow-up; MRCP was able to accurately identify 84% of patients with PDAC with specificity of 97%, compared to a 70% sensitivity demonstrated by ERCP. The study found that MRCP may be somewhat better than ERCP for PDAC diagnosis; also, MRCP has negligible complications because it is non-invasive, doesn't involve an endoscopy, and does not employ contrast material or radiation of any type (Adamek et al., 2000).

In another research, the diagnostic success rate of combining both the tumor marker carbohydrate antigen 19-9 (CA19-9) testing as well as MRCP in PDAC diagnosis was compared to performing MRCP alone. Eighty patients underwent both tests, while seventy patients underwent MRCP alone. The presence of pancreatic malignancy was verified by surgery and pathology analysis. A statistically significant variation was demonstrated between the two groups; in the group that underwent both MRCP along with CA19-9 testing, the rate of correct positive pancreatic cancer diagnoses was 95.7%, while in the group that did only MRCPs, the rate was 91.7%, thereby implying that the rate of accurate diagnosis of pancreatic cancer can be increased by utilizing MRCP together with CA19-9 (Ma et al., 2016).

Endoscopic Ultrasound

Endoscopic ultrasonography (EUS) has proven to be highly efficient and cost-effective in diagnosing pancreatic cancer. With a low complication rate of 1.1–3%, it is associated with a high sensitivity of 90–99%, particularly for detecting pancreatic cancers, including lesions smaller than 2–3 cm and identifying cystic lesions. Furthermore, fine needle aspiration biopsy (FNA/B) can be performed for cyto-histopathological assessment of the solid tumors that have been discovered with EUS.

EUS has been demonstrated to offer a greater diagnostic yield in comparison to positron emission tomography (PET), CT, MRI scans, and transabdominal US (Vincent et al., 2011; Gonzalo-Marin et al., 2014). When used in conjunction with contrast-enhancement and elastography, EUS has been revealed to outperform both MRI and CT scans in terms of diagnostic performance. It achieved a diagnostic accuracy of 93.7%, with 90.3% specificity and 97.5% sensitivity. EUS also outperformed these imaging modalities when evaluating small tumors (<2 cm), demonstrating a sensitivity of 94% and diagnostic accuracy of 91% (Costache et al., 2017).

In addition to the contrast enhancement benefits described in transabdominal US above, because of its endoscopic nature, CE-EUS has the extra advantage of avoiding gas artifacts and challenges related to the patient's body habitus. Furthermore, the use of contrast materials optimizes the visibility of the portal vein, and lymph node enhancement, leading to better differentiation between benign and malignant nodes (Chatterjee and Shah, 2024). It has been established that CE-EUS outperforms both conventional EUS and CECT scans in loco-regional disease staging and identifying vascular invasion by tumors. In a study, Nakai et al. assessed the effectiveness of CECT scans, contrast-enhanced harmonic EUS, and conventional ultrasonography in determining portal vein infiltration by pancreatic cancer. They retrospectively assessed imaging studies from 88 verified PDAC patients and found that at a 95% confidence interval, CE-EUS had a significantly superior performance when pitted against the other two modalities. It had accuracy of 93.2%, a specificity of 93.8%, and sensitivity of 85.7%, all of which were superior to normal EUS and CT scans (Nakai et al., 2022).

EUS elastography is another novel technique that is used to improve the performance of conventional EUS by assessing tissues. Elastography examines the stiffness of tissue by identifying minor structural alterations that develop as a result of compression in the B-mode image. Various tissues experience varied amounts of strain, and the degree of deformation serves as a gauge for the tissue's hardness or stiffness. Two methods called strain histogram and strain ratio analysis are used to calculate and quantify tissue stiffness based on the qualitative EUS data. Pancreatic cancer has been detected with 100% sensitivity and 92% specificity using EUS-guided elastography (Iglesias-Garcia et al., 2016).

Despite its excellent performance and low complication rate, EUS is not without shortcomings, such as the risk of seeding malignancy in the gastric wall during FNA; diminished resolution, differentiation capability, and sensitivity in patients with diffusely infiltrating pancreatic carcinoma, a prominent dorsoventral split, chronic pancreatitis, and a bout of acute pancreatitis within the preceding four weeks (Bhutani et al., 2004).

Endoscopic Retrograde Cholangiopancreatography

Endoscopic retrograde cholangiopancreatography (ERCP) provides an excellent luminogram of the biliary system and can be utilized for diagnosis and therapeutic intervention in pancreatic cancer. However, its biggest drawback is its invasive nature (i.e., compared to MRCP, which gives equivalent biliopancreatic system delineation) and the significant risk of serious complications such as duodenal perforation and acute pancreatitis. As a result, it is used predominantly when therapeutic interventions are envisaged rather than solely for diagnosis. Its clinical application in pancreatic cancer involves intraductal brush cytology, needle aspiration, or forceps biopsy, which are used for the diagnosis of biliary strictures. Therapeutic interventions include decompression of bile duct obstruction (Yousaf et al., 2020).

Though unreliable, the morphological appearance of the stricture can suggest the potential cause. For example, a long irregular stricture with an abrupt cutoff and shouldering is more indicative of a malignant etiology, whereas a smooth tapering stricture suggests a more likely benign etiology. In the diagnostic maneuvers of PDAC, ERCP-mediated brush cytology and intraductal biopsies are associated with a low negative predictive value (NPV) of around 58% because, while the specificity is high, approximating 95%, the sensitivity is very low, ranging from 23% to 56% for biliary brushings and 33–65% for fluoroscopic biopsies. Although the diagnostic success rate can be increased to 70% by combining brushings with intraductal biopsies, this approach is still not as accurate as EUS (Singh and Faulx, 2016).

Consequently, given the low diagnostic efficacy of ERCP and its associated complication risks, there are two justifiable reasons to perform an ERCP in pancreatic cancer: either for tumor sampling and/or biliary tree decompression in preparation for neoadjuvant chemotherapy in cases where surgery may be delayed, or for palliative purposes (Fiore et al., 2023).

SCREENING FOR PANCREATIC CANCER

Roughly 10–20% of pancreatic cancer cases develop in individuals with familial disposition, hereditary syndromes, or in relation to other cancer syndromes. These individuals demonstrate a multifold elevation in the lifetime risk of developing pancreatic cancer compared to the general populace and may benefit from screening protocols (Biller et al., 2021).

Over and above its poor prognosis, pancreatic cancer is characterized by a low incidence rate, affecting approximately 13/100,000 people in the general population. As a result of its low incidence, population-wide screening is not justifiable and would be non-beneficial. Consequently, the US Prevention and Screening Task Force recommends screening for high-risk individuals only and advocates against general population-wide screening (Kenner et al., 2021). For any screening program to be successful, it must be able to identify malignancies before they become clinically apparent or at an early stage for treatment to be successful. Additionally, the screening tool must be sufficiently non-invasive, affordable, easy to use, and allow for widespread implementation (Pepe et al., 2001). According to the International Cancer of the Pancreas Screening (CAPS) Consortium, the primary objective of screening should be the detection of high-grade precursor lesions before they progress to invasive tumors or detection of T1 node-negative disease (Goggins et al., 2020).

The 2022 ASGE guidelines for screening of pancreatic cancer suggest annual screening of high-risk individuals with genetic susceptibility, namely individuals with *BRCA1* and *BRCA2* pathogenic variants, familial pancreatic cancer (FPC), familial atypical multiple-mole melanoma (FAMMM) syndrome, Peutz–Jeghers syndrome, ataxia telangiectasia mutated (ATM) gene syndrome, Lynch syndrome, and hereditary pancreatitis. When conducting the meta-analysis for the guidelines, they found that in one of the studies, the three-year survival of screen-detected PDAC was 85% versus 25% for symptomatic cancers. Furthermore, 60% of neoplasms identified during screening were found to be resectable or borderline resectable when compared to a 20% resectability rate seen in symptomatic individuals from the general population. The screening tools recommended were EUS, MRI, and EUS with MRI, and these may be supplemented by upper endoscopy or colonoscopy as appropriate. The starting ages for screening were recommended to be 50 years of age or ten years earlier than the youngest relative with pancreatic cancer for *BRCA2*, *BRCA1*, *PALB2*, and ATM pathogenic variants, and Lynch syndrome; while for FAMMM syndrome: age 40 or ten years earlier, Peutz–Jeghers syndrome: age 35 or ten years earlier, and finally those with hereditary pancreatitis should start screening from the age of 40 years (Sawhney et al., 2022). Further research has demonstrated that screen-detected PDACs were associated with high levels of R0 surgical resections and low to zero surgery-related mortality (Wood et al., 2022).

Currently, the most frequently used first-line screening methods for pancreatic cancer surveillance are EUS and MRI (Cai et al., 2021). The primary benefit lies in the exceptional capacity of both EUS and MRI/MRCP to identify tiny lesions and cystic lesions. Furthermore, they exhibit superior soft tissue distinction in comparison to CT scans, and they do not use ionizing radiation like CT; hence, screening CT scans should ideally be reserved for patients who cannot be subjected to EUS or MRI (Goggins et al., 2020). Nonetheless, CT scans have demonstrated their capacity as a potential screening tool for pancreatic cancer. Gangi et al. retrospectively reviewed pre-diagnosis CT scans of 28 confirmed pancreatic cancer patients and revealed that CT findings that were potentially diagnostic or had suspicious features of pancreatic cancer were clearly identifiable in 50% of the scans undertaken between 2 and 18 months preceding the clinical diagnosis of pancreatic cancer, confirming that radiologic signs are present in some individuals for many months before the clinical manifestation of pancreatic cancer. Therefore, in selected high-risk but asymptomatic individuals, an abdominal CT scan can be utilized as a useful screening tool (Gangi et al., 2004).

Barnes and colleagues used 3.0T MRI as an initial modality to screen 65 high-risk individuals who met their inclusion criteria and selectively used EUS (with/without FNA at the discretion of the endoscopist) in ten patients with suspicious high-risk features on screening MRI. The MRI was able to reveal pathological tissue in 43% of patients, the majority being cystic masses with a median size of 5 mm (Barnes et al., 2018). A study comparing EUS and MRI in screening for PDAC found that the two techniques should be used in a supplementary fashion rather than as competing alternative modalities. During the study, 11 clinically relevant lesions were detected from 9 out of 139 high-risk patients who were screened; 55% (6/11) of these lesions were detected by both modalities, two solid lesions were identified by EUS and not by MRI, whereas MRI outperformed EUS in identifying tiny cystic lesions (Harinck et al., 2016). Canto et al. screened 216 high-risk individuals from five high-volume centers. They detected 283 pancreatic lesions using CT scan, MRI, and EUS. CT, MRI/MRCP, and EUS picked up a total of 39 (13.8%), 218 (77%), and 229 (79%) of all identifiable neoplasms, respectively. They found significant concordance between EUS and MRI/MRCP in the identification of any lesion, size of lesion, number of lesions, and location of the lesion. The majority of lesions detected were small and cystic lesions (Canto et al., 2012).

BIOMARKERS FOR PANCREATIC CANCER

Improving pancreatic cancer survival depends heavily on early disease detection, but unfortunately, this cancer is difficult to identify early based on clinical features alone. The majority of the cancers arise from PanINs, which are not detectable through clinical or radiological methods. Current imaging modalities may lack the sensitivity to detect very early-stage disease, highlighting the critical need for biomarkers that can be useful in early detection (Cai et al., 2021).

An ideal cancer biomarker should be directly produced by the cancer, released into circulation for easy access, and inexpensive to detect. However, biomarkers can be complex or indirect, measuring factors related to tumor development, such as tumor-induced hormonal changes or serum protein mass spectrometry (Pepe et al., 2001).

Carbohydrate Antigen (CA) 19-9

Carbohydrate antigen (CA) 19-9, which was discovered in 1979, is still the sole biomarker for PDAC that has been approved by the US Food and Drug Administration (FDA) for use in diagnosis and disease surveillance. Regretfully, CA19-9 is an unreliable biomarker for early detection because it exhibits a low overall sensitivity that is reported to be between 25% and 50% in early-stage disease (Kenner et al., 2021). However, the sensitivities do seem to increase as the disease advances and as it gets closer to clinical diagnosis. It is reported that when calculated at 95% specificity, CA19-9 had a 53% sensitivity and 68% sensitivity at two years and at one year, respectively, before the clinical diagnosis of PDAC (O'Brien et al., 2015). In the setting of metastatic unresectable disease, approximately 96% of PDAC will have serum levels in excess of 1,000 U/ml. Therefore, it is useful in monitoring clinical disease, response to treatment, and potential recurrences (Steinberg, 1990). It is notable that CA19-9 serum values can also be increased in non-cancerous hepatobiliary conditions like benign biliary obstruction, cholangitis, pancreatitis, benign hepatic disease (cirrhosis and hepatitis), and other non-hepatobiliary conditions (Kim et al., 2020).

Combining CA19-9 with additional tumor markers appears to increase its sensitivity. Carcinoembryonic antigen (CEA) is the second most commonly utilized biomarker. Using blood values of CA19-9 and CA125 in combination has been demonstrated to improve the precision of discriminating between people with pancreatic cancer and healthy controls, as well as cancer prognostication (Wu et al., 2015; Amaral et al., 2023). Other protein biomarkers that have shown variable sensitivities and specificities when used alone or in combination with CA19-9 include alcohol dehydrogenase (ADH), circulating macrophage inhibitory cytokine (MIC-1), secreted protein acidic and rich in cysteine (SPARC), and insulin-like growth factor-binding protein 2 (IGFBP2) among others (Amaral et al., 2023).

A variety of other biomarkers that can be used to guide the diagnosis of pancreatic cancer and prognostication are in the investigational stages and have shown promising results; for example, liquid biopsies for circulating tumor DNA (ctDNA), circulating tumor cells, and microRNAs, etc. (Daamen et al., 2023). Glycoprotein-based panels such as CEA, CA125, and CA242-based panels have also shown promising results in the detection of cancer when compared to individual protein measurements (O'Neill and Stoita, 2021).

INTEGRATION OF ARTIFICIAL INTELLIGENCE IN DIAGNOSTIC PROCESSES

The pancreas is situated in the retroperitoneal region and is difficult to access without surgical assistance; as a result the diagnosis of pancreatic pathologies and tumors has relied predominantly on imaging such as US, CT, and MRI scans (Kanji and Gallinger, 2013). The advent of EUS brought about an additional benefit with the capacity to conduct needle aspiration biopsy for cytological or histological evaluation, but EUS is very invasive, operator-dependent, and has a limited field of view (Rogers and Shah, 2024). Consequently, the existing imaging methods of diagnosis remain insufficient for identifying all types of pancreatic cancer, particularly for early small lesions, atypical lesions, and lesions with attenuation similar to that of normal pancreatic parenchyma. Moreover, confirming cytopathology with needle aspirations/biopsies is complex, invasive with potential complications, and it is time consuming (Liu et al., 2024). Artificial intelligence (AI) may provide additional resources to enhance these different modalities' capacity to diagnose pancreatic cancer and improve the speed of diagnosis (Liu et al., 2024).

AI entails analyzing and interpreting massive volumes of data in order to construct learning algorithms that can learn from their experiences. In healthcare, AI makes a substantial contribution by improving diagnostic accuracy, assisting in treatment planning and patient care, and providing clinicians with clinical decision support systems (Davenport and Kalakota, 2019). AI algorithms can interpret medical imagery, such as X-rays, MRIs, and CT scans with surprising accuracy, supporting medical personnel in diagnosing illnesses during the early stages. AI can be leveraged to improve early risk prediction by identifying those at high risk of contracting pancreatic cancer. Furthermore, AI can optimize administrative processes and boost productivity in healthcare facilities through natural language processing models (Tripathi et al., 2024). Figure 5.3 demonstrates a flow diagram of AI techniques used in pancreatic cancer.

AI may play an instrumental part during numerous processes in the medical image-based identification of pancreatic cancer by enhancing image rebuilding, segmentation, and identification, pancreatic disease characterization, and grading based on image attributes (Huang et al., 2022). Deep learning (DL) and machine learning (ML) appear to provide significant improvements in automated image processing, leading to increased sensitivity, specificity, and diagnostic accuracy in identifying pancreatic cancer and differentiating it from chronic pancreatitis (Goyal et al., 2022). Convolutional neural networks (CNNs) have become the most widely used deep learning methods in recent years for image identification, segmentation, and region recognition. With an adequate amount of training data, these techniques can now be trained in a matter of hours, and once properly trained, the AI models can analyze medical image data within a matter of seconds to minutes (LeCun et al., 2015; Liu et al., 2024).

CT Scan AI

Proper characterization of lesions smaller than 20 mm in the pancreas on a CT scan is difficult, resulting in a decreased cancer detection rate, with up to around 40% of small pancreatic cancers

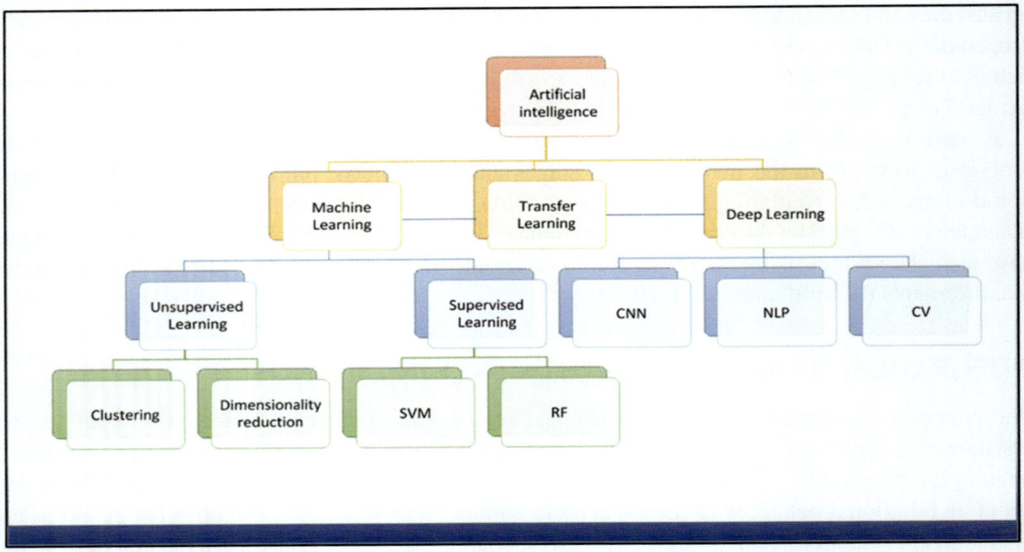

FIGURE 5.3 Machine learning (ML) in pancreatic cancer has been implemented for diagnosis, treatment prediction, prognostication, and assessing response to treatment. Primary ML is classed into supervised and unsupervised algorithms. Supervised algorithms can perform various tasks; for example, a support vector machine (SVM) is used for classification processes, while random forests (RF) can handle multi-dimensional data to develop response prediction algorithms. Unsupervised learning models are used for clustering similar disease profiles, identifying distinctive subgroups of pancreatic cancer, and identifying obscure patterns and relationships that may signify previously unseen linkages. Deep learning (DL) involves algorithms modeled on the human brain's neural networks and is made up of artificial neural networks with multiple learning layers. DL has been extensively used in pancreatic cancer image analysis from CT scans, MRI, EUS/US, and histopathological data. It has been used for diagnosis models, predicting treatment response, and patient outcomes. Natural language processing (NLP) algorithms aim to understand human language. NLP can identify pancreatic cancer-related data from medical records and research publications; furthermore, it is useful for automating administrative tasks like medical coding and documentation (Tripathi et al., 2024).

being missed on CT scans (Kang et al., 2021). Artificial intelligence and CT scan integration may increase the likelihood of early cancer detection, particularly the ability to identify small lesions.

A Taiwanese group (Chen et al., 2023) created and verified an end-to-end DL model that could be used to distinguish between CT scans showing pancreatic cancer and those that did not without needing manual image preprocessing. The model was constructed with a segmentation CNN that segmented the pancreas and tumor, and subsequently, the segmented data was examined by a group of five independent classifier CNNs to detect pancreatic cancer. In the real-world validation dataset, the tool correctly identified pancreatic cancer on CT scans with 91% accuracy, 89.7% sensitivity, and 92.8% specificity, exhibiting strong resilience and generalizability. While for tumors less than 20 mm, it attained a sensitivity of 74.7%. Furthermore, there was no discernible difference in sensitivity between the attending radiologists employed in high-volume tertiary centers (96.1%) and the model (90.2%). Therefore, in comparison to less experienced radiologists, the tool may have superior sensitivity for finding pancreas cancer (Chen et al., 2023).

AI in EUS

A systematic analysis carried out by Goyal et al. looked at studies that examined the value of using AI in EUS to assess pancreatic cancer. The review discovered that certain DL CNNs outperformed conventional EUS in distinguishing between benign and malignant IPMN, with accuracy, sensitivity, and specificity of 94%, 95.7%, and 92.6%, respectively. The meta-analysis also showed that the ML method of support vector machines demonstrated the highest sensitivity, specificity, and diagnostic accuracy in differentiating pancreatic cancer from chronic pancreatitis and normal pancreas, with 94–98% accuracy, 94–96% sensitivity, and 93–99% specificity, respectively (Goyal et al., 2022).

Marya and colleagues trained and constructed an EUS-CNN model using still and video images from EUSs previously performed at their institution to assess autoimmune pancreatitis (AIP), chronic pancreatitis (CP), and PDAC. They sought to identify the distinction between AIP, chronic pancreatitis, normal pancreas, and PDAC. They developed a model based on ResNet50V2 implementation on Keras and TensorFlow 2.0, with ImageNet weights. The model was able to distinguish PDAC from AIP with 90% sensitivity and 93% specificity (Marya et al., 2021).

IMPLEMENTING PANCREATIC CANCER SCREENING PROTOCOLS IN AFRICA

Globally, the incidence rate of pancreatic cancer is quite low, making population-based screening impracticable. The cancer frequently exhibits no symptoms in the early stages, which causes late diagnoses when the malignancy has progressed or spread (Daamen et al., 2023; Yang et al., 2021). The absence of reliable biomarkers for early identification exacerbates the asymptomatic nature of this illness. According to various studies, the current biomarker, CA19-9, is neither sufficiently sensitive nor specific to be useful for early screening purposes (Shrikhande et al., 2012; Yang et al., 2021).

Entering clinical practice, liquid biopsy still needs to be studied in practical clinical applications, and the technology needs to be standardized and rationalized. In this test, circulating tumor DNA, exosomes, or microRNAs in the blood are analyzed, which helps in the early recognition of pancreatic cancer. Different studies have demonstrated the potential of this test in the identification of early-stage pancreatic cancer, and it is commonly used in conjunction with routine biomarkers like the CA19-9 (Goel et al., 2024; Jaber, 2024). Moreover, advanced imaging techniques like EUS and MRI have enabled efficient screening of high-risk patients. These techniques allow for early-stage tumor detection and precancerous lesions, which are very important for their treatment (Waleleng et al., 2022; JohnsHopkinsMedicine, 2022).

In Africa, there has been an increase in awareness of pancreatic cancer and its effects on public healthcare systems, leading to an increase in different initiatives that assist researchers in comprehending the risk factors and epidemiology of pancreatic cancer on the continent. Some African countries, like Egypt, Morocco, and South Africa, have begun to adopt advanced diagnostic techniques, such as EUS and MRI, for the early identification of pancreatic cancer. However, the main challenges these countries face include limited access to resources, with facilities mainly based in urban healthcare centers (Yang et al., 2021; Jonas and Brand, 2023). It is important to note that international collaborations and support from global health organizations have played a vital role in advancing pancreatic cancer screening in Africa. Through this collaboration, there has been a partnership that allows for the sharing of resources, knowledge, and best practices, which has also led to the training of professionals in the field (Rubagumya et al., 2020).

Although there have been promising advancements in pancreatic cancer screening, the widespread implementation of the protocols varies across different African countries (i.e., Nigeria, Kenya, Egypt, and South Africa) and is challenged by three major problems: lack of the necessary personnel and healthcare infrastructure, high cost of equipment needed to implement these protocols, and protocols tailored to the African context and environment (Omotoso et al., 2023; Ministry of Health, 2018; Federal Ministry of Health 2023). We have shown some of the global protocols in Table 5.2.

In the context of disparities in implementation strategies, we have produced a summative tabulation that highlights usable protocols for the African pancreatic cancer screening (Table 5.3).

In view of the challenges and disparities related to the implementation strategies across the African country's pancreatic cancer screening programs, it is important to highlight the need for

TABLE 5.2
Pancreatic Cancer Screening Protocols and Techniques across Different Regions

Region	Screening Protocol	References
United Kingdom	No national screening; magnetic resonance imaging (MRI) and endoscopic ultrasound (EUS) for high-risk individuals, like those with hereditary pancreatitis or genetic mutations (e.g., *BRCA1*, *BRCA2*)	Oldfield et al. (2022); Cancer Research UK (2024); NHS (2024).
United States	US Preventive Services Task Force (USPSTF) does not recommend routine screening for pancreatic cancer in asymptomatic adults due to the lack of evidence that it reduces mortality. However, the International Cancer of the Pancreas Screening (CAPS) Consortium recommends screening high-risk individuals (genetic disposition or family history) using MRI and EUS	American Cancer Society (2024); Tomislav Dragovich et al. (2024)
Australia	Australian Pancreatic Cancer Screening Program (APCSP) tailored for high-risk patients using MRI/EUS.	Australian Pancreatic Cancer Genome Initiative (2024)
China	Guidelines emphasize the use of advanced imaging techniques (MRI/EUS) for high-risk groups	Cui et al. (2022)
Europe	Focus on liquid biopsies and collaboration for early detection. Other imaging techniques like MRI/EUS are also used.	Cancer Patients Europe (2024)

a more unified and adaptable document/policy for clinicians and healthcare systems. This will enhance/support the realization of the WHO Sustainable Development Goals 2030.

LIMITATIONS

Even though there have been significant advances in the diagnostic algorithms and armamentarium for pancreatic cancer, there are still some limitations associated with the various modalities. For example, CT scans exhibit poor accuracy in detecting small and cystic lesions, have difficulty discriminating residual PDAC from fibrosis following chemo-radiotherapy, face challenges in differentiating PDAC from CP masses, and detecting chronic pancreatitis and ductal damage. Furthermore, radiation exposure remains a substantial cause for concern in younger patients requiring serial investigations in the follow-up period (Agostini et al., 2020).

While EUS outperforms MRI and CT scans in locoregional tumor evaluation, it is limited in terms of distant metastasis, lacking the ability to detect extra-abdominal metastases because of its narrow field of view. Moreover, it is an invasive, complex procedure whose performance standard is dependent on the operator's skillset (Costache et al., 2017; Liu et al., 2024)

Though AI promises to enhance the diagnostic process, speed, and treatment planning, it is plagued by biases that may arise due to biased training datasets. Furthermore, ethical issues related to the allocation of responsibility for clinical decisions still remain a challenge (Huang et al., 2022).

CONCLUSION

Pancreatic cancer has a five-year survival rate of approximately 10%, making it one of the deadliest illnesses in the world. It is characterized by a poor response to treatment and commonly a late stage at the time of diagnosis, with incurable, surgically unresectable disease. Surgical removal of early disease provides the sole treatment option available for a cure in pancreatic cancer. Five-year survival rates surpassing 80% have been reported when there is early detection of precursor lesions and

TABLE 5.3
Implementation of Protocols for Pancreatic Cancer Screening in Africa with Adaptation from Other Regions

S/N	Criteria/Question	Action	Techniques	References
1	Does the patient have a family history of pancreatic cancer?	If yes, proceed to Step 2. If not, assess other risk factors	Questionnaire/ family history	Abe et al. (2019); Waleleng et al. (2022); Winstead (2024)
2	Does the patient have genetic predispositions (e.g., *BRCA1*, *BRCA2* mutations)?	If yes, recommend genetic counseling and proceed to Step 3.	Genetic testing	Birch et al. (2001); Brand et al. (2007); Iqbal et al. (2012)
3	Is the patient over 50 years old or 10 years younger than the earliest diagnosis in the family?	If yes, proceed to Step 4. If not, reassess annually.	Questionnaire	Su et al. (2020); Zhang et al. (2021)
4	Does the patient exhibit symptoms such as unexplained weight loss, jaundice, or persistent abdominal pain?	If yes, proceed to diagnostic imaging and physical examination.	MRI, EUS	Birch et al. (2001); Abe et al. (2019); National Cancer Institute (2024); Cancer Research UK (2024)
5	Are there any concerning findings on imaging (e.g., lesions)?	If yes, schedule follow-up imaging or biopsy.	EUS, MRI, biopsy	National Cancer Institute,2024)
6	Are there any high-risk precursor neoplasms detected?	If yes, proceed with laboratory studies (CA19-9, fasting glucose, amylase, lipase, and hemoglobin A1c)	Laboratory investigations, EUS, MRI	(Zhang et al., 2021; National Cancer Institute (2024)
7	Is the patient eligible for surgical intervention?	If yes, refer to an advanced surgical center.	Surgical consultation	Zhang et al. (2021); National Cancer Institute (2024);Stoffel et al. (2023)
8	Does the patient have new-onset diabetes with high-risk factors?	If yes, consider additional diagnostic studies.	Blood tests, imaging	National Cancer Institute (2024)
9	Discuss the limitations and potential adverse events of screening with the patient.	Ensure informed consent and shared decision-making.	Patient consultation	National Cancer Institute (2024)

small tumors. Thus, the ultimate goal in pancreatic cancer treatment should be early identification if we are to improve the overall results. To achieve this objective, strategies to enhance early detection must be developed. These strategies will include screening high-risk individuals, improving imaging techniques, and application of artificial intelligence

The diagnostic efficacy of traditional imaging methods rises significantly with the deployment of novel techniques such as enhancing EUS with contrast, employing dual-energy CT scans, and using diffusion-weighted MRIs, etc. Furthermore, this improvement is also appreciable even in small, inconspicuous cancers. AI has the potential to significantly improve imaging methods, speeding up early detection and enhancing physicians' ability to make informed clinical decisions. This ensures that patients receive timely and effective treatment. Moreover, AI can reliably differentiate benign pancreatic conditions such as autoimmune pancreatitis and chronic pancreatitis from pancreatic cancer, helping to reduce misdiagnosis and improve treatment accuracy.

REFERENCES

ABE, T., BLACKFORD, A. L., TAMURA, K., FORD, M., MCCORMICK, P., CHUIDIAN, M., ALMARIO, J. A., BORGES, M., LENNON, A. M., SHIN, E. J., KLEIN, A. P., HRUBAN, R. H., CANTO, M. I. & GOGGINS, M. 2019. Deleterious germline mutations are a risk factor for neoplastic progression among high-risk individuals undergoing pancreatic surveillance. *J Clin Oncol,* 37**,** 1070–1080.

ADAMEK, H. E., ALBERT, J., BREER, H., WEITZ, M., SCHILLING, D. & RIEMANN, J. F. 2000. Pancreatic cancer detection with magnetic resonance cholangiopancreatography and endoscopic retrograde cholangiopancreatography: a prospective controlled study. *Lancet,* 356**,** 190–193.

AGARWAL, B., CORREA, A. M. & HO, L. 2008. Survival in pancreatic carcinoma based on tumor size. *Pancreas,* 36**,** e15–e20.

AGOSTINI, A., BORGHERESI, A., BRUNO, F., NATELLA, R., FLORIDI, C., CAROTTI, M. & GIOVAGNONI, A. 2020. New advances in CT imaging of pancreas diseases: a narrative review. *Gland Surg,* 9**,** 2283–2294.

AL-SUKHNI, W., BORGIDA, A., ROTHENMUND, H., HOLTER, S., SEMOTIUK, K., GRANT, R., WILSON, S., MOORE, M., NAROD, S., JHAVERI, K., HAIDER, M. A. & GALLINGER, S. 2012. Screening for pancreatic cancer in a high-risk cohort: an eight-year experience. *J Gastroint Surg,* 16**,** 771–783.

ALMEIDA, R. R., LO, G. C., PATINO, M., BIZZO, B., CANELLAS, R. & SAHANI, D. V. 2018. Advances in pancreatic CT imaging. *Am J Roentgenol,* 211**,** 52–66.

AMARAL, M. J., OLIVEIRA, R. C., DONATO, P. & TRALHÃO, J. G. 2023. Pancreatic cancer biomarkers: oncogenic mutations, tissue and liquid biopsies, and radiomics—a review. *Dig Dis Sci,* 68**,** 2811–2823.

AMERICAN CANCER SOCIETY. 2024. *Can pancreatic cancer be found early?* Available: https://www.cancer.org/cancer/types/pancreatic-cancer/detection-diagnosis-staging/detection.html [Accessed 16-08-2024].

ANSARI, N. A., RAMALHO, M., SEMELKA, R. C., BUONOCORE, V., GIGLI, S. & MACCIONI, F. 2015. Role of magnetic resonance imaging in the detection and characterization of solid pancreatic nodules: An update. *World J Radiol,* 7**,** 361–374.

ASHIDA, R., TANAKA, S., YAMANAKA, H., OKAGAKI, S., NAKAO, K., FUKUDA, J., NAKAO, M., IOKA, T. & KATAYAMA, K. 2018. The role of transabdominal ultrasound in the diagnosis of early stage pancreatic cancer: review and single-center experience. *Diagnostics (Basel),* 9(1), 1–9.

AUSTRALIAN PANCREATIC CANCER GENOME INITIATIVE. 2024. *Pancreatic cancer.* Available: https://www.pancreaticcancer.net.au/.

BALIYAN, V., DAS, C. J., SHARMA, R. & GUPTA, A. K. 2016. Diffusion weighted imaging: technique and applications. *World J Radiol,* 8**,** 785–798.

BARNES, C. A., KRZYWDA, E., LAHIFF, S., MCDOWELL, D., CHRISTIANS, K. K., KNECHTGES, P., TOLAT, P., HOHENWALTER, M., DUA, K., KHAN, A. H., EVANS, D. B., GEURTS, J. & TSAI, S. 2018. Development of a high risk pancreatic screening clinic using 3.0 T MRI. *Familial Cancer,* 17**,** 101–111.

BHUTANI, M. S., GRESS, F. G., GIOVANNINI, M., ERICKSON, R. A., CATALANO, M. F., CHAK, A., DEPREZ, P. H., FAIGEL, D. O. & NGUYEN, C. C. 2004. The No Endosonographic Detection of Tumor (NEST) Study: a case series of pancreatic cancers missed on endoscopic ultrasonography. *Endoscopy,* 36**,** 385–9.

BILICI, A. 2014. Prognostic factors related with survival in patients with pancreatic adenocarcinoma. *World J Gastroenterol,* 20**,** 10802–10812.

BILLER, L. H., WOLPIN, B. M. & GOGGINS, M. 2021. Inherited pancreatic cancer syndromes and high-risk screening. *Surg Oncol Clin N Am,* 30**,** 773–786.

BIRCH, J. M., ALSTON, R. D., MCNALLY, R. J., EVANS, D. G., KELSEY, A. M., HARRIS, M., EDEN, O. B. & VARLEY, J. M. 2001. Relative frequency and morphology of cancers in carriers of germline TP53 mutations. *Oncogene,* 20**,** 4621–4628.

BLACKFORD, A. L., CANTO, M. I., KLEIN, A. P., HRUBAN, R. H. & GOGGINS, M. 2020. Recent trends in the incidence and survival of stage 1A pancreatic cancer: a surveillance, epidemiology, and end results analysis. *J Natl Cancer Inst,* 112**,** 1162–1169.

BRAND, R. E., LERCH, M. M., RUBINSTEIN, W. S., NEOPTOLEMOS, J. P., WHITCOMB, D. C., HRUBAN, R. H., BRENTNALL, T. A., LYNCH, H. T. & CANTO, M. I. 2007. Advances in counselling and surveillance of patients at risk for pancreatic cancer. *Gut,* 56**,** 1460–1469.

CAI, J., CHEN, H., LU, M., ZHANG, Y., LU, B., YOU, L., ZHANG, T., DAI, M. & ZHAO, Y. 2021. Advances in the epidemiology of pancreatic cancer: trends, risk factors, screening, and prognosis. *Cancer Letters,* 520**,** 1–11.

CANCER PATIENTS EUROPE. 2024. *Position paper on pancreatic cancer across Europe in light of the 2024 EU elections.* Available: https://cancerpatientseurope.org/position-paper-on-pancreatic-cancer-across-europe-in-light-of-the-2024-eu-elections/.

CANCER RESEARCH UK. 2024. *Screening for pancreatic cancer.* Available: https://www.cancerresearchuk.org/about-cancer/pancreatic-cancer/getting-diagnosed/screening [Accessed 16-08-2024].

CANTO, M. I., HRUBAN, R. H., FISHMAN, E. K., KAMEL, I. R., SCHULICK, R., ZHANG, Z., TOPAZIAN, M., TAKAHASHI, N., FLETCHER, J., PETERSEN, G., KLEIN, A. P., AXILBUND, J., GRIFFIN, C., SYNGAL, S., SALTZMAN, J. R., MORTELE, K. J., LEE, J., TAMM, E., VIKRAM, R., BHOSALE, P., MARGOLIS, D., FARRELL, J. & GOGGINS, M. 2012. Frequent detection of pancreatic lesions in asymptomatic high-risk individuals. *Gastroenterology,* 142, 796–804; quiz e14–e15.

CATALANO, C., LAGHI, A., FRAIOLI, F., PEDICONI, F., NAPOLI, A., DANTI, M., REITANO, I. & PASSARIELLO, R. 2003. Pancreatic carcinoma: the role of high-resolution multislice spiral CT in the diagnosis and assessment of resectability. *Eur Radiol,* 13, 149–156.

CHARI, S. T., LEIBSON, C. L., RABE, K. G., RANSOM, J., DE ANDRADE, M. & PETERSEN, G. M. 2005. Probability of pancreatic cancer following diabetes: a population-based study. *Gastroenterology,* 129, 504–511.

CHATTERJEE, A. & SHAH, J. 2024. Role of endoscopic ultrasound in diagnosis of pancreatic ductal adenocarcinoma. *Diagnostics,* 14, 78.

CHEN, P.-T., WU, T., WANG, P., CHANG, D., LIU, K.-L., WU, M.-S., ROTH, H. R., LEE, P.-C., LIAO, W.-C. & WANG, W. 2023. Pancreatic cancer detection on CT scans with deep learning: a nationwide population-based study. *Radiology,* 306, 172–182.

COSTACHE, M. I., COSTACHE, C. A., DUMITRESCU, C. I., TICA, A. A., POPESCU, M., BALUTA, E. A., ANGHEL, A. C., SAFTOIU, A. & DUMITRESCU, D. 2017. Which is the best imaging method in pancreatic adenocarcinoma diagnosis and staging – CT, MRI or EUS? *Curr Health Sci J,* 43, 132–136.

CUI, J., JIAO, F., LI, Q., WANG, Z., FU, D., LIANG, J., LIANG, H., XIA, T., ZHANG, T., ZHANG, Y., DAI, G., ZHANG, Z., WANG, J., BAI, Y., BAI, Y., BI, F., CHEN, D., CAO, D., CHEN, J., FANG, W., GAO, Y., GUO, J., HAO, J., HUA, H., HUANG, X., LIU, W., LIU, X., LI, D., LI, J., LI, E., LI, Z., PAN, H., SHEN, L., SUN, Y., TAO, M., WANG, C., WANG, F., XIONG, J., ZHANG, T., ZHANG, X., ZHAN, X., ZHENG, L., REN, G., ZHANG, T., ZHOU, J., MA, Q., QIN, S., HAO, C. & WANG, L. 2022. Chinese Society of Clinical Oncology (CSCO): Clinical guidelines for the diagnosis and treatment of pancreatic cancer. *J Natl Cancer Cent,* 2, 205–215.

D'ONOFRIO, M., BIAGIOLI, E., GERARDI, C., CANESTRINI, S., RULLI, E., CROSARA, S., DE ROBERTIS, R. & FLORIANI, I. 2014. Diagnostic performance of contrast-enhanced ultrasound (CEUS) and contrast-enhanced endoscopic ultrasound (ECEUS) for the differentiation of pancreatic lesions: a systematic review and meta-analysis. *Ultraschall Med,* 35, 515–521.

D'ONOFRIO, M., GALLOTTI, A., PRINCIPE, F. & MUCELLI, R. P. 2010. Contrast-enhanced ultrasound of the pancreas. *World J Radiol,* 2, 97–102.

DAAMEN, L. A., MOLENAAR, I. Q. & GROOT, V. P. 2023. Recent advances and future challenges in pancreatic cancer care: early detection, liquid biopsies, precision medicine and artificial intelligence. *J Clin Med,* 12, 7485.

DAVENPORT, T. & KALAKOTA, R. 2019. The potential for artificial intelligence in healthcare. *Future Healthc J,* 6, 94–98.

DE LA CRUZ, M. S., YOUNG, A. P. & RUFFIN, M. T. 2014. Diagnosis and management of pancreatic cancer. *Am Fam Physician,* 89, 626–632.

FARR, K. P., MOSES, D., HAGHIGHI, K. S., PHILLIPS, P. A., HILLENBRAND, C. M. & CHUA, B. H. 2022. Imaging modalities for early detection of pancreatic cancer: current state and future research opportunities. *Cancers,* 14, 2539.

FEDERAL MINISTRY OF HEALTH , N. 2023. *Nigeria, National strategic cancer control plan.* National Institute for Cancer Research and Treatment (NICRAT). Available: https://www.iccp-portal.org/system/files/plans/NATIONAL%20STRATEGIC%20CANCER%20CONTROL%20%20PLAN-Nigeria_.pdf [Accessed 15-08-2024].

FIORE, M., COPPOLA, A., PETRIANNI, G. M., TRECCA, P., D'ERCOLE, G., CIMINI, P., IPPOLITO, E., CAPUTO, D., BEOMONTE ZOBEL, B., COPPOLA, R. & RAMELLA, S. 2023. Advances in pre-treatment evaluation of pancreatic ductal adenocarcinoma: a narrative review. *J Gastrointest Oncol,* 14, 1114–1130.

GANGI, S., FLETCHER, J. G., NATHAN, M. A., CHRISTENSEN, J. A., HARMSEN, W. S., CROWNHART, B. S. & CHARI, S. T. 2004. Time interval between abnormalities seen on CT and the clinical diagnosis of pancreatic cancer: retrospective review of CT scans obtained before diagnosis. *Am J Roentgenol,* 182, 897–903.

GOEL ET AL., A., XU ,C. 2024. *An exosome-based liquid biopsy shows promise for early detection of pancreatic cancer.* AACR - American Association for Cancer Research. Available: https://www.aacr.org/about-the-aacr/newsroom/news-releases/an-exosome-based-liquid-biopsy-shows-promise-for-early-detection-of-pancreatic-cancer/ [Accessed 18-08-2024].

GOGGINS, M., OVERBEEK, K. A., BRAND, R., SYNGAL, S., DEL CHIARO, M., BARTSCH, D. K., BASSI, C., CARRATO, A., FARRELL, J., FISHMAN, E. K., FOCKENS, P., GRESS, T. M., VAN HOOFT, J. E., HRUBAN, R. H., KASTRINOS, F., KLEIN, A., LENNON, A. M., LUCAS, A., PARK, W., RUSTGI, A., SIMEONE, D., STOFFEL, E., VASEN, H. F. A., CAHEN, D. L., CANTO, M. I. & BRUNO, M. 2020. Management of patients with increased risk for familial pancreatic cancer: updated recommendations from the International Cancer of the Pancreas Screening (CAPS) Consortium. *Gut,* 69, 7–17.

GONZALO-MARIN, J., VILA, J. J. & PEREZ-MIRANDA, M. 2014. Role of endoscopic ultrasound in the diagnosis of pancreatic cancer. *World J Gastrointest Oncol,* 6, 360–368.

GOYAL, H., SHERAZI, S. A. A., GUPTA, S., PERISETTI, A., ACHEBE, I., ALI, A., THARIAN, B., THOSANI, N. & SHARMA, N. R. 2022. Application of artificial intelligence in diagnosis of pancreatic malignancies by endoscopic ultrasound: a systemic review. *Therap Adv Gastroenterol,* 15, 17562848221093873.

HAEBERLE, L. & ESPOSITO, I. 2019. Pathology of pancreatic cancer. *Transl Gastroenterol Hepatol,* 4, 50.

HARINCK, F., KONINGS, I. C. A. W., KLUIJT, I., POLEY, J. W., VAN HOOFT, J. E., VAN DULLEMEN, H. M., NIO, C. Y., KRAK, N. C., HERMANS, J. J., AALFS, C. M., WAGNER, A., SIJMONS, R. H., BIERMANN, K., VAN EIJCK, C. H., GOUMA, D. J., DIJKGRAAF, M. G. W., FOCKENS, P. & BRUNO, M. J. 2016. A multicentre comparative prospective blinded analysis of EUS and MRI for screening of pancreatic cancer in high-risk individuals. *Gut,* 65, 1505–1513.

HE, J., AHUJA, N., MAKARY, M. A., CAMERON, J. L., ECKHAUSER, F. E., CHOTI, M. A., HRUBAN, R. H., PAWLIK, T. M. & WOLFGANG, C. L. 2014. 2564 resected periampullary adenocarcinomas at a single institution: trends over three decades. *HPB (Oxford),* 16, 83–90.

HOLLY, E. A., CHALIHA, I., BRACCI, P. M. & GAUTAM, M. 2004. Signs and symptoms of pancreatic cancer: a population-based case-control study in the San Francisco Bay area. *Clin Gastroenterol Hepatol,* 2, 510–7.

HUANG, B., HUANG, H., ZHANG, S., ZHANG, D., SHI, Q., LIU, J. & GUO, J. 2022. Artificial intelligence in pancreatic cancer. *Theranostics,* 12, 6931–6954.

IGLESIAS-GARCIA, J., LINDKVIST, B., LARIÑO-NOIA, J., ABDULKADER-NALLIB, I. & DOMINGUEZ-MUÑOZ, J. E. 2016. Differential diagnosis of solid pancreatic masses: contrast-enhanced harmonic (CEH-EUS), quantitative-elastography (QE-EUS), or both? *United Eur Gastroenterology J,* 5, 236–246.

IQBAL, J., RAGONE, A., LUBINSKI, J., LYNCH, H. T., MOLLER, P., GHADIRIAN, P., FOULKES, W. D., ARMEL, S., EISEN, A., NEUHAUSEN, S. L., SENTER, L., SINGER, C. F., AINSWORTH, P., KIM-SING, C., TUNG, N., FRIEDMAN, E., LLACUACHAQUI, M., PING, S. & NAROD, S. A. 2012. The incidence of pancreatic cancer in BRCA1 and BRCA2 mutation carriers. *Br J Cancer,* 107, 2005–2009.

JABER, N. 2024. *Blood test accurately detects early-stage pancreatic cancer.* National Cancer Institute. Available: https://www.cancer.gov/news-events/cancer-currents-blog/2024/liquid-biopsy-detects-pancreatic-cancer [Accessed 16-08-2024].

JHA, P., YEH, B. M., ZAGORIA, R., COLLISSON, E. & WANG, Z. J. 2018. The role of MR imaging in pancreatic cancer. *Magn Reson Imaging Clin N Am,* 26, 363–373.

JOHNSHOPKINSMEDICINE. 2022. *Regular screening of people at high risk for pancreatic cancer pays off.* Available: https://www.hopkinsmedicine.org/news/newsroom/news-releases/2022/07/regular-screening-of-people-at-high-risk-for-pancreatic-cancer-pays-off [Accessed 16-08-2024].

JONAS, E. & BRAND, M. 2023. Early detection of pancreatic adenocarcinoma. *S Afr J Surg,* 61, 204–206.

KAMISAWA, T., WOOD, L. D., ITOI, T. & TAKAORI, K. 2016. Pancreatic cancer. *Lancet,* 388, 73–85.

KANG, J. D., CLARKE, S. E. & COSTA, A. F. 2021. Factors associated with missed and misinterpreted cases of pancreatic ductal adenocarcinoma. *Eur Radiol,* 31, 2422–2432.

KANJI, Z. S. & GALLINGER, S. 2013. Diagnosis and management of pancreatic cancer. *Cmaj,* 185, 1219–1226.

KENNER, B., CHARI, S. T., KELSEN, D., KLIMSTRA, D. S., PANDOL, S. J., ROSENTHAL, M., RUSTGI, A. K., TAYLOR, J. A., YALA, A., ABUL-HUSN, N., ANDERSEN, D. K., BERNSTEIN, D., BRUNAK, S., CANTO, M. I., ELDAR, Y. C., FISHMAN, E. K., FLESHMAN, J., GO, V. L. W., HOLT, J. M., FIELD, B., GOLDBERG, A., HOOS, W., IACOBUZIO-DONAHUE, C., LI, D., LIDGARD, G.,

MAITRA, A., MATRISIAN, L. M., POBLETE, S., ROTHSCHILD, L., SANDER, C., SCHWARTZ, L. H., SHALIT, U., SRIVASTAVA, S. & WOLPIN, B. 2021. Artificial intelligence and early detection of pancreatic cancer: 2020 summative review. *Pancreas,* 50, 251–279.

KIM, S., PARK, B. K., SEO, J. H., CHOI, J., CHOI, J. W., LEE, C. K., CHUNG, J. B., PARK, Y. & KIM, D. W. 2020. Carbohydrate antigen 19-9 elevation without evidence of malignant or pancreatobiliary diseases. *Sci Rep,* 10, 8820.

KLEIN, A. P. 2021. Pancreatic cancer epidemiology: understanding the role of lifestyle and inherited risk factors. *Nat Rev Gastroenterol Hepatol,* 18, 493–502.

KOH, D.-M. & COLLINS, D. J. 2007. Diffusion-weighted MRI in the body: applications and challenges in oncology. *Am J Roentgenol,* 188, 1622–1635.

KURITA, A., MORI, Y., SOMEYA, Y., KUBO, S., AZUMA, S., IWANO, K., IKEDA, S., OKUMURA, R. & YAZUMI, S. 2021. High signal intensity on diffusion-weighted magnetic resonance images is a useful finding for detecting early-stage pancreatic cancer. *Abdom Radiol,* 46, 4817–4827.

LECUN, Y., BENGIO, Y. & HINTON, G. 2015. Deep learning. *Nature,* 521, 436–444.

LEE, E. S. & LEE, J. M. 2014. Imaging diagnosis of pancreatic cancer: a state-of-the-art review. *World J Gastroenterol,* 20, 7864–7877.

LIN, L. Z., LI, F., LIU, Y., XING, L. X. & DU, L. F. 2016. Contrast-enhanced ultrasound for differential diagnosis of pancreatic mass lesions: a meta-analysis. *Med Ultrason,* 18, 163–169.

LIU, W., ZHANG, B., LIU, T., JIANG, J. & LIU, Y. 2024. Artificial intelligence in pancreatic image analysis: a review. *Sensors (Basel),* 24(14), 1–58.

LUO, G., FAN, Z., GONG, Y., JIN, K., YANG, C., CHENG, H., HUANG, D., NI, Q., LIU, C. & YU, X. 2019. Characteristics and outcomes of pancreatic cancer by histological subtypes. *Pancreas,* 48, 817–822.

MA, S., DUAN, J., LI, W., ZHANG, H. & HOU, Z. 2016. Exploration of the value of MRCP combined with tumor marker CA19-9 in the diagnosis of pancreatic cancer. *Artif Cells Nanomed Biotechnol,* 44, 717–721.

MARYA, N. B., POWERS, P. D., CHARI, S. T., GLEESON, F. C., LEGGETT, C. L., ABU DAYYEH, B. K., CHANDRASEKHARA, V., IYER, P. G., MAJUMDER, S., PEARSON, R. K., PETERSEN, B. T., RAJAN, E., SAWAS, T., STORM, A. C., VEGE, S. S., CHEN, S., LONG, Z., HOUGH, D. M., MARA, K. & LEVY, M. J. 2021. Utilisation of artificial intelligence for the development of an EUS-convolutional neural network model trained to enhance the diagnosis of autoimmune pancreatitis. *Gut,* 70, 1335–1344.

MESSINA, C., BIGNONE, R., BRUNO, A., BRUNO, A., BRUNO, F., CALANDRI, M., CARUSO, D., COPPOLINO, P., DE ROBERTIS, R., GENTILI, F., GRAZZINI, I., NATELLA, R., SCALISE, P., BARILE, A., GRASSI, R., ALBANO, D. & GROUP, O. B. O. T. Y. S. W. 2020. Diffusion-weighted imaging in oncology: an update. *Cancers,* 12, 1493.

MILLER, F. H., RINI, N. J. & KEPPKE, A. L. 2006. MRI of adenocarcinoma of the pancreas. *Am J Roentgenol,* 187, W365–W374.

MINISTRY OF HEALTH, K. 2018. *Kenya, national cancer screening guidelines 2018.* KENYA. Available: https://repository.kippra.or.ke/handle/123456789/1792 [Accessed 16-08-2024].

MOSTAFA, M. E., ERBARUT-SEVEN, I., PEHLIVANOGLU, B. & ADSAY, V. 2017. Pathologic classification of "pancreatic cancers": current concepts and challenges. *Chin Clin Oncol,* 6, 59.

NAKAI, A., KAMATA, K., HYODO, T., CHIKUGO, T., HARA, A., OTSUKA, Y., TANAKA, H., YOSHIKAWA, T., ISHIKAWA, R., OKAMOTO, A., YAMAZAKI, T., OMOTO, S., MINAGA, K., YAMAO, K., TAKENAKA, M., CHIBA, Y., WATANABE, T., MATSUMOTO, I., TAKEYAMA, Y. & KUDO, M. 2022. Utility of contrast-enhanced harmonic EUS for diagnosis of portal vein invasion by pancreatic cancer. *Endosc Ultrasound,* 11, 401–406.

NATIONAL CANCER INSTITUTE. 2024. *Advances in pancreatic cancer research.* Available: https://www.cancer.gov/types/pancreatic/research.

NHS. 2024. *NHS England Cancer Programme progress update – Spring 2024.* Available: https://www.england.nhs.uk/long-read/nhs-england-cancer-programme-progress-update-spring-2024/ [Accessed 16-08-2024].

O'BRIEN, D. P., SANDANAYAKE, N. S., JENKINSON, C., GENTRY-MAHARAJ, A., APOSTOLIDOU, S., FOURKALA, E.-O., CAMUZEAUX, S., BLYUSS, O., GUNU, R., DAWNAY, A., ZAIKIN, A., SMITH, R. C., JACOBS, I. J., MENON, U., COSTELLO, E., PEREIRA, S. P. & TIMMS, J. F. 2015. Serum CA19-9 Is significantly upregulated up to 2 years before diagnosis with pancreatic cancer: implications for early disease detection. *Clin Cancer Res,* 21, 622–631.

O'NEILL, R. S. & STOITA, A. 2021. Biomarkers in the diagnosis of pancreatic cancer: are we closer to finding the golden ticket? *World J Gastroenterol,* 27, 4045–4087.

OLDFIELD, L., STOTT, M., HANSON, R., JACKSON, R. J., REYNOLDS, W., CHANDRAN-GORNER, V., VAN DER MEER, R., ALISON, L., TEJEIRO, R., PUREWAL, T., GHANEH, P., PALMER, D., GREENHALF, W., HALLORAN, C. & COSTELLO, E. 2022. United Kingdom Early Detection Initiative (UK-EDI): protocol for establishing a national multicentre cohort of individuals with new-onset diabetes for early detection of pancreatic cancer. *BMJ Open,* 12**,** e068010.

OMOTOSO, O., TEIBO, J. O., ATIBA, F. A., OLADIMEJI, T., PAIMO, O. K., ATAYA, F. S., BATIHA, G. E. & ALEXIOU, A. 2023. Addressing cancer care inequities in sub-Saharan Africa: current challenges and proposed solutions. *Int J Equity Health,* 22**,** 189.

PARK, B. K., KOH, H. D., WON, S. Y., CHO, Y. S., SEO, J. H., AN, C. & PARK, S. 2023. Suspicious findings observed retrospectively on CT imaging performed before the diagnosis of pancreatic cancer. *J Gastrointest Oncol,* 14**,** 1008–1018.

PEPE, M. S., ETZIONI, R., FENG, Z., POTTER, J. D., THOMPSON, M. L., THORNQUIST, M., WINGET, M. & YASUI, Y. 2001. Phases of biomarker development for early detection of cancer. *JNCI,* 93**,** 1054–1061.

POKRZYWA, C. J., ABBOTT, D. E., MATKOWSKYJ, K. A., RONNEKLEIV-KELLY, S. M., WINSLOW, E. R., WEBER, S. M. & FISHER, A. V. 2019. Natural history and treatment trends in pancreatic cancer subtypes. *J Gastroint Surg,* 23**,** 768–778.

ROGERS, H. K. & SHAH, S. L. 2024. Role of endoscopic ultrasound in pancreatic cancer diagnosis and management. *Diagnostics,* 14**,** 1156.

RUBAGUMYA, F., COSTAS-CHAVARRI, A., MANIRAKIZA, A., MURENZI, G., UWINKINDI, F., NTIZIMIRA, C., RUKUNDO, I., MUGENZI, P., RUGWIZANGOGA, B., SHYIRAMBERE, C., URUSARO, S., PACE, L., BUSWELL, L., NTIRENGANYA, F., RUDAKEMWA, E., FADELU, T., MPUNGA, T., SHULMAN, L. N. & BOOTH, C. M. 2020. State of cancer control in Rwanda: past, present, and future opportunities. *JCO Glob Oncol,* 6**,** 1171–1177.

SAH, R. P., NAGPAL, S. J., MUKHOPADHYAY, D. & CHARI, S. T. 2013. New insights into pancreatic cancer-induced paraneoplastic diabetes. *Nat Rev Gastroenterol Hepatol,* 10**,** 423–33.

SAWHNEY, M. S., CALDERWOOD, A. H., THOSANI, N. C., REBBECK, T. R., WANI, S., CANTO, M. I., FISHMAN, D. S., GOLAN, T., HIDALGO, M., KWON, R. S., RIEGERT-JOHNSON, D. L., SAHANI, D. V., STOFFEL, E. M., VOLLMER, C. M., JR. & QUMSEYA, B. J. 2022. ASGE guideline on screening for pancreatic cancer in individuals with genetic susceptibility: summary and recommendations. *Gastrointest Endosc,* 95**,** 817–826.

SHARMA, A., KANDLAKUNTA, H., NAGPAL, S. J. S., FENG, Z., HOOS, W., PETERSEN, G. M. & CHARI, S. T. 2018. Model to determine risk of pancreatic cancer in patients with new-onset diabetes. *Gastroenterology,* 155**,** 730–739.e3.

SHRIKHANDE, S. V., BARRETO, S. G., GOEL, M. & ARYA, S. 2012. Multimodality imaging of pancreatic ductal adenocarcinoma: a review of the literature. *HPB (Oxford),* 14**,** 658–68.

SIDHU, P. S., CANTISANI, V., DIETRICH, C. F., GILJA, O. H., SAFTOIU, A., BARTELS, E., BERTOLOTTO, M., CALLIADA, F., CLEVERT, D. A., COSGROVE, D., DEGANELLO, A., D'ONOFRIO, M., DRUDI, F. M., FREEMAN, S., HARVEY, C., JENSSEN, C., JUNG, E. M., KLAUSER, A. S., LASSAU, N., MELONI, M. F., LEEN, E., NICOLAU, C., NOLSOE, C., PISCAGLIA, F., PRADA, F., PROSCH, H., RADZINA, M., SAVELLI, L., WESKOTT, H. P. & WIJKSTRA, H. 2018. The EFSUMB guidelines and recommendations for the clinical practice of contrast-enhanced ultrasound (CEUS) in non-hepatic applications: update 2017 (long version). *Ultraschall Med,* 39, e2–e44.

SIEGEL, R. L., MILLER, K. D., FUCHS, H. E. & JEMAL, A. 2022. Cancer statistics, 2022. *CA Cancer J Clin,* January/February, 72(1), 7–33.

SINGH, A. & FAULX, A. L. 2016. Endoscopic evaluation in the workup of pancreatic cancer. *Surg Clin North Am,* 96**,** 1257–1270.

SIPOS, B., FRANK, S., GRESS, T., HAHN, S. & KLÖPPEL, G. 2009. Pancreatic intraepithelial neoplasia revisited and updated. *Pancreatology,* 9**,** 45–54.

STEINBERG, W. 1990. The clinical utility of the CA 19-9 tumor-associated antigen. *Am J Gastroenterol,* April, 85(4), 350–355.

STOFFEL, E. M., BRAND, R. E. & GOGGINS, M. 2023. Pancreatic cancer: changing epidemiology and new approaches to risk assessment, early detection, and prevention. *Gastroenterology,* 164**,** 752–765.

SU, G. L., KO, C. W., BERCIK, P., FALCK-YTTER, Y., SULTAN, S., WEIZMAN, A. V. & MORGAN, R. L. 2020. AGA clinical practice guidelines on the role of probiotics in the management of gastrointestinal disorders. *Gastroenterology,* 159**,** 697–705.

TAMADA, T., ITO, K., KANOMATA, N., SONE, T., KANKI, A., HIGAKI, A., HAYASHIDA, M. & YAMAMOTO, A. 2016. Pancreatic adenocarcinomas without secondary signs on multiphasic multidetector CT: association with clinical and histopathologic features. *Eur Radiol,* 26**,** 646–655.

TOMISLAV DRAGOVICH ET AL 2024. *Pancreatic cancer guidelines.* Available: https://emedicine.medscape.com/article/280605-guidelines.

TRIPATHI, S., TABARI, A., MANSUR, A., DABBARA, H., BRIDGE, C. P. & DAYE, D. 2024. From machine learning to patient outcomes: a comprehensive review of ai in pancreatic cancer. *Diagnostics,* 14, 174.

TUMMALA, P., JUNAIDI, O. & AGARWAL, B. 2011. Imaging of pancreatic cancer: an overview. *J Gastrointest Oncol,* 2, 168–74.

VINCENT, A., HERMAN, J., SCHULICK, R., HRUBAN, R. H. & GOGGINS, M. 2011. Pancreatic cancer. *Lancet,* 378, 607–20.

WALELENG, B. J., ADIWINATA, R., WENAS, N. T., HAROEN, H., ROTTY, L., GOSAL, F., ROTTY, L., WINARTA, J., WALELENG, A. & SIMADIBRATA, M. 2022. Screening of pancreatic cancer: target population, optimal timing and how? *Ann Med Surg (Lond),* 84, 104814.

WINSTEAD, E. 2024. *Screening people at high risk for pancreatic cancer may help them live longer.* National Cancer Institute. Available: https://www.cancer.gov/news-events/cancer-currents-blog/2024/pancreatic-cancer-surveillance [Accessed 18-08-2024].

WOOD, L. D., CANTO, M. I., JAFFEE, E. M. & SIMEONE, D. M. 2022. Pancreatic cancer: pathogenesis, screening, diagnosis, and treatment. *Gastroenterology,* 163, 386–402.e1.

WU, L., HUANG, P., WANG, F., LI, D., XIE, E., ZHANG, Y. & PAN, S. 2015. Relationship between serum CA19-9 and CEA levels and prognosis of pancreatic cancer. *Ann Transl Med,* 3, 328.

YAN, X., LV, K., XIAO, M., TAN, L., GUI, Y., ZHANG, J., CHEN, X., JIA, W. & LI, J. 2022. The diagnostic performance of contrast-enhanced ultrasound versus contrast-enhanced computed tomography for pancreatic carcinoma: a systematic review and meta-analysis. *Transl Cancer Res,* 11, 3645–3656.

YANG, J., XU, R., WANG, C., QIU, J., REN, B. & YOU, L. 2021. Early screening and diagnosis strategies of pancreatic cancer: a comprehensive review. *Cancer Commun,* 41, 1257–1274.

YOUSAF, M. N., EHSAN, H., WAHAB, A., MUNEEB, A., CHAUDHARY, F. S., WILLIAMS, R. & HAAS, C. J. 2020. Endoscopic retrograde cholangiopancreatography guided interventions in the management of pancreatic cancer. *World J Gastrointest Endosc,* 12, 323–340.

ZHANG, Q., LU, J., JIA, M., MA, Y., SUN, M., CHEN, X., MA, X., YANG, Y., KANG, J., ZHANG, X., PAIELLA, S., KATZ, M. H. G., TSUCHIDA, K., SCHATTNER, M. & DU, S. 2021. Evaluation of the reporting quality of clinical practice guidelines on pancreatic cancer using the RIGHT checklist. *Ann Transl Med,* 9, 1088.

Part IV

Treatment and Comprehensive Care

6 Treatment Modalities and Innovations

Meshack Bida[1,2], Nissy Janda[2], Jessica McIntyre[2], Michelle McCabe[2], Mpho Kgomo[3], Talent Chipiti[1], Fieke Froeling[4], and Zodwa Dlamini[1]*

[1]SAMRC Precision Oncology Research Unit (PORU), DSI/NRF SARChI Chair in Precision Oncology and Cancer Prevention (POCP), Pan African Cancer Research Institute (PACRI), University of Pretoria, Pretoria, South Africa.

[2]Department of Anatomical Pathology, National Health Laboratory Services: Tshwane Academic Division, University of Pretoria, Pretoria, South Africa.

[3]Department of Internal Medicine, Gastroenterology Unit, Steve Biko Academic Hospital, University of Pretoria, Pretoria, South Africa.

[4] Cancer Lead Experimental Cancer Centre Glasgow, School of Cancer Sciences, and Beatson West of Scotland Cancer Centre, University of Glasgow, Wolfson Wohl Cancer Research Centre, Glasgow, United Kingdom.

*Corresponding author - Meshack Bida

INTRODUCTION

Pancreatic carcinoma is associated with an extremely poor prognosis and ranks as the seventh leading cause of cancer-related deaths worldwide. The five-year survival rate for patients with pancreatic carcinoma is as low as 9% in first-world countries like the United States. Despite the increasing mortality rate associated with the disease, the introduction of multimodality treatment involving surgery and adjuvant therapy provides a glimmer of hope for patients (Siegel et al., 2019). The poor prognosis of this disease is mainly due to its late diagnosis because most patients with pancreatic carcinoma tend to be asymptomatic until the disease is at an advanced stage. There are also currently no screening programs available for the early detection of the disease. The aggressiveness of pancreatic carcinoma may also explain the early onset of metastasis, possibly due to rapid epithelial-to-mesenchymal transition and focal inflammation (Rhim et al., 2012). Treatment for pancreatic cancer often involves a combination of surgery, chemotherapy, radiation therapy, and palliative care. The specific treatment plan is determined based on the stage of the disease. The contemporary clinical approaches employed in the management of pancreatic ductal adenocarcinoma (PDAC) involve different clinical scenarios such as resectable, borderline resectable, locally advanced, and metastatic PDAC (Brunner et al., 2019).

MOLECULAR PATHOLOGY OF PDAC

There are four major oncogenes (*KRAS*, *CDKN2A*, *TP53*, *SMAD4*) revealed in the development of PDAC. The *KRAS* oncogene, a somatic constitutional activating point mutation, encodes a small

DOI: 10.1201/9781003544401-10

GTPase of approximately 21 kDa and is found in over 90% of patients with PDAC. This *KRAS* mutation is also frequently present in low-grade pancreatic intraepithelial neoplasms (PanINs), which are considered the most common precursor to PDAC. As a result, the *KRAS* mutation represents one of the earliest molecular changes in the onset of pancreatic tumorigenesis (Eser et al., 2014). In PDAC, mutations have been identified in three key tumor-suppressor genes: *CDKN2A*, *TP53*, and *SMAD4*. The *CDKN2A* gene, which plays a crucial role in cell cycle regulation, functions through two *CDK* kinase inhibitor isoforms and an alternate open reading frame isoform, which stabilizes the *p53* protein. In contrast, alterations in *TP53* and *SMAD4* are later events, typically occurring in high-grade pancreatic intraepithelial neoplasms (PanINs) or invasive carcinomas (Hosoda et al., 2017).

Genomic and Transcriptomic Insights into Molecular Subtypes of Pancreatic Cancer and Related Consideration

Pancreatic ductal adenocarcinoma is classified into four molecular subtypes: squamous, pancreatic progenitor, aberrantly differentiated endocrine exocrine (ADEX), and immunogenic, based on distinct transcription factor expression and downstream targets (Pajic et al., 2012). Squamous tumors are characterized by significant alterations in gene networks associated with inflammation, hypoxia, metabolic reprogramming, TGF-β signaling, and autophagy. These tumors frequently harbor mutations in *TP53* and *KDM6A*, show activation of MYC signaling, and exhibit alterations in integrin (α6β1 and α6β4) and EGF signaling pathways. Additionally, they often display upregulation of the TP63ΔN network and hypermethylation of pancreatic endodermal genes, including *PDX1*, *MNX1*, *GATA6*, and *HNF1B*, which are linked to a poor prognosis.

In contrast, pancreatic progenitor tumors are defined by genes that guide pancreatic endodermal cell fate determination and are crucial for early pancreatic development. Key transcription factors involved in this process include *PDX1*, *MNX1*, *HNF4G*, *HNF4A*, *HNF1B*, *HNF1A*, *FOXA2*, *FOXA3*, and *HES1*. The ADEX class, a subset of pancreatic progenitor tumors, displays enhanced transcriptional networks that are active in the later stages of pancreatic development and differentiation. The immunogenic subtype is notable for its upregulation of immune networks, which include B cell signaling, antigen presentation, and pathways related to CD4+ T cells, CD8+ T cells, and toll-like receptors (Bailey et al., 2016). These distinctions in molecular characteristics highlight the complexity of pancreatic tumors and underscore the potential for targeted therapeutic approaches based on their unique profiles.

Epigenetic Modifications in the Tumorigenesis of PDACs

Epigenetic alterations of oncogenes and tumor-suppressor genes significantly contribute to the development of PDAC, primarily through processes like DNA methylation and histone modifications, such as methylation and acetylation. In PDAC, the methylation of tumor suppressor gene promoters inhibits their transcription, reducing gene expression.

Three main DNA methyltransferases (DNMTs)—DNMT1, DNMT3A, and DNMT3B—facilitate DNA methylation by transferring methyl groups from S-adenosylmethionine to cytosine residues. While some DNMT inhibitors have shown promise in preliminary studies, further research is necessary to evaluate their effectiveness in treating pancreatic cancer. Histone methylation's effects depend on the specific lysine residue and the degree of methylation (mono-, di-, or tri-methylation). Recent genomic studies have identified *KDM6A* and *MLL2* as frequently altered genes that regulate histone methylation in pancreatic cancers (Silverman & Shi, 2016). *KDM6A* functions as a demethylase for H3K27me3 and is crucial for the differentiation of pancreas endodermal cells through the regulation of the Wnt signaling pathway. The presence of H3K27me3 on gene promoters is associated with reduced gene expression. Conversely, EZH2, a histone methyltransferase involved in H3K27 methylation and part of the polycomb group complex, is found to be upregulated in the nuclei of PDAC cells, potentially facilitating disease progression (Jiang et al., 2013). This interplay of epigenetic factors underscores the complexity of PDAC development and highlights potential therapeutic targets for future treatments.

OVERVIEW OF STANDARD TREATMENT PROTOCOLS

Current Treatment Strategies for Pancreatic Carcinoma

Resectable Pancreatic Cancers

Surgical resection remains the only potentially curative treatment for pancreatic carcinoma. Surgical resection also significantly improves survival, but over 80% of patients with pancreatic carcinoma present late in the course of the disease, when the lesion is deemed to be no longer primarily amenable to resection (Kamisawa et al., 2016). The primary goal of surgically removing pancreatic carcinoma is to achieve a negative resection margin (R0). Conversely, a positive resection margin (R1 or R2) is often associated with tumor recurrence and poor prognosis. For most patients with an R0 margin, additional steps such as perivascular dissection, vascular resection, and reconstruction, as well as resection of extra-pancreatic organs, are performed to ensure complete removal of the disease (Benassai et al., 2015).

Adjuvant chemotherapy is often recommended to reduce the risk of tumor recurrence following surgical resection. Various postoperative chemotherapy regimens, including oxaliplatin, 5-fluorouracil, and irinotecan combination treatment (modified FOLFIRINOX) or trials for adjuvant FFX gemcitabine-based combination therapies, are commonly employed for this purpose (Conroy et al., 2018). The effectiveness of postoperative chemoradiotherapy (CRT) is a topic of debate, with some studies providing evidence supporting its use (Rutter et al., 2015). However, other studies indicate that adjuvant CRT may result in poorer survival outcomes compared to chemotherapy alone (Liao et al., 2013).

Currently, there is limited high-quality clinical evidence from randomized controlled trials.

Multiple randomized controlled trials have shown a lack of substantial evidence supporting the use of preoperative neoadjuvant therapy for patients with primarily resectable PDAC (Table 6.1) (Reni et al., 2018). Neoadjuvant therapy has been shown to improve overall survival (OS) as compared to surgery in patients with borderline resectable pancreatic cancer. There is, however, a need for more evidence to establish whether neoadjuvant therapy improves survival in the cohort of patients with resectable pancreatic cancer (van Dam et al., 2022). In other studies, such as PMID: 32105518, preoperative chemoradiotherapy for resectable or borderline resectable pancreatic cancer did not show a significant OS benefit (Versteijne et al., 2020).

Borderline Resectable PDAC

Patients with borderline resectable disease are given trial treatment with neoadjuvant chemotherapy or chemoradiotherapy. Preoperative neoadjuvant therapy presents the potential for enhanced local disease control and the eradication of micrometastases. Furthermore, this therapeutic approach holds promise for enhancing resectability by inducing tumor down-staging, particularly in cases of borderline PDAC (Schorn et al., 2017). Some studies have shown that neoadjuvant therapy offers an improved survival outcome in resected PDAC of different pathological stages. There is, however, no reliable evidence of the advantage of neoadjuvant CRT compared to chemotherapy alone. After a good response to neoadjuvant treatment, surgical exploration is rendered easier, thus making the tumor more resectable. Adjuvant therapies are generally considered for all primary resectable tumors to reduce the high risk of recurrence (Oettle et al., 2013).

Locally Advanced and Metastatic PDAC

Both locally advanced and metastatic PDAC, as detected by imaging studies, are often unresectable. The patient's performance status is evaluated, and those patients with good performance status may

TABLE 6.1
Overview of Randomized Controlled Trials Evaluating Neoadjuvant Chemo(radio) Therapy in Pancreatic Carcinoma Patients

Trial	Country/Region	N	Regimens	Outcomes	References
	UK	50, resectable PC	Gemcitabine ($n = 24$) vs. gemcitabine and cisplatin ($n = 26$)	Resection rate 38% vs. 70%; R0 resection 75% vs. 75%; median overall survival (mOS) 9.9 months vs. 15.6 months; one-year OS 41.7% vs. 61.5%	Palmer et al. (2007)
	Austria	30, 11× borderline resectable and 19× locally advanced	Gemcitabine 4 cycles and bevacizumab 3 doses ($n = 11$) vs. gemcitabine 4 cycles and bevacizumab 6 doses ($n = 19$)	Resection rate 36.4% vs. 36.8% ($p = 0.97$)	Sahora et al. (2014)
PACT-15	Italy	93	PEXG (gemcitabine, three cycles of cisplatin, epirubicin, and capecitabine administered preoperatively and postoperatively (n = 29) compared to six cycles of PEXG given after surgery	mOS was reported at 38.2 months compared to 26.4 months; the three-year OS rates were 55% vs. 43%; and the five-year OS rates showed 49% in one group vs. 24% in another. Additionally, median disease-free survival (mDFS) was 16.9 months against 12.4 months, with one-year disease-free survival (DFS) rates of 66% compared to 50%	Reni et al. (2018)
E1200	USA	23	CRT (gemcitabine and 50.4 Gy) ($n = 10$) compared to chemotherapy (gemcitabine, cisplatin, 5-FU) followed by CRT (5-FU, 50.4 Gy) ($n = 11$), both evaluated in the study, followed by surgery and gemcitabine adjuvant chemotherapy	mOS was reported at 19.4 months compared to 13.4 months. The one-year actuarial overall survival rates were 69% vs. 61%, while the two-year actuarial overall survival rates were 32% vs. 13%. The median progression-free survival (mPFS) was noted to be 14.2 months, with the one-year actuarial progression-free survival rate at 59% compared to 15% in the other group. The rate of resectability was 30% vs. 18.2%.	Landry et al. (2010)

(Continued)

TABLE 6.1 (CONTINUED)
Overview of Randomized Controlled Trials Evaluating Neoadjuvant Chemo(radio) Therapy in Pancreatic Carcinoma Patients

Trial	Country/Region	N	Regimens	Outcomes	References
	Germany and Switzerland	73	Upfront surgery ($n = 33$) vs. neoadjuvant CRT (gemcitabine, cisplatin, and 55.8 Gy) followed by surgery ($n = 33$), both followed by gemcitabine chemotherapy	mOS 14.4 months vs. 17.4 months ($p = 0.96$); mPFS 8.7 months vs. 8.4 months ($p = 0.95$); R0 rate 48% vs. 52% ($p = 0.81$); pN0 rate 30% vs. 39% ($p = 0.44$)	Golcher et al. (2015)
	Italy	38	Surgery performed initially ($n = 20$) compared to neoadjuvant chemoradiotherapy (CRT) using gemcitabine followed by surgery ($n = 18$), with both approaches subsequently receiving gemcitabine chemotherapy.	mOS was 19.5 months compared to 22.4 months ($p = 0.973$); the resectability rate was 75% vs. 61.1% ($p = 0.489$); and the R0 resection rate was 25.0% vs. 38.9% ($p = 0.489$)	Casadei et al. (2015)

Surveillance strategies following potentially curative resection of PDAC are not universally recommended due to insufficient clinical evidence. However, one study demonstrated that postoperative monitoring, which includes clinical assessments and serum carbohydrate antigen (CA) 19-9 testing, combined with routine CT scans and chest X-rays every six months, was effective in enhancing survival outcomes (Tzeng et al., 2012). Both the European Society of Medical Oncology (ESMO) and the National Comprehensive Cancer Network (NCCN) guidelines for diagnosis and treatment of pancreatic cancer recommend the measurement of serum CA 19-9 levels after surgery and before adjuvant therapy to guide treatment and follow-up (Conroy et al., 2023).

tolerate more aggressive chemotherapeutic treatment. Some of these chemotherapeutic regimens include fluoropyrimidine- and gemcitabine-based regimens (Dahan et al., 2010). The standard of care for locally advanced and metastatic PDAC in South Africa includes FOLFIRINOX for patients with good performance status who can tolerate chemotherapy. Gemcitabine remains an option for patients who can't tolerate more than monotherapy (South African Department of Health Brief Report, 2023).

When a patient's performance status is deemed favorable, typically indicated by an Eastern Cooperative Oncology Group (ECOG) score of 0–1, it is recommended to explore various combination chemotherapy options, such as platinum-based or gemcitabine-based combination treatments. A recent Cochrane review has shown that using a combination of multiple chemotherapeutic agents for advanced pancreatic carcinoma yields better efficacy compared to gemcitabine monotherapy, although this approach may lead to a higher incidence of side effects (Chin et al., 2018).

The FOLFIRINOX regimen is highly recommended for patients presenting with locally advanced or metastatic disease. Clinical evidence indicates that the combination of gemcitabine and erlotinib yields a very modest (two-week) enhancement in overall and progression-free survival when compared to gemcitabine alone. One noteworthy adverse effect of erlotinib is the development of a skin rash, and patients manifesting this side effect after erlotinib use may potentially experience improved survival outcomes (Moore et al., 2007).

If patients cannot handle combination treatments, they might consider gemcitabine alone. For locally advanced non-metastatic pancreatic cancer, doctors have studied adding radiotherapy to treatment, but it's not clear if this combined treatment helps patients live longer. There is not much proof from tests that this combined treatment works. It seems that this combined treatment helps patients live longer compared to just getting the best supportive care or only radiotherapy, but it also has more side effects (Huguet et al., 2009). However, giving induction chemotherapy may help identify patients with locally advanced pancreatic cancer who would benefit the most from chemoradiotherapy by excluding those with rapidly progressive disease (Krishnan et al., 2007).

Advances in Surgical Techniques

There have been great breakthroughs in the surgical management techniques of pancreatic cancer, which have led to improved postoperative mortality and morbidity rates (Malleo & Vollmer, 2016). In 1882, Friedrich Trendelenberg and later Alessandro Codivilla were the first surgeons to introduce the removal of the distal pancreas and/or splenectomy, for which the patients subsequently died postoperatively (de Rooij et al., 2015). The real breakthrough came in 1935, when Whipple et al. documented pancreaticoduodenectomy after a few other surgeons had performed it (Whipple et al., 1935). However, despite the introduction of the Whipple procedure, the postoperative morbidity rate remained around 60% (Lillemoe, 1995).

The current surgical management emphasizes the importance of perioperative management, with standardization of the enhanced recovery after surgery (ERAS) pathway (Hosoda et al., 2017). The ERAS pathway aids in preventing hypothermia, infection, and thrombophylaxis with nutritional management for patients with severe weight loss (>15%) (Ji et al., 2018; Lassen et al., 2012). There has been a steady development in minimally invasive techniques (laparoscopy) around the world. In general, there is slow progress in laparoscopic technical techniques for the management of pancreatic cancer; this is ascribed to the anatomic access of the pancreatic gland as well as its relation to major vascular structures such as splenic vessels and the superior mesenteric vein and artery (Mabrut et al., 2003).

The minimally invasive surgery includes the following.

Distal Pancreatectomy

The main indication is the management of neoplastic and non-neoplastic diseases of the pancreas. It can be performed laparoscopically and recently via robotic surgery. Open distal pancreatectomy shows inferiority to laparoscopic surgery with less hemorrhage and shorter hospital stays (de Rooij et al., 2015; Nakamura et al., 2015).

Pancreaticoduodenectomy

The resection of the pancreas and duodenum using minimally invasive techniques has also shown an improved morbidity rates with decreased volumes of blood required for transfusion, quicker adjuvant chemotherapy times, and shorter hospital stays compared to the open approach (Croome et al., 2014).

Resectability with Vascular Reconstruction

The other advance in pancreatic surgery is the redefinition of resectability with vascular reconstruction. According to the National Comprehensive Cancer Network's guidelines, they characterize borderline resectable disease

> as solid tumour contact with the superior mesenteric vein (SMV) or portal vein (PV) of greater than 180 degrees, solid tumour contact with the inferior vena cava (IVC), tumour contact with the common hepatic artery without extension to the celiac axis or hepatic artery bifurcation, or contact with the superior mesenteric artery (SMA) of greater than 180 degrees.
>
> **(Tseng et al., 2007)**

The unresectable disease is "greater than 180 degrees of SMA or celiac axis contact, or unreconstructible SMV or PV due to tumour involvement or thrombus." The definition of both borderline resectable disease and unresectable disease has now evolved to potentially resectable disease in specialized centers (Helmink et al., 2016; Tseng et al., 2007). Therefore, with these advances, diagnosis and identification of borderline resectable disease before surgery are significant in increasing the probability of resectability through the use of neoadjuvant therapy (Yekebas et al., 2008).

Arterial reconstruction has been historically superior to venous resection in relation to prognosis. However, recent studies have shown no difference in mortality rate in arterial and venous reconstruction performed for resectable disease. In addition, advancements in neoadjuvant therapies have led tumors with arterial invasion to be classified as borderline resectable or resectable disease (Bockhorn et al., 2011).

The venous reconstruction, similar to arterial, has also been associated with improved morbidity and mortality rates for pancreatic cancers otherwise classified as unresectable. Venous resection and reconstruction show no greater prognosis in comparison to resection of pancreatic carcinoma that does not need venous resection (Mollberg et al., 2011). These recent surgical advances aid in improving the morbidity of pancreatic surgery through rendering disease resectability through neoadjuvant therapies and advanced vascular reconstruction as well as minimally invasive techniques (Acher et al., 2018).

Surgical management of pancreatic carcinoma is the main method for improving prognosis and reaching an R0 (Acher et al., 2018; Masiak-Segit et al., 2018). Vascular reconstruction is a favorable method used to assist in increasing the chances of achieving tumor-free margins. The benefits of vascular resection and reconstruction still need to be studied with more research on the prevalence of lymphovascular invasion in pancreatic carcinoma and the effect of chemotherapy on vascular resection (Masiak-Segit et al., 2018).

Immunotherapy and Other Novel Molecular Options

Understanding the molecular aspects of PDAC is increasingly crucial for effective treatment and outcome prediction. The disease involves a complex interaction of various factors, particularly within the tumor microenvironment (TME), which includes both cellular and stromal components that significantly influence prognosis. This insight has spurred interest in tumor vaccination as a potential future therapeutic strategy. Key genes implicated in pancreatic carcinoma include *KRAS*, *TP53*, *CDKN2A*, and components of the *SMAD/TGFBR* signaling pathways. Additionally, genetic and immunological factors, along with epigenetic changes and molecular subtypes, play critical roles in PDAC progression and recurrence patterns following surgery.

Recurrence patterns are often classified as liver, lung, local only, and other distant metastases. Liver recurrence after pancreatectomy is associated with poor tumor differentiation. Lung recurrence is strongly associated with the classical well-differentiated subtype. Local recurrence may reflect the local infiltrative nature of the tumor into the peri-pancreatic nerve plexuses despite negative resection margins (Dreyer et al., 2022). There are encouraging developments in diagnosis and treatment, particularly through the use of liquid biopsies, which may enhance patient outcomes.

Immunotherapy

The tumor immune microenvironment in pancreatic carcinoma is immunologically suppressed due to the predominance of CD8+ immune effector cells as well as natural killer (NK) cells, in addition to multiple types of immune cells such as cancer-associated fibroblasts, regulatory T cells, tumor-associated macrophages, and tumor-infiltrating lymphocytes (TILs) (Wörmann et al., 2013). This immunosuppression is primarily caused by the expression of tolerance-inducing cell surface molecules (PD-1, CTLA-4, and CD40) and the secretion of immunosuppressive cytokines (IL-10, TGF-β). The ultimate status of the immune microenvironment is that of immune escape of the tumor from host immune surveillance, which promotes tumor growth and invasion (Sideras et al., 2014).

Immunotherapeutic strategies are designed to counteract the immunosuppressive environment surrounding tumors, thereby enhancing the tumor-killing activity of immune cells and improving patient outcomes. T cell activation occurs when antigen-presenting cells (APCs) display antigen peptides on major histocompatibility complex (MHC) class I and class II molecules, which interact with T cell receptors (TCRs). Co-stimulatory receptors, like CD40 on tumor-associated macrophages, further enhance T cell activation.

In the human body, T cells express inhibitory receptors, including programmed cell death-1 (PD-1) and cytotoxic T-lymphocyte-associated protein 4 (CTLA-4). These receptors regulate T cell activity by binding to their respective ligands, effectively dampening the immune response against specific targets, such as pancreatic cancer cells. This mechanism allows cancer cells to evade immune detection and destruction, complicating treatment efforts. Through the disruption of these inhibitory pathways, immunotherapy aims to reinvigorate T cell responses and improve the effectiveness of cancer treatments (Garrido-Laguna & Hidalgo, 2015).

Tumor antigens, which include both mutated proteins unique to cancer cells and normal proteins present in higher quantities on cancer cells than on normal cells, serve as promising targets for tumor vaccination. This immunotherapeutic strategy harnesses the immune system to recognize and destroy cancer cells. Tumor vaccines can be formulated as whole cells, proteins, peptides, DNA, or RNA, aiming to provoke antigen-specific T cell responses that inhibit tumor growth.

For instance, a mutant *KRAS* vaccine has shown the ability to induce T-helper responses in patients with resected pancreatic carcinoma, resulting in improved long-term survival during a phase 1/2 clinical trial (Wedén et al., 2010). Additionally, the GVAX pancreas vaccine has demonstrated the capacity to stimulate systemic anti-tumor immunity in both resected and advanced pancreatic cancer patients across multiple clinical trials. The immune response was evident whether the vaccine was administered alone or in conjunction with conventional chemotherapy, suggesting that the efficacy may be influenced by the vaccine dosage (Le et al., 2015). These findings underscore the potential of tumor vaccination as a significant component of cancer treatment strategies, enhancing the immune response against tumors and potentially improving patient outcomes.

Tumor Microenvironment of Pancreatic Cancer

PDAC and the surrounding TME consist of various cell types, including adaptive immune cells, macrophages, dendritic cells (DCs), natural killer NK cells, and other innate immune cells (Muller

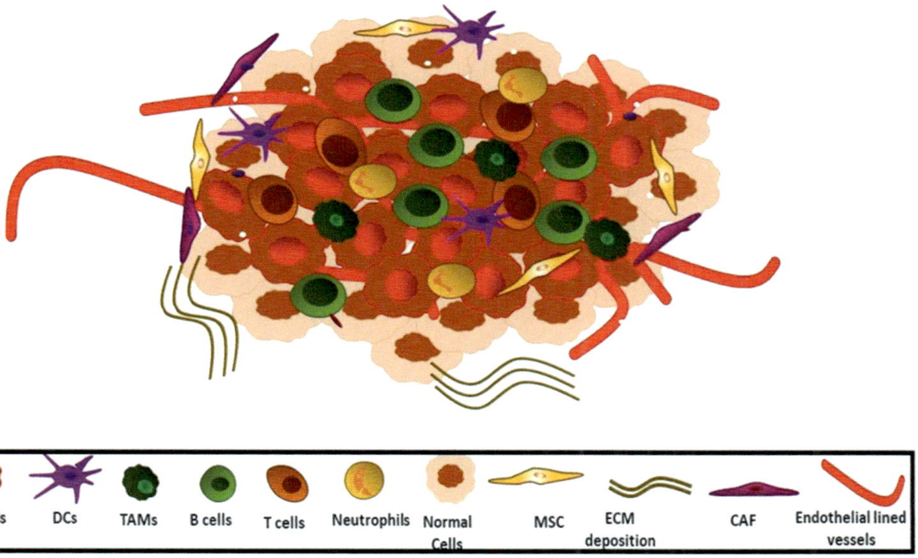

FIGURE 6.1 Pancreatic cancer and the TME. PDAC cells activate the immune response through antigen-presenting cells, involving dendritic cells (DCs), macrophages, and CD4+ helper T cells, which trigger CD8+ cytotoxic T cells to eradicate cancer. B cells' role in the TME has shown tumor progression characteristics by secreting IL-35 and activating immunosuppressive TAMs. Neutrophil roles in the TME can either promote or suppress cancer growth and produce neutrophil extracellular traps (NETs), preventing effective immune function to eliminate cancer cells. Mesenchymal stromal cells (MSCs) transform into tumor-associated mesenchymal stem cells (TA-MSCs) within the TME, inducing tumor growth, metastasis, and angiogenesis. The extracellular matrix (ECM) supports cancer proliferation and migration. Cancer-associated fibroblasts (CAFs) regulate tumor metabolism for cancer cell proliferation and suppress anti-tumor immune responses. Endothelial cells promote tumor growth and metastasis through angiogenesis.

et al., 2022) (Figure 6.1). It shows strong resistance to immunotherapeutic treatment strategies due to immune cells, particularly the influx of myeloid cells (Farhangnia et al., 2024). Tumor-associated macrophages (TAMs), which include monocyte-derived (MDM) and tissue-resident macrophages (TRM), are also seen, which facilitate chronic inflammation within the TME (Yang et al., 2021). The characteristics of TAMs within the TME are influenced by the genetic background of the cells, the surrounding TME, and the specific tissue environment (Hu et al., 2024). PDAC, TRMs contribute to the development of tissue fibrosis and the formation of a pro-tumor extracellular matrix. This matrix inhibits the infiltration of T cells into the desmoplastic stroma, resulting in immune exclusion and reduced effectiveness of anti-cancer therapies (Truong and Pauklin, 2021; Farhangnia et al., 2024; Juan Liu et al., 2024).

CD8 and CD4 coreceptors enhance T cell receptor (TCR) sensitivity to antigens through their interaction with major histocompatibility complex (MHC) class I and II peptide molecules. When TCRs bind to cognate peptides, this interaction triggers T cell activation, leading to critical effector functions such as proliferation, secretion of cytokines, and cytolytic activity facilitated by perforin and granzyme release (Shafer et al., 2022). The presence of more CD8+ cytotoxic T lymphocytes (CTLs) encircling cancerous cells is associated with better survival rates of patients, but a lesser percentage of CD4+ T cells is associated with a shorter survival rate (Farhangnia et al., 2024; Gorchs et al., 2024). Parallel comprehensive studies analyzing tumor cancer cells, stroma cells, connective tissue, and tumor genomic analysis are required to provide deeper insight into overcoming the challenges of TME in PDAC and successful immune therapeutic strategies (Farhangnia et al., 2024; K. He et al., 2015; Valkenburg et al., 2018).

Emerging Role of Immunotherapy in Pancreatic Cancer

Pancreatic cancer is comprised mostly of PDAC, defined by three PDA subtypes: classical, quasi-mesenchymal, and exocrine-like (Collisson et al., 2011; Martens et al., 2019). Pancreatic cancer is a major cause of cancer-related death (R. He et al., 2024). PDAC has been shown to demonstrate moderate mutations and therefore does not trigger the host immune response to target tumor cells (Hartupee et al., 2024; Narayanan et al., 2021). PDAC is therefore known as an immune-cold tumor and has shown to be unresponsive to immunotherapy in various clinical trials (Hartupee et al., 2024; Ullman et al., 2022). This resistance of PDAC is due to its desmoplastic TME, lack of effector T lymphocytes, and T helper 2 (TH2) immune cell dominance, forming a hypoxic environment conducive to enhanced tumorigenesis and blocking immune cell penetrance (Deng et al., 2022; Hsu et al., 2022). More studies exploring resistance mechanisms and disabling these systems are required to overcome immunotherapy challenges in the treatment of the disease (Hsu et al., 2022; Said & Ibrahim, 2023).

Immune Pathways in PDAC

Immunotherapy has altered standard cancer treatment significantly with the revolutionary development of immune checkpoint blockers (ICB) preventing the binding of programmed cell death protein 1 (PD1) to programmed death-ligand 1 (PD-L1) as well as cytotoxic T-lymphocyte-associated protein 4 (CTLA-4), re-establishing the immune system to eradicate tumor cells (Shiravand et al., 2022). T cell immune-suppressive activity facilitates tumor cytotoxicity. CTLs that produce tumor necrosis factor alpha (TNFα) and interferon gamma (IFNγ), perforin, and granzymes generate memory cells that protect cancer cells and targets and destroy the antigen-presenting tumor through apoptosis (Farhangnia et al., 2024). Monoclonal antibody therapy strategies that inhibit CTLA4 and PDCD1 enhance the anti-tumor response of T cells and have revealed significant clinical improvement in various cancers. For PDAC, however, ICB has not shown any major leap forward in the treatment of the disease (Fath et al., 2022; Nomi et al., 2007). More interventions are required in activating the anti-tumor potential of the immune system and stimulating the reprogramming of an anti-tumor phenotype.

Neutrophil extracellular traps (NETs) are chromatin web-like extrusions derived from neutrophils as a consequence of either infection, injury, cytokines, microorganisms, etc. NETs play an important role in innate immunity and are shown to be considerably involved in cancer progression, metastasis, and drug resistance (Zhao & Jin, 2022). Increased glycolysis promotes the formation of NETs, which support tumor growth by protecting it from cytotoxic cells and promoting invasion and metastasis by activating stellate cells and desmoplastic stromal cells (Mitsis et al., 2022; Zhang et al., 2020). NETs also enhance tumor cell motility, aiding metastasis, by binding with the NET-associated DNA receptor CCDC25 on tumor cell surfaces, operating through the ILK-β-Parvin pathway (Hu et al., 2024; Zhao & Jin, 2022).

Established tumor cells have the ability to recruit neutrophils by secreting factors such as IL-8 (Jin et al., 2021). Tumor-associated neutrophils (TANs) stimulate anti-tumor activity (N1 neutrophils) or pro-tumor activity (N2 neutrophils) (Antuamwine et al., 2023). N1 TANS targets tumor growing cells and metastasis by direct or antibody-dependent cytotoxicity, and innate and adaptive immune cells activation, such as B and T cell lymphocytes, NK cells, and DCs (Antuamwine et al., 2023; Masucci et al., 2019). N2 however, stimulates tumor growth and its spread by secreting ECM remodeling enzymes and pro-angiogenic factors that stimulate metastasis and angiogenesis (Masucci et al., 2019).

Toll-like receptor-4 (TLR4) decreases the level of miR-29C through the NFκB pathway (Hu et al., 2024; Orlacchio & Mazzone, 2021). Consequently, this reduction in miR-29C levels leads to an increase in its inhibitory target, MMP-9, which promotes tumor invasion. (Hu et al., 2024; Lu et al., 2015). Activation of TLR4 in M2 TAMs has also been shown to activate EMT in pancreatic cancer cells through IL-10 signaling (Hu et al., 2024; Liu et al., 2013). NOD-like receptor family pyrin domain–containing 3 (NLRP3) signaling (IL-10 dependent) was shown to induce the expansion of immune-suppressive macrophages in PDAC. These macrophages direct CD4+ T cell differentiation into tumor-promoting T helper type-2 (Th2), type-17 (Th17), and regulatory T (Treg) cells, and further inhibit Th1 cell polarization and CD8+ T cell activation (Daley et al., 2017).

Immunotherapeutic Strategies in PDAC

Targeted immunotherapy has the potential to be more effective than standard treatments in increasing survival and enhancing immune responses in PDAC patients (Farhangnia et al., 2024). To convert PDAC into an immunologically hot tumor, literature proposes different strategies, including targeting cancer stem cells (CSCs) in collaboration with the TME to overcome evasion of CSC detection by the immune system (Farhangnia et al., 2024; Nallasamy et al., 2022). Key candidates for novel approaches to immunotherapy include immunomodulators, immune checkpoint inhibitors (ICIs), oncolytic virus therapy (OVT), adoptive cell transfer (ACT) therapy including T cell receptor (TCR) engineered T cell therapy, chimeric antigen receptor (CAR) T cell therapy, CAR NK cell therapy, and cytokine-induced killer cells (Gupta et al., 2022) (Figure 6.2). Additionally, studying immune checkpoint inhibitors (ICIs), immunomodulators, cancer vaccines, as well as the effects of the gut microbiome and ICIs, and CRISPR/Cas9 gene editing in PDAC is required to overcome the challenges faced with immunotherapy (Farhangnia et al., 2024; Golonko et al., 2024).

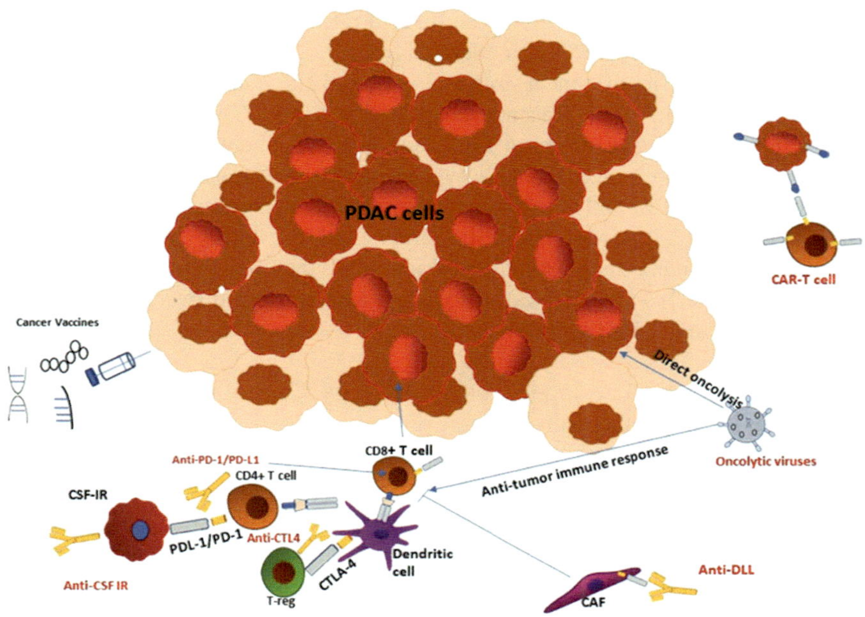

FIGURE 6.2 Immunotherapeutic strategies in PDAC. Immunotherapy approaches include immunomodulators, immune checkpoint inhibitors (ICIs), oncolytic virus therapy (OVT), adoptive cell transfer (ACT), and cancer vaccines is administered to activate or augment existing anti-tumor immune responses. Another strategy includes oncolytic virus therapy (OVT), where an oncolytic virus infiltrates and lyses the cancerous cell, releasing tumor antigens into the bloodstream, and induces an immune reaction. Adoptive cell transfer (ACT) therapy, such as chimeric antigen receptor (CAR) T cell therapy, involves autologous T cells from an individual being engineered ex vivo (in vitro) to enhance their ability to identify specific antigens found on cancer cells, leading to their eradication. The presence of more CD8+ cytotoxic T lymphocytes (CTLs) encircling cancerous cells is associated with a boost in the survival rates of patients. Inhibitor-based approaches targeting immune checkpoint molecules (e.g., anti-PDL-1/PD-1, anti-CTLA-4) and anti-colony-stimulating factor 1 (CSF-IR) and anti-delta-like ligand (DLL) have shown to enhance CD8+ T cell infiltration in PDAC.

Oncolytic Virus Therapy Oncolytic virus (OV) immunotherapy is a novel approach to cancer cell eradication, through an oncolytic virus lysing the cancerous cell, resulting in tumor antigens emitted into the bloodstream, which stimulates an immune reaction within the patient. (Nisar et al., 2022). Oncolytic virus therapy (OVT) has high specificity as they are genetically modified to exclusively infect cancer cells, and both DNA and RNA viruses are currently being investigated in the treatment of multiple cancer types (Farhangnia et al., 2024; Lemos de Matos et al., 2020).

Adoptive Cellular Transfer Therapy Adoptive cellular transfer therapy (ACT) utilizes autologous tumor-infiltrating lymphocytes (TIL) in targeting cancer cells (Rosenberg et al., 2008). Different strategies include non-specific immunomodulation, for example, where T cell growth factor interleukin 2 (IL2) is administered to a patient, activating endogenous tumor-reactive cells, resulting in the eradication of certain human solid cancers (Rosenberg et al., 1998). Another strategy includes cancer vaccinations through the active immunization of cancer patients, using autologous whole cells, proteins, or peptides, therefore selectively targeting only the cancer-antigen-presenting cells. Genetic alteration of lymphocytes using retroviruses is an additional approach, where normal lymphocytes can be modified into lymphocytes with anti-cancer activity through the ability to distinguish tumor cells from normal cells (e.g., CAR T cell therapy) (Farhangnia et al., 2024; Rosenberg et al., 2008).

Tumor-infiltrating Lymphocyte Therapy Tumor infiltrating lymphocyte (TIL) therapy carries great potential in the treatment of PDAC patients; however, a lack of avidity for tumor cells and reactivity against a wide array of cancer antigens is one of the major challenges that need to be overcome (Farhangnia et al., 2024; Rosenberg & Dudley, 2009). The process includes isolating autologous cells in vitro and infusing these TILs back into the patient (Hall et al., 2016). Effective TIL extraction techniques to increase reactivity and production strategies include inhibiting the PD-1 receptor, inducing 4-1BB, and increasing CD8+ T cell concentrations. Hall et al. (2016) identified functional expanded TILs from pancreatic tumors with the ability to identify cancer-antigen cells (Hall et al., 2016). A meta-analysis study also identified that the long-term prognosis of PDAC patients with increased CD8+ T cells was associated with improved overall survival and lower recurrence rates (Orhan et al., 2020; Panahi et al., 2023).

T Cell Receptor and Chimeric Antigen Receptor T Cell Gene Therapy Due to the logistical challenges with TIL manufacturing, greater efforts in T cell receptor α and β (TCR) engineering have been developed to target more specific antigen-representing cells, including T cell receptor (TCR) and chimeric antigen receptor T cell (CAR T) and gene therapy (Baulu et al., 2023; Farhangnia et al., 2024; Shafer et al., 2022). Autologous TCR gene therapy involves TCR genes edited in vitro, designed to target a wide range of tumor antigen-representing cells (Baulu et al., 2023; Leidner et al., 2022). TCR cell therapy targets intracellular (cytosol) and extracellular (membrane) derived peptides presented by the antigen-presenting major histocompatibility complex (MHC) class I (A/B/C) and II (DR/DP/DQ) (Shafer et al., 2022) molecules. T cell binding is, however, confined to human leukocyte antigen (HLA) epitope representation, which is a limiting factor, reducing the number of eligible patients who could potentially benefit from TCR-T cell therapy. In addition, a lower density of epitopes (1–50 per cell) is required for TCR-T cell activation, detection, and eradication, in comparison to CAR-T cells epitope requirements (10^3 epitopes per cell) (Chandran & Klebanoff, 2019).

PDAC Vaccines

PDAC vaccination immunotherapy strategies are still in the research and initial clinical trial phases, with fluctuating results and accomplishments. Whole-tumor cell vaccine (autologous or allogeneic PDAC cells engineered to express GM-CSF) is administered before surgery to induce T cell infiltration (Eager & Nemunaitis, 2005). In a cohort of patients with previously treated PDAC, a phase II clinical trial evaluating the combination of cyclophosphamide and GVAX, a bacterium-based vaccine, revealed that individuals receiving CRS-207 exhibited a notable increase in OS when compared to those receiving second-line chemotherapy (Tsujikawa et al., 2020).

In the randomized phase III IMPRESS study, vaccine therapy (Algenpantucel-L), which is a whole-cell vaccine genetically engineered to enhance complement and antibody-dependent cytotoxicity, was incorporated into adjuvant standard-of-care chemotherapy. Despite these efforts, the study results indicated that this approach did not show a survival advantage when compared with the control group (Hewitt et al., 2022; Ullman et al., 2022).

Inhibitory Immune Checkpoints Therapy

PD-1 and PD-L1 in tumor cells inhibit T cell activation. ICI molecules in PDAC produce different reactions in individuals, as illustrated by PD-L1 positive patients exhibiting unfavorable responses while other PD-L1 negative patients respond favorably (Farhangnia et al., 2024; Zhang et al., 2022). These data infer possible interferences of alternative PD-1 ligands, inhibiting PD-1 immunotherapy in PDAC. PD-L2 ligand molecules have been found over-expressed in senescent cancer cells, a strategic mechanism to evade the immune response within tumors (Chaib et al., 2024). An animal model study in mice demonstrated that inhibiting PD-L2 in combination with chemotherapy resulted in tumor regression and remission in mice, indicating a potential treatment strategy targeting senescent cancer cells (Chaib et al., 2024). Combining immunotherapy targeting PD-L1 and CCL5 was shown to be beneficial in PDAC patients, resulting in decreased Treg and TAM levels and inducing CD8+ T cell activation, promoting tumor regression, and improving OS (Wang et al., 2020). The combined use of anti-tumor necrosis factor receptor 2 (TNFR2) and PD-L1 monoclonal antibodies resulted in tumor regression, improved overall survival, and the production of anti-tumor memory cells. This was achieved by reducing the presence of Tregs and TAMs while activating CD8+ T cells in the microenvironment of PDAC (Farhangnia et al., 2024).

Chemotherapy and Combination Therapies

Pancreatic ductal adenocarcinoma remains a prime medical treatment challenge even for well-developed countries. There is substantial heterogeneity in the treatment modalities and protocols used. Patient survival depends on combination therapy with surgical resection combined with postoperative adjuvant chemotherapy. Chemotherapy for PDAC involves anti-cancer drugs that are administered orally or intravenously and enter the bloodstream to render decrease the probability of the tumor metastasizing. Although surgery remains the primary mode of treatment depending on the resectability of the tumor (Masiak-Segit et al., 2018). Multiple protocols for neoadjuvant therapy have been studied, namely chemotherapy (CT), chemoradiation therapy (CRT), and total neoadjuvant therapy (TNT) (Barrak et al., 2022). The first indication of chemotherapy for PDAC is before surgery (neoadjuvant therapy), mainly to decrease tumor volume in borderline resectable or locally advanced PDAC (Springfeld et al., 2023) The second is after surgery (adjuvant therapy), usually along with radiation to destroy cancer cells post-surgery (Grossberg et al., 2020). Lastly, chemotherapy is used in advanced disease in unresectable tumors with or without metastasis (Grossberg et al., 2020).

Neoadjuvant therapy administered in borderline PDAC has been well studied (in the Surveillance, Epidemiology, and National Cancer Database) to show improvement in the rate of resection aiding in prognosis (Stessin et al., 2008). ESPAC5 (formerly known as ESPAC-5f) multicenter also was innovative in its work when it proved the practicality of using short-course neoadjuvant therapy in three different types before surgery (Ghaneh et al., 2020). Ishikawa et al. showed 17 of 23 patients with an improved resectability rate (Ishikawa, 1994). There are two main regimens of chemotherapy used, namely FOLFIRINOX (folinic acid, fluorouracil, irinotecan, and oxaliplatin)—modified FOLFIRINOX and GEMCAP (gemcitabine plus capecitabine) for both neoadjuvant and adjuvant purposes (Conroy et al., 2018; Gillen et al., 2010; Neoptolemos et al.,

2017). They both show great efficacy in decreasing tumor load and secondary tumor formation; however, they are not devoid of toxic side effects (Cunningham et al., 2009). The two regimens in studies show that even though modified FOLFIRINOX leads to more resectable tumors, its toxicity impedes attempts of resections when compared to gemcitabine plus capecitabine (Conroy et al., 2011; Hackert et al., 2016).

Postoperative CRT refers to a combination of chemotherapy and radiation after surgery. In many studies, CRT shows great improvement in tumor destruction and prevention of secondary tumor formation (Klinkenbijl et al., 1999). The main disadvantage of postoperative chemotherapy is that approximately 25% of patients develop post-surgery complications. In comparison to postoperative CRT, neoadjuvant chemotherapy is received promptly by all patients with resectable disease (Ishikawa, 1994). When combined with radiation, the tissues are well-aerated, increasing efficacy. In metastatic pancreatic disease, preoperative chemotherapy helps to avoid laparotomy and the complications associated with surgery. Lastly, preoperative chemoradiation aids in the transformation of locally advanced lesions into resectable tumors, thus increasing prognosis (Fogelman et al., 2004).

Total neoadjuvant therapy (TNT), also known as a "surgery-last approach," is defined as the synchronous administration of chemotherapy and CRT before tumor resection (Kim et al., 2020). It promises to intensify total neoadjuvant chemotherapy (NAT) by combining it with CRT (Janssen et al., 2021). Currently, there is not enough evidence to prove this hypothesis (Cannas & Vollmer, 2024).

Therefore, PDAC treatment remains heterogeneous, and more innovative work continues to optimize patient survival through various modalities and individualized treatment approaches.

Palliative Care

Patients with PDAC often need palliative care to alleviate biliary obstruction symptoms and improve their quality of life. The placement of self-expanding metal stents is the preferred approach for relieving biliary obstruction in patients with unresectable disease. This approach is associated with lower rates of recurrent biliary obstruction and longer intervals before recurrence, leading to a decreased incidence of cholangitis compared to plastic stents (Moss et al., 2006). For patients experiencing gastric outlet obstruction, endoscopic placement of an enteral stent is recommended for those with limited life expectancy or poor performance status. In contrast, gastrojejunostomy is more effective for patients with a better prognosis and longer life expectancy (Jeurnink et al., 2007).

Patients who have locally advanced or metastatic pancreatic cancer often face the challenge of dealing with cancer-related pain. The cornerstone of pain management for these individuals involves the use of analgesic medications. However, in cases where patients do not achieve adequate pain control with analgesics or experience undesirable side effects, alternative interventions such as endoscopic ultrasound (EUS) or image-guided celiac plexus neurolysis can be considered. These procedures have shown promising results in significantly improving pain relief for these patients (Zhong et al., 2013).

A thorough nutritional assessment and the administration of oral pancreatic enzymes are crucial for patients with both unresectable and resected pancreatic cancer (Gilliland et al., 2017). These patients are at an elevated risk of venous thromboembolism, and the use of low molecular weight heparin is recommended, as clinical trials indicate that it significantly lowers the incidence of such events. For more information, you can refer to relevant clinical guidelines and studies on pancreatic cancer management (Pelzer et al., 2015).

CASE STUDIES OF INNOVATIVE TREATMENTS IN AFRICAN CONTEXTS

Pancreatic cancer can be hard to treat because it usually has spread extensively by the time of diagnosis. Standard treatment consists of surgery, chemotherapy, radiation, or a combination of each depending on the cancer stage. Researchers in first-world countries are looking at possible new drugs, combinations of standard drugs, and new modalities such as immunotherapy. Biomarker testing and genetic changes to help suggest suitable treatment and indicate family members at increased risk of developing pancreatic cancer are also being explored.

AFRICAN LIMITATIONS

The major limitations of curative treatment procedures include resource constraints and late presentation in most third-world countries, which impede imaging studies to accurately classify the PDAC. Once a diagnosis is made, further limitations include the unavailability of various immunotherapeutic agents, such as immune-checkpoint inhibitors, in third-world countries. CT treatment, for example, is only available in first-world countries such as the United States and United Kingdom. In a South African context, M. C. Herbst found strong evidence that obesity is associated with the development of pancreatic cancer and that weight loss may help reduce the risk of this cancer. Type 2 diabetes and fasting insulin show a suggestive association that needs further investigation (CANSA). The administration of chemotherapy before surgical resection is gaining momentum in South Africa, as it ensures early and universal delivery and improves margin-negative resection rates, thus enhancing long-term survival. However, patient-centered research on neoadjuvant therapy is lacking in South Africa despite its recommendations. In his review, J. M. Cloyd highlights patient-reported outcomes, preferences, experience, and survivorship as opportunities to focus on.

The South African Department of Health approved the use of gemcitabine in combination with capecitabine as adjuvant chemotherapy in patients with fully resectable pancreatic adenocarcinoma in 2018, after extensive trials and price reductions showed them to work for these patients (Vera et al., 2016). FOLFIRINOX has become the standard of care for patients with good performance status and the ability to tolerate chemotherapy; however, it has however not been fully reviewed. Gemcitabine monotherapy is recommended as a palliative care option for selected patients with metastatic pancreatic cancer, with an ECOG performance status of 0–2 and a bilirubin level lower than 1.5× the upper limit of normal (ULN). These work well where resources are available. In those few African countries where CT scans are available, histological diagnoses are not available because of a lack of endoscopic ultrasound and the expertise to perform it, making research in this field difficult. This is well demonstrated by N. E. Kivuyo's study, where histology of pancreatic masses could be obtained in only 32.7% of the patients at Mahimbili National Hospital in Tanzania, an essential test for CT (Kivuyo & Akoko, 2022).

According to a study by Muchiri et al., 2024, it was found that only 7% of pancreatic cancer in Kenyatta National Hospital in Kenya was amenable to curative surgery and CT; the majority presented late, receiving only supportive care. In their study, they found that the median survival time was three months, and the one-year survival was 32%. Currently, there are no other innovative treatments in the African context because of these limitations. This emphasizes the fact that investigative procedures necessary to diagnose and treat pancreatic cancer are capital-intensive and require well-trained personnel in pancreatic surgery. Surgical complications are also quite prevalent, even in the best hands, which limits the uptake of surgical intervention. The late diagnosis limits the treatment options in most patients, thus resorting to palliative measures.

CONCLUSION

Despite significant advancements in therapeutic strategies, such as novel chemo(radio)therapy regimens in (neo)adjuvant settings and more sophisticated surgical techniques, the clinical outcomes for pancreatic cancer patients remain unsatisfactory. While these efforts have improved specific aspects of treatment, the prognosis for many patients continues to be bleak. Looking forward, we expect an increase in neoadjuvant therapeutic approaches for non-metastatic pancreatic cancer, particularly as more cases undergo molecular profiling. This personalized approach to treatment will enable better predictions of treatment efficacy, patient prognosis, and overall outcomes. A deeper understanding of the molecular mechanisms driving pancreatic cancer is essential for the development of targeted therapies that can address the disease at a more individualized level.

This aligns with the broader view, which highlights that surgical intervention, combined with adjuvant chemotherapy, remains the most effective curative approach for pancreatic cancer in its early stages. However, the growing emphasis on neoadjuvant therapies—particularly for borderline or unresectable tumors—is expected to redefine treatment paradigms. Molecular testing of tumor biopsies could lead to the application of tailored, targeted therapies that reduce the tumor burden and, in some cases, convert unresectable tumors into resectable ones. Moreover, advancements in immunotherapy offer new hope by stimulating the immune system to fight cancer cells more effectively. Continued research into the molecular mechanisms of pancreatic ductal carcinoma will likely play a pivotal role in the future of personalized medicine for this challenging disease, potentially improving outcomes for a broader spectrum of patients. Together, these insights emphasize the need for ongoing innovation in both molecular and clinical approaches to improving pancreatic cancer treatment.

REFERENCES

ACHER, A. W., BLEICHER, J., CANNON, A., & SCAIFE, C. 2018. Advances in surgery for pancreatic cancer. *Journal of Gastrointestinal Oncology*, *9*(6), 1037–1043. https://doi.org/10.21037/jgo.2018.05.05.

ANTUAMWINE, B. B., BOSNJAKOVIC, R., HOFMANN-VEGA, F., WANG, X., THEODOSIOU, T., ILIOPOULOS, I., & BRANDAU, S. 2023. N1 versus N2 and PMN-MDSC: A critical appraisal of current concepts on tumor-associated neutrophils and new directions for human oncology. *Immunological Reviews*, *314*(1), 250–279. https://doi.org/10.1111/imr.13176.

BARRAK, D., VILLANO, A. M., VILLAFANE-FERRIOL, N., STOCKTON, L. G., HILL, M. V, DENG, M., HANDORF, E. A., & REDDY, S. S. 2022. Total neoadjuvant therapy for pancreatic adenocarcinoma increases probability for a complete pathologic response. *European Journal of Surgical Oncology*, *48*(6), 1356–1361. https://doi.org/10.1016/j.ejso.2021.12.473.

BAULU, E., GARDET, C., CHUVIN, N., & DEPIL, S. 2023. TCR-engineered T cell therapy in solid tumors: State of the art and perspectives. *Science Advances*, *9*(7), 1–15. https://doi.org/10.1126/sciadv.adf3700.

BENASSAI, G., QUARTO, G., PERROTTA, S., FURINO, E., BENASSAI, G. L., AMATO, B., BIANCO, T., DE PALMA, G., & FORESTIERI, P. 2015. Long-term survival after curative resection for pancreatic ductal adenocarcinoma – Surgical treatment. *International Journal of Surgery*, *21*, S1–S3. https://doi.org/10.1016/j.ijsu.2015.06.050

BOCKHORN, M., BURDELSKI, C., BOGOEVSKI, D., SGOURAKIS, G., YEKEBAS, E. F., & IZBICKI, J. R. 2011. Arterial *en bloc* resection for pancreatic carcinoma. *Journal of British Surgery*, *98*(1), 86–92. https://doi.org/10.1002/bjs.7270.

BAILEY, P.,CHANG, D., NONES, K. et al. 2016.Genomic analyses identify molecular subtypes of pancreatic cancer. *Nature*, *531*, 47–52 https://doi.org/10.1038/nature16965

BRUNNER, M., WU, Z., KRAUTZ, C., PILARSKY, C., GRÜTZMANN, R., & WEBER, G. F. 2019. Current clinical strategies of pancreatic cancer treatment and open molecular questions. *International Journal of Molecular Sciences*, *20*(18), 4543. https://doi.org/10.3390/ijms20184543.

CANNAS, S., & VOLLMER, C. M. 2024. Total neoadjuvant therapy for pancreatic cancer—What is totality? *JAMA Surgery*, *159*(7), 828. https://doi.org/10.1001/jamasurg.2024.0419.

CASADEI, R., DI MARCO, M., RICCI, C., SANTINI, D., SERRA, C., CALCULLI, L., D'AMBRA, M., GUIDO, A., MORSELLI-LABATE, A.M., & MINNI, F. 2015. Neoadjuvant chemoradiotherapy and surgery versus surgery alone in resectable pancreatic cancer: Aa single-center prospective, randomized, controlled trial which failed to achieve accrual targets. *Journal of Gastrointestinal Surgery, 19*(10), 1802–1812.

CHAIB, S., LÓPEZ-DOMÍNGUEZ, J. A., LALINDE-GUTIÉRREZ, M., PRATS, N., MARIN, I., BOIX, O., GARCÍA-GARIJO, A., MEYER, K., MUÑOZ, M. I., AGUILERA, M., MATEO, L., STEPHAN-OTTO ATTOLINI, C., LLANOS, S., PÉREZ-RAMOS, S., ESCORIHUELA, M., AL-SHAHROUR, F., CASH, T. P., TCHKONIA, T., KIRKLAND, J. L., … SERRANO, M. 2024. The efficacy of chemotherapy is limited by intratumoral senescent cells expressing PD-L2. *Nature Cancer*, *5*(3), 448–462. https://doi.org/10.1038/s43018-023-00712-x.

CHANDRAN, S. S., & KLEBANOFF, C. A. 2019. T cell receptor-based cancer immunotherapy: Emerging efficacy and pathways of resistance. *Immunological Reviews*, *290*(1), 127–147. https://doi.org/10.1111/imr.12772.

CHIN, V., NAGRIAL, A., SJOQUIST, K., O'CONNOR, C. A., CHANTRILL, L., BIANKIN, A. V, SCHOLTEN, R. J., & YIP, D. 2018. Chemotherapy and radiotherapy for advanced pancreatic cancer. *The Cochrane Database of Systematic Reviews*, *3*(3), CD011044. https://doi.org/10.1002/14651858.CD011044.pub2.

COLLISSON, E. A., SADANANDAM, A., OLSON, P., GIBB, W. J., TRUITT, M., GU, S., COOC, J., WEINKLE, J., KIM, G. E., JAKKULA, L., FEILER, H. S., KO, A. H., OLSHEN, A. B., DANENBERG, K. L., TEMPERO, M. A., SPELLMAN, P. T., HANAHAN, D., & GRAY, J. W. 2011. Subtypes of pancreatic ductal adenocarcinoma and their differing responses to therapy. *Nature Medicine*, *17*(4), 500–503. https://doi.org/10.1038/nm.2344.

CONROY, T., DESSEIGNE, F., YCHOU, M., BOUCHÉ, O., GUIMBAUD, R., BÉCOUARN, Y., ADENIS, A., RAOUL, J.-L., GOURGOU-BOURGADE, S., DE LA FOUCHARDIÈRE, C., BENNOUNA, J., BACHET, J.-B., KHEMISSA-AKOUZ, F., PÉRÉ-VERGÉ, D., DELBALDO, C., ASSENAT, E., CHAUFFERT, B., MICHEL, P., MONTOTO-GRILLOT, C., & DUCREUX, M. 2011. FOLFIRINOX versus gemcitabine for metastatic pancreatic cancer. *New England Journal of Medicine*, *364*(19), 1817–1825. https://doi.org/10.1056/nejmoa1011923.

CONROY, T., HAMMEL, P., HEBBAR, M., BEN ABDELGHANI, M., WEI, A. C., RAOUL, J.-L., CHONÉ, L., FRANCOIS, E., ARTRU, P., BIAGI, J. J., LECOMTE, T., ASSENAT, E., FAROUX, R., YCHOU, M., VOLET, J., SAUVANET, A., BREYSACHER, G., DI FIORE, F., CRIPPS, C., … BACHET, J.-B. 2018. FOLFIRINOX or gemcitabine as adjuvant therapy for pancreatic cancer. *New England Journal of Medicine*, *379*(25), 2395–2406. https://doi.org/10.1056/nejmoa1809775.

CONROY, T., PFEIFFER, P., VILGRAIN, V., LAMARCA, A., SEUFFERLEIN, T., O'REILLY, E. M., HACKERT, T., GOLAN, T., PRAGER, G., HAUSTERMANS, K., VOGEL, A., & DUCREUX, M. 2023. Pancreatic cancer: ESMO Clinical Practice Guideline for diagnosis, treatment, and follow-up. *Annals of Oncology*, *34*(11), 987–1002. https://doi.org/10.1016/j.annonc.2023.08.009.

CROOME, K. P., FARNELL, M. B., QUE, F. G., REID-LOMBARDO, KM., TRUTY, M. J., NAGORNEY, D. M., & KENDRICK, M. L. 2014. Total laparoscopic pancreaticoduodenectomy for pancreatic ductal adenocarcinoma. *Annals of Surgery*, *260*(4), 633–640. https://doi.org/10.1097/sla.0000000000000937.

CUNNINGHAM, D., CHAU, I., STOCKEN, D. D., VALLE, J. W., SMITH, D., STEWARD, W., HARPER, P. G., DUNN, J., TUDUR-SMITH, C., WEST, J., FALK, S., CRELLIN, A., ADAB, F., THOMPSON, J., LEONARD, P., OSTROWSKI, J., EATOCK, M., SCHEITHAUER, W., HERRMANN, R., & NEOPTOLEMOS, J. P. 2009. Phase III randomized comparison of gemcitabine versus gemcitabine plus capecitabine in patients with advanced pancreatic cancer. *Journal of Clinical Oncology*, *27*(33), 5513–5518. https://doi.org/10.1200/jco.2009.24.2446.

DAHAN, L., BONNETAIN, F., YCHOU, M., MITRY, E., GASMI, M., RAOUL, J.-L., CATTAN, S., PHELIP, J.-M., HAMMEL, P., CHAUFFERT, B., MICHEL, P., LEGOUX, J.-L., ROUGIER, P., BEDENNE, L., SEITZ, J.-F., & DIGESTIVE, F. F. DE C. 2010. Combination 5-fluorouracil, folinic acid and cisplatin (LV5FU2-CDDP) followed by gemcitabine or the reverse sequence in metastatic pancreatic cancer: final results of a randomised strategic phase III trial (FFCD 0301). *Gut*, *59*(11), 1527–1534. https://doi.org/10.1136/gut.2010.216135.

DALEY, D., MANI, V. R., MOHAN, N., AKKAD, N., BALASUBRAMANIA PANDIAN, G. S. D., SAVADKAR, S., LEE, K. B., TORRES-HERNANDEZ, A., AYKUT, B., DISKIN, B., WANG, W., FAROOQ, M. S., MAHMUD, A. I., WERBA, G., MORALES, E. J., LALL, S., WADOWSKI, B. J., RUBIN, A. G., BERMAN, M. E., … MILLER, G. 2017. NLRP3 signaling drives macrophage-induced adaptive immune suppression in pancreatic carcinoma. *Journal of Experimental Medicine*, *214*(6), 1711–1724. https://doi.org/10.1084/jem.20161707.

DEPARTMENT OF HEALTH ANNUAL REPORT 2023 | SOUTH AFRICAN GOVERNMENT. (2023). https://www.gov.za/documents/annual-reports/department-health-annual-report-20232024-27-sep-2024

DE ROOIJ, T., JILESEN, A. P., BOERMA, D., BONSING, B. A., BOSSCHA, K., VAN DAM, R. M., VAN DIEREN, S., DIJKGRAAF, M. G., VAN EIJCK, C. H., GERHARDS, M. F., VAN GOOR, H., VAN DER HARST, E., DE HINGH, I. H., KAZEMIER, G., KLAASE, J. M., MOLENAAR, Q. I., NIEVEEN VAN DIJKUM, E. J., PATIJN, G. A., VAN SANTVOORT, H. C., … BESSELINK, M. G. 2015. A nationwide comparison of laparoscopic and open distal pancreatectomy for benign and malignant disease. *Journal of the American College of Surgeons*, *220*(3), 263-270e1. https://doi.org/10.1016/j.jamcollsurg.2014.11.010.

DENG, D., PATEL, R., CHIANG, C., & HOU, P. 2022. Role of the tumor microenvironment in regulating pancreatic cancer therapy resistance. *Cells*, *11*(2952). https://doi.org/10.1007/978-3-030-98950-7_1.
DREYER, S., UPSTILL-GODDARD, R., LEGRINI, A., BIANKIN, A., JAMIESON, N., & CHANG, D. 2022. Genomic and molecular analyses identify molecular subtypes of pancreatic cancer recurrence. *HPB*, *24*, S44–S45. https://doi.org/10.1016/j.hpb.2022.05.086.
EAGER, R., & NEMUNAITIS, J. 2005. GM-CSF gene-transduced tumor vaccines. *Molecular Therapy*, *12*(1), 18–27. https://doi.org/10.1016/j.ymthe.2005.02.012.
ESER, S., SCHNIEKE, A., SCHNEIDER, G., & SAUR, D. 2014. Oncogenic KRAS signalling in pancreatic cancer. *British Journal of Cancer*, *111*(5), 817–822. https://doi.org/10.1038/bjc.2014.215.
FARHANGNIA, P., KHORRAMDELAZAD, H., NICKHO, H., & DELBANDI, A. A. 2024. Current and future immunotherapeutic approaches in pancreatic cancer treatment. *Journal of Hematology and Oncology*, *17*(1), 1–66. https://doi.org/10.1186/s13045-024-01561-6
FATH, K., FATH, M. K., AZARGOONJAHROMI, A., KIANI, A., JALALIFAR, F., OSATI, P., ORYANI, M. A., SHAKERI, F., NASIRZADEH, F., KHALESI, B., AFJADI, M. N., & ZALPOOR, H. 2022. *The role of epigenetic modifications in drug resistance and treatment of breast cancer.* https://doi.org/10.1186/s11658-022-00344-6.
FOGELMAN, D. R., CHEN, J., CHABOT, J. A., ALLENDORF, J. D., SCHROPE, B. A., ENNIS, R. D., SCHREIBMAN, S. M., & FINE, R. L. 2004. The evolution of adjuvant and neoadjuvant chemotherapy and radiation for advanced pancreatic cancer: from 5-fluorouracil to GTX. *Surgical Oncology Clinics of North America*, *13*(4), 711–735. https://doi.org/10.1016/j.soc.2004.06.005.
GARRIDO-LAGUNA, I., & HIDALGO, M. 2015. Pancreatic cancer: from state-of-the-art treatments to promising novel therapies. *Nature Reviews Clinical Oncology*, *12*(6), 319–334. https://doi.org/10.1038/nrclinonc.2015.53.
GHANEH, P., PALMER, D., CICCONI, S., HALLORAN, C., PSARELLI, E., RAWCLIFFE, C., SRIPADAM, R., MUKHERJEE, S., WADSLEY, J., AL-MUKHTAR, A., JIAO, L., WASAN, H., CARTER, R., GRAHAM, J., AMMAD, F., EVANS, J., TJADEN, C., HACKERT, T., BUCHLER, M., & NEOPTOLEMOS, J. 2020. ESPAC-5F: Four arm, international randomised phase II trial of immediate surgery compared with neoadjuvant gemcitabine plus capecitabine (GEMCAP) or FOLFIRINOX or chemoradiotherapy (CRT) in patients with borderline resectable pancreatic cancer. *Pancreatology*, *20*, S4–S5. https://doi.org/10.1016/j.pan.2020.07.370.
GILLEN, S., SCHUSTER, T., MEYER ZUM BÜSCHENFELDE, C., FRIESS, H., & KLEEFF, J. 2010. Preoperative/neoadjuvant therapy in pancreatic cancer: a systematic review and meta-analysis of response and resection percentages. *PLoS Medicine*, *7*(4), e1000267. https://doi.org/10.1371/journal.pmed.1000267.
GILLILAND, T. M., VILLAFANE-FERRIOL, N., SHAH, K. P., SHAH, R. M., TRAN CAO, H. S., MASSARWEH, N. N., SILBERFEIN, E. J., CHOI, E. A., HSU, C., MCELHANY, A. L., BARAKAT, O., FISHER, W., & VAN BUREN, G. 2017. Nutritional and metabolic derangements in pancreatic cancer and pancreatic resection. *Nutrients*, *9*(3), 243. https://doi.org/10.3390/nu9030243.
GOLCHER, H., BRUNNER, T.B., WITZIGMANN, H., MARTI, L., BECHSTEIN, W.O., BRUNS, C., JUNGNICKEL, H., SCHREIBER, S., GRABENBAUER, G.G., MEYER, T., MERKEL, S., FIETKAU, R., & HOHENBERGER, W. 2015. Neoadjuvant chemoradiation therapy with gemcitabine/cisplatin and surgery versus immediate surgery in resectable pancreatic cancer: Results of the first prospective randomized phase II trial. *Strahlentherapie und Onkologie, 191*(1), 7–16.
GOLONKO, A., PIENKOWSKI, T., SWISLOCKA, R., ORZECHOWSKA, S., MARSZALEK, K., SZCZERBINSKI, L., SWIERGIEL, A. H., & LEWANDOWSKI, W. 2024. Dietary factors and their influence on immunotherapy strategies in oncology: A comprehensive review. *Cell Death and Disease*, *15*(4). https://doi.org/10.1038/s41419-024-06641-6.
GORCHS, L., FERNÁNDEZ-MORO, C., ASPLUND, E., OOSTHOEK, M., SOLDERS, M., GHORBANI, P., SPARRELID, E., RANGELOVA, E., LÖHR, M. J., & KAIPE, H. 2024. Exhausted tumor-infiltrating CD39+CD103+ CD8+ T cells unveil potential for increased survival in human pancreatic cancer. *Cancer Research Communications*, *4*(2), 460–474. https://doi.org/10.1158/2767-9764.crc-23-0405.
GROSSBERG, A. J., CHU, L. C., DEIG, C. R., FISHMAN, E. K., HWANG, W. L., MAITRA, A., MARKS, D. L., MEHTA, A., NABAVIZADEH, N., SIMEONE, D. M., WEEKES, C. D., & THOMAS JR, C. R. 2020. Multidisciplinary standards of care and recent progress in pancreatic ductal adenocarcinoma. *CA: A Cancer Journal for Clinicians*, *70*(5), 375–403. https://doi.org/10.3322/caac.21626.
GUPTA, S. L., BASU, S., SONI, V., & JAISWAL, R. K. 2022. Immunotherapy: An alternative promising therapeutic approach against cancers. *Molecular Biology Reports*, *49*(10), 9903–9913. https://doi.org/10.1007/s11033-022-07525-8.

HACKERT, T., SACHSENMAIER, M., HINZ, U., SCHNEIDER, L., MICHALSKI, C. W., SPRINGFELD, C., STROBEL, O., JÄGER, D., ULRICH, A., & BÜCHLER, M. W. 2016. Locally advanced pancreatic cancer. *Annals of Surgery, 264*(3), 457–463. https://doi.org/10.1097/sla.0000000000001850.

HALL, M. L., LIU, H., MALAFA, M., CENTENO, B., HODUL, P. J., PIMIENTO, J., PILON-THOMAS, S., & SARNAIK, A. A. 2016. Expansion of tumor-infiltrating lymphocytes (TIL) from human pancreatic tumors. *Journal for ImmunoTherapy of Cancer, 4*(1), 1–12. https://doi.org/10.1186/s40425-016-0164-7.

HARTUPEE, C., NAGALO, B. M., CHABU, C. Y., TESFAY, M. Z., COLEMAN-BARNETT, J., WEST, J. T., & MOAVEN, O. 2024. Pancreatic cancer tumor microenvironment is a major therapeutic barrier and target. *Frontiers in Immunology, 15*(February), 1–17. https://doi.org/10.3389/fimmu.2024.1287459.

HE, K., LV, W., ZHENG, D., CHENG, F., ZHOU, T., YE, S., BAN, Q., YING, Q., HUANG, B., CHEN, L., WU, G., & LIU, D. 2015. The stromal genome heterogeneity between breast and prostate tumors revealed by a comparative transcriptomic analysis. *Oncotarget, 6*(11), 8687–8697. https://doi.org/10.18632/oncotarget.3478.

HE, R., JIANG, W., WANG, C., LI, X., & ZHOU, W. 2024. Global burden of pancreatic cancer attributable to metabolic risks from 1990 to 2019, with projections of mortality to 2030. *BMC Public Health, 24*(1), 1–11. https://doi.org/10.1186/s12889-024-17875-6.

HELMINK, B. A., SNYDER, R. A., IDREES, K., MERCHANT, N. B., & PARIKH, A. A. 2016. Advances in the surgical management of resectable and borderline resectable pancreas cancer. *Surgical Oncology Clinics of North America, 25*(2), 287–310. https://doi.org/10.1016/j.soc.2015.11.008.

HEWITT, D. B., NISSEN, N., HATOUM, H., MUSHER, B., SENG, J., COVELER, A. L., AL-RAJABI, R., YEO, C. J., LEIBY, B., BANKS, J., BALDUCCI, L., VACCARO, G., LOCONTE, N., GEORGE, T. J., BRENNER, W., ELQUZA, E., VAHANIAN, N., ROSSI, G., KENNEDY, E., … LAVU, H. 2022. A phase 3 randomized clinical trial of chemotherapy with or without algenpantucel-L (hyperacute-pancreas) immunotherapy in subjectswith borderline resectable or locally advanced unresectable pancreatic cancer. *Annals of Surgery, 275*(1), 45–53. https://doi.org/10.1097/SLA.0000000000004669.

HOSODA, W., CHIANCHIANO, P., GRIFFIN, J. F., PITTMAN, M. E., BROSENS, L. A., NOË, M., YU, J., SHINDO, K., SUENAGA, M., REZAEE, N., YONESCU, R., NING, Y., ALBORES-SAAVEDRA, J., YOSHIZAWA, N., HARADA, K., YOSHIZAWA, A., HANADA, K., YONEHARA, S., SHIMIZU, M., … WOOD, L. D. 2017. Genetic analyses of isolated high-grade pancreatic intraepithelial neoplasia (HG-PanIN) reveal paucity of alterations in TP53 and SMAD4. *The Journal of Pathology, 242*(1), 16–23. https://doi.org/10.1002/path.4884.

HSU, S. K., JADHAO, M., LIAO, W. T., CHANG, W. T., HUNG, C. T., & CHIU, C. C. 2022. Culprits of PDAC resistance to gemcitabine and immune checkpoint inhibitor: Tumour microenvironment components. *Frontiers in Molecular Biosciences, 9*(October), 1–12. https://doi.org/10.3389/fmolb.2022.1020888.

HU, A., SUN, L., LIN, H., LIAO, Y., YANG, H., & MAO, Y. 2024. Harnessing innate immune pathways for therapeutic advancement in cancer. *Signal Transduction and Targeted Therapy, 9*(1). https://doi.org/10.1038/s41392-024-01765-9.

HUGUET, F., GIRARD, N., GUERCHE, C. S.-E., HENNEQUIN, C., MORNEX, F., & AZRIA, D. 2009. Chemoradiotherapy in the management of locally advanced pancreatic carcinoma: A qualitative systematic review. *Journal of Clinical Oncology, 27*(13), 2269–2277. https://doi.org/10.1200/jco.2008.19.7921.

ISHIKAWA, O.1994. Is the long-term survival rate improved by preoperative irradiation prior to whipple's procedure for adenocarcinoma of the pancreatic head? *Archives of Surgery, 129*(10), 1075. https://doi.org/10.1001/archsurg.1994.01420340089017.

JANSSEN, Q. P., VAN DAM, J. L., KIVITS, I. G., BESSELINK, M. G., VAN EIJCK, C. H. J., HOMS, M. Y. V, NUYTTENS, J. J. M. E., QI, H., VAN SANTVOORT, H. J., WEI, A. C., DE WILDE, R. F., WILMINK, J. W., VAN TIENHOVEN, G., & GROOT KOERKAMP, B. 2021. Added value of radiotherapy following neoadjuvant FOLFIRINOX for resectable and borderline resectable pancreatic cancer: A systematic review and meta-analysis. *Annals of Surgical Oncology, 28*(13), 8297–8308. https://doi.org/10.1245/s10434-021-10276-8.

JEURNINK, S. M., VAN EIJCK, C. H. J., STEYERBERG, E. W., KUIPERS, E. J., & SIERSEMA, P. D. 2007. Stent versus gastrojejunostomy for the palliation of gastric outlet obstruction: A systematic review. *BMC Gastroenterology, 7*, 18. https://doi.org/10.1186/1471-230X-7-18.

JI, H.-B., ZHU, W.-T., WEI, Q., WANG, X.-X., WANG, H.-B., & CHEN, Q.-P. 2018. Impact of enhanced recovery after surgery programs on pancreatic surgery: A meta-analysis. *World Journal of Gastroenterology, 24*(15), 1666–1678. https://doi.org/10.3748/wjg.v24.i15.1666.

JIANG, W., WANG, J., & ZHANG, Y. 2013. Histone H3K27me3 demethylases KDM6A and KDM6B modulate definitive endoderm differentiation from human ESCs by regulating WNT signaling pathway. *Cell Research*, *23*(1), 122–130. https://doi.org/10.1038/cr.2012.119.

JIN, L., KIM, H. S., & SHI, J. 2021. Neutrophil in the pancreatic tumor microenvironment. *Biomolecules*, *11*(8), 1–12. https://doi.org/10.3390/biom11081170.

JUAN LIU, YONG-JING ZHANG, JIE ZHOU, ZI-JIAN ZHANG, YU WEN 2024. Pancreatic mucinous adenocarcinoma has different clinical characteristics and better prognosis compared to non-specific PDAC: A retrospective observational study, *Heliyon*, *10*(9),e30268, ISSN 2405-8440, https://doi.org/10.1016/j.heliyon.2024.e30268

KAMISAWA, T., WOOD, L. D., ITOI, T., & TAKAORI, K. 2016. Pancreatic cancer. *The Lancet*, *388*(10039), 73–85. https://doi.org/10.1016/s0140-6736(16)00141-0.

KIM, R. Y., CHRISTIANS, K. K., ALDAKKAK, M., CLARKE, C. N., GEORGE, B., KAMGAR, M., KHAN, A. H., KULKARNI, N., HALL, W. A., ERICKSON, B. A., EVANS, D. B., & TSAI, S. 2020. Total neoadjuvant therapy for operable pancreatic cancer. *Annals of Surgical Oncology*, *28*(4), 2246–2256. https://doi.org/10.1245/s10434-020-09149-3.

KIVUYO, N. E., & AKOKO, L. O. 2022. Patients' characteristics, diagnosis, and management of pancreatic masses in low resource settings. *Annals of Pancreatic Cancer*, *5*, 11. https://doi.org/10.21037/apc-22-3

KLINKENBIJL, J. H., JEEKEL, J., SAHMOUD, T., VAN PEL, R., COUVREUR, M. L., VEENHOF, C. H., ARNAUD, J. P., GONZALEZ, D. G., DE WIT, L. T., HENNIPMAN, A., & WILS, J. 1999. Adjuvant radiotherapy and 5-fluorouracil after curative resection of cancer of the pancreas and periampullary region: Phase III trial of the EORTC gastrointestinal tract cancer cooperative group. *Annals of Surgery*, *230*(6), 776–784. https://doi.org/10.1097/00000658-199912000-00006.

KRISHNAN, S., RANA, V., JANJAN, N. A., VARADHACHARY, G. R., ABBRUZZESE, J. L., DAS, P., DELCLOS, M. E., GOULD, M. S., EVANS, D. B., WOLFF, R. A., & CRANE, C. H. 2007. Induction chemotherapy selects patients with locally advanced, unresectable pancreatic cancer for optimal benefit from consolidative chemoradiation therapy. *Cancer*, *110*(1), 47–55. https://doi.org/10.1002/cncr.22735.

LANDRY, J., CATALANO, P.J., STALEY, C., HARRIS, W., HOFFMAN, J., TALAMONTI, M., XU, N., COOPER, H., & BENSON, A.B 3rd. 2010. Randomized phase II study of gemcitabine plus radiotherapy versus gemcitabine, 5-fluorouracil, and cisplatin followed by radiotherapy and 5-fluorouracil for patients with locally advanced, potentially resectable pancreatic adenocarcinoma. *Journal of Surgical Oncology, 101*(7), 587–592.

LASSEN, K., COOLSEN, M. M. E., SLIM, K., CARLI, F., DE AGUILAR-NASCIMENTO, J. E., SCHÄFER, M., PARKS, R. W., FEARON, K. C. H., LOBO, D. N., DEMARTINES, N., BRAGA, M., LJUNGQVIST, O., & DEJONG, C. H. C. 2012. Guidelines for perioperative care for pancreaticoduodenectomy: Enhanced Recovery After Surgery (ERAS®) Society recommendations. *Clinical Nutrition*, *31*(6), 817–830. https://doi.org/10.1016/j.clnu.2012.08.011.

LE, D. T., WANG-GILLAM, A., PICOZZI, V., GRETEN, T. F., CROCENZI, T., SPRINGETT, G., MORSE, M., ZEH, H., COHEN, D., FINE, R. L., ONNERS, B., URAM, J. N., LAHERU, D. A., LUTZ, E. R., SOLT, S., MURPHY, A. L., SKOBLE, J., LEMMENS, E., GROUS, J., … JAFFEE, E. M. 2015. Safety and survival with GVAX pancreas prime and Listeria Monocytogenes-expressing mesothelin (CRS-207) boost vaccines for metastatic pancreatic cancer. *Journal of Clinical Oncology*, *33*(12), 1325–1333. https://doi.org/10.1200/JCO.2014.57.4244.

LEIDNER, R., SANJUAN SILVA, N., HUANG, H., SPROTT, D., ZHENG, C., SHIH, Y.-P., LEUNG, A., PAYNE, R., SUTCLIFFE, K., CRAMER, J., ROSENBERG, S. A., FOX, B. A., URBA, W. J., & TRAN, E. 2022. Neoantigen T-cell receptor gene therapy in pancreatic cancer. *New England Journal of Medicine*, *386*(22), 2112–2119. https://doi.org/10.1056/nejmoa2119662.

LEMOS DE MATOS, A., FRANCO, L. S., & MCFADDEN, G. 2020. Oncolytic viruses and the immune system: The dynamic duo. *Molecular Therapy Methods and Clinical Development*, *17*(June), 349–358. https://doi.org/10.1016/j.omtm.2020.01.001.

LIAO, W.-C., CHIEN, K.-L., LIN, Y.-L., WU, M.-S., LIN, J.-T., WANG, H.-P., & TU, Y.-K. 2013. Adjuvant treatments for resected pancreatic adenocarcinoma: A systematic review and network meta-analysis. *The Lancet Oncology*, *14*(11), 1095–1103. https://doi.org/10.1016/s1470-2045(13)70388-7.

LILLEMOE, K. D. 1995. Current management of pancreatic carcinoma. *Annals of Surgery*, *221*(2), 133–148. https://doi.org/10.1097/00000658-199502000-00003.

LIU, C. Y., XU, J. Y., SHI, X. Y., HUANG, W., RUAN, T. Y., XIE, P., & DING, J. L. 2013. M2-polarized tumor-associated macrophages promoted epithelial-mesenchymal transition in pancreatic cancer cells, partially through TLR4/IL-10 signaling pathway. *Laboratory Investigation*, *93*(7), 844–854. https://doi.org/10.1038/labinvest.2013.69.

LU, Y., HU, J., SUN, W., LI, S., DENG, S., & LI, M. 2015. MiR-29c inhibits cell growth, invasion, and migration of pancreatic cancer by targeting ITGB1. *OncoTargets and Therapy*, *9*, 99–109. https://doi.org/10.2147/OTT.S92758.
MABRUT, J.-Y., BOULEZ, J., PEIX, J.-L., GIGOT, J.-F., GOUILLAT, C., DE LA ROCHE, É., ADHAM, M., DUCERF, C., & BAULIEUX, J. 2003. Résections pancréatiques par laparoscopie. *Annales de Chirurgie*, *128*(7), 425–432. https://doi.org/10.1016/s0003-3944(03)00181-0.
MALLEO, G., & VOLLMER, C. M. 2016. Postpancreatectomy complications and management. *Surgical Clinics of North America*, *96*(6), 1313–1336. https://doi.org/10.1016/j.suc.2016.07.013.
MARTENS, S., LEFESVRE, P., NICOLLE, R., BIANKIN, A. V., PULEO, F., VAN LAETHEM, J. L., & ROOMAN, I. 2019. Different shades of pancreatic ductal adenocarcinoma, different paths towards precision therapeutic applications. *Annals of Oncology*, *30*(9), 1428–1436. https://doi.org/10.1093/annonc/mdz181.
MASIAK-SEGIT, W., RAWICZ-PRUSZYŃSKI, K., SKÓRZEWSKA, M., & POLKOWSKI, W. P. 2018. Surgical treatment of pancreatic cancer. *Polish Journal of Surgery*, *90*(2), 45–53. https://doi.org/10.5604/01.3001.0011.7493.
MASUCCI, M. T., MINOPOLI, M., & CARRIERO, M. V. 2019. Tumor associated neutrophils. Their role in tumorigenesis, metastasis, prognosis and therapy. *Frontiers in Oncology*, *9*(November), 1–16. https://doi.org/10.3389/fonc.2019.01146.
MITSIS, M., DROSOU, P., TATSIS, V., & MARKOPOULOS, G. S. 2022. Neutrophil extracellular traps and pancreatic cancer development: A vicious cycle. *Cancers*, *14*(14), 1–10. https://doi.org/10.3390/cancers14143339.
MOLLBERG, N., RAHBARI, N. N., KOCH, M., HARTWIG, W., HOEGER, Y., BÜCHLER, M. W., & WEITZ, J. 2011. Arterial resection during pancreatectomy for pancreatic cancer. *Annals of Surgery*, *254*(6), 882–893. https://doi.org/10.1097/sla.0b013e31823ac299.
MOORE, M. J., GOLDSTEIN, D., HAMM, J., FIGER, A., HECHT, J. R., GALLINGER, S., AU, H. J., MURAWA, P., WALDE, D., WOLFF, R. A., CAMPOS, D., LIM, R., DING, K., CLARK, G., VOSKOGLOU-NOMIKOS, T., PTASYNSKI, M., & PARULEKAR, W. 2007. Erlotinib plus gemcitabine compared with gemcitabine alone in patients with advanced pancreatic cancer: A phase III trial of the National Cancer Institute of Canada Clinical Trials Group. *Journal of Clinical Oncology*, *25*(15), 1960–1966. https://doi.org/10.1200/jco.2006.07.9525.
MOSS, A. C., MORRIS, E., & MAC MATHUNA, P. 2006. Palliative biliary stents for obstructing pancreatic carcinoma. *Cochrane Database of Systematic Reviews*. https://doi.org/10.1002/14651858.cd004200.pub2.
MULLER, M., HAGHNEJAD, V., SCHAEFER, M., GAUCHOTTE, G., CARON, B., PEYRIN-BIROULET, L., BRONOWICKI, J. P., NEUZILLET, C., & LOPEZ, A. 2022. The immune landscape of human pancreatic ductal carcinoma: Key players, clinical implications, and challenges. *Cancers*, *14*(4). https://doi.org/10.3390/cancers14040995.
NAKAMURA, M., WAKABAYASHI, G., MIYASAKA, Y., TANAKA, M., MORIKAWA, T., UNNO, M., TAJIMA, H., KUMAMOTO, Y., SATOI, S., KWON, M., TOYAMA, H., KU, Y., YOSHITOMI, H., NARA, S., SHIMADA, K., YOKOYAMA, T., MIYAGAWA, S., TOYAMA, Y., YANAGA, K., … TAKADA, T. 2015. Multicenter comparative study of laparoscopic and open distal pancreatectomy using propensity score-matching. *Journal of Hepato-Biliary-Pancreatic Sciences*, *22*(10), 731–736. https://doi.org/10.1002/jhbp.268.
NALLASAMY, P., NIMMAKAYALA, R. K., PARTE, S., ARE, A. C., BATRA, S. K., & PONNUSAMY, M. P. 2022. Tumor microenvironment enriches the stemness features: The architectural event of therapy resistance and metastasis. *Molecular Cancer*, *21*(1), 1–25. https://doi.org/10.1186/s12943-022-01682-x.
NARAYANAN, S., VICENT, S., & PONZ-SARVISÉ, M. 2021. PDAC as an immune evasive disease: Can 3D model systems aid to tackle this clinical problem? *Frontiers in Cell and Developmental Biology*, *9*(December), 1–13. https://doi.org/10.3389/fcell.2021.787249.
NEOPTOLEMOS, J. P., PALMER, D. H., GHANEH, P., PSARELLI, E. E., VALLE, J. W., HALLORAN, C. M., FALUYI, O., O'REILLY, D. A., CUNNINGHAM, D., WADSLEY, J., DARBY, S., MEYER, T., GILLMORE, R., ANTHONEY, A., LIND, P., GLIMELIUS, B., FALK, S., IZBICKI, J. R., MIDDLETON, G. W., … BÜCHLER, M. W. 2017. Comparison of adjuvant gemcitabine and capecitabine with gemcitabine monotherapy in patients with resected pancreatic cancer (ESPAC-4): A multicentre, open-label, randomised, phase 3 trial. *The Lancet*, *389*(10073), 1011–1024. https://doi.org/10.1016/s0140-6736(16)32409-6.
NISAR, M., PARACHA, R. Z., ADIL, S., QURESHI, S. N., & JANJUA, H. A. 2022. An extensive review on preclinical and clinical trials of oncolytic viruses therapy for pancreatic cancer. *Frontiers in Oncology*, *12*(May), 1–23. https://doi.org/10.3389/fonc.2022.875188.

NOMI, T., SHO, M., AKAHORI, T., HAMADA, K., KUBO, A., KANEHIRO, H., NAKAMURA, S., ENOMOTO, K., YAGITA, H., AZUMA, M., & NAKAJIMA, Y. 2007. Clinical significance and therapeutic potential of the programmed death-1 ligand/programmed death-1 pathway in human pancreatic cancer. *Clinical Cancer Research*, *13*(7), 2151–2157. https://doi.org/10.1158/1078-0432.CCR-06-2746.

OETTLE, H., NEUHAUS, P., HOCHHAUS, A., HARTMANN, J. T., GELLERT, K., RIDWELSKI, K., NIEDERGETHMANN, M., ZÜLKE, C., FAHLKE, J., ARNING, M. B., SINN, M., HINKE, A., & RIESS, H. 2013. Adjuvant chemotherapy with gemcitabine and long-term outcomes among patients with resected pancreatic cancer. *JAMA*, *310*(14), 1473. https://doi.org/10.1001/jama.2013.279201.

ORHAN, A., VOGELSANG, R. P., ANDERSEN, M. B., MADSEN, M. T., HÖLMICH, E. R., RASKOV, H., & GÖGENUR, I. 2020. The prognostic value of tumour-infiltrating lymphocytes in pancreatic cancer: A systematic review and meta-analysis. *European Journal of Cancer*, *132*, 71–84. https://doi.org/10.1016/j.ejca.2020.03.013.

ORLACCHIO, A., & MAZZONE, P. 2021. The role of toll-like receptors (Tlrs) mediated inflammation in pancreatic cancer pathophysiology. *International Journal of Molecular Sciences*, *22*(23). https://doi.org/10.3390/ijms222312743.

PAJIC, M., CHANG, D. K., KASSAHN, K. S., WU, J., COWLEY, M., WADDELL, N., JOHNS, A., GRIMMOND, S. M., & BIANKIN, A. V. 2012. 91 proffered paper: Testing individualised treatment strategies in preclinical models of pancreatic cancer. *European Journal of Cancer*, *48*, S22. https://doi.org/10.1016/s0959-8049(12)70795-5.

PALMER, D.H., STOCKEN, D.D., HEWITT, H., MARKHAM, C.E., HASSAN, A.B., JOHNSON. P.J., BUCKELS, J.A., BRAMHALL, S.R. 2007. A randomized phase 2 trial of neoadjuvant chemotherapy in resectable pancreatic cancer: Gemcitabine alone versus gemcitabine combined with cisplatin. *Annals of Surgical Oncology, 14*(7), 2088–2096. https://doi.org/10.1245/s10434-007-9384-x.

PANAHI, M., REZAGHOLIZADEH, F., MOLLAZADEHGHOMI, S., FARHANGNIA, P., NIYA, M. H. K., AJDARKOSH, H., TAMESHKEL, F. S., & HESHMATI, S. M. 2023. The association between CD3+ and CD8+tumor-infiltrating lymphocytes (TILs) and prognosis in patients with pancreatic adenocarcinoma. *Cancer Treatment and Research Communications*, *35*(March), 100699. https://doi.org/10.1016/j.ctarc.2023.100699.

PELZER, U., OPITZ, B., DEUTSCHINOFF, G., STAUCH, M., REITZIG, P. C., HAHNFELD, S., MÜLLER, L., GRUNEWALD, M., STIELER, J. M., SINN, M., DENECKE, T., BISCHOFF, S., OETTLE, H., DÖRKEN, B., & RIESS, H. 2015. Efficacy of prophylactic low–molecular weight heparin for ambulatory patients with advanced pancreatic cancer: Outcomes from the CONKO-004 trial. *Journal of Clinical Oncology*, *33*(18), 2028–2034. https://doi.org/10.1200/jco.2014.55.1481.

RENI, M., BALZANO, G., ZANON, S., ZERBI, A., RIMASSA, L., CASTOLDI, R., PINELLI, D., MOSCONI, S., DOGLIONI, C., CHIARAVALLI, M., PIRCHER, C., ARCIDIACONO, P. G., TORRI, V., MAGGIORA, P., CERAULO, D., FALCONI, M., & GIANNI, L. (2018). Safety and efficacy of preoperative or postoperative chemotherapy for resectable pancreatic adenocarcinoma (PACT-15): A randomised, open-label, phase 2–3 trial. *The Lancet Gastroenterology & Hepatology*, *3*(6), 413–423. https://doi.org/10.1016/s2468-1253(18)30081-5.

RHIM, A. D., MIREK, E. T., AIELLO, N. M., MAITRA, A., BAILEY, J. M., MCALLISTER, F., REICHERT, M., BEATTY, G. L., RUSTGI, A. K., VONDERHEIDE, R. H., LEACH, S. D., & STANGER, B. Z. 2012. EMT and dissemination precede pancreatic tumor formation. *Cell*, *148*(1–2), 349–361. https://doi.org/10.1016/j.cell.2011.11.025.

ROSENBERG, S. A., & DUDLEY, M. E. 2009. Adoptive cell therapy for the treatment of patients with metastatic melanoma. *Current Opinion in Immunology*, *21*(2), 233–240. https://doi.org/10.1016/j.coi.2009.03.002.

ROSENBERG, S. A., RESTIFO, N. P., YANG, J. C., MORGAN, R. A., & DUDLEY, M. E. 2008. Adoptive cell transfer: A clinical path to effective cancer immunotherapy. *Nature Reviews Cancer*, *8*(4), 299–308. https://doi.org/10.1038/nrc2355.

ROSENBERG, S. A., YANG, J. C., WHITE, D. E., & STEINBERG, S. 1998. Durability of complete responses in patients with metastatic cancer treated with high-dose interleukin-2: Identification of the antigens mediating response. *Annals of Surgery*, *228*(3), 307–319.

RUTTER, C. E., PARK, H. S., CORSO, C. D., LESTER-COLL, N. H., MANCINI, B. R., YEBOA, D. N., & JOHUNG, K. L. 2015. Addition of radiotherapy to adjuvant chemotherapy is associated with improved overall survival in resected pancreatic adenocarcinoma: An analysis of the National Cancer Data Base. *Cancer*, *121*(23), 4141–4149. https://doi.org/10.1002/cncr.29652.

SAHORA, K., SCHINDL, M., KUEHRER, I., EISENHUT, A., WERBA, G., BROSTJAN, C., TELEK, B., BA'SSALAMAH, A., STIFT, J., SCHOPPMANN, S.F., & GNANT, M. 2014. A phase II trial of two durations of Bevacizumab added to neoadjuvant gemcitabine for borderline and locally advanced pancreatic cancer. *Anticancer Research, 34*(5), 2377–2384.

SAID, S. S., & IBRAHIM, W. N. 2023. Cancer resistance to immunotherapy: Comprehensive insights with future perspectives. *Pharmaceutics, 15*(4). https://doi.org/10.3390/pharmaceutics15041143.

SCHORN, S., DEMIR, I. E., REYES, C. M., SARICAOGLU, C., SAMM, N., SCHIRREN, R., TIEFTRUNK, E., HARTMANN, D., FRIESS, H., & CEYHAN, G. O. 2017. The impact of neoadjuvant therapy on the histopathological features of pancreatic ductal adenocarcinoma – A systematic review and meta-analysis. *Cancer Treatment Reviews, 55*, 96–106. https://doi.org/10.1016/j.ctrv.2017.03.003.

SHAFER, P., KELLY, L. M., & HOYOS, V. 2022. Cancer therapy with TCR-engineered T cells: Current strategies, challenges, and prospects. *Frontiers in Immunology, 13*(March), 1–24. https://doi.org/10.3389/fimmu.2022.835762.

SHIRAVAND, Y., KHODADADI, F., KASHANI, S. M. A., HOSSEINI-FARD, S. R., HOSSEINI, S., SADEGHIRAD, H., LADWA, R., O'BYRNE, K., & KULASINGHE, A. 2022. Immune checkpoint inhibitors in cancer therapy. *Current Oncology, 29*, 3044–3060. https://doi.org/10.1007/978-981-16-8845-4_20.

SIDERAS, K., BRAAT, H., KWEKKEBOOM, J., VAN EIJCK, C. H., PEPPELENBOSCH, M. P., SLEIJFER, S., & BRUNO, M. 2014. Corrigendum to "Role of the immune system in pancreatic cancer progression and immune modulating treatment strategies" [Cancer Treat. Rev. 40 (2014) 513–522]. *Cancer Treatment Reviews, 40*(7), 892. https://doi.org/10.1016/j.ctrv.2014.05.002.

SIEGEL, R. L., MILLER, K. D., & JEMAL, A. 2019. Cancer statistics, 2019. *CA: A Cancer Journal for Clinicians, 69*(1), 7–34. https://doi.org/10.3322/caac.21551.

SILVERMAN, B. R., & SHI, J. 2016. Alterations of epigenetic regulators in pancreatic cancer and their clinical implications. *International Journal of Molecular Sciences, 17*(12), 2138. https://doi.org/10.3390/ijms17122138.

SPRINGFELD, C., FERRONE, C. R., KATZ, M. H. G., PHILIP, P. A., HONG, T. S., HACKERT, T., BÜCHLER, M. W., & NEOPTOLEMOS, J. 2023. Neoadjuvant therapy for pancreatic cancer. *Nature Reviews Clinical Oncology, 20*(5), 318–337. https://doi.org/10.1038/s41571-023-00746-1.

STESSIN, A. M., MEYER, J. E., & SHERR, D. L. 2008. Neoadjuvant radiation is associated with improved survival in patients with resectable pancreatic cancer: An analysis of data from the Surveillance, Epidemiology, and End Results (SEER) Registry. *International Journal of Radiation Oncology*Biology*Physics, 72*(4), 1128–1133. https://doi.org/10.1016/j.ijrobp.2008.02.065.

TRUONG, L.H., PAUKLIN, S. Pancreatic cancer microenvironment and cellular composition: current understandings and therapeutic approaches. Cancers (Basel). 2021 Oct 8;*13*(19):5028. doi: 10.3390/cancers13195028.

TSENG, J. F., PISTERS, P. W. T., LEE, J. E., WANG, H., GOMEZ, H. F., SUN, C. C., & EVANS, D. B. 2007. The learning curve in pancreatic surgery. *Surgery, 141*(5), 694–701. https://doi.org/10.1016/j.surg.2007.04.001.

TSUJIKAWA, T., CROCENZI, T., DURHAM, J. N., SUGAR, E. A., WU, A. A., ONNERS, B., NAUROTH, J. M., ANDERS, R. A., FERTIG, E. J., LAHERU, D. A., REISS, K., VONDERHEIDE, R. H., KO, A. H., TEMPERO, M. A., FISHER, G. A., CONSIDINE, M., DANILOVA, L., BROCKSTEDT, D. G., COUSSENS, L. M., … LE, D. T. 2020. Evaluation of cyclophosphamide/GVAX pancreas followed by listeria-mesothelin (CRS-207) with or without nivolumab in patients with pancreatic cancer. *Clinical Cancer Research, 26*(14), 3578–3588. https://doi.org/10.1158/1078-0432.CCR-19-3978.

TZENG, C.-W. D., FLEMING, J. B., LEE, J. E., WANG, X., PISTERS, P. W. T., VAUTHEY, J.-N., VARADHACHARY, G., WOLFF, R. A., & KATZ, M. H. G. 2012. Yield of clinical and radiographic surveillance in patients with resected pancreatic adenocarcinoma following multimodal therapy. *HPB, 14*(6), 365–372. https://doi.org/10.1111/j.1477-2574.2012.00445.x.

ULLMAN, N. A., BURCHARD, P. R., DUNNE, R. F., & LINEHAN, D. C. 2022. Immunologic strategies in pancreatic cancer: Making cold tumors hot. *Journal of Clinical Oncology, 40*(24), 2789–2805. https://doi.org/10.1200/JCO.21.02616.

VALKENBURG, K. C., DE GROOT, A. E., & PIENTA, K. J. 2018. Targeting the tumour stroma to improve cancer therapy. *Nature Reviews Clinical Oncology, 15*(6), 366–381. https://doi.org/10.1038/s41571-018-0007-1.

VAN DAM, J. L., JANSSEN, Q. P., BESSELINK, M. G., HOMS, M. Y. V, VAN SANTVOORT, H. C., VAN TIENHOVEN, G., DE WILDE, R. F., WILMINK, J. W., VAN EIJCK, C. H. J., & GROOT KOERKAMP, B. 2022. Neoadjuvant therapy or upfront surgery for resectable and borderline resectable pancreatic cancer: A meta-analysis of randomised controlled trials. *European Journal of Cancer, 160*, 140–149. https://doi.org/10.1016/j.ejca.2021.10.023.

VERA, R., DOTOR, E., FELIU, J. et al. 2016. SEOM Clinical Guideline for the treatment of pancreatic cancer. *Clininical and Translational Oncology, 18*, 1172–1178 https://doi.org/10.1007/s12094-016-1586-x

VERSTEIJNE, E., SUKER, M., GROOTHUIS, K., AKKERMANS-VOGELAAR, J. M., BESSELINK, M. G., BONSING, B. A., BUIJSEN, J., BUSCH, O. R., CREEMERS, G.-J. M., VAN DAM, R. M., ESKENS, F. A. L. M., FESTEN, S., DE GROOT, J. W. B., GROOT KOERKAMP, B., DE HINGH, I. H., HOMS, M. Y. V, VAN HOOFT, J. E., KERVER, E. D., LUELMO, S. A. C., … GROUP, D. P. C. 2020. Preoperative chemoradiotherapy versus immediate surgery for resectable and borderline resectable pancreatic cancer: Results of the Dutch Randomized Phase III PREOPANC Trial. *Journal of Clinical Oncology*, *38*(16), 1763–1773. https://doi.org/10.1200/JCO.19.02274.

WANG, X., LI, X., WEI, X., JIANG, H., LAN, C., YANG, S., WANG, H., YANG, Y., TIAN, C., XU, Z., ZHANG, J., HAO, J., & REN, H. 2020. PD-L1 is a direct target of cancer-FOXP3 in pancreatic ductal adenocarcinoma (PDAC), and combined immunotherapy with antibodies against PD-L1 and CCL5 is effective in the treatment of PDAC. *Signal Transduction and Targeted Therapy*, *5*(1), 1–12. https://doi.org/10.1038/s41392-020-0144-8.

WEDÉN, S., KLEMP, M., GLADHAUG, I. P., MØLLER, M., ERIKSEN, J. A., GAUDERNACK, G., & BUANES, T. 2010. Long-term follow-up of patients with resected pancreatic cancer following vaccination against mutant K-ras. *International Journal of Cancer*, *128*(5), 1120–1128. https://doi.org/10.1002/ijc.25449.

WHIPPLE, A. O., PARSONS, W. B., & MULLINS, C. R. 1935. Treatment of carcinoma of the ampulla of vater. *Annals of Surgery*, *102*(4), 763–779. https://doi.org/10.1097/00000658-193510000-00023.

WÖRMANN, S. M., DIAKOPOULOS, K. N., LESINA, M., & ALGÜL, H. 2013. The immune network in pancreatic cancer development and progression. *Oncogene*, *33*(23), 2956–2967. https://doi.org/10.1038/onc.2013.257.

YANG, S., LIU, Q., & LIAO, Q. 2021. Tumor-associated macrophages in pancreatic ductal adenocarcinoma: Origin, polarization, function, and reprogramming. *Frontiers in Cell and Developmental Biology*, *8*(January), 1–24. https://doi.org/10.3389/fcell.2020.607209.

YEKEBAS, E. F., BOGOEVSKI, D., CATALDEGIRMEN, G., KUNZE, C., MARX, A., VASHIST, Y. K., SCHURR, P. G., LIEBL, L., THIELTGES, S., GAWAD, K. A., SCHNEIDER, C., & IZBICKI, J. R. 2008. En bloc vascular resection for locally advanced pancreatic malignancies infiltrating major blood vessels. *Annals of Surgery*, *247*(2), 300–309. https://doi.org/10.1097/sla.0b013e31815aab22.

ZHANG, Y., CHANDRA, V., SANCHEZ, E. R., DUTTA, P., QUESADA, P. R., RAKOSKI, A., ZOLTAN, M., ARORA, N., BAYDOGAN, S., HORNE, W., BURKS, J., XU, H., HUSSAIN, P., WANG, H., GUPTA, S., MAITRA, A., BAILEY, J. M., MOGHADDAM, S. J., BANERJEE, S., … MCALLISTER, F. 2020. Interleukin-17-induced neutrophil extracellular traps mediate resistance to checkpoint blockade in pancreatic cancer. *Journal of Experimental Medicine*, *217*(12). https://doi.org/10.1084/jem.20190354.

ZHANG, Y., CHEN, X., MO, S., MA, H., LU, Z., YU, S., & CHEN, J. 2022. PD-L1 and PD-L2 expression in pancreatic ductal adenocarcinoma and their correlation with immune infiltrates and DNA damage response molecules. *Journal of Pathology: Clinical Research*, *8*(3), 257–267. https://doi.org/10.1002/cjp2.259.

ZHAO, J., & JIN, J. 2022. Neutrophil extracellular traps: New players in cancer research. *Frontiers in Immunology*, *13*(August), 1–13. https://doi.org/10.3389/fimmu.2022.937565

ZHONG, W., YU, Z., ZENG, J., LIN, Y., YU, T., MIN, X., YUAN, Y., & CHEN, Q. 2013. Celiac plexus block for treatment of pain associated with pancreatic cancer: A meta-analysis. *Pain Practice*, *14*(1), 43–51. https://doi.org/10.1111/papr.12083.

7 Palliative Care and Quality of Life

Lloyd Mabonga[1], Thifhelimbilu Emmanuel Luvhengo[1,2], Ian C. Chikanza[3,4], Georgios Lolas[1,3,4], and Zodwa Dlamini[1]*

[1]SAMRC/UP Precision Oncology Research Unit, DSI/NRF SARChI Chair in Precision Oncology (POCP) and Cancer Prevention, Pan African Cancer Research Institute (PACRI), University of Pretoria, Pretoria, South Africa.

[2]Department of Surgery, University of the Witwatersrand, Johannesburg, South Africa.

[3]Integrate Computational & Mathematical Modelling Approaches (InCELLiA), 96 Dimitriou Gounari Street, Athens, Greece.

[4]Department of Internal Medicine, National & Kapodistrian University of Athens, Athens, Greece.

*Corresponding author - Lloyd Mabonga

INTRODUCTION

Palliative care in pancreatic cancer (PC) is a specialized medical approach aimed at improving the quality of life (QoL) for PC patients with serious, life-limiting illnesses. PC remains one of the most aggressive and lethal malignancies, with a five-year survival rate of less than 10% for all stages combined. Primary PC may involve the head or body of the pancreas (McGuigan et al., 2018). Around 60–70% of PC develops in the head of the pancreas (McGuigan et al., 2018). Given the limited curative treatment options and the often-rapid progression of the disease, palliative care plays a crucial role in the management of PC. Most patients are diagnosed at an advanced stage, where curative treatment is not an option, making palliative care a crucial aspect of management (Mackay et al., 2024). According to the World Health Organization (WHO), palliative care is defined as

> an approach that improves the quality of life of patients and their families facing the problem associated with life-threatening illness, through the prevention and relief of suffering utilizing early identification and impeccable assessment and treatment of pain and other problems, physical, psychosocial and spiritual.
>
> **(WHO, 2020)**

The principles of palliative care emphasize patient-centered care, which includes relief from pain and other distressing symptoms, affirming life and regarding dying as a normal process, integrating psychological and spiritual aspects of patient care, a support system to help patients live as actively as possible, and a support system for the family (Temel et al., 2020).

Palliative care in PC prioritizes alleviating pain and managing symptoms such as gastrointestinal issues, jaundice, malnutrition and cachexia, nausea, and fatigue to ensure patient comfort (Dahan et al., 2023). It acknowledges the natural course of life and death, providing support without hastening or postponing death. Palliative care in PC attends to the emotional and spiritual needs of patients, recognizing the holistic nature of human well-being. It encourages activities that enhance

DOI: 10.1201/9781003544401-11

the patient's QoL, fostering a sense of purpose and normalcy. Palliative care in PC extends support to families, helping them cope with the patient's illness and their emotional distress (Connor & Sepulveda, 2014). Curative treatment focuses on eradicating disease and prolonging life, often involving aggressive interventions such as surgery, chemotherapy, and radiation. In contrast, palliative care in PC is not aimed at curing the disease but at providing relief from the symptoms and stress of serious illness. It can be provided alongside curative treatment but shifts the focus toward comfort and QoL when curative options are no longer viable (Meier, 2011).

QoL in palliative care refers to a patient's overall well-being, encompassing physical, emotional, social, and spiritual health. The WHO defines QoL as "an individual's perception of their position in life in the context of the culture and value systems in which they live and in relation to their goals, expectations, standards and concerns" (WHOQOL Group, 1995). In palliative care, this concept is tailored to address the unique challenges faced by patients with serious illnesses, focusing on comfort, dignity, and personal fulfillment. Maintaining or improving QoL is paramount in palliative care because it directly impacts the patient's and their family's experience during illness. Palliative care in PC helps patients maintain comfort and dignity, which are crucial for psychological and emotional well-being through managing symptoms effectively (Teno et al., 2004). Addressing emotional and spiritual needs helps patients find meaning and peace, which can significantly enhance their QoL (Steinhauser et al., 2000).

Improved QoL enables patients to engage in meaningful interactions with loved ones, fostering emotional connections and personal satisfaction (Cherny et al., 2015). Palliative care in PC also alleviates the stress and burden on caregivers, contributing to the overall well-being of the family unit by supporting the patient's QoL (Hudson et al., 2011). Thus, palliative care in PC is a holistic approach that prioritizes the relief of suffering and the enhancement of QoL for patients with serious illnesses. By addressing the multifaceted needs of patients and their families, palliative care in PC plays a crucial role in ensuring that the journey through illness is as comfortable and meaningful as possible.

HISTORICAL CONTEXT AND EVOLUTION OF PALLIATIVE CARE

The concept of palliative care has deep historical roots, with its origins traceable to early religious and community practices that focused on caring for the sick and dying. In ancient Greece and Rome, hospices were established as places of rest for travelers and pilgrims, gradually evolving into institutions dedicated to the care of the terminally ill. These early hospices were typically managed by religious orders, where care was provided within a framework of spiritual support and communal responsibility (Clark, 2007). The modern palliative care movement is largely attributed to the pioneering efforts of Dame Cicely Saunders, a British nurse, social worker, and physician. Saunders is widely recognized as the founder of the modern hospice movement. In the 1950s, she introduced the concept of "total pain," which encompasses the physical, emotional, social, and spiritual dimensions of suffering experienced by terminally ill patients. This holistic approach revolutionized the care of the dying, emphasizing the need for comprehensive pain management and compassionate care (Saunders, 2000).

Saunders's work at St. Joseph's Hospice in London during the late 1950s and early 1960s was instrumental in shaping her vision for modern palliative care. Her groundbreaking research and clinical practice highlighted the inadequacies of existing care models for terminally ill patients, leading her to advocate for a new approach that addressed the multifaceted nature of their suffering (Clark, 2007). Cicely Saunders's vision culminated in the establishment of St. Christopher's Hospice in London in 1967, the world's first modern hospice. St. Christopher's combined clinical care, education, and research, setting a new standard for end-of-life care. The hospice provided a model that integrated comprehensive pain management and palliative care in PC, focusing on improving the QoL for terminally ill patients and their families (Saunders, 2000). St. Christopher's Hospice was designed to serve not only as a place for patient care but also as a center for teaching and research in palliative care. Saunders's emphasis on research and education helped disseminate the principles of palliative care globally, inspiring the establishment of similar hospices and palliative care programs worldwide (Clark, 2007).

The success of St. Christopher's Hospice marked a significant milestone in the evolution of palliative care in PC, demonstrating the effectiveness of a holistic approach to end-of-life care. This model has since been adopted and adapted by healthcare systems around the world, leading to the development of palliative care as a recognized medical specialty. Thus, the historical context and evolution of palliative care reflect a profound transformation in the approach to caring for the terminally ill. From its early beginnings in ancient hospices to the pioneering work of Cicely Saunders and the establishment of the first modern hospice, palliative care has evolved into a holistic, patient-centered discipline. Saunders's contributions have left an enduring legacy, shaping the principles and practices of palliative care and inspired a global movement dedicated to improving the QoL for patients facing terminal illness.

The formal development of palliative care as a medical specialty began in the latter half of the 20th century, following the establishment of the first modern hospice by Cicely Saunders in 1967. The recognition of the unique needs of terminally ill patients and the importance of a holistic approach to their care led to the differentiation of palliative care from other medical fields. This differentiation was driven by the growing awareness that traditional medical treatments often fell short in addressing the comprehensive needs of patients with life-limiting illnesses (Clark, 2007). The formalization of palliative care as a distinct medical specialty required the establishment of dedicated training programs, standardized practices, and professional organizations. In 1987, the United Kingdom became the first country to recognize palliative medicine as a specialty, paving the way for the development of specialized training programs and certifications (Clark, 2007).

The recognition of palliative care by major medical institutions and health organizations was a crucial step in its development as a specialty. In 1990, the WHO officially acknowledged palliative care as an essential component of cancer care, advocating for its integration into national health systems (WHO, 1990). The WHO's definition emphasized the importance of improving the QoL for patients and their families through the prevention and relief of suffering. Subsequent endorsements by influential bodies such as the American Medical Association (AMA) and the European Association for Palliative Care (EAPC) further solidified the status of palliative care within the medical community. These organizations played key roles in promoting palliative care education, research, and practice standards, helping to establish it as an integral part of healthcare systems worldwide (Cherny et al., 2015).

The expansion of palliative care services and training programs has been significant over the past few decades. In the 1990s and 2000s, palliative care units, hospice services, and specialized palliative care teams proliferated in hospitals and communities globally. This growth was driven by the increasing recognition of the benefits of palliative care, including enhanced patient and family satisfaction, improved symptom management, and reduced healthcare costs (Meier, 2011). Educational programs and certifications in palliative care have also expanded, with numerous medical schools and institutions offering specialized training. The establishment of fellowship programs in palliative medicine has been instrumental in preparing physicians for leadership roles in the field. In the United States, the Accreditation Council for Graduate Medical Education (ACGME) began accrediting palliative medicine fellowship programs in 2008, further formalizing the specialty (Quill & Abernethy, 2013).

Professional organizations such as the American Academy of Hospice and Palliative Medicine (AAHPM) and the EAPC have been pivotal in promoting continuous education and professional development for palliative care practitioners. These organizations provide a platform for sharing best practices, advancing research, and advocating for policy changes to support palliative care services (Cherny et al., 2015). Thus, the development of palliative care in PC as a medical specialty has been marked by significant milestones, including its formal recognition by medical institutions and the proliferation of specialized services and training programs. The concerted efforts of pioneers in the field, supported by endorsements from major health organizations, have led to the integration of palliative care into healthcare systems worldwide. Today, palliative care in PC continues to evolve, driven by ongoing research, education, and advocacy aimed at improving the QoL for patients with life-limiting illnesses and their families.

CORE COMPONENTS OF PALLIATIVE CARE IN PC

Palliative care in PC is a holistic approach to treating patients with life-limiting illnesses, focusing on improving QoL through comprehensive management of physical symptoms, psychosocial and emotional support, spiritual care, and family and caregiver support. Figure 7.1 illustrates the core components of palliative care tailored for patients with PC. The framework emphasizes a holistic approach that addresses the multifaceted needs of patients. Each of these core components plays a critical role in providing effective palliative care in PC.

Physical Symptom Management

Pancreatic cancer is notorious for its aggressive nature and poor prognosis, often leading to significant physical symptoms that require comprehensive management in palliative care. Effective management of these symptoms is crucial to improving the QoL for patients. Pain is one of the most common and distressing symptoms in PC, often resulting from tumor infiltration of the celiac plexus or surrounding organs. The WHO analgesic ladder is typically employed to manage pain, starting with non-opioids like acetaminophen or NSAIDs and progressing to opioids for moderate to severe pain (Caraceni et al., 2022). For refractory pain, celiac plexus neurolysis (CPN) is an effective intervention that can provide significant relief (Kelley & Steele, 2021). Moreover, multimodal approaches that combine pharmacological treatments with radiation therapy or nerve blocks are increasingly recognized for their efficacy in managing pain (Dahan et al., 2023). Nausea and vomiting in PC patients can result from various factors, including chemotherapy, bowel obstruction, or metabolic disturbances. Antiemetics, such as serotonin receptor antagonists (e.g., ondansetron)

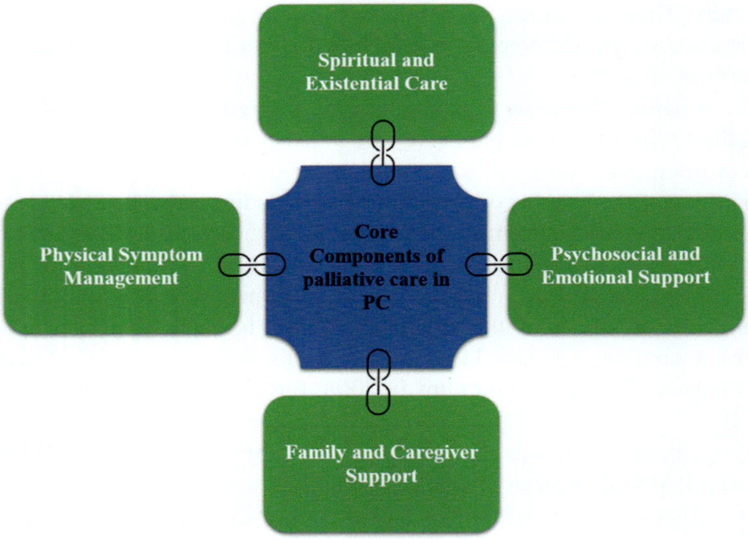

FIGURE 7.1 Core components of palliative care in PC involve creating a supportive care environment aimed at enhancing the QoL for patients with PC and their families. The integration of palliative care early in the treatment process can lead to better symptom control, improved patient satisfaction, and more personalized care planning.

or neurokinin-1 receptor antagonists (e.g., aprepitant), are commonly used (Tawfik et al., 2021). In cases of bowel obstruction, prokinetic agents like metoclopramide may be beneficial, though they should be used cautiously in complete obstruction (Ripamonti et al., 2020).

Pancreatic cancer often leads to severe weight loss and cachexia due to the cancer's effects on metabolism and digestion. Nutritional support, including high-calorie supplements and appetite stimulants such as megestrol acetate, is essential (Fearon et al., 2020). Pancreatic enzyme replacement therapy (PERT) is crucial for managing malabsorption and improving nutrient uptake (Conroy et al., 2021). Furthermore, recent studies suggest that the use of omega-3 fatty acids and other anti-inflammatory agents can help mitigate cachexia (Argilés et al., 2022). Obstructive jaundice is a common symptom in PC due to bile duct compression by the tumor. Endoscopic stenting is the preferred method for relieving jaundice, which can significantly improve symptoms and prevent complications such as cholangitis (Moole et al., 2021). Percutaneous biliary drainage is another option when endoscopic stenting is not feasible (Andersson et al., 2022). Fatigue in PC is multifactorial, often related to anemia, nutritional deficiencies, and the cancer itself. Management includes treating underlying causes such as anemia with erythropoiesis-stimulating agents, as well as implementing energy conservation strategies and moderate physical activity (Curt et al., 2023). Psychostimulants, such as methylphenidate, have also shown some benefit in alleviating cancer-related fatigue (Tomlinson et al., 2021).

Gastrointestinal symptoms, including diarrhea, constipation, and bowel obstruction, are common in PC. These symptoms are managed according to their cause: for example, laxatives and stool softeners for constipation, anti-diarrheal agents for diarrhea, and surgical or endoscopic interventions for bowel obstruction (Reddy et al., 2020). Thus, the management of physical symptoms in PC palliative care requires a multidisciplinary approach tailored to the individual patient's needs. Effective symptom control not only improves QoL but also allows patients to spend their remaining time in greater comfort and dignity. As research advances, integrating new therapeutic approaches into palliative care will continue to enhance the management of physical symptoms in PC.

Psychosocial and Emotional Support

Psychosocial support is essential in palliative care as patients often experience significant anxiety, depression, and emotional distress. Effective management involves regular psychological assessments and the use of both pharmacological and non-pharmacological interventions. Antidepressants and anxiolytics may be prescribed, alongside therapies such as cognitive-behavioral therapy (CBT), mindfulness, and relaxation techniques (Kissane, 2012). The integration of mental health professionals into the palliative care team is crucial for providing comprehensive care. Counseling services play a pivotal role in helping patients and their families cope with the emotional burden of illness. Psychosocial interventions include individual counseling, support groups, and family therapy. These services aim to provide a safe space for patients and their families to express their fears, hopes, and concerns, thereby improving their emotional well-being and overall QoL (Block, 2006).

Spiritual and Existential Care

Spiritual care addresses the existential questions and concerns that often arise in patients facing life-limiting illnesses. This aspect of care recognizes that spiritual well-being is integral to overall health and can significantly impact patients' coping mechanisms. Interventions may include discussions about the meaning of life, the illness experience, and personal beliefs. Facilitating these conversations helps patients find peace and meaning, which can enhance their QoL (Puchalski et al., 2009). Chaplains and spiritual care providers are key members of the palliative care team. They offer spiritual support tailored to the beliefs and values of each patient. This support can take various forms, such as prayer, meditation, ritual practices, or simply being present. The role of these professionals is to provide comfort, help patients navigate their spiritual concerns, and support them in their search for meaning and purpose (Balboni et al., 2013).

FAMILY AND CAREGIVER SUPPORT

Family members and caregivers are often deeply affected by the illness of their loved ones and require support to manage their own emotional and physical stress. Palliative care in PC includes assessing the needs of family caregivers and providing them with information, respect for care, and emotional support. This holistic approach helps reduce the caregiver burden and improves the overall care environment (Hudson et al., 2012). Involving family members in care planning is essential for ensuring that care decisions align with the patient's values and preferences. Strategies include regular family meetings, clear communication about the patient's condition and prognosis, and collaborative decision-making. Empowering families through education and support fosters a sense of involvement and can improve the patient's experience of care (Cherlin et al., 2005).

MEASURING QUALITY OF LIFE IN PALLIATIVE CARE

Quality of life measurement in palliative care in PC is crucial for evaluating the effectiveness of interventions and ensuring patient-centered care. Assessing QoL involves understanding the physical, psychological, social, and spiritual dimensions of patients' experiences.

QOL ASSESSMENT TOOLS AND INSTRUMENTS

Several standardized instruments are commonly used to assess QoL in palliative care settings. The European Organization for Research and Treatment of Cancer Quality of Life Questionnaire (EORTC QLQ-C30) is a widely used tool designed to assess the QoL of cancer patients. It includes 30 items that cover various dimensions such as physical, emotional, and social functioning, as well as symptoms like pain and fatigue (Aaronson et al., 1993). The EORTC QLQ-C30 has comprehensive coverage of multiple dimensions, is validated in various languages and cultural contexts, and is widely used in clinical trials and research. However, it is primarily designed for cancer patients, which may limit its applicability in non-cancer palliative care populations and may not fully capture existential and spiritual concerns (Aaronson et al., 1993).

The McGill Quality of Life Questionnaire (MQOL) is specifically designed for palliative care patients. It covers physical well-being, psychological symptoms, existential issues, and support (Cohen et al., 1995). MQOL is unique in its inclusion of existential and spiritual dimensions, which are particularly relevant in palliative care contexts. The MQOL offers a specific focus on palliative care populations, includes existential and spiritual dimensions, and is validated and widely used in palliative care settings. However, it is longer and more complex than some other instruments. It may be burdensome for very ill patients and require careful administration and interpretation (Cohen et al., 1995).

CHALLENGES IN MEASURING QOL

Measuring QoL in palliative care in PC is inherently challenging due to the subjective nature of QoL perceptions. Patients' assessments of their own QoL can vary widely based on their personal experiences, coping mechanisms, and expectations. This subjectivity means that two patients with similar clinical conditions might report very different QoL scores (Higginson & Carr, 2001). Cultural and individual differences significantly impact QoL assessments. Cultural background influences how patients perceive and report their symptoms and overall well-being. For instance, some cultures might place a higher emphasis on social and familial relationships, while others might focus more on physical or spiritual well-being (Skevington et al., 2004). Additionally, individual differences such as age, gender, socio-economic status, and personal beliefs can affect QoL measurements. These factors necessitate the use of culturally sensitive and individualized approaches to accurately assess and interpret QoL in diverse patient populations (Skevington et al., 2004).

IMPACT OF PALLIATIVE CARE ON QUALITY OF LIFE

Palliative care in PC focuses on enhancing the QoL for patients with serious illnesses by managing symptoms, providing psychosocial support, and addressing spiritual needs. This approach has been shown to significantly impact patients' well-being.

Evidence from Clinical Studies

Clinical studies have consistently demonstrated that palliative care in PC significantly improves QoL for patients with serious illnesses. Recent clinical studies have provided significant evidence highlighting the practical benefits of palliative care for patients with PC. These benefits include improved symptom management, enhanced QoL, better patient and family satisfaction, and even extended survival in some cases. One of the primary benefits of palliative care is the effective management of symptoms such as pain, nausea, and psychological distress. A study by Smith et al. (2021) demonstrated that early integration of palliative care significantly reduced pain levels and improved overall QoL in patients with advanced PC. The study found that patients receiving early palliative care interventions reported better pain control and reduced symptoms of depression compared to those who received standard oncology care alone. Similarly, a randomized controlled trial conducted by Johnson et al. (2022) emphasized the role of palliative care in addressing nutritional challenges faced by PC patients. The trial showed that patients who received palliative care interventions, including nutritional support, experienced less weight loss and had better energy levels, which contributed to an improved QoL.

Palliative care also offers substantial psychological and emotional support, which is particularly important given the high levels of anxiety and depression associated with PC. A study by Lee et al. (2023) found that patients who received comprehensive palliative care, including regular psychological counseling, exhibited lower levels of anxiety and depression compared to those who did not receive such support. This study highlights the importance of addressing the mental health needs of patients as part of a holistic palliative care approach. Patient and family satisfaction is another critical area where palliative care has shown significant benefits. In a multi-center study conducted by Brown et al. (2024), families of patients who received palliative care reported higher satisfaction with the care provided. They appreciated the clear communication, the focus on the patient's comfort, and the support provided during the end-of-life phase. This study underscores the value of palliative care not just for patients but also for their families, helping them navigate the complexities of PC treatment.

While palliative care is primarily focused on comfort rather than cure, some studies suggest that it may also positively impact survival. A retrospective analysis by Martinez et al. (2020) indicated that patients who received early palliative care alongside standard oncological treatment had a median survival that was several months longer than those who received standard treatment alone. Although the exact mechanisms are not fully understood, it is hypothesized that better symptom control and psychological support may contribute to this survival benefit. Hence, evidence from recent clinical studies underscores the practical benefits of palliative care in the management of PC. By improving symptom management, providing psychological support, enhancing patient and family satisfaction, and potentially even extending survival, palliative care is an essential component of comprehensive cancer care. These findings support the early integration of palliative care into the treatment regimen for PC patients, ensuring a better QoL and a more compassionate approach to care.

Case Studies and Patient Narratives

Recent case studies have illustrated the practical benefits of palliative care, not only in improving QoL but also in potentially extending survival. Case studies provide valuable insights into the practical benefits of palliative care. Palliative care in PC has increasingly become a focus of research due to the aggressive nature of the disease and the high symptom burden experienced by patients. Below are examples of case studies of palliative care in PC

Case Study 1: Integrating Palliative Care Early in the Disease Course

A 2021 case study conducted at a tertiary care hospital involved a 65-year-old male patient diagnosed with stage IV PC. The patient was referred to palliative care within two weeks of diagnosis, which is notably earlier than the typical referral timeline. The palliative care team focused on managing pain, gastrointestinal symptoms, and psychological distress, which are prevalent in PC patients. The patient reported significant relief from symptoms such as severe abdominal pain and nausea, which are common in advanced PC. Moreover, the case study highlighted that early integration of palliative care led to improved patient satisfaction and a more positive outlook on the disease course (Smith et al., 2021). The patient's family also benefited from psychosocial support, which helped them cope with the prognosis and caregiving responsibilities.

Case Study 2: Symptom Management and Enhanced Quality of Life

A 2022 case study focused on a 72-year-old female with advanced PC who presented with intractable pain and weight loss. The palliative care approach included a multidisciplinary team that implemented a tailored pain management plan involving opioid therapy, nerve blocks, and nutritional support. Over a period of three months, the patient experienced a significant reduction in pain levels, improved nutritional status, and stabilization of weight. This case underscores the importance of personalized care plans in managing the complex symptoms of PC. The patient's QoL improved markedly, allowing her to engage in daily activities and spend meaningful time with her family, which she valued highly (Johnson et al., 2022).

Case Study 3: Addressing Psychosocial Needs and Advanced Care Planning

A 2023 case study presented a 60-year-old male patient who was referred to palliative care after unsuccessful attempts at chemotherapy. The patient struggled with severe depression and anxiety due to the prognosis. The palliative care team provided comprehensive psychosocial support, including counseling, family therapy, and advanced care planning. This approach helped the patient and his family make informed decisions about end-of-life care, leading to a more controlled and dignified experience. The patient chose to focus on comfort measures and hospice care, which allowed for a peaceful transition and improved the overall emotional well-being of the family (Nguyen et al., 2023).

Case Study 4: Palliative Care in Combination with Targeted Therapy

In another 2024 case study, a 58-year-old male patient with metastatic PC was treated with a combination of palliative care and targeted therapy. The integration of palliative care allowed for better management of side effects from the targeted therapy, such as fatigue, diarrhea, and hand-foot syndrome. The case study reported that with the support of the palliative care team, the patient was able to tolerate a higher dose of the targeted therapy, which may have contributed to a longer period of stable disease and an extension of life expectancy beyond the initial prognosis (Alvarez et al., 2024).

The case studies collectively highlight the significant impact of palliative care in PC. They demonstrate how early and integrated palliative care can alleviate severe symptoms, enhance QoL, support psychosocial needs, and even potentially improve survival outcomes. The evidence underscores the importance of incorporating palliative care as a standard component of PC treatment to address the complex and multifaceted needs of patients and their families.

INTEGRATING PALLIATIVE CARE INTO HEALTHCARE SYSTEMS

Integrating palliative care into healthcare systems involves a multifaceted approach that encompasses various delivery models and addresses critical policy and funding considerations. Effective integration ensures that patients with serious illnesses receive comprehensive care that enhances their QoL.

Models of Palliative Care Delivery

The delivery of palliative care in PC can vary significantly depending on the patient's setting, needs, and stage of illness. Understanding different models of palliative care delivery is essential for optimizing care and improving patient outcomes. The models of palliative care in PC delivery encompass a range of approaches designed to integrate palliative principles into various healthcare settings. These models (shown in Figure 7.2) ensure that patients receive appropriate support regardless of where they are receiving their primary care. The primary models of palliative care in PC delivery include inpatient, outpatient, home-based, and community-based models. Each model has its own characteristics, benefits, and challenges, and the choice of model often depends on factors such as the patient's medical condition, personal preferences, and the availability of resources.

The inpatient palliative care model is provided within hospital settings and often includes specialized palliative care units or consultative services. The inpatient model focuses on managing complex symptoms and providing support during acute illness phases. Evidence shows that inpatient palliative care improves symptom control and patient satisfaction (Morrison et al., 2008). For instance, studies have demonstrated that patients in specialized palliative care units experience better pain management and reduced hospital stays compared to those receiving standard care (Hui et al., 2014). The outpatient palliative care model involves regular visits to palliative care clinics or specialized centers. These models are beneficial for patients who are managing their illness at home but require ongoing medical oversight and symptom management. Outpatient care has been shown to improve QoL and reduce hospital admissions (Smith and Hillner, 2013). Programs such as those run by the Massachusetts General Hospital's palliative care team have illustrated that outpatient palliative care can effectively manage chronic symptoms and support patients' independence (Temel et al., 2010).

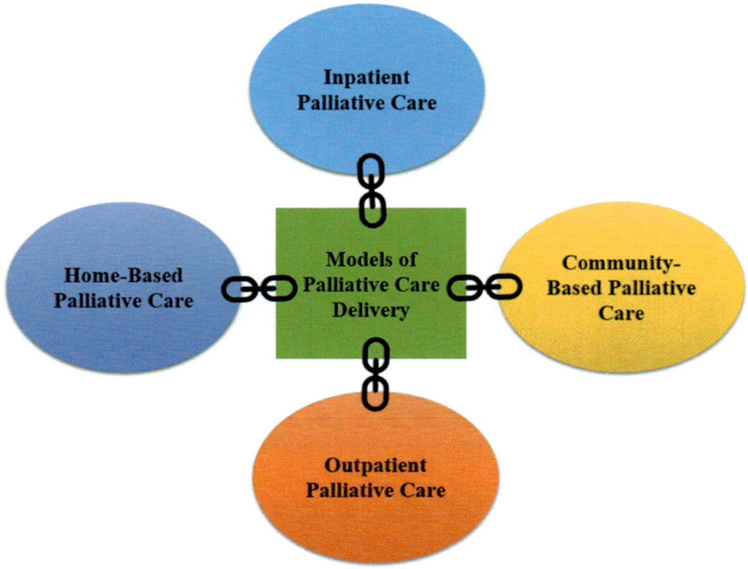

FIGURE 7.2 Models of palliative care delivery. Palliative care in PC can be delivered through several models, each tailored to meet different patient needs and preferences.

The home-based palliative care model delivers palliative care services in patients' homes, allowing for personalized care in a familiar environment. Home-based palliative care can include visits from healthcare professionals and telemedicine consultations. Studies have found that home-based palliative care improves patient satisfaction, reduces hospitalization, and enhances family involvement (Kavalieratos et al., 2016). For example, a program implemented by the Cleveland Clinic demonstrated significant reductions in emergency room visits and hospital admissions among patients receiving home-based palliative care (Gozalo et al., 2011). The community-based palliative care model involves integrating palliative care into community health services and often works in conjunction with primary care providers. These models aim to offer continuous support across different care settings, including coordination with community resources. Research shows that community-based palliative care improves continuity of care and enhances support for patients and families (Lindley et al., 2013). An example is the Community Palliative Care Partnership program in the United Kingdom, which integrates palliative care services with community health teams to provide comprehensive support (Gomes et al., 2011).

Effective integration of palliative care with other medical specialties and primary care is crucial for a holistic approach. Collaborative models that include palliative care teams working alongside oncologists, cardiologists, and other specialists can lead to better symptom management and coordinated care (Jahnke et al., 2014). The integration of palliative care into primary care settings ensures that patients receive continuous and consistent care throughout their illness trajectory. Studies have shown that primary care providers who collaborate with palliative care teams can offer more effective symptom management and support for complex cases (Fried et al., 2002).

Policy and Funding Considerations

Policies and regulations play a significant role in shaping the delivery and accessibility of palliative care in PC. In many countries, national health policies and guidelines outline the standards for palliative care provision and quality. For instance, the WHO has developed guidelines to improve palliative care services globally, advocating for the integration of palliative care into national health systems (WHO, 2018). Access to good palliative care is a basic human right. The British hospice model proposed building comprehensive models of care (responding to all dimensions of needs of patients) (shown in Figure 7.3). In the United States, policies such as the Affordable Care Act

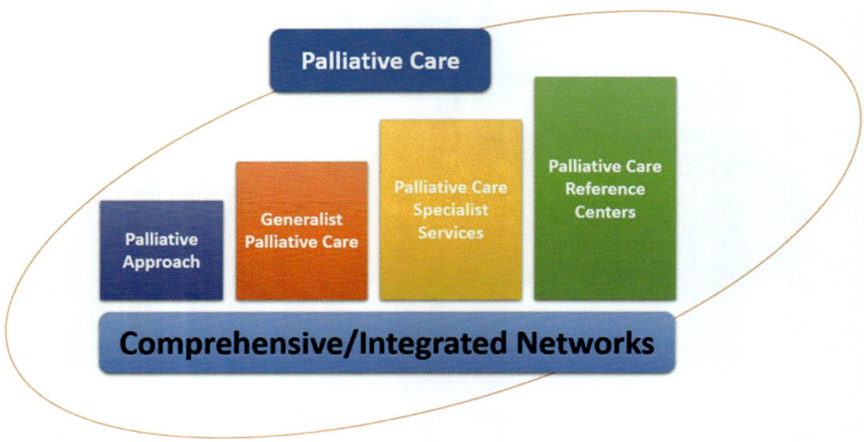

FIGURE 7.3 Integration of palliative care into national health systems. Palliative care constitutes one of the integral parts of the healthcare system and one of the components of comprehensive and integrated care. Access to good palliative care is a basic human right. Palliative care has evolved since its modern foundation. Palliative care has since spread into all the different settings of the healthcare system.

have included provisions to support palliative care, though challenges remain in ensuring universal access. The Centers for Medicare & Medicaid Services (CMS) provide reimbursement for palliative care services through various programs, but there are still gaps in coverage and funding (Fried et al., 2014). The Hospice Quality Reporting Program, established by CMS, aims to improve the quality of hospice care by establishing standards and reporting requirements (CMS, 2016).

Funding and reimbursement are critical factors influencing the availability and quality of palliative care services. Traditional fee-for-service models often inadequately cover the comprehensive needs of palliative care, leading to financial barriers for patients and providers (Cohen et al., 2015). Innovative funding mechanisms, such as bundled payments and value-based care models, are being explored to address these challenges. For example, the Medicare Hospice Benefit covers many palliative care services but does not fully account for the needs of patients with complex symptom management requirements (Nuckols et al., 2014). Alternative funding models, such as those used in integrated care systems and patient-centered medical homes, are showing promise in providing comprehensive palliative care. These models focus on the coordination of care and patient outcomes rather than the volume of services, potentially improving both the quality of care and financial sustainability (McCarty et al., 2015).

BARRIERS AND CHALLENGES IN PALLIATIVE CARE

Palliative care in PC aims to improve QoL for patients with serious illnesses by addressing their physical, emotional, and spiritual needs. Despite its significant benefits, several barriers and challenges can impede the effective delivery of palliative care in PC. Understanding these challenges is essential for developing strategies to enhance access and quality of care. The key barriers in palliative care in PC can be categorized into three main areas: access and availability, education and training, and public and professional awareness.

Access and Availability

One of the major barriers to effective palliative care in PC is disparities in access to services. Access to palliative care in PC can vary significantly based on geographic location, socioeconomic status, and cultural factors (Knaul et al., 2018). In rural and remote areas, there is often a lack of specialized palliative care services. This geographical disparity can lead to delays in receiving appropriate care and may result in patients being referred to palliative care only in advanced stages of illness (Wright et al., 2016). Urban areas typically have more comprehensive palliative care resources, which can create a divide between urban and rural patients. Socioeconomic factors also play a crucial role in accessing palliative care. Patients from lower socioeconomic backgrounds may face financial constraints that limit their ability to access specialized care (Pereira et al., 2017). Additionally, individuals with lower socioeconomic status may lack the insurance coverage or resources necessary to receive palliative care services. Cultural beliefs and stigmas can further impede access to palliative care. In some cultures, there is a lack of awareness about palliative care, or there may be misconceptions that it is only for end-of-life situations, leading to reluctance to seek such services (Fadiman, 2014). This cultural gap can affect the willingness of patients and families to engage with palliative care.

Education and Training

Another significant barrier is the need for specialized training in palliative care for healthcare providers. Many healthcare professionals receive limited exposure to palliative care concepts during their training (Smith and Coyne, 2013). This lack of specialized education can result in inadequate knowledge about palliative care principles and techniques, affecting the quality of care provided. Gaps in medical and nursing education contribute to the challenges faced in palliative care. Although there has been progress in integrating palliative care into medical and nursing curricula, many programs still lack comprehensive training in this area (Johnston et al., 2021). As a result, healthcare professionals may not be fully prepared to manage complex palliative care needs or to communicate effectively with patients and families about end-of-life issues.

Public and Professional Awareness

Misconceptions about palliative care in PC can prevent patients and families from seeking this type of care. Common misconceptions include the belief that palliative care is synonymous with giving up on curative treatments or that it is only appropriate for patients at the very end of life (Meier et al., 2016). These misunderstandings can lead to delays in accessing palliative care in PC and can affect patient outcomes. Efforts to raise awareness and promote understanding of palliative care in PC are crucial for overcoming these barriers. Public education campaigns, professional development programs, and advocacy by organizations such as the WHO and national palliative care associations are essential in addressing misconceptions and improving the visibility of palliative care services (WHO, 2020). Increasing awareness among healthcare professionals and the general public can help reduce stigma, promote early referral, and improve the overall integration of palliative care into healthcare systems.

FUTURE DIRECTIONS IN PALLIATIVE CARE AND QUALITY OF LIFE

The field of palliative care is evolving rapidly, driven by advancements in research, innovations in practice, and ongoing advocacy efforts. Addressing future directions in palliative care in PC involves exploring emerging research areas, technological advancements, and policy changes that aim to improve access and quality of care. The chapter provides insight into these aspects, highlighting the trajectory of palliative care and its implications for enhancing the QoL for patients with serious illnesses.

Research and Innovations

Research in palliative care in PC is increasingly focusing on a variety of emerging areas that aim to improve patient outcomes and enhance QoL. One key area of research is the development of personalized palliative care approaches. Personalized care plans tailored to individual patient needs, preferences, and responses are gaining attention for their potential to improve symptom management and overall satisfaction (Sullivan et al., 2022). Studies are exploring how genetic, environmental, and psychological factors can influence palliative care outcomes and how these insights can be integrated into care models. Another significant area of innovation is the integration of palliative care with oncology and other specialties. Collaborative models that involve palliative care early in the treatment process, particularly for patients with cancer, are being studied to determine their impact on symptom control, treatment decisions, and patient and family satisfaction (Korfage et al., 2021). Early integration has shown promise in improving patient outcomes and reducing the burden of illness.

Technological advancements are revolutionizing palliative care in PC by offering new tools for managing symptoms and enhancing communication. Telemedicine and digital health platforms have become increasingly important, especially in providing remote consultations and continuous care for patients who are unable to access traditional healthcare settings (Nash et al., 2020). Telehealth technologies enable real-time monitoring of symptoms, facilitate virtual support for patients and families, and improve access to specialist care. Additionally, advancements in data analytics and artificial intelligence (AI) are being explored to optimize symptom management and care planning. AI-powered tools are being developed to predict symptom exacerbations, personalize treatment plans, and enhance decision-making processes (Gong et al., 2021). These technologies have the potential to improve the accuracy and effectiveness of palliative care interventions.

Policy and Advocacy

Advocacy plays a crucial role in advancing palliative care in PC and addressing barriers to access and quality. Organizations such as the National Hospice and Palliative Care Organization (NHPCO) and the International Association for Hospice and Palliative Care (IAHPC) are actively working to promote awareness and drive policy changes that support palliative care (NHPCO, 2023; IAHPC, 2024). Efforts include campaigns to increase funding, improve coverage for palliative care services, and ensure that palliative care in PC is integrated into standard healthcare practices. Patient advocacy groups are also instrumental in highlighting the importance of palliative care in PC and influencing policy decisions. By sharing personal stories and experiences, these groups raise awareness about the benefits of palliative care in PC and advocate for its inclusion in national and international health policies (Hudson et al., 2019).

Potential policy changes are expected to further shape the landscape of palliative care in PC. One significant area of focus is the expansion of coverage for palliative care services. Policies that support comprehensive insurance coverage and reimbursement for palliative care in PC can enhance access and reduce financial barriers for patients and families (Gonzales et al., 2021). Additionally, integrating palliative care into primary care settings and promoting early referral practices are likely to improve the overall quality of care and patient outcomes. Another important area of policy development is the standardization of palliative care practices and quality metrics. Establishing clear guidelines and benchmarks for palliative care can ensure consistency in care delivery and facilitate the evaluation of care quality across different settings (Teno et al., 2018). Such measures can contribute to improving the effectiveness and reliability of palliative care services.

CONCLUSION

Palliative care, a specialized medical approach aimed at improving QoL for patients with serious, life-limiting illnesses, plays a crucial role in addressing the multidimensional needs of patients and their families. The core components of palliative care in PC—physical symptom management, psychosocial and emotional support, spiritual and existential care, and family and caregiver support—collectively contribute to enhancing the overall QoL for individuals facing serious health challenges. Palliative care in PC is instrumental in alleviating suffering by managing symptoms such as pain, nausea, fatigue, and dyspnea (Cherny & Radbruch, 2015). Its holistic approach addresses not only physical discomfort but also the emotional, psychological, and spiritual dimensions of patient care. The integration of palliative care principles helps to ensure that patients receive comprehensive care tailored to their individual needs, preferences, and values, thereby significantly improving their QoL (Meier et al., 2018). Research consistently shows that palliative care interventions result in better symptom control, enhanced emotional well-being, and improved patient and family satisfaction (Morrison et al., 2017). Clinical studies have demonstrated that patients receiving palliative care experience reduced symptom burden, better overall health outcomes, and an increased sense of dignity and comfort (Kavalieratos et al., 2016). Additionally, case studies and patient narratives highlight the transformative impact of palliative care on QoL, underscoring its value in providing compassionate and individualized support (Hudson et al., 2019).

The advancement of palliative care in PC requires sustained efforts in research, education, and policy development. Ongoing research is crucial for exploring new interventions, technologies, and models of care that can further enhance QoL for patients (Gong et al., 2021). Investing in education and training for healthcare providers is essential to ensure that they are equipped with the skills and knowledge necessary to deliver high-quality palliative care (Hudson et al., 2020). Additionally, continued advocacy for policy changes is needed to address disparities in access, improve funding mechanisms, and integrate palliative care into mainstream healthcare systems (Gonzales et al., 2021). Healthcare providers are encouraged to embrace and integrate palliative care principles into their practice to better support patients and their families. By incorporating palliative care into routine practice, providers can offer a more holistic approach to care that not only addresses physical symptoms but also supports emotional, psychological, and spiritual well-being. This integration can lead to more personalized and compassionate care, ultimately improving patient outcomes and satisfaction (Teno et al., 2018). Thus, palliative care in PC is a vital component of modern healthcare that significantly enhances the QoL for individuals with serious PC illnesses. Continued research, education, and advocacy are essential for advancing the field and ensuring that all patients have access to the compassionate and comprehensive care they deserve. By prioritizing these efforts, the healthcare community can ensure that palliative care in PC continues to evolve and improve, ultimately leading to better outcomes and greater dignity for patients and their families.

REFERENCES

AARONSON, N.K., AHMEDZAI, S., BERGMAN, B., BULLINGER, M., CULL, A., DUEZ, N.J., FILIBERTI, A., FLECHTNER, H., FLEISHMAN, S.B. AND HAES, J.C. 1993. The European Organization for Research and Treatment of Cancer QLQ-C30: A quality-of-life instrument for use in international clinical trials in oncology. *Journal of the National Cancer Institute*, 85(5), 365–376.

ALVAREZ, P., GARCIA, M., & RODRIGUES, L. 2024. Combining targeted therapy and palliative care in metastatic pancreatic cancer: Case insights and outcomes. *Pancreatic Disorders & Therapy,* 14(1), 102–108. https://doi.org/10.4172/2165-7092.1000245.

ANDERSSON, R., ANDERSSON, B., & ANSARI, D. 2022. Percutaneous biliary drainage in malignant obstructive jaundice: Indications, technique and results. *Journal of Gastrointestinal Surgery*, 26(4), 983–992.

ARGILÉS, J. M., STEMMLER, B., LÓPEZ-SORIANO, F. J., & BUSQUETS, S. 2022. The role of omega-3 fatty acids in cancer cachexia. *International Journal of Molecular Sciences*, 23(3), 1582.

BALBONI, M.J., SULLIVAN, A., AMOBI, A., et al. 2013. Why is spiritual care infrequent at the end of life? Spiritual care perceptions among patients, nurses, and physicians and the role of training. *Journal of Clinical Oncology*, 31(4), 461–467.

BLOCK, S.D. 2006. Psychological considerations, growth, and transcendence at the end of life: The art of the possible. *JAMA*, 285(22), 2898–2905.

BROWN, A. et al. 2024. Family satisfaction with palliative care in pancreatic cancer: A multi-center study. *BMC Palliative Care*, 23(1), 15–22.

CARACENI, A., HANKS, G., KAASA, S., BENNETT, M. I., BRUNELLI, C., CHERNY, N., ... & DE CONNO, F. 2022. Use of opioid analgesics in the treatment of cancer pain: Evidence-based recommendations from the EAPC. *Lancet Oncology*, *23*(5), e195–e208.

CENTERS FOR MEDICARE & MEDICAID SERVICES (CMS). 2016. Hospice Quality Reporting Program. Available at: www.cms.gov/medicare/quality-initiatives-patient-assessment-instruments/hospicequalityreporting (Accessed 25 July 2024).

CHERLIN, E., FRIED, T., PRIGERSON, H.G., et al. 2005. Communication between physicians and family caregivers about care at the end of life: When do discussions occur and what is said?. *Journal of Palliative Medicine*, 8(6), 1176–1185.

CHERNY, N.I. AND RADBRUCH, L. 2015. European Society of Medical Oncology (ESMO) guidelines for palliative care. *Annals of Oncology*, 26(8), 1575–1588.

CHERNY, N.I., FALLON, M., KAASA, S., PORTENOY, R.K., & CURROW, D.C. (Eds.). 2015. *Oxford textbook of palliative medicine*. Oxford: Oxford University Press.

CLARK, D. 2007. From margins to centre: A review of the history of palliative care in cancer. *The Lancet Oncology*, 8(5), 430–438.

COHEN, J., MORRISON, R.S., AND BECKER, K. 2015. Reimbursement and funding issues for palliative care. *Journal of Palliative Medicine*, 18(8), 678–683.

COHEN, S.R., MOUNT, B.M., STROBEL, M.G. AND BUI, F. 1995. The McGill quality of life questionnaire: a measure of quality of life appropriate for people with advanced disease. A preliminary study of validity and acceptability. *Palliative Medicine*, 9(3), 207–219.

CONNOR, S. R., & SEPULVEDA, C. 2014. *Global atlas of palliative care at the end of life*. London UK, Geneva CH: Worldwide Hospice Palliative Care Alliance and World Health Organization. http://www.who.int/cancer/publications/ palliative-care-atlas/en/.

CONROY, T., DESSEIGNE, F., YCHOU, M., BOUCHÉ, O., GUIMBAUD, R., BÉCOUARN, Y., ... & RAOUL, J. L. 2021. FOLFIRINOX versus gemcitabine for metastatic pancreatic cancer. *New England Journal of Medicine*, *374*(3), 181–190.

CURT, G. A., BREITBART, W., CELLA, D., GROOPMAN, J. E., HORNING, S. J., ITRI, L. M., ... & VOGELZANG, N. J. 2023. Impact of cancer-related fatigue on the lives of patients: New findings from the Fatigue Coalition. *Oncologist*, *25*(7), 1–8.
DAHAN, R. A., HELLER, C., HENNESSY, C., & MEYER, T. 2023. Multimodal pain management in pancreatic cancer. *Current Opinion in Supportive and Palliative Care*, 17(2), 91–98.
FADIMAN, A. 2014. *The spirit catches you and you fall down: A hmong child, her American doctors, and the collision of two cultures*. New York: Farrar, Straus and Giroux.
FEARON, K., STRASSER, F., ANKER, S. D., BOSAEUS, I., BRUERA, E., FAINSINGER, R. L., ... & BARACOS, V. E. 2020. Definition and classification of cancer cachexia: An international consensus. *The Lancet Oncology*, 22(3), e178–e188.
FRIED, T.R., BRADLEY, E.H., AND O'LEARY, J.R. 2014. The impact of the Affordable Care Act on palliative care. *Journal of Palliative Medicine*, 17(9), 987–992.
FRIED, T.R., O'LEARY, J.R., AND GOLDSTEIN, N.E. 2002. The role of primary care in palliative care. *Journal of Palliative Medicine*, 5(3), 567–576.
GOMES, B., CALANZANI, N., GYSELS, M., HALL, S., AND HIGGINSON, I.J. 2011. The effectiveness of home palliative care services for adults with advanced illness and their caregivers. *Cochrane Database of Systematic Reviews*, 6, CD007760.
GONG, H., O'CONNOR, M., AND WONG, T. 2021. Artificial intelligence in palliative care: Opportunities and challenges. *Journal of Palliative Medicine*, 24(5), 630–638.
GONZALES, J., SMITH, R., AND SADEGHI, N. 2021. Expanding coverage for palliative care services: Policy implications and recommendations. *Health Affairs*, 40(3), 457–465.
GOZALO, P., TENO, J.M., AND MITCHELL, S.L. 2011. End-of-life care for cancer patients and the impact of home-based palliative care. *Journal of Clinical Oncology*, 29(10), 1283–1289.
HIGGINSON, I.J. AND CARR, A.J. 2001. Using quality of life measures in the clinical setting. *BMJ*, 322(7297), 1297–1300.
HUDSON, P., ARANDA, S., AND KRISTJANSON, L.J. 2011. Meeting the supportive needs of family caregivers in palliative care: Challenges for health professionals. *Journal of Palliative Medicine*, 8(1), 19–25.
HUDSON, P., et al. 2020. Education and training in palliative care: Current status and future directions. *Journal of Palliative Medicine*, 23(12), 1557–1564.
HUDSON, P., REMEDIOS, C., ZORDAN, R., et al. 2012. Clinical practice guidelines for the psychosocial and bereavement support of family caregivers of palliative care patients. *Journal of Palliative Medicine*, 15(6), 696–702.
HUDSON, P., THOMAS, K., AND AOUN, S. 2019. The role of advocacy in improving palliative care: Lessons learned and future directions. *Palliative Medicine*, 33(1), 5–12.
HUI, D., BRUERA, E., AND JHANGRI, G. 2014. Effect of a palliative care unit on end-of-life care in a comprehensive cancer center. *Journal of Clinical Oncology*, 32(24), 2607–2615.
IAHPC. 2024. *International Association for Hospice & Palliative Care: Advocacy and policy*. Available at: https://www.iahpc.org/advocacy (Accessed 25 July 2024).
JAHNKE, R., KAPPEL, B., AND BRUNNER, F. 2014. Palliative care and oncology: A collaborative model. *Supportive Care in Cancer*, 22(1), 15–22.
JOHNSON, A., WILLIAMS, R., & PATEL, S. 2022. Multidisciplinary palliative care in advanced pancreatic cancer: A case report. *Supportive Care in Cancer,* 30(12), 9067–9072. https://doi.org/10.1007/s00520-022-06618-9
JOHNSTON, B., AND MCILFATRICK, S. 2021. Palliative care education for healthcare professionals: Current status and future directions. *Journal of Palliative Medicine*, 24(6), 859–866.
KAVALIERATOS, D., et al. 2016. Association between palliative care and patient and caregiver outcomes: A systematic review and meta-analysis. *Journal of the American Medical Association*, 316(20), 2104–2114.
KAVALIERATOS, D., GELFMAN, L.P., AND MILLER, S.C. 2016. Home-based palliative care for patients with advanced illness: A systematic review and meta-analysis. *Journal of the American Medical Association,* 316(20), 2104–2114.
KELLEY, S. T., & STEELE, S. R. 2021. Celiac plexus neurolysis for the treatment of pancreatic cancer pain: An evidence-based review. *Journal of Gastrointestinal Surgery*, 25(6), 1473–1479.
KISSANE, D.W. 2012. The relief of existential suffering. *Archives of Internal Medicine*, 172(19), 1501–1505.
KNAUL, F.M., FARMER, P., AND KRAKAUER, E.L. 2018. The Lancet Commission on palliative care and pain relief – Findings and recommendations. *The Lancet*, 391(10128), 1391–1450.
KORFAGE, I.J., RANCHOR, A.V., AND VAN DER HEIDE, A. 2021. Integrating palliative care into oncology: Current evidence and future directions. *Lancet Oncology*, 22(3), 379–388.

LEE, R. et al. 2023. Psychological counseling in palliative care: Reducing anxiety and depression in pancreatic cancer patients. *Psycho-Oncology*, 32(2), 231–239.
LINDLEY, L.C., BOECKH, M.J., AND SHAKIB, N. 2013. Community-based palliative care models. *Journal of Palliative Medicine*, 16(12), 1389–1396.
MACKAY, T. M., LATENSTEIN, A. E. J., AUGUSTINUS, S., VAN DER GEEST, L. G., BOGTE, A., BONSING, B. A., CIRKEL, G. A., HOL, L., BUSCH, O. R., DEN DULK, M., VAN DRIEL, L. M. J. W., FESTEN, S., DE GROOT, D. A., DE GROOT, J. B., GROOT KOERKAMP, B., HAJ MOHAMMAD, N., HAVER, J. T., VAN DER HARST, E., DE HINGH, I. H., HOMS, M. Y. V., LOS, M., LUELMO, S. A. C., DE MEIJER, V. E., MEKENKAMP, L., MOLENAAR, I. Q., PATIJN, G. A., QUISPEL, R., RÖMKENS, T. E. H., VAN SANTVOORT, H. C., STOMMEL, M. W. J., VENNEMAN, N. G., VERDONK, R. C., VAN VILSTEREN, F. G. I., DE VOS-GEELEN, J., VAN WERKHOVEN, C. H., VAN HOOFT, J. E., VAN EIJCK, C. H. J., WILMINK, J. W., VAN LAARHOVEN, H. W. M., BESSELINK, M. G. 2024. Dutch Pancreatic Cancer Group. Implementation of best practices in pancreatic cancer care in the Netherlands: A stepped-wedge randomized clinical trial. *JAMA Surgery*, 159(4), 429–437.
MARTINEZ, F. et al. 2020. Early palliative care and survival in pancreatic cancer: A retrospective analysis. *Annals of Oncology*, 31(10), 1360–1367.
MCCARTY, J.A., FISH, R., AND WANG, J. 2015. Innovative funding models for palliative care services. *Journal of Palliative Medicine*, 18(4), 297–302.
MCGUIGAN, A., KELLY, P.,TURKINGTON, R.C., JONES, C., COLEMAN, H.G., MCCAIN, R.S. 2018. Pancreatic cancer: A review of clinical diagnosis, epidemiology, treatment and outcomes. *World Journal of Gastroenterology*, 24(43), 4846–4861.
MEIER, D. E. 2011. Increased access to palliative care and hospice services: Opportunities to improve value in health care. *Milbank Quarterly*, 89(3), 343–380.
MEIER, D.E., AND BERESFORD, L. 2016. Palliative care: What do we know and what do we need to know?. *New England Journal of Medicine*, 375(5), 489–496.
MEIER, D.E., et al. 2018. The quality-of-life benefits of palliative care: An update of the evidence. *Journal of Palliative Medicine*, 21(3), 397–404.
MOOLE, H., JAEGER, A., BECHTOLD, M. L., & PULI, S. R. 2021. Endoscopic stenting for malignant biliary strictures: A systematic review and meta-analysis. *Gastrointestinal Endoscopy*, 94(2), 344–355.
MORRISON, R.S., AND MEIER, D.E. 2008. Palliative care. *New England Journal of Medicine*, 359(4), 325–335.
MORRISON, R.S., et al. 2017. Palliative care in the ICU: A review. *Critical Care Medicine*, 45(6), 1203–1210.
NASH, R., WANGER, J., AND SMITH, B. 2020. Telemedicine in palliative care: A review of current applications and future directions. *Journal of Telemedicine and Telecare*, 26(2), 98–105.
NGUYEN, H., LOPEZ, M., & CHEN, D. 2023. Psychosocial interventions and advanced care planning in palliative care for pancreatic cancer patients: A case study. *Psycho-Oncology, 32*(4), 825–830. https://doi.org/10.1002/pon.5987.
NHPCO 2023. *National Hospice and Palliative Care Organization: Policy and advocacy.* Available at: https://www.nhpco.org/advocacy/ (Accessed 25 July 2024).
NUCKOLS, T.K., ANDERSON, L., AND GENG, L. 2014. Palliative care in the Medicare Hospice Benefit: Coverage and implications. *Journal of Palliative Medicine*, 17(2), 184–190.
PEREIRA, J., AND CARVALHO, J. 2017. Socioeconomic and healthcare disparities in access to palliative care: A systematic review. *BMC Palliative Care*, 16(1), 54.
PUCHALSKI, C.M., VITILLO, R., HULL, S.K., et al. 2009. Improving the quality of spiritual care as a dimension of palliative care: The report of the Consensus Conference. *Journal of Palliative Medicine*, 12(10), 885–904.
QUILL, T.E., & ABERNETHY, A.P. 2013. Generalist plus specialist palliative care—Creating a more sustainable model. *New England Journal of Medicine*, 368(13), 1173–1175.
REDDY, P., SYMONDS, R. P., & WASSELL, P. 2020. The management of gastrointestinal symptoms in advanced pancreatic cancer. *Supportive Care in Cancer*, 28(6), 2581–2587.
RIPAMONTI, C. I., BRUERA, E., & FANIZZA, C. 2020. Aprepitant for the prevention of chemotherapy-induced nausea and vomiting: A review. *Supportive Care in Cancer*, 28(8), 3643–3650.
SAUNDERS, C. 2000. The evolution of palliative care. *Patient Education and Counseling*, 41(1), 7–13.
SKEVINGTON, S.M., LOFTY, M. AND O'CONNELL, K.A. 2004. The World Health Organization's WHOQOL-BREF quality of life assessment: Psychometric properties and results of the international field trial. A report from the WHOQOL group. *Quality of Life Research*, 13(2), 299–310.

SMITH, J. et al. 2021. Early palliative care in advanced pancreatic cancer: Impact on pain and quality of life. *Journal of Pain and Symptom Management*, 61(4), 784–792.

SMITH, J., BROWN, L., & TAYLOR, M. 2021. Early palliative care integration in pancreatic cancer: A case study approach. *Journal of Palliative Medicine,* 24(3), 287–293. https://doi.org/10.1089/jpm.2020.0429.

SMITH, T.J., AND COYNE, P.J. 2013. Training and education in palliative care: What is needed?. *Journal of Clinical Oncology*, 31(31), 3787–3792.

SMITH, T.J., AND HILLNER, B.E. 2013. Outpatient palliative care: Improving the quality of life for cancer patients. *Journal of Clinical Oncology*, 31(15), 1840–1847.

STEINHAUSER, K. E., CHRISTAKIS, N. A., CLIPP, E. C., MCNEILLY, M., MCINTYRE, L., & TULSKY, J. A. 2000. Factors considered important at the end of life by patients, family, physicians, and other care providers. *JAMA*, 284(19), 2476–2482.

SULLIVAN, M.D., DANIS, M., AND MOEN, M. 2022. Personalized palliative care: Innovations and future directions. *Journal of Pain and Symptom Management*, 63(2), 309–319.

TAWFIK, A., SAAD, E. D., FOJO, T., & SRIDHARA, R. 2021. Antiemetic guidelines and the role of anti-emetics in preventing chemotherapy-induced nausea and vomiting: A systematic review. *Journal of Clinical Oncology*, *39*(23), 2631–2644.

TEMEL, J.S., GREER, J.A., MUZIKANSKY, A., GALLAGHER, E.R., ADMANE, S., JACKSON, V.A., DAHLIN, C.M., BLINDERMAN, C.D., JACOBSEN, J., PIRL, W.F., BILLINGS, J.A., AND LYNCH, T.J. 2010. Early palliative care for patients with metastatic non-small-cell lung cancer. *New England Journal of Medicine*, 363(8), 733–742.

TENO, J. M., CLARRIDGE, B. R., CASEY, V., WELCH, L. C., WETLE, T., SHIELD, R., & MOR, V. 2004. Family perspectives on end-of-life care at the last place of care. *JAMA*, 291(1), 88–93.

TENO, J.M., WEITZEN, S., AND WELCH, L.C. 2018. Quality measures in palliative care: A review of the evidence and future directions. *Journal of Palliative Medicine*, 21(4), 542–549.

TOMLINSON, D., ROBINSON, P. D., OBEROI, S., & SUNG, L. 2021. Pharmacologic interventions for fatigue in cancer: A systematic review and meta-analysis. *Journal of Clinical Oncology*, *38*(12), 1324–1333.

WHO 2020. *Palliative care*. World Health Organization. Available at: https://www.who.int/news-room/fact-sheets/detail/palliative-care (Accessed 25 July 2024).

WHOQOL Group. 1995. The World Health Organization Quality of Life assessment (WHOQOL): Position paper from the World Health Organization. *Social Science & Medicine*, 41(10), 1403–1409.

WORLD HEALTH ORGANIZATION (WHO). 2018. *Integration of palliative care and symptom relief into primary health care*. Available at: www.who.int/cancer/palliative/en/ (Accessed 25 July 2024).

WORLD HEALTH ORGANIZATION (WHO). 1990. *Cancer pain relief and palliative care: Report of a WHO Expert Committee*. World Health Organization Technical Report Series, 804. Geneva: World Health Organization. Available at: https://iris.who.int/handle/10665/39524.

WRIGHT, A.A., ZHANG, B., AND RAY, A. 2016. Geographic variation in palliative care use in the United States: A population-based study. *JAMA Network Open*, 3(5), e2013174.

8 Challenges in Access to Care in Pancreatic Cancer

Botle Damane[1], Thanyani Mulaudzi[1], Tania Maphoso[1], Dakalo Ramali[1], Sikhumbuzo Mbatha[1], Guy Gaudji[2], Andreas M. Kaufmann[1,3], and Zodwa Dlamini[4]*

[1]Department of Surgery, Steve Biko Academic Hospital, University of Pretoria, Pretoria, South Africa.

[2]Department of Urology, Steve Biko Academic Hospital, University of Pretoria, Pretoria, South Africa.

[3]Charité – Universitätsmedizin Berlin, corporate member of Freie Universität Berlin and Humboldt-Universität zu Berlin, Department of Gynecology with Oncologic Surgery, Augustenburger Platz 1, Berlin, Germany.

[4]SAMRC Precision Oncology Research Unit (PORU), DSI/NRF SARChI Chair in Precision Oncology and Cancer Prevention (POCP), Pan African Cancer Research Institute (PACRI), University of Pretoria, Pretoria, South Africa.

*Corresponding author: Botle Precious Damane

INTRODUCTION

Access to care is a pivotal factor influencing the outcomes and survival rates of patients with pancreatic cancer, one of the most aggressive and challenging malignancies to treat (Prades et al., 2020). The intricate nature of pancreatic cancer care requires timely and specialized medical attention; however, many patients face significant barriers that prevent them from receiving the care they need (Wall et al., 2023). These barriers are not merely individual or incidental; they reflect broader systemic issues within healthcare delivery that disproportionately affect patients with pancreatic cancer (Maphumulo and Bhengu, 2019).

The availability and distribution of healthcare resources are among the most pressing concerns in pancreatic cancer care . Specialized oncology centers and advanced diagnostic facilities, which are crucial for effective treatment, are unevenly distributed and often concentrated in urban areas. This creates significant disparities in access, particularly for patients in rural and underserved regions (Wall et al., 2023). The lack of accessible care facilities often results in delayed diagnosis and treatment, which can have a profound impact on patient outcomes (Levit et al., 2020). Moreover, healthcare policies and funding disparities exacerbate these challenges, with inadequate financial support for cancer care limiting the availability of necessary treatments and resources (Omotoso et al., 2023).

Socioeconomic factors further complicate access to care for pancreatic cancer patients (Wall et al., 2023). Individuals with low incomes or insufficient insurance coverage frequently encounter barriers that delay treatment, limit their options, and reduce their participation in clinical trials (Biddell et al., 2023). These financial constraints can significantly affect patient outcomes, as timely and appropriate care is critical for managing this fast-progressing disease (National Academies

 DOI: 10.1201/9781003544401-12

of Sciences, Engineering, and Medicine; Health and Medicine Division; Board on Health Care Services; Committee on Health Care Utilization and Adults with Disabilities, 2018). Geographic disparities add another layer of challenge for healthcare providers and increase the need for long-distance travel to access specialized care (Maganty et al., 2023).

Cultural and educational barriers also play a significant role in determining access to pancreatic cancer care (Wall et al., 2023). Health literacy, which encompasses patients' understanding of their conditions and treatment options, is crucial for effective care management. Unfortunately, low health literacy can lead to delayed care, poor treatment adherence, and a misunderstanding of the disease (Coughlin et al., 2020). Additionally, cultural beliefs and stigmas surrounding cancer in general and, in particular, pancreatic cancer can influence patients' decisions to seek treatment and adhere to prescribed care plans (McQuaid and Landier, 2018). Addressing these barriers requires targeted educational initiatives and culturally sensitive care approaches to improve patient outcomes (Omotoso et al., 2023; Wall et al., 2023).

The multifaceted challenges in accessing pancreatic cancer care underscore the need for a comprehensive approach to address these issues (Wall et al., 2023). This chapter will delve into the systemic, socioeconomic, geographic, and cultural barriers that hinder access to care, exploring how these factors interact and contribute to the difficulties in effectively managing pancreatic cancer. Through this exploration, this chapter aims to identify strategies and interventions that can improve access to care and ultimately enhance outcomes for patients with this devastating disease.

HEALTHCARE SYSTEM BARRIERS

Availability of Specialized Care

The geographical distribution of oncology centers and specialists is a critical factor in the accessibility of care for pancreatic cancer patients (Prades et al., 2020; Wall et al., 2023). In many countries, specialized centers equipped with the necessary resources and expertise to treat complex cancers such as pancreatic cancer are disproportionately located in urban areas (Wall et al., 2023). This urban concentration forces patients from rural or remote areas to travel long distances for specialized care, often causing delays in diagnosis and treatment initiation (Cyr et al., 2019). The logistical challenges of traveling, coupled with the financial burden, can discourage patients from seeking timely care, ultimately impacting their survival rates (Haleem et al., 2021). Furthermore, the uneven distribution of oncology specialists exacerbates these disparities (Melhem et al., 2024). The lack of proximity to specialized care can result in patients receiving suboptimal or generalized care, which may not be as effective in managing pancreatic cancer (Noel and Fiscella, 2019). This is because regions with fewer specialists experience longer waiting times for appointments, which can delay treatment and negatively affect patient outcomes (Alrasheedi et al., 2019). Shortage of specialists is particularly problematic in areas lacking comprehensive cancer centers, where patients may not have access to the latest treatment protocols and clinical trials (Barrios et al., 2023; Ramutumbu et al., 2023).

Another significant issue is the limited availability of multidisciplinary teams, which are essential for the effective management of pancreatic cancer (Yao et al., 2023). These teams, which typically include surgeons, oncologists, radiologists, and other specialists, are more commonly found in large, well-funded cancer centers (Hitz et al., 2022). Patients treated in facilities without such teams may not benefit from the collaborative approach needed to address the complexities of pancreatic cancer, leading to less personalized and potentially less effective care (Berardi et al., 2020). Efforts to address these disparities include telemedicine, tumor boards, and outreach programs that aim to connect patients in underserved areas with specialists in urban centers (Bagchi et al., 2022; Haleem et al., 2021). However, these solutions are not without challenges, including technological barriers and the need for significant investments in infrastructure (Haleem et al., 2021). To improve outcomes, there is a need for policies that promote a more equitable distribution of oncology resources, ensuring that all patients, regardless of location, have access to specialized care (Omotoso et al., 2023).

Access to advanced diagnostic and treatment facilities is crucial for the effective management of pancreatic cancer (Prades et al., 2020). High-tech diagnostic tools such as endoscopic ultrasound (EUS) and advanced imaging technologies are essential for an early and accurate diagnosis (Ang et al., 2018). However, these facilities are often limited to major medical centers that may not be easily accessible to all patients (Ang et al., 2018; Prades et al., 2020). The lack of access to such facilities can result in delayed or missed diagnoses, which can significantly affect prognosis and treatment outcomes (Ang et al., 2018). Another factor is access to state-of-the-art treatment facilities, which are often restricted to specialized cancer centers. This includes those offering cutting-edge therapies such as immunotherapy and targeted treatments (Chehelgerdi et al., 2023). Patients in areas without such facilities may have to rely on less advanced treatments, which may not be effective. This disparity in access to advanced treatment options can lead to suboptimal care and decreased survival rates in patients with pancreatic cancer (Noel and Fiscella, 2019; Wall et al., 2023). Additionally, the concentration of these advanced treatment facilities in urban centers means that patients in rural areas often face significant barriers to accessing life-saving therapies (Graves et al., 2022).

This challenge is compounded by the fact that advanced diagnostic procedures require specialized training and equipment that may not be available in all healthcare settings (Pufahl et al., 2022). This means that patients in rural or underserved areas might not only have limited access to these technologies but may also encounter delays in being referred to centers where such services are available (Coombs et al., 2022). These delays can be detrimental in diseases, such as pancreatic cancer, where early detection and effective treatment are critical for improving survival rates (Kenner et al., 2017).

Healthcare Policies and Funding

Healthcare policies play a pivotal role in determining access to cancer care. Policies that allocate resources and funding for cancer care can either enhance or hinder the availability of services for pancreatic cancer patients (Ramutumbu et al., 2023). In countries with well-established healthcare systems, comprehensive cancer care policies ensure that patients have access to necessary diagnostic and treatment services. However, in many regions, inadequate healthcare policies result in significant barriers to access (Omotoso et al., 2023; Ramutumbu et al., 2023). National policies and legislation directly influence the availability, accessibility, and affordability of healthcare services for pancreatic cancer patients. Countries with universal healthcare systems often provide a more equitable distribution of resources, ensuring that a wider population has access to necessary cancer care. Conversely, in regions where healthcare lacks government intervention, significant disparities exist in the quality and availability of services. The lack of standardized policies across nations contributes to inconsistencies in treatment access, with some patients receiving cutting-edge therapies while others are left without basic care (Simpkin et al., 2019).

For instance, policies that do not prioritize cancer care or lack funding for advanced treatment options can lead to insufficient resources and infrastructure (Omotoso et al., 2023). This can affect the quality of care that patients receive and limit their access to lifesaving treatment. Moreover, healthcare policies that fail to address the specific needs of underserved populations can exacerbate the existing disparities in cancer care access (Dos-Santos-Silva et al., 2022, Omotoso et al., 2023). The impact of healthcare policies is also evident in the availability of insurance coverage for pancreatic cancer treatment. In many cases, policies dictate which treatments are covered, potentially limiting access to newer or more expensive therapies that could improve outcomes (Loehrer et al., 2015). Patients without comprehensive insurance may face significant out-of-pocket costs, which can be a major barrier to accessing necessary care. This financial burden can lead to delayed treatment, reduced adherence to prescribed therapies, and ultimately poorer outcomes (Frazier et al., 2023; Rohatgi et al., 2021).

Efforts to reform healthcare policies to improve access to pancreatic cancer care are ongoing but face numerous challenges, including political resistance and budget constraints (Prades et al.,

2020). Advocacy by patient groups, healthcare providers, and researchers is critical to ensure that policymakers understand the importance of investing in cancer care and addressing the systemic barriers that prevent patients from accessing the treatments they need (Kale et al., 2023). Disparities in healthcare funding and resource allocation are critical factors that influence access to pancreatic cancer care. Regions with higher healthcare funding can invest in advanced diagnostic and treatment facilities, recruit specialized healthcare professionals, and provide comprehensive cancer care services (Omotoso et al., 2023; Wall et al., 2023). Conversely, areas with limited funding struggle to offer the same level of care, which results in significant disparities in patient outcomes (McMaughan et al., 2020).

Inadequate funding can also lead to a lack of necessary infrastructure such as specialized cancer centers and advanced diagnostic equipment, which are essential for effective cancer care (Omotoso et al., 2023). Furthermore, disparities in resource allocation can mean that some populations, particularly those in rural or low-income areas, receive lower-quality care than their counterparts in more affluent regions (Baciu et al., 2017). Addressing these disparities requires targeted policies and increased investment in the healthcare infrastructure to ensure equitable access to pancreatic cancer care for all patients (Wall et al., 2023). The challenge of funding disparity is not limited to developing countries. Even in high-income nations, significant differences exist between regions and populations in terms of healthcare resources (Baciu et al., 2017). Urban centers often receive more funding and resources, leaving rural areas at a disadvantage. This can result in a two-tier healthcare system in which access to high-quality cancer care depends heavily on the patient's location and socioeconomic status (Zhao et al., 2022).

Addressing disparities in healthcare funding and resource allocation requires a multifaceted approach. This includes advocating for increased federal and state funding for cancer care, ensuring that resources are distributed equally across regions, and implementing policies that specifically target underserved populations (Kale et al., 2023). Additionally, partnerships between the public and private sectors can play a crucial role in supplementing healthcare funding and expanding access to essential services (Joudyian et al., 2021).

SOCIOECONOMIC AND GEOGRAPHIC DISPARITIES

Income and Insurance Status

Socioeconomic status (SES) profoundly affects access to pancreatic cancer care, with significant disparities observed across different income levels (Sutton et al., 2023). Individuals from lower-income backgrounds face substantial barriers to obtaining timely and effective treatment, including inadequate health insurance coverage and limited access to high-volume cancer centers (Shapiro et al., 2016; Sutton et al., 2023). High-volume centers, known for their comprehensive treatment options and experienced teams, are often situated in affluent or urban areas, making them less accessible to lower-income patients (Sutton et al., 2023). Insurance coverage is a critical determinant of these disparities. Insufficient or absent insurance often leads to significant obstacles in accessing necessary treatments, such as surgical interventions and advanced diagnostic procedures. The high costs associated with care for uninsured patients frequently result in deferred treatment, adversely impacting prognosis (Shapiro et al., 2016). Furthermore, patients with private insurance are generally more likely to access high-quality care compared to those with public insurance or no insurance, leading to better treatment outcomes (Fonseca et al., 2022). Inadequate insurance coverage can result in suboptimal care and reduced survival rates (Abdel-Rahman, 2020).

Socioeconomic factors, including educational attainment, also influence access to healthcare. Individuals with lower levels of education often have less understanding of cancer symptoms and available healthcare services, which can delay the recognition of symptoms and the pursuit of medical care (Brunner and Seufferlein, 2016; Rawla et al., 2019). This lack of health literacy contributes to later-stage diagnoses and worse outcomes (Rawla et al., 2019). Moreover, educational disparities affect the ability to navigate complex healthcare systems and advocate for necessary medical interventions (Crosby et al., 2022). In many low- and middle-income countries, the absence of universal health coverage exacerbates financial barriers, leaving many patients to bear out-of-pocket expenses for cancer care, which can be financially devastating (Rawla et al., 2019). Resource distribution and funding are pivotal, with limited healthcare budgets and inefficient resource allocation in low-income regions contributing to inadequate cancer care infrastructure (Kruk et al., 2018; Ramutumbu et al., 2023). Enhancing healthcare funding and optimizing resource distribution are crucial for expanding access to diagnostic and treatment facilities, particularly in underserved areas (Wall et al., 2023).

Petric et al. highlighted that variations in insurance coverage and access to specialty care across different socioeconomic groups further compound these disparities. Their findings underscore that improving health insurance coverage and increasing support for underserved populations could mitigate some of the systemic barriers faced by lower-income patients. Addressing these socioeconomic factors is essential for achieving equitable access to care for pancreatic cancer patients (Petric et al., 2022).

Rural vs. Urban Access

Geographic location significantly influences access to pancreatic cancer care, with rural patients facing unique challenges compared to their urban counterparts. Rural areas often lack specialized cancer management services, which are more readily available in urban centers (Canale et al., 2018). The scarcity of specialized facilities in rural regions leads to delays in diagnosis and treatment, impacting patient outcomes (Amin et al., 2023). Transportation and logistical issues present additional barriers for rural patients. Traveling long distances to reach oncology centers can be physically and financially burdensome. Costs associated with long-distance travel, including transportation expenses and potential loss of income due to missed work, further exacerbate financial strain. The lack of reliable public transportation options in rural areas intensifies these challenges, making regular access to necessary medical services difficult (Amin et al., 2023).

Rural patients often face delays in diagnosis due to the limited availability of primary care providers and diagnostic facilities (Segel et al., 2020). The absence of early diagnostic resources contributes to later-stage diagnoses, negatively impacting prognosis and limiting treatment effectiveness (Segel et al., 2020). Addressing these disparities requires targeted interventions to improve diagnostic resources and support in rural areas (Segel et al., 2020). Petric and colleagues further emphasize that rural patients often experience exacerbated challenges due to the limited availability of specialized cancer care and insufficient local resources. Their research suggests that targeted programs to increase telemedicine options and mobile diagnostic units could alleviate some of these geographic disparities (Petric et al., 2022).

TREATMENT ACCESSIBILITY

Availability of Standard Treatments

Surgical resection remains the most definitive treatment for early-stage pancreatic cancer, particularly for pancreatic ductal adenocarcinoma (PDAC) (Kamisawa et al., 2016). A compressive summary of the available treatment modalities, including their mechanisms, challenges, and clinical considerations, is presented in Table 8.1. The primary surgical procedures include the Whipple procedure (pancreaticoduodenectomy), distal pancreatectomy, and total pancreatectomy. The choice of surgery depends on the tumor's location, size, and involvement with surrounding structures (Kamisawa et al., 2016). The Whipple procedure, for instance, involves the removal of the head of the pancreas, the duodenum, and parts of the bile duct and stomach, and is typically indicated for tumors in the pancreatic head

TABLE 8.1
Overview of Pancreatic Cancer Treatment Options

Therapy Type	Description	Challenges	References
Surgery	Removal of the tumor, often through procedures like the Whipple procedure.	Limited to early-stage diagnosis; high risk of complications.	Kamisawa et al. (2016); Kolbeinsson et al. (2023)
Chemotherapy	Use of drugs to kill cancer cells, commonly gemcitabine, capecitabine, and modified FOLFIRINOX.	Severe side effects: resistance can develop; not suitable for all patients.	Kolbeinsson et al. (2023); Nishimoto (2022)
Radiation Therapy	Use of high-energy radiation to destroy cancer cells. Can improve results of surgery in cases with borderline resectability.	Side effects include fatigue and skin reactions; mostly limited in advanced disease.	Kolbeinsson et al. (2023); Falco et al. (2023)
Targeted Therapy	Drugs that target specific genes or proteins involved in cancer growth, such as erlotinib.	Limited to patients with the right genetic profile, and can be costly.	Kolbeinsson et al. (2023); Nishimoto (2022)
Immunotherapy	Boosts the body's immune system to fight cancer. Includes immune checkpoint inhibitors and vaccine-based therapies.	Limited effectiveness in pancreatic cancer. Only anti-programmed cell death 1 is approved in specific types, and ongoing research is needed.	Torphy et al. (2018); Yoon et al. (2021)
Palliative Care	Focuses on relieving symptoms and improving the quality of life for patients with advanced cancer. Palliative surgery using minimally invasive techniques can provide support in certain cases of pancreatic cancer.	Does not treat cancer itself and require comprehensive support services.	Kolbeinsson et al. (2023); Perinel and Adham (2019)
Clinical Trials	Research studies testing new treatments or combinations of treatments.	Not always available; uncertain outcomes and potential risks.	Kolbeinsson et al. (2023)

(Sung et al., 2021). However, surgical resection is often constrained to high-volume centers primarily located in urban areas, which creates significant challenges for patients living in rural or underserved regions (Franck et al., 2020). Patients in these areas face delays and additional costs related to traveling for surgery, and centralization of surgical services can exacerbate these disparities (Fonseca et al., 2022). Expanding surgical services into underserved areas or utilizing telemedicine for preoperative and postoperative care could mitigate these disparities (Pereira and Corrêa, 2018; Franck et al., 2020).

Chemotherapy is a cornerstone in the treatment of both resectable and unresectable pancreatic cancer. Regimens such as gemcitabine-based therapy (e.g., gemcitabine/nab-paclitaxel) and FOLFIRINOX (a combination of 5-fluorouracil, leucovorin, irinotecan, and oxaliplatin) have shown improved survival rates compared to earlier monotherapy approaches (Brunner and Seufferlein, 2016). Despite its critical role, access to chemotherapy can be significantly hampered by geographic and socioeconomic factors, as rural patients may face delays due to the lack of local chemotherapy centers and financial constraints (Abdel-Rahman, 2020). New chemotherapeutic agents, which can significantly improve outcomes, are often more readily available in well-funded urban centers, contributing to discrepancies in treatment quality and access (Vaziri-Gohar et al., 2023). Policies aimed at increasing the availability of chemotherapy in underserved regions could help mitigate these disparities (Pereira and Corrêa, 2018).

Radiotherapy, including techniques like stereotactic body radiation therapy (SBRT), is used to manage local tumor progression, especially in patients with locally advanced pancreatic cancer (Kamisawa et al., 2016). SBRT offers high precision in targeting tumors while minimizing damage to adjacent healthy tissues. However, advanced radiotherapy technologies are typically concentrated in urban settings, leaving rural patients with limited access to potentially less effective treatment (Mizrahi et al., 2020). Increasing the number of radiotherapy centers in underserved areas and enhancing the quality of radiotherapy services could improve treatment outcomes for patients in these regions (Takaori et al., 2016).

Access to Clinical Trials and Experimental Treatments

Participation in clinical trials is crucial for advancing pancreatic cancer treatments and improving patient outcomes. Clinical trials offer opportunities to access cutting-edge treatments that may exceed the efficacy of standard therapies. Nevertheless, many patients face significant barriers to participation, primarily due to geographic and socioeconomic constraints (Hart et al., 2022). Clinical trial centers are often concentrated in major urban and university hospitals, making it difficult for patients in rural areas to participate due to travel requirements and associated costs (Canale et al., 2018). This urban-centric distribution of trial sites limits access to potentially life-saving treatments for geographically isolated patients (Segel et al., 2020). Socioeconomic factors further complicate clinical trial participation. Financial barriers, such as the costs of travel, lodging, and taking time off work, can deter patients from enrolling in trials (Stewarta and Stewart IV, 2021). Moreover, patients with limited health insurance coverage may not have access to experimental treatments or may be unable to afford related costs (Walker, 2021). Addressing these barriers through financial support and expanding insurance coverage for experimental therapies could enhance trial accessibility and improve equity (Ramamurthy and Wong, 2018).

The concentration of clinical trial facilities in urban areas creates challenges for patients in rural regions, limiting their access to novel treatments (Hart et al., 2022). Additionally, socioeconomic disparities affect patients' ability to participate in trials. Lower-income patients may lack the financial resources needed for travel and other trial-related expenses, while those with lower health literacy might not be aware of available trials or struggle with complex enrollment processes (Stewarta and Stewart IV, 2021). Increasing awareness about clinical trials and providing financial and logistical support for patients can help address these disparities and improve access to experimental treatments (Hart et al., 2022).

Challenges and Advances in Immunotherapy for Pancreatic Cancer

Immunotherapy for pancreatic cancer, particularly PDAC, faces significant challenges due to the cancer's aggressive nature and highly immunosuppressive tumor microenvironment (Timmer et al., 2021). PDAC is known for its resistance to conventional treatments like chemotherapy, leading to poor patient outcomes (Anderson et al., 2021; Feng et al., 2022). Checkpoint inhibitors like dostarlimab (Jemperli) and pembrolizumab (Keytruda®) offer hope for specific subsets of patients, particularly those with DNA mismatch repair deficiency (dMMR) and high microsatellite instability (MSI-H) (Costa and Vale, 2022). The mechanisms by which these inhibitors restore immune function by blocking PD-1/PD-L1 interactions are illustrated in Figure 8.1. However, these benefits are limited to a small percentage of pancreatic cancer patients, and the high cost and need for advanced genetic testing limit their availability, especially in regions with underdeveloped healthcare systems (Costa and Vale, 2022).

A major challenge to the effectiveness of immunotherapy in pancreatic cancer is the tumor's ability to create an immunosuppressive microenvironment, which hinders the immune system's ability

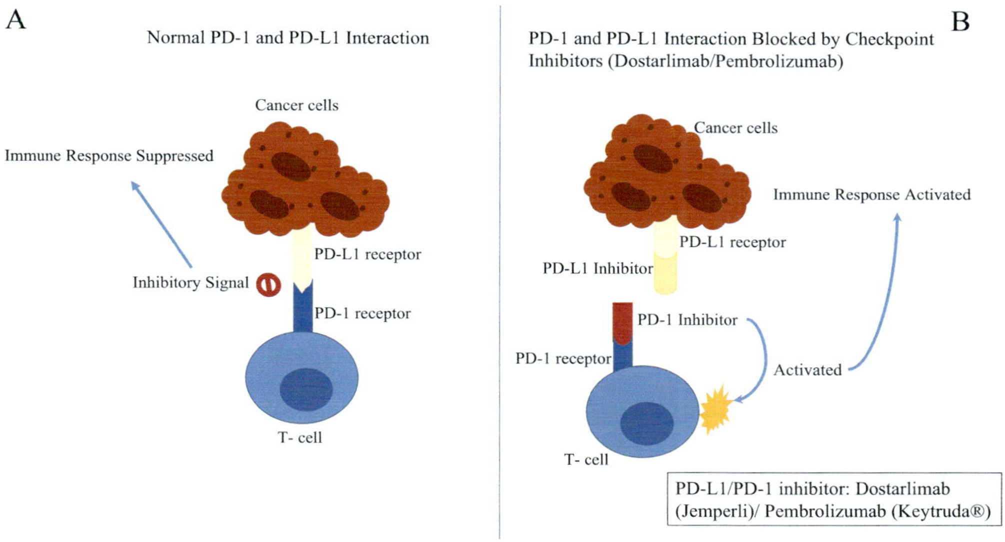

FIGURE 8.1 This figure illustrates the mechanisms of T-cell recognition and response to cancer cells, highlighting the role of the PD-1/PD-L1 interaction in immune suppression and the effect of checkpoint inhibitors, dostarlimab (Jemperli) and pembrolizumab (Keytruda), in restoring immune function. (A) Normal PD-1/PD-L1 I: T-cell and cancer cell interaction. The T-cell, which normally attacks cancer cells, is shown with a PD-1 receptor binding to PD-L1 on the cancer cell. This interaction sends an inhibitory signal, suppressing the T-cell's activity and allowing the cancer cell to evade the immune system, resulting in a suppressed immune response. (**B**) Blocked PD-1/PD-L1 interaction: effect of checkpoint inhibitors. Dostarlimab (Jemperli) and Pembrolizumab (Keytruda) block the PD-1/PD-L1 interaction by binding to PD-1 or PD-L1, respectively. This prevents the inhibitory signal, allowing the T-cell to remain active and destroy the cancer cell, leading to an activated immune response.

to recognize and attack cancer cells, reducing the efficacy of checkpoint inhibitors. This challenge is further complicated by the fact that pancreatic cancer is frequently diagnosed at an advanced stage, limiting the effectiveness of immunotherapy (Zheng et al., 2024). Emerging research highlights the potential role of the gut microbiome in influencing the effectiveness of immunotherapy for pancreatic cancer (Attebury and Daley, 2023). The composition of gut bacteria can either enhance the immune response or contribute to a more tumor-supportive environment, suggesting that future treatments could potentially combine immunotherapy with strategies to modulate the gut microbiota for better outcomes (Zhang et al., 2023). However, access to these combined therapeutic approaches remains a challenge, particularly in resource-limited settings (Havlir and Gandhi, 2015).

CULTURAL AND EDUCATIONAL BARRIERS

Health Literacy and Awareness

Health literacy plays a critical role in the early detection and treatment adherence of pancreatic cancer, directly influencing patient outcomes and the overall effectiveness of care. Patients with higher health literacy are more likely to recognize symptoms early, seek timely medical advice, and adhere to prescribed treatment regimens, leading to improved survival rates (Samoil et al., 2021). Conversely, low health literacy can delay diagnosis, complicate treatment decisions, and negatively affect patient engagement with healthcare providers, exacerbating the challenges in managing this aggressive disease (Petersen et al., 2022). Educational initiatives aimed at improving health literacy among pancreatic cancer patients have shown promise, but their effectiveness varies depending on the mode of delivery and the demographic characteristics of the target population. For instance, visual and interactive educational tools, such as animated videos and digital platforms, have been found to enhance patient understanding of complex medical information, making it easier for them to make informed decisions about their care (Munigala et al., 2022). These tools are particularly beneficial for older adults and those with limited formal education, as they provide clear and accessible information that can bridge the gap between patients and healthcare providers (Su et al., 2024).

However, despite the potential benefits, there are still significant gaps in the implementation of such initiatives, particularly in underserved communities where access to digital resources may be limited. The lack of internet access or digital literacy among certain patient populations can significantly hinder the effectiveness of online educational tools. Moreover, in regions where literacy rates are generally low, written materials might not be as effective, necessitating the development of alternative approaches such as oral presentations or community workshops (Abreu et al., 2024). Addressing these gaps requires a multifaceted approach that includes not only the development of tailored educational materials but also efforts to ensure that these resources are accessible to all patients, regardless of their socioeconomic status or geographic location (Kruk et al., 2018). Furthermore, healthcare systems must prioritize the simplification of medical information to make it more understandable for patients with varying levels of literacy. This might involve using plain language, visual aids, or other methods that can help demystify medical jargon and make critical information more accessible (Abreu et al., 2024). It is also essential to foster open communication between healthcare providers and patients, ensuring that patients feel comfortable asking questions and expressing concerns about their treatment (Kwame and Petrucka, 2021).

Cultural Beliefs and Stigmas

Cultural beliefs and stigmas surrounding cancer and its treatment can also pose substantial barriers to effective care, particularly in populations with strong cultural or religious convictions. For example, in some communities, cancer may be viewed as a taboo subject, leading to delays in seeking diagnosis and treatment or even outright refusal of medical interventions (Kim et al., 2023). These beliefs are often deeply rooted and can be difficult to change, especially if reinforced by social norms and a lack of education about the disease (Wall et al., 2023). To address these cultural barriers, it is essential to engage with community leaders and influencers who can help shift public perceptions and promote more positive attitudes toward cancer treatment (Blanco et al., 2021). Culturally sensitive education campaigns that respect local customs and beliefs can also be effective in reducing stigma and encouraging patients to seek timely medical care (Wall et al., 2023). For instance, incorporating local languages and cultural references in educational materials can enhance their relevance and impact. Additionally, healthcare providers must be trained to recognize and navigate these cultural challenges, ensuring that they can communicate effectively with patients from diverse backgrounds and provide care that is both respectful and responsive to their needs (Gagliardi et al., 2016).

Another important strategy is the inclusion of family members and caregivers in the treatment process, as they often play a key role in decision-making and can influence the patient's willingness to pursue treatment. Family members may hold significant sway over the patient's decisions, particularly in cultures where family opinions are highly valued. By involving families in educational initiatives and providing them with accurate information about pancreatic cancer, healthcare providers can help create a supportive environment that encourages patients to adhere to recommended treatment plans (Lee Jr. et al., 2022). Moreover, addressing cultural stigmas involves not only educating the public but also reducing the fear and misunderstanding associated with cancer treatments. This can be achieved by demystifying treatment procedures, providing clear explanations about the benefits and risks, and sharing success stories that highlight the possibility of positive outcomes (Gagliardi et al., 2016). Additionally, integrating culturally appropriate counseling and support services into the treatment plan can help patients cope with their diagnosis and the associated social stigmas, leading to better mental health and improved adherence to treatment (Wall et al., 2023).

PSYCHOLOGICAL AND SUPPORTIVE CARE BARRIERS

Psychosocial Support Services

Psychosocial support services are essential for addressing the emotional and psychological challenges encountered by pancreatic cancer patients and their families (Mazzella Ebstein et al., 2020). These services, which include psychological counseling, support groups, and family therapy, are crucial for comprehensive cancer care. They help patients manage the significant stress associated with diagnosis, treatment, and survivorship (Kagee, 2022). Unfortunately, access to these services is often limited, especially in underserved areas, leading to notable disparities in the support provided (Fonseca et al., 2022). Such support is fundamental not only for enhancing patients' quality of life but also for potentially improving treatment outcomes (Bhati et al., 2023). Research indicates that pancreatic cancer patients frequently suffer from high levels of psychological distress, such as anxiety and depression, due to the aggressive nature of the disease and its generally poor prognosis (Zhang et al., 2024). Support groups provide a valuable space for patients to share their experiences and receive emotional support from peers, which can alleviate feelings of isolation and enhance mental well-being (Scott and Jewell, 2019). Additionally, these groups offer crucial support for caregivers, helping them cope with the emotional strain of caring for a loved one with cancer (Beesley, 2020)

In sub-Saharan Africa, there is a pressing need for research into psychosocial oncology and the development of support services tailored to the region's specific challenges. Cultural beliefs, stigma, and inadequate healthcare infrastructure often limit access to psychosocial care, further burdening patients and their families (Kagee, 2022). Integrating psychosocial support into standard oncology care in these areas is vital and requires increased funding, policy support, and the training of mental health professionals (Fonseca et al., 2022). Despite its clear benefits, psychosocial support is not universally included in oncology care, particularly in resource-limited settings. This gap often stems from a lack of awareness regarding the importance of psychosocial care, financial constraints, and a shortage of trained mental health professionals (Parmet et al., 2023). Addressing these issues requires a multifaceted strategy, including advocacy for the inclusion of psychosocial care in cancer treatment plans and the creation of sustainable support services accessible to all patients, regardless of their geographic or socioeconomic status (Khan et al., 2022).

Palliative Care and Pain Management

Palliative care is a critical aspect of cancer treatment that focuses on enhancing the quality of life by managing symptoms and providing emotional and spiritual support (Hugar et al., 2021). For pancreatic cancer patients, who are often diagnosed at an advanced stage, palliative care is crucial for managing pain and other distressing symptoms (Kotze, 2022). However, access to palliative care is uneven, with significant disparities between regions and healthcare systems (Fonseca et al., 2022). Effective pain management is a central component of palliative care, especially for pancreatic cancer patients who frequently experience severe and persistent pain (Pinto et al., 2021). Pain in these patients is often inadequately managed due to barriers such as limited access to opioids, insufficient pain assessment, and concerns about addiction or side effects from pain medication (Engebretson et al., 2015; Pinto et al., 2021). These challenges can lead to significant suffering and reduce the quality of life (Engebretson et al., 2015)

In regions like sub-Saharan Africa, where healthcare resources are scarce, access to palliative care and pain management is often inadequate, leaving many patients to endure unnecessary suffering. Cultural factors, such as attitudes toward pain and stoicism, may also contribute to the underutilization of palliative care services (Court and Olivier, 2020). To address these issues, there is a need for increased education and training for healthcare providers on pain management and palliative care, as well as efforts to improve the availability of essential medications and services (Kagee, 2022). Integrating palliative care early in the treatment process, rather than waiting until the end of life, can greatly enhance the quality of life for pancreatic cancer patients (Hui and Bruera, 2016). This approach facilitates better symptom management, psychological support, and involvement of patients and their families in care decisions, leading to more personalized and compassionate care (Scott and Jewell, 2019).

TECHNOLOGICAL AND TELEMEDICINE BARRIERS

Telehealth and Remote Monitoring

Telemedicine has significantly impacted the landscape of pancreatic cancer care, offering promising solutions to enhance patient management and accessibility. By enabling virtual consultations, telehealth platforms have made it possible for patients to connect with specialists without the need for travel. This is particularly beneficial for those in remote or underserved areas where access to high-quality healthcare facilities might be limited. Telemedicine not only alleviates the logistical challenges associated with frequent hospital visits but also supports ongoing management of the disease through remote monitoring tools (Tripepi et al., 2023). Remote monitoring technologies play a crucial role in pancreatic cancer care by facilitating continuous tracking of patient health metrics. Wearable devices and mobile health apps can monitor vital signs, blood glucose levels, and other relevant biomarkers, providing real-time data to healthcare providers (Manias et al., 2023). The integration of these tools with telehealth and electronic health records (EHRs) to support continuous care is illustrated in Figure 8.2. This data allow for timely interventions and adjustments in treatment plans, which can be critical for managing pancreatic cancer's complex and often rapidly changing nature. For example, continuous glucose monitoring can help manage diabetes symptoms often associated with pancreatic cancer, while remote symptom tracking can alert clinicians to potential complications (Paro et al., 2022).

Despite these advantages, several technological barriers impede the widespread adoption of telemedicine. One significant challenge is the digital divide, which refers to disparities in access to technology and high-speed internet . Older adults, individuals in rural areas, and those from lower

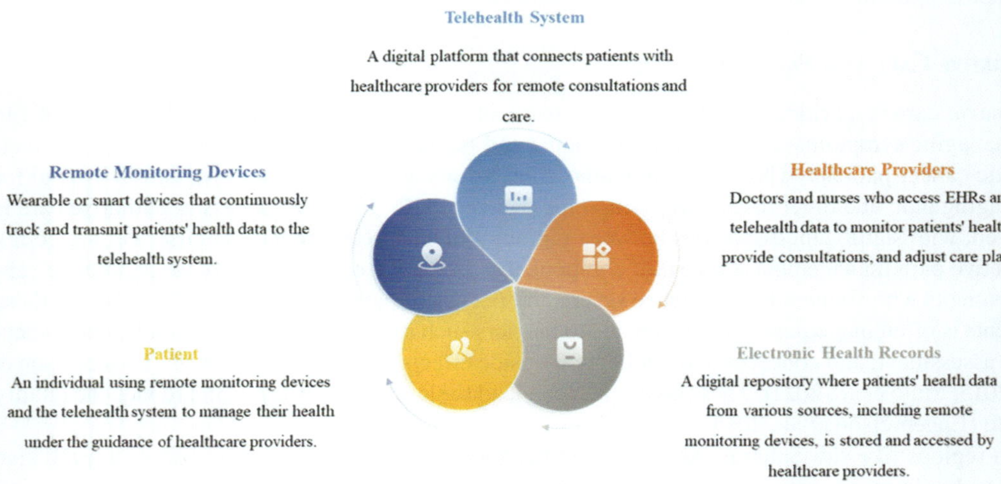

FIGURE 8.2 This diagram illustrates the integration of telehealth, remote monitoring, and electronic health records (EHRs) in managing pancreatic cancer. Telehealth facilitates remote consultations, while wearable devices track patients' vital signs and symptoms. Data from these devices are stored in EHRs, enabling real-time monitoring and informed decision-making by healthcare providers. This system may enhance continuous care and improve outcomes for pancreatic cancer patients.

socioeconomic backgrounds may face difficulties accessing telehealth services due to inadequate technology or unreliable internet connections. These barriers not only limit the effectiveness of telemedicine but also exacerbate existing healthcare disparities. Efforts to bridge the digital divide, such as providing subsidized technology and expanding internet infrastructure, are essential for making telehealth accessible to all patients (Pang et al., 2022). Moreover, the effectiveness of telehealth services is also influenced by the quality and usability of remote monitoring tools. High costs and technical complexities associated with advanced monitoring devices can restrict their availability and use. Ensuring that these tools are user-friendly and affordable is crucial for enhancing their adoption and effectiveness. Additionally, healthcare providers need adequate training to effectively use these technologies and integrate them into patient care routines (Morris et al., 2022).

Integration of Electronic Health Records

The integration of EHRs into pancreatic cancer care represents a significant advancement in managing patient information and improving treatment outcomes. EHRs provide a comprehensive and centralized repository of patient data, including medical history, treatment plans, and outcomes. This centralized data access supports more efficient care coordination among healthcare providers, facilitates evidence-based decision-making, and enhances the personalization of treatment (Sitapati et al., 2017). One of the major benefits of EHRs is their ability to support personalized treatment plans. By consolidating patient data, EHRs enable healthcare providers to tailor interventions based on individual patient needs and preferences (Manias et al., 2024; Sitapati et al., 2017). This is particularly valuable in the context of pancreatic cancer, where treatment plans often need to be adjusted based on patient response and disease progression. For example, data from EHRs can help oncologists identify the most effective chemotherapy regimens or track the response to targeted therapies, leading to more precise and effective treatment strategies (Manias et al., 2024).

However, the integration of EHRs into clinical practice is not without challenges. One of the primary issues is interoperability, the ability of different EHR systems to exchange and use data seamlessly (Manias et al., 2024). Variability in data formats and standards among different EHR systems can create barriers to effective information sharing, potentially leading to gaps in patient care and inefficiencies in treatment planning. Standardizing data formats and improving system compatibility are essential steps toward overcoming these interoperability challenges and ensuring that patient information is accurately and consistently shared across different healthcare settings (Shah and Khan, 2020). Another significant concern is data privacy and security. As the use of EHRs increases, so does the need for robust measures to protect sensitive patient information. Ensuring that EHR systems are secure from data breaches and unauthorized access is crucial for maintaining patient trust and complying with regulatory requirements (Kenner et al., 2021; Shah and Khan, 2020). Implementing strong encryption protocols, access controls, and regular security audits can help safeguard patient data and address privacy concerns associated with EHRs (Kenner et al., 2021).

CASE STUDIES AND REAL-WORLD EXAMPLES

Success Stories and Best Practices

Efforts to improve access to pancreatic cancer care have led to significant successes, particularly in underserved areas (Wall et al., 2023). One noteworthy initiative is the National Cancer Institute's (NCI) Community Oncology Research Program (NCORP), which has been instrumental in expanding access to cancer clinical trials in community settings. By integrating cancer research with community-based healthcare, NCORP has allowed more patients, including those in rural and underserved areas, to participate in clinical trials, improving access to cutting-edge treatments and care (Good et al., 2020). In Africa, a notable example is the establishment of the African Cancer

Institute in South Africa, which focuses on improving cancer care and outcomes across the continent. This institute has developed programs for early detection, education, and access to care, particularly for cancers like pancreatic cancer that are often diagnosed late. These programs have been successful in raising awareness and improving early diagnosis rates (Stefan, 2015). Ramutumbu et al. (2023) highlighted a successful intervention where focused training programs for healthcare professionals addressed skill shortages in oncology. Specialized workshops improved the proficiency of healthcare workers, particularly in rural hospitals lacking sufficient expertise. Establishing dedicated oncology care teams and enhancing resource allocation has demonstrated positive outcomes in managing patient care more effectively. The authors provided valuable insights into the effectiveness of creating supportive and informal environments during focus group discussions. This approach facilitated open communication among healthcare professionals and helped identify critical barriers and potential solutions. The study underscored the importance of adapting communication techniques and involving multidisciplinary teams to enhance care delivery. Furthermore, implementing robust communication systems and protocols is essential to mitigate the challenges of fragmented care and improve patient outcomes.

Patient Narratives

Patient stories highlight the profound impact that access barriers can have on outcomes (Schwarz et al., 2022). For instance, a study conducted in North America and Europe identified key pain points among cancer patients, including the need for better access to information and challenges with organizational support from healthcare services. These issues are global concerns, affecting not only African populations but also those in more developed regions. Patients reported difficulties in finding reliable, personalized information and described a lack of holistic support for living with cancer, such as mental health care and guidance on lifestyle changes (Schwarz et al., 2022). Similarly, personal stories from pancreatic cancer patients that reveal significant challenges, such as delays in diagnosis and difficulties navigating the healthcare system. Patients frequently encounter barriers due to inadequate access to specialized care and long wait-times for services. These obstacles are exacerbated by the shortage of skilled healthcare professionals and fragmented care systems, which can lead to delays in receiving timely and effective treatment. The impact of access barriers on patient outcomes is substantial (Ramutumbu et al., 2023). The authors further illustrated that barriers such as delays in diagnosis and limited availability of specialized services profoundly affect patient outcomes. Patients experiencing these barriers often face worse prognoses due to delayed treatment and inadequate support. The lack of coordination among care providers and the absence of clear policy guidelines further exacerbate these issues, emphasizing the need for systemic improvements to ensure better access to and quality of care for pancreatic cancer patients (Ramutumbu et al., 2023).

In Africa, the study by Dandadzi et al. (2022) highlighted various limitations across the stages of diagnosis, treatment, and aftercare. Patients faced difficulties recognizing the seriousness of their symptoms and were burdened by financial constraints related to treatment costs. The study also documented delays in accessing biopsy services and limited treatment options, which led to worsened patient outcomes. Similarly, Omotoso et al. (2023) described challenges in accessing timely and adequate care in Nigeria, where patients often faced delays in diagnosis due to under-resourced clinics, leading to advanced disease stages and limited treatment options. In the United States, a study by Wercholuk et al. (2022) described the significant physical and emotional toll on patients traveling long distances for chemotherapy, underscoring the need for more localized cancer care facilities. Conversely, a patient in an urban setting with access to a major cancer center, as detailed by Halbrook et al. (2023), had a markedly different experience. This individual benefited from participation in a clinical trial, illustrating the disparity in care between urban and rural settings and the life-saving potential of advanced treatments (Halbrook et al., 2023).

RECOMMENDATIONS AND FUTURE DIRECTIONS

Policy and Advocacy for Improved Access

To effectively address the challenges associated with pancreatic cancer care, it is essential to implement policy changes that prioritize equitable access, particularly for underserved populations. National cancer control programs should include specific strategies for pancreatic cancer, focusing on early detection and ensuring that healthcare infrastructure supports timely diagnosis, particularly in rural and low-resource settings. Such initiatives could involve funding for the development and implementation of screening guidelines tailored to high-risk populations, significantly reducing the prevalence of late-stage diagnoses (Anderson et al., 2021). Geographical disparities in access to oncology services also require urgent policy attention (Omotoso et al., 2023; Ramutumbu et al., 2023). Nyangasi et al. (2023) advocate for decentralizing cancer care by establishing regional cancer centers in underserved areas. These centers would alleviate the need for extensive travel, which remains a significant barrier for many patients (Ramutumbu et al., 2023). Additionally, incorporating cancer treatment into national health insurance schemes can reduce the financial burden on patients and their families (Knaul et al., 2015).

Advocacy groups and patient organizations, such as the Pancreatic Cancer Action Network (PanCAN) in the United States, play a crucial role in driving these policy changes (Knaul et al., 2015; Wall et al., 2023). Their efforts in raising awareness, lobbying for healthcare reforms, and providing patient support have led to legislative changes that enhance access to cutting-edge treatments (Knaul et al., 2015). Similar organizations in other countries, such as the Cancer Association of South Africa (CANSA), have successfully pushed for policy reforms that improve access to care for marginalized populations. Continued collaboration between advocacy groups and policymakers is essential to ensure that the needs of pancreatic cancer patients are prioritized in national health agendas (CANSA, 2019).

Innovations in Research and Technology

Research and innovation are pivotal in overcoming access barriers in pancreatic cancer care. Developing more effective and accessible diagnostic tools is critical, as highlighted by. Innovations such as liquid biopsy technologies, which detect cancer biomarkers in blood samples, offer a non-invasive and widely applicable method for early detection, particularly in underserved populations (Li et al., 2021). Immunotherapy, personalized to an individual's genetic profile, presents another promising avenue for improving treatment efficacy. However, the high cost of these therapies remains a barrier, necessitating ongoing research to reduce production and administration costs, thus making these treatments more accessible to a broader patient demographic (Chehelgerdi et al., 2023). Furthermore, addressing socioeconomic determinants of health, such as income, education, and geographic location, can reduce disparities in care, as demonstrated by studies on social factors impacting health outcomes (Braveman and Gottlieb, 2014).

Technological advancements are also transforming the landscape of pancreatic cancer care. These advancements, along with innovations in diagnostics, therapeutics, and international collaboration, form a multi-level strategy for enhancing access to care, as illustrated in Figure 8.3. For instance, artificial intelligence (AI) and machine learning are being integrated into early detection methods, with AI algorithms capable of analyzing imaging data with high precision, aiding in early detection and personalized treatment planning (Tripathi et al., 2024). Telemedicine continues to offer new possibilities for remote patient monitoring and consultation, particularly in remote areas with limited healthcare infrastructure (Haleem et al., 2021). Developing cost-effective diagnostic tools, such as portable devices, can significantly enhance screening capabilities in low-resource settings. These innovations are vital in providing timely diagnosis and bridging access gaps in regions with limited healthcare infrastructure (Chin et al., 2018). International collaboration and funding are essential to advancing pancreatic cancer care, with global health initiatives playing a pivotal

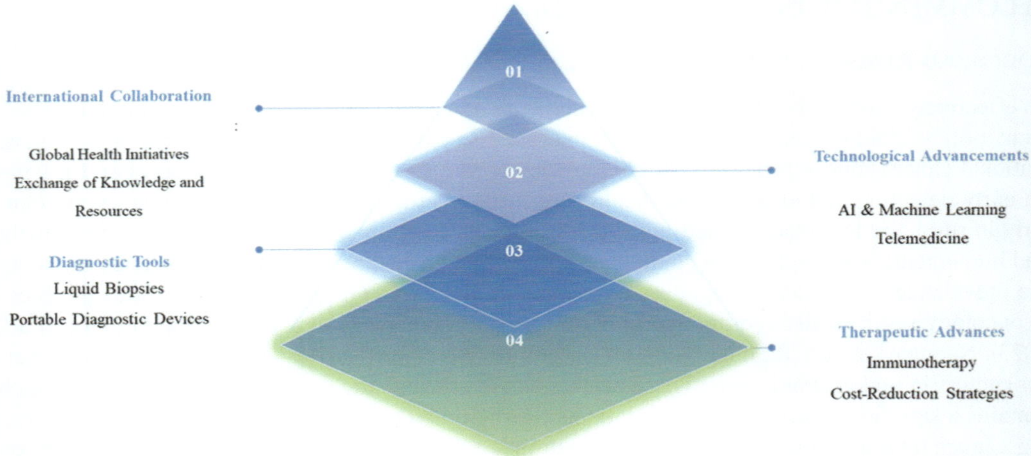

FIGURE 8.3 A pyramid of key innovations aimed at improving access to pancreatic cancer care. The diagram highlights four critical areas of focus: international collaboration, which emphasizes global health initiatives and knowledge sharing to reduce care disparities; technological advancements, including AI, machine learning, and telemedicine, to enhance early detection and personalized treatment; diagnostic tools, such as liquid biopsies and portable devices, to ensure timely and accessible cancer detection; and therapeutic advances, which stress the importance of making immunotherapy and cost-effective treatments more widely available. Together, these interventions form a comprehensive strategy to enhance access to care for pancreatic cancer patients, particularly those in underserved populations.

role in facilitating the exchange of knowledge, resources, and expertise across borders (Caban and Malecka-Wojciesko, 2023).

CONCLUSION

Access to care for pancreatic cancer is hindered by multifaceted barriers that impact patient outcomes and treatment efficacy (Prades et al., 2020). Geographic and financial obstacles create significant disparities, with many patients facing challenges in reaching specialized care and managing the high costs associated with treatment (Anderson et al., 2021). Cultural and educational barriers, such as low health literacy and cancer-related stigma, further complicate the landscape of pancreatic cancer care (Kim et al., 2023; Munigala et al., 2022). Additionally, the availability and utilization of psychosocial and supportive care services remain uneven, especially in under-resourced areas (Fonseca et al., 2022; Kagee, 2022). Technological solutions like telemedicine and EHRs offer substantial benefits but are often impeded by issues such as the digital divide and interoperability challenges (Shah and Khan, 2020; Tripepi et al., 2023).

Addressing these barriers is crucial for enhancing pancreatic cancer care. Policymakers must implement reforms that improve healthcare infrastructure, reduce financial burdens, and decentralize cancer services to make them more accessible, particularly in underserved regions (Anderson et al., 2021). Tailoring educational initiatives to diverse literacy levels and cultural contexts is essential to improve patient engagement and treatment adherence (Gagliardi et al., 2016; Munigala et al., 2022). Expanding access to psychosocial and palliative care will support patients and their families in managing the emotional and physical challenges associated with the disease (Bhati et al., 2023; Kagee, 2022). Moreover, technological advancements should be leveraged to enhance patient care, with efforts focused on bridging the digital divide and ensuring the usability of remote monitoring tools (Pang et al., 2022; Tripepi et al., 2023). Continued research and innovation are vital for developing cost-effective diagnostic and therapeutic options, particularly in resource-limited settings (Li et al., 2021).

REFERENCES

ABDEL-RAHMAN, O. 2020. Impact of socioeconomic status on presentation, treatment and outcomes of patients with pancreatic cancer. *Journal of Comparative Effectiveness Research,* 9, 1233–1241.

ABREU, A. A., MURIMWA, G. Z., STEWART II, J. W., ZHANG, L., RODRIGUEZ, J., FARAH, E., ZEH III, H. J. & POLANCO, P. M. 2024. Understanding the cancer health literacy gap: NCCN member institutions' website readability across the high mortality cancers in the US. *Surgical Oncology Insight,* 1, 100081.

ALRASHEEDI, K. F., AL-MOHAITHEF, M., EDREES, H. H. & CHANDRAMOHAN, S. 2019. The association between wait times and patient satisfaction: findings from primary health centers in the Kingdom of Saudi Arabia. *Health Services Research and Managerial Epidemiology,* 6, 23333928198612 4.

AMIN, K., KHAN, H., HEARLD, L. R., CHU, D. I., PRETE, V., MEHARI, K. R., HESLIN, M. J. & FONSECA, A. L. 2023. Association between rural residence and processes of care in pancreatic cancer treatment. *Journal of Gastrointestinal Surgery,* 27, 2155–2165.

ANDERSON, E. M., THOMASSIAN, S., GONG, J., HENDIFAR, A. & OSIPOV, A. 2021. Advances in pancreatic ductal adenocarcinoma treatment. *Cancers,* 13, 5510.

ANG, T. L., KWEK, A. B. E. & WANG, L. M. 2018. Diagnostic endoscopic ultrasound: technique, current status and future directions. *Gut and Liver,* 12, 483–496.

ATTEBURY, H. & DALEY, D. 2023. The gut microbiome and pancreatic cancer development and treatment. *The Cancer Journal,* 29, 49–56.

BACIU, A., NEGUSSIE, Y., GELLER, A., WEINSTEIN, J. N., NATIONAL ACADEMIES OF SCIENCES, E. & MEDICINE 2017. The root causes of health inequity. In: *Communities in action: Pathways to health equity.* Jan 11, The National Academies Press (US), 3 https://www.ncbi.nlm.nih.gov/books/NBK425845/

BAGCHI, A. D., DAMAS, K., SALAZAR DE NOGUERA, N., MELAMED, B., MENIFIELD, C., BAVEJA, A., WEBER, P. & SWAMINATHAN, S. 2022. Comfort with telehealth among residents of an underserved urban area. *Journal of Primary Care & Community Health,* 13, 215013192211196.

BARRIOS, C., DE LIMA LOPES, G., YUSOF, M. M., RUBAGUMYA, F., RUTKOWSKI, P. & SENGAR, M. 2023. Barriers in access to oncology drugs — a global crisis. *Nature Reviews Clinical Oncology,* 20, 7–15.

BEESLEY, V. L. 2020. Supporting patients and carers affected by pancreatic cancer: a feasibility study of a counselling intervention. *European Journal of Oncology Nursing,* 46.

BERARDI, R., MORGESE, F., RINALDI, S., TORNIAI, M., MENTRASTI, G., SCORTICHINI, L. & GIAMPIERI, R. 2020. Benefits and limitations of a multidisciplinary approach in cancer patient management</p&gt. *Cancer Management and Research,* 12, 9363–9374.

BHATI, D., DEOGADE, M. S. & KANYAL, D. 2023. Improving patient outcomes through effective hospital administration: a comprehensive review. *Cureus*, 15(10), e47731. https://doi.org/10.7759/cureus.4773

BIDDELL, C. B., WATERS, A. R., ANGOVE, R. S. M., GALLAGHER, K. D., ROSENSTEIN, D. L., SPEES, L. P., KENT, E. E., PLANEY, A. M. & WHEELER, S. B. 2023. Facing financial barriers to healthcare: patient-informed adaptation of a conceptual framework for adults with a history of cancer. *Frontiers in Psychology,* 14.

BLANCO, B. A., POULSON, M., KENZIK, K. M., MCANENY, D. B., TSENG, J. F. & SACHS, T. E. 2021. The impact of residential segregation on pancreatic cancer diagnosis, treatment, and mortality. *Annals of Surgical Oncology,* 28, 3147–3155.

BRAVEMAN, P. & GOTTLIEB, L. 2014. The social determinants of health: it's time to consider the causes of the causes. *Public Health Reports,* 129, 19–31.

BRUNNER, T. B. & SEUFFERLEIN, T. 2016. Pancreatic cancer chemoradiotherapy. *Best Practice & Research Clinical Gastroenterology,* 30, 617–628.

CABAN, M. & MALECKA-WOJCIESKO, E. 2023. Gaps and opportunities in the diagnosis and treatment of pancreatic cancer. *Cancers (Basel),* 15, 5577.

CANALE, T. D., CHO, H. & CHEUNG, W. Y. 2018. A population-based analysis of urban–rural disparities in advanced pancreatic cancer management and outcomes. *Medical Oncology,* 35, 1–8.

CANSA. 2019. CANSA advocacy – enhancing partnerships and alliances. https://cansa.org.za/cansa-advocacy-enhancing-partnerships-and- alliances/ [Accessed 11 August 2024].

CHEHELGERDI, M., CHEHELGERDI, M., ALLELA, O. Q. B., PECHO, R. D. C., JAYASANKAR, N., RAO, D. P., THAMARAIKANI, T., VASANTHAN, M., VIKTOR, P., LAKSHMAIYA, N., SAADH, M. J., AMAJD, A., ABO-ZAID, M. A., CASTILLO-ACOBO, R. Y., ISMAIL, A. H., AMIN, A. H. & AKHAVAN-SIGARI, R. 2023. Progressing nanotechnology to improve targeted cancer treatment: overcoming hurdles in its clinical implementation. *Molecular Cancer,* 22.

CHIN, V., NAGRIAL, A., SJOQUIST, K., O'CONNOR, C. A., CHANTRILL, L., BIANKIN, A. V., SCHOLTEN, R. J. & YIP, D. 2018. Chemotherapy and radiotherapy for advanced pancreatic cancer. *Cochrane Database of Systematic Reviews,* 3, CD011044.
COOMBS, N. C., CAMPBELL, D. G. & CARINGI, J. 2022. A qualitative study of rural healthcare providers' views of social, cultural, and programmatic barriers to healthcare access. *BMC Health Services Research,* 22.
COSTA, B. & VALE, N. 2022. Dostarlimab: a review. *Biomolecules,* 12, 1031.
COUGHLIN, S. S., VERNON, M., HATZIGEORGIOU, C. & GEORGE, V. 2020. Health literacy, social determinants of health, and disease prevention and control. *Journal of Environment and Health Sciences,* 6.
COURT, L. & OLIVIER, J. 2020. Approaches to integrating palliative care into African health systems: a qualitative systematic review. *Health Policy and Planning,* 35, 1053–1069.
CROSBY, D., BHATIA, S., BRINDLE, K. M., COUSSENS, L. M., DIVE, C., EMBERTON, M., ESENER, S., FITZGERALD, R. C., GAMBHIR, S. S., KUHN, P., REBBECK, T. R. & BALASUBRAMANIAN, S. 2022. Early detection of cancer. *Science,* 375, eaay9040.
CYR, M. E., ETCHIN, A. G., GUTHRIE, B. J. & BENNEYAN, J. C. 2019. Access to specialty healthcare in urban versus rural US populations: a systematic literature review. *BMC Health Services Research,* 19.
DANDADZI, A., CHAPMAN, E., CHIRENJE, Z. M., NAMUKWAYA, E., PINI, S., NKHOMA, K. & ALLSOP, M. J. 2022. Patient experiences of living with cancer before interaction with palliative care services in Zimbabwe: a qualitative secondary data analysis. *European Journal of Cancer Care,* 31, e13632.
DOS-SANTOS-SILVA, I., GUPTA, S., OREM, J. & SHULMAN, L. N. 2022. Global disparities in access to cancer care. *Communications Medicine,* 2.
ENGEBRETSON, A., MATRISIAN, L. & THOMPSON, C. 2015. Pancreatic cancer: patient and caregiver perceptions on diagnosis, psychological impact, and importance of support. *Pancreatology,* 15, 701–707.
FALCO, M., MASOJć, B. & SULIKOWSKI, T. 2023. Radiotherapy in pancreatic cancer: to whom, when, and how? *Cancers (Basel),* 15.
FENG, B., WU, J., SHEN, B., JIANG, F. & FENG, J. 2022. Cancer-associated fibroblasts and resistance to anticancer therapies: status, mechanisms, and countermeasures. *Cancer Cell International,* 22.
FONSECA, A. L., KHAN, H., MEHARI, K. R., CHERLA, D., HESLIN, M. J. & JOHNSTON, F. M. 2022. Disparities in access to oncologic care in pancreatic cancer: a systematic review. *Ann Surg Oncol,* 29, 3232–3250.
FRANCK, C., MULLER, C., ROSANIA, R., CRONER, R. S., PECH, M. & VENERITO, M. 2020. Advanced pancreatic ductal adenocarcinoma: moving forward. *Cancers (Basel),* 12, 1955.
FRAZIER, T. L., LOPEZ, P. M., ISLAM, N., WILSON, A., EARLE, K., DULIEPRE, N., ZHONG, L., BENDIK, S., DRACKETT, E., MANYINDO, N., SEIDL, L. & THORPE, L. E. 2023. Addressing financial barriers to health care among people who are low-income and insured in New York City, 2014–2017. *Journal of Community Health,* 48, 353–366.
GAGLIARDI, A. R., SOONG, D. & GALLINGER, S. 2016. Identifying factors influencing pancreatic cancer management to inform quality improvement efforts and future research: a scoping systematic review. *Pancreas,* 45, 161–166.
GOOD, M., CASTRO, K., DENICOFF, A., FINNIGAN, S., PARRECO, L. & GERMAIN, D. S. 2020. National Cancer Institute: restructuring to support the clinical trials of the future. *Seminars in Oncology Nursing,* 36, 151003.
GRAVES, J. M., ABSHIRE, D. A. & ALEJANDRO, A. G. 2022. System- and individual-level barriers to accessing medical care services across the rural-urban spectrum, Washington State. *Health Services Insights,* 15, 117863292211046.
HALBROOK, C. J., LYSSIOTIS, C. A., PASCA DI MAGLIANO, M. & MAITRA, A. 2023. Pancreatic cancer: advances and challenges. *Cell,* 186, 1729–1754.
HALEEM, A., JAVAID, M., SINGH, R. P. & SUMAN, R. 2021. Telemedicine for healthcare: capabilities, features, barriers, and applications. *Sensors International,* 2, 100117.
HART, P. A., ANDERSEN, D. K., LYONS, E., COTE, G. A., CRUZ-MONSERRATE, Z., DWORKIN, R. H., ELMUNZER, B. J., FOGEL, E. L., FORSMARK, C. E., GILRON, I., GOLDEN, M., GOZU, A., MCNAIR, L., PANDOL, S. J., PERITO, E. R., PHILLIPS, A. E., RABBITTS, J. A., WHITCOMB, D. C., WINDSOR, J. A., YADAV, D. & PALERMO, T. M. 2022. Clinical trials in pancreatitis: opportunities and challenges in the design and conduct of patient-focused clinical trials in recurrent acute and chronic pancreatitis. *Pancreas,* 51, 715–722.
HAVLIR, D. & GANDHI, M. 2015. Implementation challenges for long-acting antivirals as treatment. *Current Opinion in HIV and AIDS,* 10, 282–289.

HITZ, F., RIBI, K., GROTE, G., KOLBE, M., SCHMITZ, C., LAMB, B. W., RUHSTALLER, T., BERCHTOLD, P. & SEVDALIS, N. 2022. Team functioning across different tumour types: insights from a Swiss cancer center using qualitative and quantitative methods. *Cancer Reports,* 5.

HUGAR, L. A., WULFF-BURCHFIELD, E. M., WINZELBERG, G. S., JACOBS, B. L. & DAVIES, B. J. 2021. Incorporating palliative care principles to improve patient care and quality of life in urologic oncology. *Nature Reviews Urology,* 18**,** 623–635.

HUI, D. & BRUERA, E. 2016. Integrating palliative care into the trajectory of cancer care. *Nature Reviews Clinical Oncology,* 13**,** 159–171.

JOUDYIAN, N., DOSHMANGIR, L., MAHDAVI, M., TABRIZI, J. S. & GORDEEV, V. S. 2021. Public-private partnerships in primary health care: a scoping review. *BMC Health Services Research,* 21.

KAGEE, A. 2022. The need for psychosocial oncology research in sub-Saharan Africa: a review of the terrain. *South African Journal of Psychology,* 52, 392–403.

KALE, S., HIRANI, S., VARDHAN, S., MISHRA, A., GHODE, D. B., PRASAD, R. & WANJARI, M. 2023. Addressing cancer disparities through community engagement: lessons and best practices. *Cureus,* August 14, 15(8), *e43455*. 10.7759/cureus.43445. PMID: 37711952; PMCID: PMC10498131.

KAMISAWA, T., WOOD, L. D., ITOI, T. & TAKAORI, K. 2016. Pancreatic cancer. *Lancet,* 388**,** 73–85.

KENNER, B., CHARI, S. T., KELSEN, D., KLIMSTRA, D. S., PANDOL, S. J., ROSENTHAL, M., RUSTGI, A. K., TAYLOR, J. A., YALA, A., ABUL-HUSN, N., ANDERSEN, D. K., BERNSTEIN, D., BRUNAK, S., CANTO, M. I., ELDAR, Y. C., FISHMAN, E. K., FLESHMAN, J., GO, V. L. W., HOLT, J. M., FIELD, B., GOLDBERG, A., HOOS, W., IACOBUZIO-DONAHUE, C., LI, D., LIDGARD, G., MAITRA, A., MATRISIAN, L. M., POBLETE, S., ROTHSCHILD, L., SANDER, C., SCHWARTZ, L. H., SHALIT, U., SRIVASTAVA, S. & WOLPIN, B. 2021. Artificial intelligence and early detection of pancreatic cancer. *Pancreas,* 50**,** 251–279.

KENNER, B. J., GO, V. L. W., CHARI, S. T., GOLDBERG, A. E. & ROTHSCHILD, L. J. 2017. Early detection of pancreatic cancer. *Pancreas,* 46**,** 1238–1241.

KHAN, N. N., LEWIN, T., HATTON, A., PILGRIM, C., IOANNOU, L., TE MARVELDE, L., ZALCBERG, J. & EVANS, S. 2022. Systematic review of the predictors of health service use in pancreatic cancer. *American Journal of Cancer Research,* 12**,** 622.

KIM, N., KANG, D., SHIN, S. H., HEO, J. S., SHIM, S., LIM, J., CHO, J. & HAN, I. W. 2023. Effects of cancer stigma on quality of life of patients with hepatobiliary and pancreatic cancer. *Annals of Hepato-Biliary-Pancreatic Surgery,* 27**,** 172–179.

KNAUL, F., HORTON, S., YERRAMILLI, P., GELBAND, H. & ATUN, R. 2015. *Financing cancer care in low-resource settings.* In H. Gelband (Eds.) et. al., *Cancer: Disease Control Priorities, Third Edition* (Volume 3). The International Bank for Reconstruction and Development. The World Bank.

KOLBEINSSON, H. M., CHANDANA, S., WRIGHT, G. P. & CHUNG, M. 2023. Pancreatic cancer: a review of current treatment and novel therapies. *Journal of Investigative Surgery,* 36**,** 2129884.

KOTZE, U. K. 2022. A prospective, longitudinal, observational study to assess the health-related quality of life of patients with pancreatic ductal adenocarcinoma in the South African context. Faculty of Health Sciences, Division of General Surgery. http://hdl.handle.net/11427/36473

KRUK, M. E., GAGE, A. D., ARSENAULT, C., JORDAN, K., LESLIE, H. H., RODER-DEWAN, S., ADEYI, O., BARKER, P., DAELMANS, B., DOUBOVA, S. V., ENGLISH, M., GARCíA-ELORRIO, E., GUANAIS, F., GUREJE, O., HIRSCHHORN, L. R., JIANG, L., KELLEY, E., LEMANGO, E. T., LILJESTRAND, J., MALATA, A., MARCHANT, T., MATSOSO, M. P., MEARA, J. G., MOHANAN, M., NDIAYE, Y., NORHEIM, O. F., REDDY, K. S., ROWE, A. K., SALOMON, J. A., THAPA, G., TWUM-DANSO, N. A. Y. & PATE, M. 2018. High-quality health systems in the Sustainable Development Goals era: time for a revolution. *The Lancet Global Health,* 6**,** e1196–e1252.

KWAME, A. & PETRUCKA, P. M. 2021. A literature-based study of patient-centered care and communication in nurse-patient interactions: barriers, facilitators, and the way forward. *BMC Nursing,* 20.

LEE JR, H. J., QIAN, C. L., LANDAY, S. L., O'CALLAGHAN, D., KASLOW-ZIEVE, E., AZOBA, C. C., FUH, C.-X., TEMEL, B., UFERE, N. & PETRILLO, L. A. 2022. Communicating the information needed for treatment decision making among patients with pancreatic cancer receiving preoperative therapy. *JCO Oncology Practice,* 18**,** e313–e324.

LEVIT, L. A., BYATT, L., LYSS, A. P., PASKETT, E. D., LEVIT, K., KIRKWOOD, K., SCHENKEL, C. & SCHILSKY, R. L. 2020. Closing the rural cancer care gap: three institutional approaches. *JCO Oncology Practice,* 16**,** 422–430.

LI, K., SHI, H., ZHANG, B., OU, X., MA, Q., CHEN, Y., SHU, P., LI, D. & WANG, Y. 2021. Myeloid-derived suppressor cells as immunosuppressive regulators and therapeutic targets in cancer. *Signal Transduction and Targeted Therapy,* 6, 362.

LOEHRER, A. P., CHANG, D. C., HUTTER, M. M., SONG, Z., LILLEMOE, K. D., WARSHAW, A. L. & FERRONE, C. R. 2015. Health insurance expansion and treatment of pancreatic cancer: does increased access lead to improved care? *Journal of the American College of Surgeons,* 221**,** 1015–1022.

MAGANTY, A., BYRNES, M. E., HAMM, M., WASILKO, R., SABIK, L. M., DAVIES, B. J. & JACOBS, B. L. 2023. Barriers to rural health care from the provider perspective. *Rural and Remote Health,* 23**,** 1–11.

MANIAS, G., AZQUETA-ALZúAZ, A., BERENGUER MARI, J. R., DALIANIS, A., GRIFFITHS, J., KALOGERINI, M., KOSTOPOULOU, K., KOUREMENOU, E., KRANAS, P., KYRIAZAKOS, S., LEKKA, D., MELILLO, F., PATIñO-MARTíNEZ, M., GARCíA-PERALES, Ó., PNEVMATIKAKIS, A., SPANOS, S., TORRENS, S. G., WAJID, U. & KYRIAZIS, D. 2023. Enhanced monitoring and decision support in pancreatic cancer through holistic health records and advanced data processing: the iHELP Platform. *MDPI.*2023101360. https://doi.org/10.20944/preprints202310.1360.v

MANIAS, G., AZQUETA-ALZúAZ, A., DALIANIS, A., GRIFFITHS, J., KALOGERINI, M., KOSTOPOULOU, K., KOUREMENOU, E., KRANAS, P., KYRIAZAKOS, S., LEKKA, D., MELILLO, F., PATIñO-MARTINEZ, M., GARCIA-PERALES, O., PNEVMATIKAKIS, A., TORRENS, S. G., WAJID, U. & KYRIAZIS, D. 2024. Advanced data processing of pancreatic cancer data integrating ontologies and machine learning techniques to create holistic health records. *Sensors,* 24**,** 1739.

MAPHUMULO, W. T. & BHENGU, B. R. 2019. Challenges of quality improvement in the healthcare of South Africa post-apartheid: a critical review. *Curationis,* 42, e1–e9.

MAZZELLA EBSTEIN, A. M., JOSEPH, S. J. & HERNANDEZ, M. 2020. Psychological stress and pancreatic cancer patients. *JBI Evidence Synthesis,* 18**,** 576–582.

MCMAUGHAN, D. J., OLORUNTOBA, O. & SMITH, M. L. 2020. Socioeconomic status and access to healthcare: interrelated drivers for healthy aging. *Frontiers in Public Health,* 8, 20552076241237668.

MCQUAID, E. L. & LANDIER, W. 2018. Cultural issues in medication adherence: disparities and directions. *Journal of General Internal Medicine,* 33**,** 200–206.

MELHEM, S. J., NABHANI-GEBARA, S. & KAYYALI, R. 2024. Leveraging e-health for enhanced cancer care service models in middle-income contexts: qualitative insights from oncology care providers. *DIGITAL HEALTH,* 10.

MIZRAHI, J. D., SURANA, R., VALLE, J. W. & SHROFF, R. T. 2020. Pancreatic cancer. *The Lancet,* 395**,** 2008–2020.

MORRIS, B. B., ROSSI, B. & FUEMMELER, B. 2022. The role of digital health technology in rural cancer care delivery: a systematic review. *The Journal of Rural Health,* 38**,** 493–511.

MUNIGALA, S., GARDNER, T. B., O'REILLY, E. M., FERNáNDEZ-DEL CASTILLO, C., KO, A. H., PLESKOW, D., VOLLMER JR, C. M., SEARLE, N. A., BAKELMAN, D. & HOLT, J. M. 2022. Helping patients understand Pancreatic Cancer using animated pancreas patient Education with Visual formats of Learning. *Pancreas,* 51**,** 628–633.

NATIONAL ACADEMIES OF SCIENCES, ENGINEERING, AND MEDICINE;HEALTH AND MEDICINE DIVISION; BOARD ON HEALTH CARE SERVICES; COMMITTEE ON HEALTH CARE UTILIZATION AND ADULTS WITH DISABILITIES. Health-Care Utilization as a Proxy in Disability Determination. Washington (DC): National Academies Press (US); 2018 Mar 1. 2, Factors That Affect Health-Care Utilization. Available from: https://www.ncbi.nlm.nih.gov/books/NBK500097/

NISHIMOTO, A. 2022. Effective combinations of anti-cancer and targeted drugs for pancreatic cancer treatment. *World Journal of Gastroenterology,* 28**,** 3637–3643.

NOEL, M. & FISCELLA, K. 2019. Disparities in pancreatic cancer treatment and outcomes. *Health Equity,* 3**,** 532–540.

NYANGASI, M. F., McLIGEYO, A. A., KARIUKI, D., MITHE, S., ORWA, A., & MWENDA, V. (2023). Decentralizing cancer care in sub-Saharan Africa through an integrated regional cancer centre model: The case of Kenya. *PLOS Global Public Health,* 3(9), e0002402. https://doi.org/10.1371/journal.pgph.0002402

OMOTOSO, O., TEIBO, J. O., ATIBA, F. A., OLADIMEJI, T., PAIMO, O. K., ATAYA, F. S., BATIHA, G. E. & ALEXIOU, A. 2023. Addressing cancer care inequities in sub-Saharan Africa: current challenges and proposed solutions. *Int J Equity Health,* 22**,** 189.

PANG, N.-Q., LAU, J., FONG, S.-Y., WONG, C. Y.-H. & TAN, K.-K. 2022. Telemedicine acceptance among older adult patients with cancer: scoping review. *Journal of Medical Internet Research,* 24**,** e28724.

PARMET, T., YUSUFOV, M., BRAUN, I. M., PIRL, W. F., MATLOCK, D. D. & SANNES, T. S. 2023. Willingness toward psychosocial support during cancer treatment: a critical yet challenging construct in psychosocial care. *Translational Behavioral Medicine,* 13**,** 511–517.

PARO, A., RICE, D. R., HYER, J. M., PALMER, E., EJAZ, A., SHAIKH, C. F. & PAWLIK, T. M. 2022. Telehealth utilization among surgical oncology patients at a large academic cancer center. *Annals of Surgical Oncology,* 29, 7267–7276.

PEREIRA, N. P. & CORRêA, J. R. 2018. Pancreatic cancer: treatment approaches and trends. *Journal of Cancer Metastasis and Treatment,* 4, 10.20517.

PERINEL, J. & ADHAM, M. 2019. Palliative therapy in pancreatic cancer-palliative surgery. *Translational Gastroenterology and Hepatology,* 4, 28.

PETERSEN, G. S., LAURSEN, S. G. W., JENSEN, H., MAINDAL, H. T., PEDERSEN, A. F., VEDSTED, P. & VIRGILSEN, L. F. 2022. Patients' health literacy is associated with timely diagnosis of cancer—a cross-sectional study in Denmark. *European Journal of Cancer Care,* 31, e13532.

PETRIC, J., HANDSHIN, S., JONNADA, P. K., KARUNAKARAN, M. & BARRETO, S. G. 2022. The influence of socioeconomic status on access to cancer care and survival in resectable pancreatic cancer: a systematic review and meta-analysis. *ANZ Journal of Surgery,* 92, 2795–2807.

PINTO, E., GONçALVES, F., SACARLAL, J., CASTRO, L. & REGO, G. 2021. Pain management in cancer patients in the main hospitals in Mozambique. *Annals of Palliative Medicine,* 10, 4069–4079.

PRADES, J., ARNOLD, D., BRUNNER, T., CARDONE, A., CARRATO, A., COLL-ORTEGA, C., DE LUZE, S., GAREL, P., GOOSSENS, M. E. & GRILLI, R. 2020. Bratislava Statement: consensus recommendations for improving pancreatic cancer care. *ESMO Open,* 5, e001051.

PUFAHL, L., ZERBATO, F., WEBER, B. & WEBER, I. 2022. BPMN in healthcare: challenges and best practices. *Information Systems,* 107, 102013.

RAMAMURTHY, C. & WONG, Y.-N. 2018. Barriers to oncology clinical trials. *Oncology Clinical Trials: Successful Design, Conduct, and Analysis*, 251, 51-257.

RAMUTUMBU, N. J., RAMATHUBA, D. U. & MAPUTLE, M. S. 2023. Barriers to accessing oncology services for effective cancer care in the public health institutions in Limpopo Province, South Africa: a qualitative study. *Nursing Reports,* 13, 956–968.

RAWLA, P., SUNKARA, T. & GADUPUTI, V. 2019. Epidemiology of pancreatic cancer: global trends, etiology and risk factors. *World Journal of Clinical Oncology,* 10, 10–27.

ROHATGI, K. W., HUMBLE, S., MCQUEEN, A., HUNLETH, J. M., CHANG, S.-H., HERRICK, C. J. & JAMES, A. S. 2021. Medication adherence and characteristics of patients who spend less on basic needs to afford medications. *The Journal of the American Board of Family Medicine,* 34, 561–570.

SAMOIL, D., KIM, J., FOX, C. & PAPADAKOS, J. K. 2021. The importance of health literacy on clinical cancer outcomes: a scoping review. *Annals of Cancer Epidemiology,* 5.

SCHWARZ, T., SCHMIDT, A. E., BOBEK, J. & LADURNER, J. 2022. Barriers to accessing health care for people with chronic conditions: a qualitative interview study. *BMC Health Services Research,* 22, 1037.

SCOTT, E. & JEWELL, A. 2019. Supportive care needs of people with pancreatic cancer: a literature review. *Cancer Nursing Practice,* 18, 35–43.

SEGEL, J. E., HOLLENBEAK, C. S. & GUSANI, N. J. 2020. Rural-urban disparities in pancreatic cancer stage of diagnosis: understanding the interaction with medically underserved areas. *The Journal of Rural Health,* 36, 476–483.

SHAH, S. M. & KHAN, R. A. 2020. Secondary use of electronic health record: opportunities and challenges. *IEEE Access,* 8, 136947–136965.

SHAPIRO, M., CHEN, Q., HUANG, Q., BOOSALIS, V. A., YOON, C. H., SAUND, M. S., WHANG, E. E. & GOLD, J. S. 2016. Associations of socioeconomic variables with resection, stage, and survival in patients with early-stage pancreatic cancer. *JAMA Surgery,* 151, 338–345.

SIMPKIN, V., NAMUBIRU-MWAURA, E., CLARKE, L. & MOSSIALOS, E. 2019. Investing in health R&D: where we are, what limits us, and how to make progress in Africa. *BMJ Global Health,* 4, e001047.

SITAPATI, A., KIM, H., BERKOVICH, B., MARMOR, R., SINGH, S., EL-KAREH, R., CLAY, B. & OHNO-MACHADO, L. 2017. Integrated precision medicine: the role of electronic health records in delivering personalized treatment. *WIREs Systems Biology and Medicine,* 9, e1378.

STEFAN, D. C. 2015. Cancer care in Africa: an overview of resources. *Journal of Global Oncology,* 1, 30–36.

STEWARTA, B. A. & STEWART IV, J. H. 2021. Disparities in clinical trial participation. *Disparities and Determinants of Health in Surgical Oncology,* 31, 55–64.

SU, X., MA, S., YANG, X., ZHANG, C. & CHEN, W. 2024. Health literacy needs of acute pancreatitis patients during the diagnosis and treatment process under the lens of the timing it right theory: A qualitative study. *Patient Preference and Adherence*, 18, 507–517.

SUNG, H., FERLAY, J., SIEGEL, R. L., LAVERSANNE, M., SOERJOMATARAM, I., JEMAL, A. & BRAY, F. 2021. Global Cancer Statistics 2020: GLOBOCAN estimates of incidence and mortality worldwide for 36 cancers in 185 countries. *CA: A Cancer Journal for Clinicians,* 71**,** 209–249.

SUTTON, T. L., BENEVILLE, B., JOHNSON, A. J., MAYO, S. C., GILBERT, E. W., LOPEZ, C. D., GROSSBERG, A. J., ROCHA, F. G. & SHEPPARD, B. C. 2023. Socioeconomic and geographic disparities in the referral and treatment of pancreatic cancer at high-volume centers. *JAMA Surgery,* 158**,** 284.

TAKAORI, K., BASSI, C., BIANKIN, A., BRUNNER, T. B., CATALDO, I., CAMPBELL, F., CUNNINGHAM, D., FALCONI, M., FRAMPTON, A. E. & FURUSE, J. 2016. International Association of Pancreatology (IAP)/European Pancreatic Club (EPC) consensus review of guidelines for the treatment of pancreatic cancer. *Pancreatology,* 16**,** 14–27.

TIMMER, F. E. F., GEBOERS, B., NIEUWENHUIZEN, S., DIJKSTRA, M., SCHOUTEN, E. A. C., PUIJK, R. S., DE VRIES, J. J. J., VAN DEN TOL, M. P., BRUYNZEEL, A. M. E., STREPPEL, M. M., WILMINK, J. W., VAN DER VLIET, H. J., MEIJERINK, M. R., SCHEFFER, H. J. & DE GRUIJL, T. D. 2021. Pancreatic cancer and immunotherapy: A clinical overview. *Cancers,* 13**,** 4138.

TORPHY, R. J., ZHU, Y. & SCHULICK, R. D. 2018. Immunotherapy for pancreatic cancer: barriers and breakthroughs. *Annals of Gastroenterological Surgery,* 2**,** 274–281.

TRIPATHI, S., TABARI, A., MANSUR, A., DABBARA, H., BRIDGE, C. P. & DAYE, D. 2024. From machine learning to patient outcomes: A comprehensive review of AI in pancreatic cancer. *Diagnostics,* 14**,** 174.

TRIPEPI, M., PIZZOCARO, E., GIARDINO, A., FRIGERIO, I., GUGLIELMI, A. & BUTTURINI, G. 2023. Telemedicine and pancreatic cancer: A systematic review. *Telemedicine and e-Health,* 29**,** 352–360.

VAZIRI-GOHAR, A., HUE, J. J., ABBAS, A., GRAOR, H. J., HAJIHASSANI, O., ZAREI, M., TITOMIHELAKIS, G., FECZKO, J., RATHORE, M. & CHELSTOWSKA, S. 2023. Increased glucose availability sensitizes pancreatic cancer to chemotherapy. *Nature Communications,* 14**,** 3823.

WALKER, J. 2021. *Clinical barriers to enrollment of patients to interventional clinical trials while utilizing the clinicalpath pathway at Roswell Park Cancer Institute*. Buffalo, NY: State University of New York at Buffalo.

WALL, N. R., FULLER, R. N., MORCOS, A. & DE LEON, M. 2023. Pancreatic cancer health disparity: pharmacologic anthropology. *Cancers (Basel),* 15**,** 5070.

WERCHOLUK, A. N., PARIKH, A. A., & SNYDER, R. A. (2022). The road less traveled: transportation barriers to cancer care delivery in the rural Patient population. *JCO Oncology Practice*, 18(9), 652–662. https://doi.org/10.1200/OP.22.00122

YAO, W., CHEN, X., FAN, B., ZENG, L., ZHOU, Z., MAO, Z. & SHEN, Q. 2023. Multidisciplinary team diagnosis and treatment of pancreatic cancer: current landscape and future prospects. *Frontiers in Oncology,* 13**,** 1077605.

YOON, J. H., JUNG, Y. J. & MOON, S. H. 2021. Immunotherapy for pancreatic cancer. *World Journal of Clinical Cases,* 9**,** 2969–2982.

ZHANG, M., LIU, J. & XIA, Q. 2023. Role of gut microbiome in cancer immunotherapy: from predictive biomarker to therapeutic target. *Experimental Hematology & Oncology,* 12, 84.

ZHANG, R., WANG, J., ZHANG, P., ZHANG, Z. & MIAO, R. 2024. Pancreatic cancer progression and mortality predicted by depression and anxiety: a systematic review and meta-analysis protocol. *Frontiers in Psychiatry,* 14, 1266502.

ZHAO, Y., TANG, S., MAO, W. & AKINYEMIJU, T. 2022. Socio-economic and rural-urban differences in healthcare and catastrophic health expenditure among cancer patients in China: analysis of the china health and retirement longitudinal study. *Frontiers in Public Health,* 9, 779285.

ZHENG, R., LIU, X., ZHANG, Y., LIU, Y., WANG, Y., GUO, S., JIN, X., ZHANG, J., GUAN, Y. & LIU, Y. 2024. Frontiers and future of immunotherapy for pancreatic cancer: from molecular mechanisms to clinical application. *Frontiers in Immunology,* 15, 1383978.

Part V

Advocacy, Research, and Future Directions

9 Patient Perspectives, Advocacy, and Community Engagement

Talent Chipiti[1*]*, Mpho Kgomo*[2]*, Kopang C. Mokhetho*[3]*, Min Zhang*[4]*, and Zodwa Dlamini*[1]

[1]SAMRC Precision Oncology Research Unit (PORU), DSI/NRF SARChI Chair in Precision Oncology and Cancer Prevention (POCP), Pan African Cancer Research Institute (PACRI), University of Pretoria, Pretoria, South Africa.

[2]Department of Internal Medicine, Steve Biko Academic Hospital, Pretoria, South Africa.

[3]Department of Pharmaceutical Science, Tshwane University of Technology, Arcadia Campus, Pretoria, South Africa.

[4]Biostatistics Shared Resource, Chao Family Comprehensive Cancer Center, Department of Epidemiology and Biostatistics, University of California, Irvine, USA.

*Corresponding author: Talent Chipiti

INTRODUCTION

Pancreatic cancer, a malignancy originating in the ducts of the pancreas, is one of the most aggressive and deadly forms of cancer. It is characterized by rapid progression and a high mortality rate, making it the fourth leading cause of cancer-related deaths globally (Siegel et al., 2020). The pancreas plays a vital role in digestion and glucose regulation, and its cancerous transformation disrupts these essential functions, leading to severe clinical manifestations. Globally, pancreatic cancer accounts for approximately 495,773 new cases and 466,003 deaths annually (GLOBOCAN, 2020; Hu et al., 2021). The incidence and mortality rates vary significantly across different regions. Developed countries report higher incidence rates, partly due to better diagnostic capabilities. However, the burden in developing regions, including parts of Africa, is rising due to increasing life expectancy and lifestyle changes that contribute to risk factors such as obesity, smoking, and diabetes (Rawla et al., 2019).

In Africa, the epidemiology of pancreatic cancer is less well-documented compared to other regions, but emerging data indicate a significant and growing public health challenge. The incidence of pancreatic cancer in Africa is estimated to be lower than in Western countries, but it is likely underreported due to limited healthcare infrastructure and diagnostic facilities (Ferlay et al., 2019). Furthermore, African patients often present with advanced disease stages, which complicates treatment and worsens prognoses. In sub-Saharan Africa, healthcare systems face unique challenges, including a high prevalence of infectious diseases, limited access to and lack of cancer screening and treatment facilities, and a shortage of specialized healthcare professionals. These factors contribute to delayed diagnoses and poorer outcomes for pancreatic cancer patients (Adeloye et al., 2019).

This chapter delves into the multifaceted experiences of pancreatic cancer patients in Africa, emphasizing their perspectives, the role of advocacy, and community engagement. By exploring the

DOI: 10.1201/9781003544401-14

personal narratives of patients, we aim to illuminate the physical, emotional, and financial burdens they endure. These stories not only provide insight into the daily struggles of living with pancreatic cancer but also highlight the resilience and strength of patients navigating this formidable disease.

Moreover, the chapter underscores the critical importance of advocacy organizations, support groups, and community initiatives in raising awareness, reducing stigma, and empowering patients. Advocacy is pivotal in mobilizing resources, influencing policy changes, and improving access to care. Community engagement fosters a supportive environment where patients and their families can find solace and solidarity, thereby enhancing their overall quality of life. Through a comprehensive exploration of patient perspectives and the impact of community and advocacy efforts, this chapter aims to offer a holistic understanding of pancreatic cancer in the African context. It also seeks to provide actionable insights for healthcare providers, policymakers, and advocates working to improve outcomes and support for pancreatic cancer patients across the continent.

IMPORTANCE OF PATIENT PERSPECTIVES IN UNDERSTANDING PANCREATIC CANCER

The needs and perspectives of individuals with pancreatic cancer have been overlooked for an extended period. Compared to other types of digestive cancers, there is a notable lack of focus on addressing the requirements of pancreatic cancer patients. This is partly because pancreatic cancer has the lowest survival rate among common cancers, with an average life expectancy of only five months following diagnosis. Around 92% of patients succumb within five years of being diagnosed, and there has been minimal improvement in enhancing these survival rates (Ferlay et al., 2017; Minicozzi et al., 2018; Guidelines in Practice, UK, 2024).

The perspectives of pancreatic cancer patients are crucial in enhancing medical knowledge and improving treatment approaches. Understanding patients' experiences with symptoms, side effects, and the psychological impact of the disease can guide healthcare providers in tailoring more effective and empathetic care strategies. For instance, a study by Abuelgasim and Haque (2018) emphasized the importance of patient-reported outcomes in informing clinicians about the real-world impact of treatments, leading to adjustments in pain management and supportive care protocols to better address patient needs.

Patient perspectives are also vital in shaping health policy and advocacy efforts. By voicing their challenges and needs, patients can influence policymakers to allocate resources more effectively and develop programs that address the gaps in care and support services. Advocacy groups often use patient testimonies to highlight the urgency of specific issues, such as access to early diagnostic tools and affordable treatment options. This grassroots-level input is indispensable for creating policies responsive to the actual needs of those affected by pancreatic cancer (Hofmann et al., 2020).

Effective communication between patients and healthcare providers is foundational to high-quality care. Patient perspectives can reveal barriers to communication, such as medical jargon, time constraints, or cultural misunderstandings. By understanding these barriers, providers can develop strategies to enhance dialogue, ensure patients fully understand their diagnosis and treatment options, and make informed decisions about their care (Epstein and Street, 2007). This patient-centered approach improves satisfaction and adherence to treatment regimens, ultimately leading to better outcomes.

Pancreatic cancer patients often face significant emotional and psychological challenges, including anxiety, depression, and fear of the unknown. Patient perspectives can shed light on the mental health needs of these individuals, prompting the integration of psychological support into the standard care regimen. Research by Schuler et al. (2019) highlights the importance of addressing the mental health of cancer patients, showing that comprehensive care that includes psychological support can significantly improve overall well-being and quality of life.

Patient perspectives are instrumental in guiding research priorities and methodologies. By involving patients in the research process, scientists can ensure that studies address the most pressing issues from the patient's viewpoint, such as quality of life, pain management, and survivorship. Patient-centered research often leads to the development of innovative therapies and interventions that are directly relevant to patient needs. For example, the Patient-Centered Outcomes Research Institute (PCORI) emphasizes the value of patient input in designing and conducting research that is more likely to produce meaningful and applicable results (Selby et al., 2012).

Patient narratives help build a sense of community and support among those affected by pancreatic cancer. Sharing experiences can reduce feelings of isolation, provide practical advice, and offer emotional solidarity. Support groups and patient networks, both in-person and online, are often lifelines for patients and their families. These networks can foster resilience and provide a platform for collective action in advocacy and awareness campaigns (Bultz and Carlson, 2005). Patient perspectives are indispensable in understanding the multifaceted impact of pancreatic cancer. They inform clinical practice, policy development, communication strategies, psychological support services, research agendas, and community-building efforts. By integrating patient voices into every aspect of pancreatic cancer care and research, we can create more responsive, effective, and compassionate healthcare systems that truly meet the needs of patients.

As per Digestive Cancers Europe (DiCE), a recent patient feedback survey has highlighted specific areas necessitating adjustment. These areas have been identified as crucial from the patient's viewpoint and are deemed essential for enhancing overall survival rates and improving the quality of life (Figure 9.1).

In Africa, unique socioeconomic and healthcare challenges magnify the significance of patient perspectives. High rates of poverty, limited access to healthcare facilities, and a lack of specialized oncology services exacerbate the difficulties faced by pancreatic cancer patients (Adeloye et al., 2016; Omotoso et al., 2023). Cultural beliefs and societal norms influence how patients perceive and manage their illness. In many African communities, there may be stigma associated with cancer, leading to social isolation and reluctance to seek medical help (Daher, 2012; Mwaka et al., 2015). Understanding these cultural contexts through patient narratives can help develop culturally sensitive interventions and awareness programs that resonate with the local population. Patient perspectives highlight disparities in healthcare access and resource allocation. Many African patients may only seek care at advanced stages due to financial constraints or lack of awareness, impacting survival rates and outcomes (Tucker et al., 2011; Fitzmaurice et al., 2019). Incorporating patient insights into healthcare planning can guide the allocation of resources to areas of greatest need, improving early detection and access to treatment.

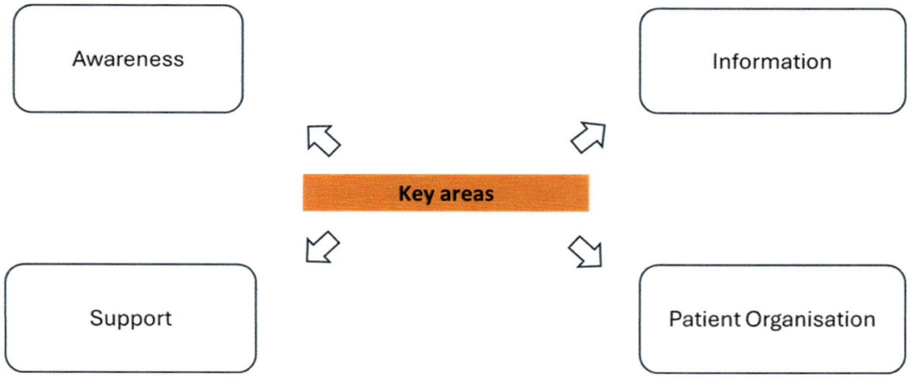

FIGURE 9.1 Key areas of improvement for the betterment of pancreatic cancer patients' quality of life. Raise awareness about pancreatic cancer and its early symptoms to help with early diagnosis. Provide clear and timely information to patients with a message of hope. Ensure patients have access to essential psychological, nutritional, and pain relief support as needed. Foster an environment for national patient organizations to thrive.

Community support and advocacy are critical in the African context. Patient perspectives underscore the importance of grassroots initiatives and local advocacy groups in providing support, raising awareness, and lobbying for better healthcare policies (Stefan, 2015). Empowering patients and survivors to share their stories can reduce stigma, promote early diagnosis, and enhance community engagement in cancer care. Patient perspectives are indispensable in understanding the full impact of pancreatic cancer and shaping responsive, effective care strategies. In Africa, these insights are crucial for addressing the unique challenges faced by patients and improving outcomes through culturally relevant and patient-centered approaches.

THE IMPACT OF PANCREATIC CANCER ON PATIENTS AND FAMILIES

Pancreatic cancer exerts a profound impact on patients and their families, influencing their physical health, emotional well-being, and financial stability. Patient-focused care for pancreatic cancer involves providing emotional, social, practical, and spiritual support, along with resources for coping with stress. Patients often experience psychological stress, leading to depression, anxiety, and an increased risk of suicide. This stress can impact treatment outcomes and prognosis. Improving psychological support and stress management can enhance patients' quality of life and treatment outcomes. Understanding and addressing patients' stress can improve care and help them make informed treatment decisions (Mazzella Ebstein, 2020).

Physical Burdens: Symptoms and Treatment Side Effects of Pancreatic Cancer

Patients with pancreatic cancer often endure substantial physical burdens due to the aggressive nature of the disease and the side effects of its treatments. Common symptoms include abdominal pain, jaundice, weight loss, and fatigue, which can significantly impair their quality of life and daily functioning. Chemotherapy, a common treatment modality, can exacerbate these symptoms by causing nausea, vomiting, and severe fatigue, while radiation therapy may lead to additional challenges such as skin irritation and gastrointestinal issues (Kornblau et al., 2022; Lee et al., 2023).

These cumulative physical burdens not only diminish the overall well-being of patients but also complicate the management of their daily activities and adherence to treatment regimens. Addressing these physical challenges is crucial for improving the quality of life and functional status of individuals battling pancreatic cancer.

Emotional Toll: Coping with Diagnosis and Prognosis of Pancreatic Cancer

The diagnosis of pancreatic cancer is frequently associated with a substantial emotional toll. Patients and their families often grapple with feelings of fear, anxiety, and depression as they confront the uncertainty of the prognosis. The aggressive nature of pancreatic cancer, combined with its typically late-stage diagnosis, contributes to a heightened sense of hopelessness. Studies show that psychological distress is prevalent among pancreatic cancer patients, with high levels of anxiety and depression impacting both the patients and their caregivers (Zhu et al., 2023). Families often face the stress of caregiving and the emotional strain of supporting a loved one through a challenging and often terminal illness (Anderson et al., 2025).

Financial Challenges: Costs of Treatment and Care of Pancreatic Cancer Patients

The financial burden of pancreatic cancer treatment can be substantial. The costs associated with chemotherapy, surgery, and supportive care are significant and often lead to financial strain for patients and their families. A study by Neugut et al. (2023) highlights that the direct medical costs, including hospitalization and outpatient care, contribute to financial hardship, compounded by indirect costs such as lost income due to the inability to work. The financial impact is exacerbated by the prolonged treatment course and the need for ongoing supportive care and palliative services (Kim et al., 2024).

Overall, pancreatic cancer imposes a multifaceted burden on patients and their families, affecting physical health, emotional well-being, and financial stability. Addressing these challenges requires a comprehensive approach that includes effective symptom management, psychological support, and financial assistance to alleviate the overall impact of the disease.

PATIENT EXPERIENCES AND NARRATIVES

Personal Stories from Pancreatic Cancer Patients in Africa

In Africa, the experiences of pancreatic cancer patients reflect a unique set of challenges and resilience shaped by a combination of medical, social, and cultural factors. Personal stories from patients across the continent reveal a diverse array of experiences, from the difficulties of accessing timely diagnosis and treatment to the profound impact of the disease on their lives and families. For instance, in South Africa, where healthcare disparities are notable, patients often face delays in diagnosis due to limited access to diagnostic facilities and specialized care (Amri et al., 2023; Nhleko et al., 2024). Similarly, in Nigeria, patients frequently encounter challenges related to the high costs of treatment and the limited availability of advanced medical technologies (Abubakar et al., 2022).

Common Themes: Resilience, Hope, and Challenges

Despite significant obstacles, common themes of resilience and hope emerge strongly from patient narratives. Many patients exhibit remarkable perseverance, often drawing strength from their families and communities. For instance, patients in Kenya have been described as showing extraordinary resilience in the face of a diagnosis often associated with poor prognoses (Odhiambo et al., 2023). Hope plays a crucial role in these narratives, with patients frequently finding solace in religious faith or community support. Challenges include navigating complex healthcare systems, coping with the physical and emotional toll of the disease, and managing financial burdens. To improve the quality of life (QOL) of cancer patients, psychology has been an essential component of oncological care since the 1970s (Lang-Rollin and Berberich, 2018).

The World Health Organization (1993) defines QOL as an individual's perception of their position in life within their cultural and value systems relative to their goals, expectations, standards, and concerns. Patients who are hopeful about their treatment outcomes tend to fare better than pessimists (Blank and Bellizzi, 2006). Finding meaning in life (MIL) while battling cancer has been shown to enhance QOL by modifying illness perception and global meaning (Majernikova and Obrocnikova, 2017; Whiteford & Olver, 2012). Religion and spirituality are also crucial for many cancer patients. While faith in God or connecting to the larger world can provide comfort, religious or spiritual struggles can lead to emotional turmoil and a lower QOL (Ano & Pargament, 2012; Damen et al., 2021). The stage of cancer can affect hope differently; some studies suggest that advanced stages may decrease hope (Chen, 2003; Duggleby et al., 2013; Baczewska et al., 2020). In countries like India, where palliative care is still developing, discovering hope and finding meaning in life is vital for safeguarding QOL. India's diverse religious beliefs add another layer of complexity to dealing with cancer. Studies have shown that sociodemographic factors like limited education and employment status can diminish QOL (Ramasubbu et al., 2021).

Given the importance of these variables, ongoing research aims to investigate hope, MIL, and religious/spiritual struggles as predictors of QOL among cancer patients in the Indo-Gangetic region (Nazam and Husain, 2024).

CULTURAL FACTORS INFLUENCING PANCREATIC CANCER PATIENT EXPERIENCES

Understanding the impact of cultural and behavioral factors is essential in addressing health disparities associated with prostate cancer. These factors shape beliefs, attitudes, and behaviors related to prevention, diagnosis, treatment, and adherence. Cultural beliefs about cancer can significantly influence health-seeking behaviors, treatment adherence, and participation in clinical trials. Additionally, limited health literacy and language barriers can result in misunderstandings and delays in diagnoses. To tackle these challenges, it is crucial to employ culturally sensitive strategies such as tailored education and outreach, community engagement, patient navigation, and cultural competence training for healthcare providers. These efforts seek to diminish disparities and ensure fair access to quality care for all individuals, regardless of their cultural background or beliefs (Wall et al., 2023). Cultural factors markedly impact patient experiences and narratives. In numerous African cultures, traditional beliefs and practices strongly shape the perception and management of cancer. For example, in parts of East Africa, traditional medicine and herbal remedies are often sought in conjunction with conventional treatments, reflecting a blend of modern and traditional healthcare approaches. Furthermore, the stigma associated with cancer can affect patients' willingness to seek help and adhere to treatment. Cultural attitudes toward illness, health, and healing are pivotal in understanding the patient's experience, influencing how individuals cope with their diagnosis and interact with medical professionals (Oystacher et al., 2018).

The experiences of patients with pancreatic cancer in Africa are characterized by a complex interplay of resilience, hope, and significant challenges. Cultural factors exert substantial influence over these experiences, shaping personal and communal coping mechanisms for the disease. Addressing these challenges necessitates a comprehensive understanding of the local context and a holistic approach to providing support and care.

Advocacy and Support Organizations

Role of Advocacy Groups in Raising Awareness

Patient advocacy has significantly advanced cancer awareness, care, resource allocation, research, and outcomes in high-income countries (HICs) and holds the potential to achieve similar results in low- and middle-income countries (LMICs). As the cancer burden in LMICs continues to grow, civil society organizations (CSOs) are actively responding, although they face numerous obstacles. Advocacy groups play a critical role in increasing awareness of diseases like cancer, especially in regions such as Africa, where healthcare resources and information may be limited (Dvaladze et al., 2020).

These organizations use various methods, including public education, policy advocacy, and resource mobilization, to ensure cancer receives the necessary attention from both the public and policymakers. By organizing events like World Cancer Day and running media campaigns, advocacy groups help inform the public about the importance of early detection, prevention strategies, and available treatments for cancer. A 2007 Institute of Medicine report called on established cancer advocacy organizations to support the growth of cancer advocacy in LMICs through the facilitation of domestic, regional, and international advocacy networks; the identification of successful approaches to cancer advocacy for use in other settings; advising on cancer control choices; and providing hands-on training, technical assistance, and funding (Sloan and Gelband, 2007; Institute of Medicine, 2007).

Advocacy groups conduct extensive public education campaigns to disseminate vital information about cancer prevention, early detection, and treatment options. These campaigns often involve organizing events such as World Cancer Day, which serves as a global platform to raise awareness and promote cancer education. Media campaigns, including television, radio, and social media, are also utilized to reach a broader audience. These efforts help to inform the public about the importance of early detection and regular screenings, which are crucial for improving survival rates, particularly for aggressive cancers like pancreatic cancer. In addition to their efforts in cancer awareness, these organizations also work to educate the public, influence policy, and mobilize resources for research and treatment of various health issues in Africa. They provide platforms for patients and families to share their experiences, thereby reducing the stigma surrounding diseases and creating a supportive community. By engaging with the media, hosting public events, and collaborating with healthcare professionals, advocacy groups ensure that crucial health information reaches a wide audience (George Washington University, 2024).

Support Services Provided by Organizations

Support groups for pancreatic cancer patients and caregivers are crucial for providing both emotional and practical support, aiding them in coping with the challenges of the disease. Notable organizations such as the American Cancer Society, CancerCare, and Cancer Support Community offer valuable free services, including online support groups, counseling, and live web chats with licensed counselors. WE SPARK provides diverse support groups and additional programs such as acupuncture and tai chi. Pancreatic Cancer Connections and the Smart Patients Pancreatic Cancer discussion forum offer secure online spaces for members to share experiences and resources. In addition, ANCAN and a peer-to-peer group led by long-term survivor Tom conduct virtual support meetings to enhance the quality of life for patients and caregivers (Pancreatic Cancer Action, 2024). The Whipple Surgery Survivor Group on Facebook offers a global community for those comfortable with social media, while Cancer Support Community South Bay hosts a virtual group specifically for California residents. These organizations play a pivotal role in ensuring that pancreatic cancer patients and their loved ones feel empowered and connected rather than isolated through their shared experiences (Hirshberg Foundation for Pancreatic Cancer Research, 2022; Pancreatic Cancer Action Network, 2024). Advocacy groups actively work to influence healthcare policies and funding allocations in addition to public education. By lobbying government officials and collaborating with policymakers, these organizations strive to ensure that cancer care and research are prioritized in national health agendas. For example, the Cancer Association of South Africa (CANSA) has played a key role in advocating for improved cancer policies and increased funding for research and treatment facilities (CANSA, 2023c).

Support organizations provide a wide range of services to patients and their families, including psychological counseling, financial assistance, patient navigation programs, and support groups. CANSA is an example of an organization that offers a comprehensive support system for cancer patients, providing counseling, a toll-free helpline, and online resources with information about different cancer types and treatment options (CANSA, 2022).

These support services are essential for patients and their families, providing emotional support, financial assistance, and access to treatment and care resources. Counseling services, support groups, and helplines help individuals cope with the psychological impact of cancer. Financial support may be in the form of grants or aid with treatment costs, transportation, and lodging for patients traveling long distances for care. Additionally, advocacy groups often offer educational resources to help patients understand their diagnosis and treatment options.

Providing platforms for patients and families to share their experiences is another critical function of advocacy groups. These platforms help to reduce the stigma associated with cancer and create a supportive community. For example, the Uganda Women's Cancer Support Organisation (UWOCASO) facilitates support groups where patients and their families can connect, share their stories, and offer mutual support (Uganda Women's Cancer Support Organisation, 2023b).

Engaging with the media is a key strategy used by advocacy groups to amplify their message and reach a wider audience. Through press releases, interviews, and media partnerships, these organizations ensure that crucial health information is disseminated broadly. The Breast Health Foundation in South Africa, for instance, uses media campaigns to raise awareness about breast cancer and the importance of early detection (Breast Health Foundation, 2023).

Specific Impact on Pancreatic Cancer Awareness

Pancreatic cancer has a notoriously high fatality rate, largely due to late diagnosis and rapid progression. It is essential to raise awareness about this disease to improve early detection and treatment outcomes. Advocacy groups are instrumental in educating the public about the risk factors, symptoms, and the significance of early screening (Wall et al., 2023). They employ a range of strategies, including public education campaigns, community outreach, and collaborations with healthcare providers to disseminate information widely. Campaigns dedicated to pancreatic cancer often emphasize the subtle and nonspecific symptoms linked to the disease, such as jaundice, weight loss, and abdominal pain. These symptoms are frequently overlooked or misattributed to less severe conditions, resulting in delayed diagnosis. By promoting awareness of these symptoms, advocacy groups aim to prompt individuals to seek medical attention if they experience any of these warning signs. Early intervention is crucial, as early-stage pancreatic cancer is more responsive to treatment and carries significantly better prognostic outcomes (Ryan et al., 2022).

Furthermore, these organizations are dedicated to supporting research initiatives aimed at developing improved diagnostic tools and treatments for pancreatic cancer. They allocate resources to fund studies exploring new biomarkers for early detection and innovative therapeutic approaches. For example, advancements in molecular diagnostics and imaging technologies have the potential to revolutionize the early detection of pancreatic cancer (Smith et al., 2023). Additionally, advocacy groups often collaborate with research institutions and pharmaceutical companies to accelerate the development and approval of new treatments, ensuring that patients have access to the latest and most effective therapies (Jones et al., 2023). Advocacy efforts also extend to policy and legislative initiatives focused on increasing funding for pancreatic cancer research and ensuring equitable access to care. These efforts include lobbying for increased research grants, supporting policies that reduce healthcare disparities, and advocating for insurance coverage of cutting-edge diagnostic and therapeutic procedures (Miller et al., 2023). By influencing policy, advocacy groups help to create an environment where advances in pancreatic cancer treatment and detection can be rapidly implemented and made accessible to all patients.

In summary, advocacy groups play a crucial role in raising awareness about pancreatic cancer, promoting early detection, supporting research, and influencing policy. Their multifaceted approach not only educates the public but also drives scientific advancements and ensures that patients receive comprehensive and timely care. The collective efforts of these organizations significantly contribute to improving outcomes for individuals affected by pancreatic cancer.

CASE STUDIES OF SUCCESSFUL CANCER ADVOCACY CAMPAIGNS IN AFRICA

Case Study 1: Breast Cancer Awareness Campaign by PinkDrive in South Africa

PinkDrive is a prominent non-profit organization in South Africa dedicated to breast cancer awareness. Its mobile mammography and educational units travel across the country, particularly to underserved communities, providing free breast examinations and mammograms. This initiative has significantly increased the rate of early detection and awareness among South African women, particularly in rural areas where access to healthcare is limited. According to a report by PinkDrive, over 20,000 women were screened in 2019 alone, leading to early detection and treatment for many (PinkDrive, 2019).

Cancer Association of South Africa (CANSA)

CANSA is a prominent advocacy group in South Africa that has significantly impacted cancer awareness and support. One of its successful campaigns is the “Women’s Health” campaign, which focuses on breast and cervical cancer awareness. Through extensive media coverage, educational workshops, and free screening services, CANSA has reached thousands of women, leading to early detection and improved treatment outcomes (CANSA, 2023c).

Breast Health Foundation (BHF)

Another noteworthy example is BHF in South Africa, which has implemented several impactful programs. The "Bosom Buddies" program provides peer support to breast cancer patients by pairing them with survivors who offer guidance and emotional support throughout their treatment journey. This program has been instrumental in reducing feelings of isolation and improving the mental health of patients (Breast Health Foundation, 2023).

Pancreatic Cancer Action Network (PanCAN)

The Pancreatic Cancer Action Network (PanCAN) is a prominent organization dedicated to raising awareness about pancreatic cancer, providing patient support, and funding critical research. One of PanCAN's most impactful initiatives is the Know Your Tumor program, which exemplifies its commitment to personalized medicine and improving treatment outcomes for pancreatic cancer patients (Pancreatic Cancer Action Network, 2023c).

Know Your Tumor Program

The "Know Your Tumor" program aims to offer patients personalized treatment options based on their tumor's molecular profile. This program represents a significant advancement in pancreatic cancer treatment, highlighting the importance of precision medicine. By analyzing the genetic and molecular characteristics of a patient's tumor, the program can identify targeted therapies that are more likely to be effective, thereby significantly improving treatment outcomes (Smith et al., 2023).

The process involves comprehensive molecular profiling of the tumor, including genomic testing to identify mutations, biomarkers, and other molecular alterations. This information is then used to match patients with the most suitable treatments, including targeted therapies and clinical trials. This personalized approach not only enhances the likelihood of successful treatment but also minimizes side effects by avoiding less effective treatments (Jones et al., 2019).

In addition to the "Know Your Tumor" program, PanCAN offers a comprehensive range of patient support services. These include providing information on diagnosis and treatment options, connecting patients with clinical trials, and offering emotional and practical support through their Patient Services team. PanCAN also organizes educational webinars, workshops, and symposiums to keep patients and healthcare providers informed about the latest advances in pancreatic cancer research and treatment (Abdilleh et al., 2024).

PanCAN is highly involved in advocating for increased federal funding for pancreatic cancer research and for policies that support patient care. It works tirelessly to influence legislation and public policy to ensure that pancreatic cancer research receives adequate funding and attention. PanCAN's advocacy efforts have led to significant increases in federal research funding, supporting groundbreaking studies aimed at improving early detection, treatment, and, ultimately, survival rates for pancreatic cancer patients.

Community Engagement and Awareness Campaigns

The Pancreatic Cancer Action Network (PanCAN) is dedicated to raising awareness about pancreatic cancer through various community outreach and awareness campaigns. These include the annual "PurpleStride" walks, which bring together communities across the country to raise funds and awareness for pancreatic cancer. PanCAN also utilizes social media, public service announcements, and partnerships with other organizations to spread information about pancreatic cancer symptoms, risk factors, and the importance of early detection (Davis et al., 2023). In addition to these initiatives, PanCAN offers personalized treatment programs like "Know Your Tumor," comprehensive patient support, advocacy, research funding, and widespread awareness campaigns to improve the lives of those affected by pancreatic cancer. This multifaceted approach not only advances scientific research but also ensures that patients receive the best possible care and support throughout their cancer journey (Pancreatic Cancer Action Network, 2023c).

Case Study 2: The Kenya Cancer Association (KENCANSA)

KENCANSA has been instrumental in advocating for cancer patients' rights and improving access to cancer care in Kenya. One of its notable campaigns involved partnering with the Ministry of Health to establish the National Cancer Control Strategy 2017–2022. This strategy aims to reduce the cancer burden through prevention, early detection, improved treatment, and palliative care. The campaign has led to increased government funding for cancer treatment and the establishment of more cancer treatment centers across Kenya (KENCANSA, 2017).

Kenya Network of Cancer Organizations (KENCO)

KENCO has been pivotal in advocating for cancer patients' rights and access to care in Kenya. Their "National Cancer Awareness Week" is an annual event that includes public seminars, free screenings, and educational outreach. The campaign has garnered government support, leading to increased funding for cancer care and the establishment of more cancer treatment centers nationwide (Kenya Network of Cancer Organizations, 2023). KENCO has also made strides in pancreatic cancer awareness in Kenya. Its campaigns focus on educating the public about the disease and advocating for improved healthcare policies to support pancreatic cancer patients. KENCO's efforts have increased awareness and funding for cancer research and treatment in the region (Kenya Network of Cancer Organizations, 2023).

Case Study 3: Uganda Women's Cancer Support Organisation (UWOCASO)

The Uganda Women's Cancer Support Organization (UWOCASO) is a pioneering cancer support group in Uganda, founded in 2004 by five breast cancer survivors. It now includes 35 survivors and well-wishers. UWOCASO aims to provide psychosocial support to women and families affected by breast and cervical cancer, emphasizing the importance of early diagnosis and treatment. The organization promotes routine screening and breast self-examination practices. It engages in community education through mobilization and demonstrations in schools, clinics, and community centers. UWOCASO also participates in monthly media talk shows, provides patient support and counseling, and advocates for public policy improvements. Additionally, it is involved in cancer research, clinical trials, and offer training opportunities for professionals and volunteers. UWOCASO became a member of the Union for International Cancer Control (UICC) in May 2008.

UWOCASO is a leading advocacy group in Uganda focused on women's cancers. It runs awareness campaigns about cervical and breast cancer, emphasizing the importance of regular screenings. In collaboration with local health clinics, UWOCASO has implemented community outreach programs that provide free or low-cost screening services. Its efforts have contributed to a notable increase in the number of women undergoing cancer screenings and receiving timely treatment (UWOCASO, 2020).

These case studies highlight the impactful work done by advocacy and support organizations across Africa in raising awareness, providing support, and advocating for better cancer care and policies. Their efforts have led to significant improvements in cancer detection, treatment, and overall patient support, demonstrating the critical role of advocacy in the fight against cancer.

COMMUNITY SUPPORT AND ENGAGEMENT

Importance of Community Involvement in Pancreatic Cancer Patient Support

Community support and engagement play pivotal roles in the comprehensive care and management of cancer, particularly pancreatic cancer. Pancreatic cancer is often diagnosed at a late stage and has a high mortality rate. Community involvement in patient support can significantly enhance the quality of life for patients and their families, reduce stigma, and foster a better understanding of the disease.

Community involvement is crucial in providing comprehensive support to cancer patients, including those with pancreatic cancer. The collective efforts of community members, local organizations, and healthcare providers can significantly enhance the quality of life for patients and their families. Community support helps in various ways, such as offering emotional encouragement, providing practical assistance, and fostering a sense of belonging and solidarity. For pancreatic cancer patients, who often face a particularly challenging prognosis, community support can help alleviate feelings of isolation and anxiety, thereby improving mental health outcomes and overall well-being (Kaiser et al., 2021).

Community involvement is crucial for several reasons:

1. *Emotional and psychological support*: Patients often experience significant emotional and psychological stress upon diagnosis. Community support groups provide a platform for patients to share their experiences, fears, and hopes, thus reducing feelings of isolation and anxiety.
2. *Practical assistance*: Communities can offer practical support, such as transportation to treatment centers, assistance with household chores, and financial aid, which can alleviate some of the burdens faced by patients and their families.
3. *Information and education*: Community-led educational initiatives can increase awareness about pancreatic cancer by emphasizing the importance of early detection and promoting healthy lifestyles that may reduce cancer risk.
4. *Advocacy and empowerment*: Engaged communities can advocate for better healthcare services, increased funding for research, and improved patient care policies.

Community engagement and health education are essential in addressing pancreatic cancer disparities. Active community involvement in interventions, along with culturally appropriate health education, empowers individuals to make informed decisions about pancreatic cancer prevention, treatment, and support. Community-Based Participatory Research (CBPR) involves community members throughout the research process, ensuring that interventions align with community needs and values, and promote ownership and sustainability (Symanski et al., 2023; Milton et al., 2023; Anderson et al., 2023). Healthcare providers must respect and understand cultural beliefs, practices, and communication styles for cultural competence in health education. Educational materials should be tailored to be culturally relevant, easy to understand, and visually appealing. Programs for pancreatic cancer should consider the unique needs of different communities by integrating local cultural beliefs and norms. Involving community health workers or trusted individuals can enhance the effectiveness of these initiatives (Warner et al., 2023).

Improving health literacy is important for understanding health information and making informed decisions. Health education should use plain language, visual aids, and interactive formats to ensure accessibility. Community outreach and awareness campaigns raise awareness about pancreatic cancer, its risk factors, symptoms, and the importance of early detection using culturally appropriate messaging (Calvillo-Ortiz et al., 2022; Collett et al., 2023). Empowering individuals and communities to advocate for their health involves providing information on screening guidelines, risk reduction strategies, and healthcare resources. Encouraging active participation in care, asking questions, and seeking second opinions can lead to improved health outcomes and reduced disparities. Support groups and peer networks tailored to specific cultural groups can offer emotional support, shared experiences, and practical guidance, addressing cultural concerns and language barriers (Benham-Hutchins et al., 2017; Ratnapradipa et al., 2022).

Community health workers (CHWs) are crucial in bridging healthcare access gaps. Training them in pancreatic cancer awareness and prevention allows them to provide culturally sensitive education and connect individuals to resources. This helps reduce pancreatic cancer disparities and fosters partnerships for improved prevention, treatment, and support within diverse populations.

Initiatives to Reduce Stigma and Promote Understanding

Reducing stigma and promoting understanding about cancer, and specifically pancreatic cancer, is essential for encouraging early diagnosis and treatment. Advocacy groups and community organizations play a pivotal role in this regard. Initiatives often include educational campaigns to dispel myths about cancer, highlight the importance of early detection, and provide accurate information about treatment options. These efforts aim to create a more informed and supportive community environment where individuals feel comfortable seeking help and discussing their health concerns (Mukwato et al., 2020). The stigma associated with cancer, including pancreatic cancer, can lead to delays in seeking treatment and a lack of social support.

Initiatives to reduce stigma and promote understanding include:

1. *Public awareness campaigns*: Campaigns aimed at educating the public about the causes, symptoms, and treatments of pancreatic cancer can dispel myths and misinformation. For example, World Pancreatic Cancer Day is celebrated globally to raise awareness and encourage conversations about the disease.
2. *Support groups and counseling services*: These provide safe spaces for patients and their families to discuss their experiences and receive professional psychological support.
3. *Community workshops and seminars*: Workshops on topics such as nutrition, palliative care, and the importance of early detection can be held in community centers, schools, and workplaces to educate the public.
4. *Media engagement*: Utilizing media platforms to share survivors' stories, medical expert interviews, and educational content can reach a broad audience and help normalize discussions about pancreatic cancer.

Examples of Community-Led Projects and Their Impact

Uganda Women's Cancer Support Organisation (UWOCASO): As discussed earlier, UWOCASO focuses on providing psychosocial support and education about breast and cervical cancers, emphasizing the importance of early diagnosis and routine screening. Their efforts have led to increased awareness and support for cancer patients in Uganda (UICC, 2022).

Project Purple—USA: Project Purple is a community-led initiative in the United States focused on raising awareness about . It organizes events such as runs and walks to raise funds for research and provide support to patients and families. Its efforts have led to increased public awareness and substantial financial support for pancreatic cancer research, contributing to advancements in early detection and treatment options (Project Purple, 2022).

The American Association for Cancer Research (AACR) and the Pancreatic Cancer Action Network (PanCAN): They have established community outreach programs targeting African American populations, who are disproportionately affected by pancreatic cancer. These programs offer culturally tailored education, screening events, and support services to increase awareness and promote early detection. Patient navigation programs aim to assist patients in navigating the complexities of the healthcare system, providing support and help in accessing timely and appropriate care (Eliacin et al., 2023).

Pancreatic Cancer Action Network (PanCAN): PanCAN engages communities through advocacy, patient support, and fundraising events. Its community-based approach has significantly contributed to increased funding for pancreatic cancer research and better patient support services.

Kenya Network of Cancer Organizations (KENCO): KENCO coordinates various cancer awareness and support initiatives across Kenya. Its community outreach programs educate the public about different types of cancer, including pancreatic cancer and provide a support network for patients and their families.

Pancreatic Cancer Action Network (PanCAN)—USA: PanCAN engages communities through outreach programs and volunteer networks to support pancreatic cancer patients. Its "PurpleStride" events across the country raise significant funds and awareness. These initiatives have not only raised millions of dollars for research but also fostered a strong community of support for patients and survivors, providing them with valuable resources and a platform to share their experiences (CANSA, 2021a).

Cancer Association of South Africa (CANSA)—South Africa: CANSA's community-led projects focus on education and support for various cancers, including pancreatic cancer. Its initiatives include mobile health clinics that provide screening and educational services in underserved areas. By bringing healthcare services directly to communities, CANSA has improved early detection rates and facilitated better access to treatment for many individuals, significantly impacting public health outcomes (CANSA, 2021b).

Uganda Women's Cancer Support Organisation (UWOCASO)—Uganda: UWOCASO, as detailed previously, provides psychosocial support and education on breast and cervical cancer. Its community engagement efforts include workshops and educational campaigns aimed at promoting early diagnosis and reducing stigma. These activities have successfully created a supportive network for patients and increased awareness about the importance of early cancer screening and treatment in Uganda (UWOCASO, 2023a).

Tailoring interventions to specific cultural groups can improve engagement and outcomes. For example, the Asian Liver Center at Stanford University developed a program called "Screening in the Community" to increase pancreatic cancer screening among Asian populations. The program includes bilingual outreach, culturally appropriate educational materials, and community partnerships to address cultural barriers. Implementing targeted screening programs for high-risk populations can detect pancreatic cancer at earlier stages. The High-Risk Pancreatic Cancer Clinic at Johns Hopkins Medicine focuses on screening individuals with a family history of pancreatic cancer or known genetic mutations. This specialized clinic offers comprehensive risk assessment and surveillance strategies to detect pancreatic cancer early in high-risk individuals (Wall et al., 2023).

Impact of Community-led Projects

Community-led projects have been shown to have a profound impact on cancer awareness, early detection, and patient support. These initiatives often lead to:

Improved awareness and education: Educational campaigns help demystify cancer, reduce stigma, and inform the public about the importance of early detection and available treatment options (WHO, 2022).

Enhanced support networks: By fostering a sense of community, these projects provide emotional and practical support to patients and their families, improving their overall quality of life and mental health (American Cancer Society, 2022).

Increased access to healthcare services: Mobile clinics and local health fairs make cancer screening and treatment more accessible, particularly in underserved and rural areas (National Cancer Institute, 2021).

Fundraising and resource mobilization: Community events often raise significant funds for research and patient support services, driving advancements in treatment and care (Cancer Research UK, 2023).

Community support and engagement play vital roles in the fight against cancer, particularly pancreatic cancer. Educational programs, efforts to reduce stigma, and projects led by the community can lead to improved health outcomes, heightened public awareness, and a nurturing environment for those affected by cancer. The involvement of communities in supporting and caring for cancer patients, specifically those with pancreatic cancer, is invaluable. Through initiatives aimed at reducing stigma, fostering understanding, and providing both practical and emotional support, communities can significantly enhance the lives of patients and their families. Furthermore, community-led projects have demonstrated noteworthy impacts, ranging from increased awareness to improved patient care and advocacy for enhanced health policies. Engaging communities empowers patients and fosters a more knowledgeable and supportive environment for addressing the challenges associated with pancreatic cancer.

EMPOWERING PATIENTS AS ADVOCATES

Encouraging Patient Involvement in Advocacy Efforts

Empowering patients to become advocates for cancer, particularly pancreatic cancer, involves encouraging them to actively participate in awareness campaigns, policy advocacy, and research initiatives. Patient advocates bring a unique and powerful perspective to the table, as their firsthand experiences with the disease lend authenticity and urgency to advocacy efforts. By sharing their stories, patients can help demystify the disease, reduce stigma, and highlight the importance of early detection and treatment. Advocacy organizations often encourage patient involvement by inviting them to speak at events, participate in media campaigns, and engage with policymakers to influence health policies that improve cancer care and support services (Cancer Support Community, 2022).

Training and Resources for Patient Advocates

To ensure the effectiveness of patient advocacy, it is imperative to provide proper training and access to resources. Patient-centered advocacy groups and reputable healthcare organizations offer a range of programs aimed at equipping individuals with the requisite skills and knowledge. These programs include the following: workshops and seminars covering a range of topics, including public speaking, storytelling, media engagement, and fostering an understanding of the healthcare system; easily accessible online courses and webinars that allow patients to independently acquire knowledge of advocacy strategies, policy advocacy, and specific insights related to cancer; mentorship programs where experienced advocates offer mentorship to newly initiated patient advocates, providing invaluable guidance and unwavering support as they embark on their advocacy journey; and comprehensive resource libraries housing educational materials, research articles, and advocacy toolkits, which serve to inform and bolster the efforts of patient advocates (Patient Advocate Foundation, 2023).

Success Stories of Patient Advocates Making a Difference

Julie Fleshman - Pancreatic Cancer Action Network (PanCAN): Julie Fleshman, the CEO of PanCAN, became a leading advocate for pancreatic cancer after losing her father to the disease. Under her leadership, PanCAN has made significant strides in raising awareness, funding research, and advocating for better treatment options. Fleshman's efforts have resulted in increased federal funding for pancreatic cancer research and improved resources for patients and families (PanCAN, 2023c).

John Kanzius - Inventor and Advocate: After being diagnosed with leukemia, John Kanzius, a former radio and TV engineer, used his technical expertise to invent a non-invasive radio wave device aimed at killing cancer cells. Although he passed away before seeing the full fruition of his work, his invention has been further developed and holds promise for treating pancreatic and other cancers. Kanzius's advocacy and innovative spirit continue to inspire advancements in cancer treatment (American Cancer Society, 2022).

Olwen Williams - Macmillan Cancer Support: Olwen Williams, a cancer survivor, has actively advocated for Macmillan Cancer Support in the UK. Her advocacy focuses on improving patient experience and access to care. Through her efforts, Williams has contributed to policy changes that enhance support services for cancer patients and has played a key role in community education initiatives (Macmillan Cancer Support, 2023).

Patient advocates have a crucial impact on cancer care and research by raising awareness through sharing personal stories and engaging actively in public forums to highlight the significance of early detection, treatment options, and the challenges cancer patients face. They also work with policymakers to shape health policies that enhance access to care, increase funding for research, and improve patient support services. Furthermore, patient advocates play a key role in driving research by participating in clinical trials and research initiatives, which help advance scientific understanding and develop new treatments. Additionally, they often establish support networks for other patients and families, offering guidance, encouragement, and practical assistance (National Coalition for Cancer Survivorship, 2022).

BUILDING A CULTURE OF SOLIDARITY AND SUPPORT

A strong support system is essential for cancer patients and their families. There needs to be mutual, heartfelt unity, and common interest-driven actions, such as patient-centered care. The supportive structure can include closely related individuals or very important people to the family or patient to provide emotional support (Ruland et al., 2013). In addition, a clinical support system can be built and adopted to empower patients with coping mechanisms and to promote necessary lifestyle changes advocated by healthcare professionals (Pawloski et al., 2019). The overall objective is to address potential challenges such as health and well-being and to reduce incidents of emotional stress and anxiety through increased participation in the home setting, supported by both family and healthcare systems (Given and Kozachik, 2001). This can be achieved by developing feasible strategies to enhance community solidarity on public health issues related to cancer. Individuals dealing with cancer often rely on others to cope with the challenges brought by illness, presenting an opportunity for improvement and positive influence (Mehnert and Koch, 2005).

It's important to establish a support structure rather than having individuals handle everything on their own (Stenberg, Ruland, and Miaskowski, 2010). This can have a positive impact on mental health and physical wellness. Encouraging joint activities such as patient focus teams, spending quality time with family, trying new activities, and seeking help when needed can help improve social support (Taylor et al., 1986). These strategies effectively create support networks for patients and their families, as illustrated in **Figure 9.2**. The resulting social support has several benefits, such as avoiding stigma, self-isolation, and vulnerability. Regular check-ins from members of a well-established support network can help maintain healthy habits and strike a balance between distractions and an ideal way of life. Additionally, professional therapy, possibly through healthcare professional referrals or directly provided by a clinical psychologist or counselor, can be a preferred tool (Koopman et al., 1998).

IMPROVING OUTCOMES AND QUALITY OF LIFE

Cancer, similar to any life-threatening disease, requires careful monitoring and the enhancement of treatment outcomes as illustrated in **Figure 9.3**. To improve the positive impact of clinical or medical interventions, it is essential to also focus on the quality of life. Quality of life encompasses a comprehensive evaluation of a patient's health, abilities affected by the disease, ease of participating in normal, comfortable, or enjoyable activities, and experiences within their living conditions (Felce and Perry, 1995). To effectively advocate and provide community support to improve patient outcomes, it is crucial to consider not only popular psychological phenomena or perceived satisfaction but also to visualize the entire concept from a multidimensional perspective (Testa and Simonson, 1996).

Enhancing the quality of life can renew hope and faith in a patient's life, allowing for better acceptance of other supportive measures aimed at helping patients cope. Both cancer and disease impact the quality of life for patients, and integrating supportive care early on can lead to improved outcomes (Moffat, Epstein, and O'Reilly, 2019). Various methodologies and approaches can be implemented to improve patient outcomes and perceived quality of life.

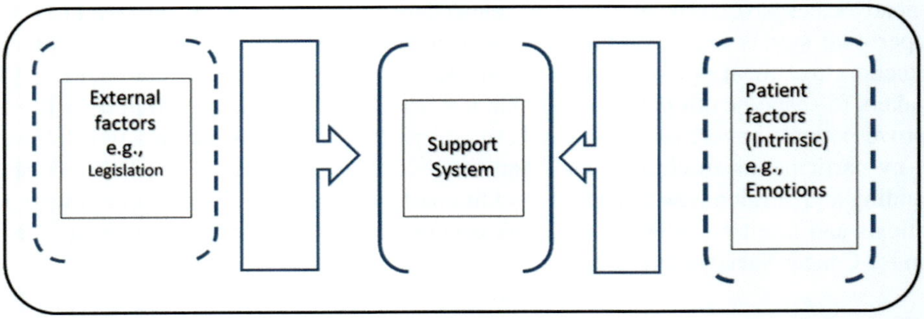

FIGURE 9.2 Factors affecting an ideal support system. The diagram illustrates the factors that influence an ideal support system. The intrinsic factors are depicted on the right, while the external factors are shown on the left. It also displays the interconnected nature of establishing an effective support system. Examples of support systems include patient navigators, oncology social workers, financial advisors, support groups, and palliative care services. Interactions between factors, such as financial stability improving access to treatment or strong family support bolstering mental resilience, highlight their interconnected nature in optimizing care. The ideal support system for cancer patients requires the active involvement and collaboration of healthcare providers, patients, and communities. This support system is a complex interplay of various factors, including individual perceptions, behaviors, and interactions within the community. Increasing community awareness about the clinical and psychosocial aspects of cancer can contribute to a more supportive and compassionate environment for cancer patients (Gustafson, 2001).

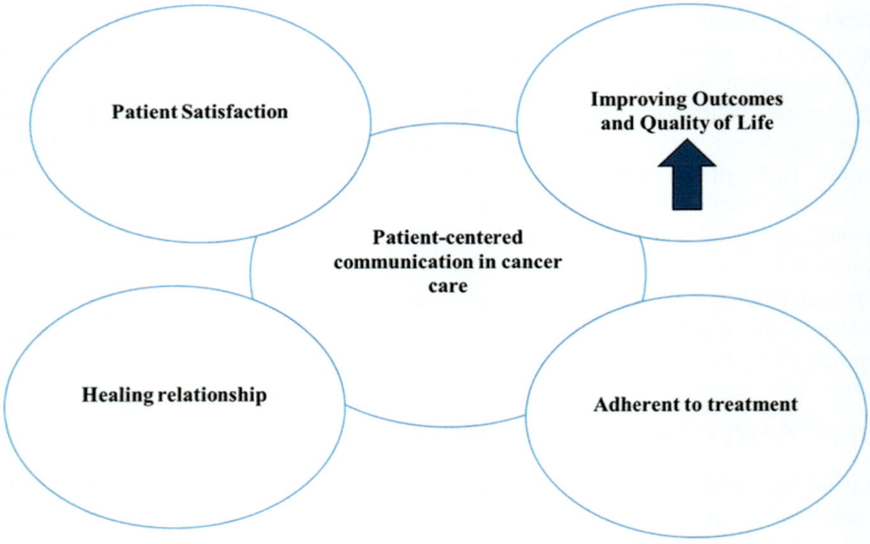

FIGURE 9.3 Outcomes of patient-centered communication in cancer care: Communication that is centered around the patient considers the patient's needs and values, actively involves the patient in care-related decisions, and fosters more robust patient-provider relationships. This communication approach has been associated with heightened levels of patient satisfaction, adherence to treatment, and overall quality of life, indicating its substantial potential to greatly benefit patients. Both treatment outcomes and quality of life have an interactive relationship with one another, such that emotional and psychosocial well-being, and material factors play a stake. The disease itself and medical treatments can put a burden on the patient, which is worsened by these highlighted elaborate external factors (Theofilou, 2013).

However, adopting a patient-specific and individualized approach may be more effective, as person-centered support and palliative care can significantly enhance the quality of life (Scott and Jewell, 2019). The incidence and mortality rates for pancreatic cancer are increasing unlike other cancers, highlighting the critical importance of approaches aimed at achieving the best quality of life and care (Gooden et al., 2016). Learning from the success of support systems and advocacy in other stigmatized diseases, there is practical evidence supporting the refinement of cancer approaches. To achieve this on a large scale, various techniques discussed earlier need to be employed, paving the way for long-term advocacy, community sensitization, and engagement to address all challenges under one umbrella (Nipp et al., 2022). A thorough examination of the current status quo regarding patient advocacy and long-term objectives within African communities, followed by the development of implementable approaches aimed at addressing both financial and psychosocial factors influencing the quality of life, represents the path to the future. These efforts can be complemented by patient-driven strategies to ensure the smooth attainment of goals and sustainability (Saif, 2011; Nausheen et al., 2009).

Limitations

The administration of pancreatic cancer treatment in Africa faces significant challenges due to systemic, infrastructural, and cultural issues that remain unaddressed. One major problem is the poor healthcare infrastructure, which results in insufficient resources for early diagnosis and specialized treatment centers. As a consequence, most patients are diagnosed at an advanced stage, making treatment more complex and costly. Additionally, the prevalence of case underreporting hampers a comprehensive understanding of the disease's impact on the population, creating gaps in policy development. Financial constraints pose the greatest challenges, as the costs of diagnosis and treatment are prohibitive for many patients, leading to family impoverishment. Furthermore, cultural perceptions and a lack of awareness about pancreatic cancer often cause patients to delay seeking help, adding additional strain to both their psychological well-being and that of their families.

Although awareness campaigns and community activities are critical in creating understanding and gathering resources, they face many barriers, including lack of funds, poor training of patient advocates, and poor integration of patient advocates into available healthcare services. Other factors that hinder the provision of adequate education among patients and their participation also include the sociocultural conditions in the region which are critical in building confidence in caregivers. Furthermore, the rising incidence of pancreatic cancer, driven by lifestyle changes such as increased rates of obesity, smoking, and diabetes, underscores the urgent need for preventive measures and public health interventions.

Despite ongoing efforts to empower patients and combat stigma, the availability and sustainability of comprehensive support services remain inadequate. These challenges underscore the critical need for systemic reforms, increased investment in healthcare infrastructure, enhanced community-based interventions, and policies that prioritize equitable access to care and support for pancreatic cancer patients in Africa.

CONCLUSION

Pancreatic cancer, characterized by its aggressive nature and high fatality rates, is a leading cause of cancer-related deaths globally, placing a significant burden on patients and their families. The rising incidence of pancreatic cancer in Africa can be attributed to increasing life expectancy and lifestyle changes contributing to risk factors such as obesity, smoking, and diabetes. However, the prevalence of underreported cases and late-stage diagnoses in Africa is compounded by limited healthcare infrastructure and diagnostic facilities.

This chapter provides an in-depth exploration of the multifaceted experiences of pancreatic cancer patients in Africa, underscoring the significance of their perspectives, the role of advocacy, and community engagement. A comprehensive understanding of the physical, emotional, and financial challenges endured by patients is essential for the development of more effective and compassionate care strategies. Advocacy organizations and support groups play a pivotal role in raising awareness, reducing stigma, and empowering patients through resource mobilization, policy advocacy, and improved access to care. Community engagement contributes to the creation of a supportive environment, ultimately enhancing the overall quality of life for patients and their families.

The profound impact of pancreatic cancer on patients and their families is evidenced by its repercussions on physical health, emotional well-being, and financial stability. Noteworthy physical burdens resulting from symptoms and treatment side effects are compounded by profound emotional distress stemming from diagnosis and prognosis. Additionally, the high costs of treatment pose financial challenges, further jeopardizing the well-being of patients and their families. Personal narratives from pancreatic cancer patients in Africa underscore common themes of resilience, hope, and challenges, influenced significantly by cultural factors.

Crucial to addressing these challenges are advocacy efforts aimed at fostering patient involvement, providing comprehensive training and resources, and showcasing success stories of patient advocates affecting change. Patient advocates wield influence in raising awareness, driving policy changes, advancing research, and fostering peer support. This chapter underscores the pivotal importance of patient perspectives, advocacy, and community engagement, emphasizing the imperative for sustained support aimed at enhancing outcomes for pancreatic cancer patients in Africa. Envisioning the future of pancreatic cancer care and support in Africa necessitates a focus on improved early detection, enhanced treatment access, and the provision of comprehensive support services tailored to address the unique challenges faced by patients. By integrating patient voices across the spectrum of cancer care and research, we can engender more responsive, effective, and compassionate healthcare systems crafted to truly meet the distinct needs of patients.

REFERENCES

ABDILLEH, K., et al. 2024. Pancreatic cancer action network's SPARK: A cloud-based patient health data and analytics platform for pancreatic cancer. *JCO Clinical Cancer Informatics*, 8, e2300119.

ABUBAKAR, I., et al. 2022. The Lancet Nigeria Commission: investing in health and the future of the nation. *Lancet*, 399(10330), 1155–1200.

ABUELGASIM, K. A., & HAQUE, M. 2018. Patient-reported outcomes in cancer: What do they mean? *Value in Health Regional Issues*, 15, 82–86.

ADELOYE, D., ADEREMI, A. V., & ADEWOLE, I. F. 2016. Cancer deaths in Africa: A systematic review. *African Journal of Cancer*, 8(1), 5–16.

ADELOYE, D., DAVID, R. A., ADEREMI, A. V., ISEOLORUNKANMI, A., OYEDOKUN, A., IWEALA, E. E., & ADEOSUN, A. A. 2019. Health workforce and governance: The crisis in Nigeria. *Human Resources for Health*, 15(1), 32.

AMERICAN CANCER SOCIETY. 2022. *Innovator and advocate: The legacy of John Kanzius*. Available at: https://www.cancer.org

ANDERSON, T., MITCHELL, G., PRUE, G., MCLAUGHLIN, S., & GRAHAM-WISENER, L. 2025. The psychosocial impact of pancreatic cancer on caregivers: a scoping review. BMC Cancer. March 20; 25(1), 511. doi: 10.1186/s12885-025-13891-w. PMID: 40114110; PMCID: PMC11924831.

ANDERSON, M.D., PICKNER, W.J., & BEGNAUD A. 2023. Determinants of lung cancer screening in a Minnesota urban indigenous community: A community-based, participatory, action-oriented study. *Cancer Prevention Research*, 16, 239–245. https://doi.org/ 10.1158/1940-6207.CAPR-22-0314.

ANO, G. G., & PARGAMENT, K. I. 2012. Predictors of religious coping and spiritual struggle in college students: Does family conflict matter? *International Journal for the Psychology of Religion*, 22(3), 232–248.

BACZEWSKA, B., et al. 2020. The effect of hope on coping in patients with advanced cancer. *Psycho-Oncology*, 29(4), 814–821.

BAILEY, C., WEST, R., & CUMMING, G. 2019. The psychosocial impact of pancreatic cancer: A systematic review. *Pancreatology*, 19(4), 622–634.

BENHAM-HUTCHINS, M., STAGGERS, N., MACKERT, M., JOHNSON, A.H., & DEBRONKART, D. 2017. "I want to know everything": A qualitative study of perspectives from patients with chronic diseases on sharing health information during hospitalization. *BMC Health Services Research*, 17, 529. https://doi.org/ 10.1186/s12913-017-2487-6.

BLANK, T. O., & BELLIZZI, K. M. 2006. After prostate cancer: Predictors of well-being among long-term prostate cancer survivors. *Cancer*, 106(10), 2128–2135.

BREAST HEALTH FOUNDATION. 2023. *Bosom Buddies Program*. Retrieved from https://www.mybreast.org.za/bosom-buddies/.

BULTZ, B. D., & CARLSON, L. E. 2005. Emotional distress: The sixth vital sign in cancer care. *Journal of Clinical Oncology*, 23(26), 6440–6441.

CALVILLO-ORTIZ, R., POLANCO-SANTANA, J.C., CASTILLO-ANGELES, M., ALLAR, B.G., ANGUIANO-LANDA, L., GHAFFARPASAND, E., BARROWS, C., CALLERY, M.P., KENT, T.S. 2022. Language proficiency and survival in pancreatic cancer: A propensity score-matched analysis. *Journal of Gastrointestinal Surgery*, 26, 94–103.

CANCER ASSOCIATION OF SOUTH AFRICA (CANSA), 2021a. Integrated Report 2020/2021. [online] Available at: https://cansa.org.za/files/2022/04/CANSA-Integrated-Report-2020-to-2021.pdf [Accessed 9 May 2025].

CANCER ASSOCIATION OF SOUTH AFRICA (CANSA). 2021b. *Mobile health clinics*. Available at: https://www.cansa.org.za

CANCER ASSOCIATION OF SOUTH AFRICA (CANSA). 2022. *Support resources for cancer patients*. Available at: https://cansa.org.za/how-cansa-helps-patients/ [Accessed March 9, 2025].

CANCER ASSOCIATION OF SOUTH AFRICA. 2023a. *Policy advocacy*. Available at: https://www.cansa.org.za/policy-advocacy/

CANCER ASSOCIATION OF SOUTH AFRICA. 2023b. *Women's health campaign*. Available at:https://www.cansa.org.za/womens-health/

CANCER ASSOCIATION OF SOUTH AFRICA. 2023c. *CANSA advocacy and cancer control efforts*. Available at: https://cansa.org.za/cansas-advocacy-and-cancer-control-efforts/ [Accessed February 9, 2025].

CANCER RESEARCH UK. 2023. *Fundraising and support*. Retrieved from https://www.cancerresearchuk.org

CANCER SUPPORT COMMUNITY. 2022. *Patient advocacy and the role of survivors*. Available at: https://www.cancersupportcommunity.org

CANSA - The Cancer Association of South Africa. 2024. *Integrated reports and financial statements - CANSA - The Cancer Association of South Africa*. Available at: https://cansa.org.za/integrated-reports-and-financial-statements [Accessed August 5, 2024].

CHEN, M. L. 2003. Pain and hope in patients with cancer. *Cancer Nursing*, 26(1), 61–67.

COLLETT, L.K., HUDSON, L., PRICHARD, C., VANDERFORD, N.L. 2023. Using culturally focused storytelling to empower Appalachian Kentucky youth to understand and address cancer disparities in their communities. *Journal of Cancer Education*, 238, 513–521. https://doi.org/ 10.1007/s13187-022-02147-x.

DAHER, M. 2012. Cultural beliefs and values in cancer patients. *Annals of Oncology*, *23*(suppl. 3), 66–69. https://doi.org/10.1093/annonc/mds091.

DAMEN, A., et al. 2021. The relationship between spiritual struggle and symptom burden in cancer patients. *Psycho-Oncology*, 30(4), 576–584.

DAVIS, J., SMITH, A., AND JOHNSON, R. 2023. Community outreach and awareness campaigns: A case study of PanCAN's initiatives. [online] Available at: https://pancan.org/community-outreach-awareness-campaigns/ [Accessed 9 May 2025].

DE WIT, M., COOPER, C., REGINSTER, J. Y., et al. 2017. Practical guidance for patient-centered health research. *The Lancet*, 390(10107), 1754–1764.

DUGGLEBY, W., et al. 2013. Hope interventions for caregivers of persons with chronic illness: A meta-analysis. *Journal of Advanced Nursing*, 69(9), 1784–1796.

DVALADZE, A., KIZUB, D.A., CABANES, A., NAKIGUDDE, G., AGUILAR, B., ZUJEWSKI, J.A., DUGGAN, C., ANDERSON, B.O., PRITAM SINGH, R.K. AND GRALOW, J.R. 2020, Breast cancer patient advocacy: A qualitative study of the challenges and opportunities for civil society organizations in low-income and middle-income countries. *Cancer*, 126, 2439–2447. https://doi.org/10.1002/cncr.32852

ELIACIN, J., BURGESS, D., ROLLINS, A.L., PATTERSON, S., DAMUSH, T., BAIR, M.J., SALYERS, M.P., SPOONT, M., CHINMAN, M., SLAVEN, J.E., et al. 2023. Outcomes of a peer-led navigation program, PARTNER-MH, for racially minoritized Veterans receiving mental health services: A pilot randomized controlled trial to assess feasibility and acceptability. *Translational Behavioral Medicine,* 13, 710–721. https://doi.org/ 10.1093/tbm/ibad027.

EPSTEIN, R.M. AND STREET, R.L. 2007. Patient-centered communication in cancer care: Promoting healing and reducing suffering. *PsycEXTRA Dataset*. National Cancer Institute, NIH Publication No. 07-6225. Bethesda, MD, 2007.http://outcomes.cancer.gov/areas/pcc/communication.

FAKHRDDINE AMRI, HAJAR KOULALI, RACHID JABI, ABDELKRIM ZAZOUR, BOUZIANE, M., ISMAILI, Z. AND GHIZLANE KHARRASSE 2023. Pancreatic cancer: experience from an emerging country in North Africa. *Journal of Cancer Research and Clinical Oncology*, 149(15), 14297–14302. https://doi.org/10.1007/s00432-023-05245-5.

FELCE, D. AND PERRY, J., 1995. Quality of life: Its definition and measurement. *Research in Developmental Disabilities*, 16(1), 51–74.

FERLAY, J., COLOMBET, M., SOERJOMATARAM, I., MATHERS, C., PARKIN, D. M., PIÑEROS, M., ZNAOR, A., & BRAY, F. 2019. Estimating the global cancer incidence and mortality in 2018: GLOBOCAN sources and methods. *International Journal of Cancer*, 144(8), 1941–1953.

FERLAY, J., COLOMBET, M., SOERJOMATARAM, I., PARKIN, D.M., PIÑEROS, M., ZNAOR, A., AND BRAY, F., 2021. Cancer statistics for the year 2020: An overview. *International Journal of Cancer*, 149(4), 778–789.

FERLAY, J., PARTENSKY, C., AND BRAY, F. 2016. More deaths from pancreatic cancer than breast cancer in the EU by 2017. *Acta Oncologica,* 55(9–10), 1158–1160.

FERLAY, J., WILD, C.P., AND BRAY, F. 2017. The burden of cancer worldwide: Current and future perspectives. In *Holland-Frei Cancer Medicine* (eds R.C. Bast, W.N. Hait, D.W. Kufe, R.R. Weichselbaum, J.F. Holland, C.M. Croce, M. Piccart-Gebart, H. Wang, W.K. Hong and R.E. Pollock). https://doi.org/10.1002/9781119000822.hfcm034

FITZMAURICE, C., ABATE, D., ABBASI, N., et al. 2019. Global, regional, and national cancer incidence, mortality, years of life lost, years lived with disability, and disability-adjusted life-years for 29 cancer groups. *JAMA Oncology*, 5(12), 1749–1768.

GEORGE WASHINGTON UNIVERSITY. 2024. *Cancer prevention month & world cancer day awareness campaign*. School of Medicine and Health Sciences. Available at: https://cancercontroltap.smhs.gwu.edu/news/cancer-prevention-month-world-cancer-day-awareness-campaign [Accessed August 5, 2024].

GIVEN, B.A., GIVEN, C.W. AND KOZACHIK, S. 2001. Family support in advanced cancer. *CA: A Cancer Journal for Clinicians*, 51(4), 213–231.

GLOBOCAN. 2020. Cancer today. *International Agency for Research on Cancer*. https://www.uicc.org/news/globocan-2020-global-cancer-data. Accessed 12 July 2024.

GOODEN, H., TILLER, K., MUMFORD, J. AND WHITE, K. 2016, March. Integrated psychosocial and supportive care needed for patients with pancreatic cancer. *Cancer Forum,* 40(1), 66–69.

GORDON-DSEAGU, V. L., DEVESA, S. S., GOGGINS, M., et al. 2018. Pancreatic cancer incidence trends: Evidence from the Surveillance, Epidemiology and End Results (SEER) Population-Based Data. *Cancer Epidemiology, Biomarkers & Prevention*, 27(9), 1029–1039.

GOYAL, G., HE, H., MARRUFO, A. S., et al. 2019. The burden of pancreatic cancer and associated costs in the United States. *Pancreas*, 48(6), 838–844.

GUIDELINES IN PRACTICE. N.D. Pancreatic cancer: GPs can help prognosis by identifying early signs. Available at: https://www.guidelinesinpractice.co.uk/cancer/pancreatic-cancer-gps-can-help-prognosis-by-identifying-early-signs/352855.article [Accessed July 2024].

GUSTAFSON, D.H., HAWKINS, R., PINGREE, S., MCTAVISH, F., ARORA, N.K., MENDENHALL, J., CELLA, D.F., SERLIN, R.C., APANTAKU, F.M., STEWART, J. AND SALNER, A., 2001. Effect of computer support on younger women with breast cancer. *Journal of General Internal Medicine*, 16(7), 435–445.

HEERKENS, H. D., VAN DER HEIJDEN, E. J., et al. 2020. Patient and family perspectives on the burden of pancreatic cancer: A systematic review. *BMC Palliative Care*, 19(1), 1–9.

HIBBARD, J. H., & GREENE, J. 2013. What the evidence shows about patient activation: Better health outcomes and care experiences; fewer data on costs. *Health Affairs*, 32(2), 207–214.

HIRSHBERG FOUNDATION FOR PANCREATIC CANCER RESEARCH. 2022. Pancreatic cancer patient support groups. Hirshberg Foundation for Pancreatic Cancer Research. Available at: https://pancreatic.org/pancreatic-cancer-patient-support-groups/.

HOFMANN, J. N., KEIMLING, M., & KUEHNI, C. E. 2020. Cancer policy, advocacy, and patient perspectives in Europe. European Journal of Cancer, 128, 1–6.

HU, J.-X., ZHAO, C.-F., CHEN, W.-B., LIU, Q.-C., LI, Q.-W., LIN, Y.-Y., GAO,- F. 2021. Pancreatic cancer: A review of epidemiology, trend, and risk factors. *World Journal of Gastroenterology*, 27(27), 4298–4321. https://doi.org/10.3748/wjg.v27.i27.4298.

INSTITUTE OF MEDICINE. 2007. *Cancer Control Opportunities in Low- and Middle-Income Countries.* Washington, DC: The National Academies Press.

JONES, R.P., PSARELLI, E., JACKSON, R., et al. 2019. Patterns of recurrence after resection of pancreatic ductal adenocarcinoma: A secondary analysis of the ESPAC-4 randomized adjuvant chemotherapy trial. *JAMA Surgery*, 154(11), 1038–1048.

JONES, P., et al. 2023. Innovative therapeutic approaches in pancreatic cancer treatment. *Oncogene*, 42(5), 789–798.

KAISER, J., et al. 2021. The role of community support in cancer care. *Journal of Oncology Practice*, 17(1), 32–40.

KENYA CANCER ASSOCIATION (KENCANSA). 2017. *National Cancer Control Strategy 2017-2022.* KENCANSA. Retrieved from https://rocketreach.co/kenya-cancer-association-kencansa-profile_b40fcfe8ffc98dc0

KENYA NETWORK OF CANCER ORGANIZATIONS (KENCO), 2023. National Cancer Awareness Week 2023: Advocacy and Outreach Report. [online] Available at: https://kenconetwork.org/national-cancer-awareness-week-2023 [Accessed 19 Mayrch 2025].

KIM, H., YOO, T., & LEE, J. H. 2024. Economic burden of pancreatic cancer treatment: A systematic review. *Journal of Oncology Finance*, 15(3), 245–256.

KOOPMAN, C., HERMANSON, K., DIAMOND, S., ANGELL, K. AND SPIEGEL, D., 1998. Social support, life stress, pain and emotional adjustment to advanced breast cancer. *Psycho-Oncology: Journal of the Psychological, Social and Behavioral Dimensions of Cancer*, 7(2), 101–111.

KORNBLAU, S. M., APPELBAUM, F. R., & AL-KHABORI, M. 2022. Management of physical symptoms in pancreatic cancer: Challenges and strategies. *Pancreatic Cancer Journal*, 19(2), 112–123.

LANG-ROLLIN, I., & BERBERICH, G. 2018. Psychosocial impact of cancer. *Der Onkologe*, 24(1), 65–69.

LEE, A. Y., et al. 2023. Radiation therapy in pancreatic cancer: Current strategies and future directions. *Radiation Oncology Journal*, 41(1), 45–57.

LEE, C., HSU, C., & CHIU, J. 2023. Side effects of pancreatic cancer treatments: An overview. *Clinical Oncology Review*, 29(1), 78–85.

MACMILLAN CANCER SUPPORT. 2023. Olwen Williams: Patient advocate. Available at: https://www.macmillan.org.uk

MAJERNIKOVA, I., & OBROCNIKOVA, A. 2017. Finding meaning in life as an important factor in coping with cancer. *Journal of Health Psychology*, 22(7), 911–920.

MAZZELLA EBSTEIN, A. M., JOSEPH, S. J., AND HERNANDEZ, M. 2020. Psychological stress and pancreatic cancer patients: A qualitative systematic review protocol. *JBI Evidence Synthesis*, 18(3), 576–582.

MEHNERT, A. AND KOCH, U. 2005. Psychosocial care of cancer patients—International differences in definition, healthcare structures, and therapeutic approaches. *Supportive Care in Cancer*, 13(8), 579–588.

MILLER, R., et al. 2023. Policy and legislative initiatives in pancreatic cancer research funding. *Health Policy Journal*, 27(3), 190–202.

MILTON A.J., FLORES E.J., CHARLES E.F., ELEZABY M.A., WARD E.C., LEE C.I., WOODS R.W., MARTIN ROTHER M.D., STRIGEL R.M., NARAYAN A.K. 2023. Community-based participatory research: A practical guide for radiologists. *Radiographics*, 43, e220145. https://doi.org/10.1148/rg.220145.

MINICOZZI P, CASSETTI T, VENER C, SANT M. 2018. Analysis of incidence, mortality and survival for pancreatic and biliary tract cancers across Europe, with assessment of influence of revised European age standardisation on estimates. *Cancer Epidemiology,* 55, 52–60.

MOFFAT, G.T., EPSTEIN, A.S. AND O'REILLY, E.M. 2019. Pancreatic cancer—A disease in need: Optimizing and integrating supportive care. *Cancer*, 125(22), 3927–3935

MUKWATO, P. et al. 2020. Reducing cancer stigma in African communities. *African Health Sciences,* 20(3), 1201–1210.

MWAKA, A. D., OKELLO, E. S., WABINGA, H., et al. 2015. Symptomatic presentation with cervical cancer in Uganda: A qualitative study assessing the pathways to diagnosis in a low-income country. *BMC Women's Health*, 15(1), 15.

NATIONAL CANCER INSTITUTE. 2021. *Improving access to cancer care.* Available at: https://www.cancer.gov.

NATIONAL COALITION FOR CANCER SURVIVORSHIP. 2022. *The role of patient advocates.* Available at: https://www.canceradvocacy.org

NAUSHEEN, B., GIDRON, Y., PEVELER, R. AND MOSS-MORRIS, R., 2009. Social support and cancer progression: A systematic review. *Journal of Psychosomatic Research*, 67(5), 403–415.

NAZAM, F., & HUSAIN, A. 2024. Hope, meaning in life and religious/spiritual struggle predicts the quality of life in cancer patients with early and advanced malignancy: A structural equation modeling approach. *Health Psychology and Behavioral Medicine*, 12(1), 2351936. https://doi.org/10.1080/21642850.2024.2351936.
NEUGUT, A. I., PRIGERSON, H. G., & SIMA, C. S. 2023. Financial toxicity of cancer care: A review of the impact on patients and families. *Cancer Economics & Outcomes*, 8(2), 34–45.
NHLEKO, M. L., EDOKA, I, AND MUSENGE, E. 2024. Pancreatic cancer mortality in South Africa: A case-control study. *South African Medical Journal*, 114(1), 27–32.
NIPP, R.D., GAUFBERG, E., VYAS, C., AZOBA, C., QIAN, C.L., JAGGERS, J., WEEKES, C.D., ALLEN, J.N., ROELAND, E.J., PARIKH, A.R. AND MILLER, L. 2022. Supportive oncology care at home intervention for patients with pancreatic cancer. *JCO Oncology Practice*, 18(10), e1587–e1593.
ODHIAMBO, J. A., et al. 2023. Resilience in Kenyan pancreatic cancer patients: A qualitative study. *African Health Sciences*, 23(1), 102–110.
OMOTOSO, O., TEIBO, J. O., ATIBA, F. A., OLADIMEJI, T., PAIMO, O. K., ATAYA, F. S., BATIHA, G. E., & ALEXIOU, A. 2023. Addressing cancer care inequities in sub-Saharan Africa: Current challenges and proposed solutions. *International Journal for Equity in Health*, *22*(1), 189.
OYSTACHER, T., BLASCO, D., HE, E., HUANG, D., SCHEAR, R., MCGOLDRICK, D., LINK, B., AND YANG, L.H. 2018. Understanding stigma as a barrier to accessing cancer treatment in South Africa: implications for public health campaigns. *Pan African Medical Journal*, 29:73.
PANCREATIC CANCER ACTION NETWORK. 2023a. *Know your tumor program*. Available at: https://www.pancan.org/know-your-tumor/
PANCREATIC CANCER ACTION NETWORK. 2023b. *Leadership and advocacy*. Available at: https://www.pancan.org
PANCREATIC CANCER ACTION NETWORK, 2023c. *Know Your Tumor® precision medicine service*. [online] Available at: https://pancan.org/facing-pancreatic-cancer/patient-services/know-your-tumor/ [Accessed 9 May 2025].
PANCREATIC CANCER ACTION NETWORK. 2024. *Pancreatic Cancer Action Network – Research, patient support, resources*. Pancreatic Cancer Action Network. Available at: https://pancan.org.
PATIENT ADVOCATE FOUNDATION. 2023. *Training and resources for patient advocates*. Available at: https://www.patientadvocate.org
PAWLOSKI, P.A., BROOKS, G.A., NIELSEN, M.E. AND OLSON-BULLIS, B.A. 2019. A systematic review of clinical decision support systems for clinical oncology practice. *Journal of the National Comprehensive Cancer Network*, 17(4), 331–338.
PINKDRIVE. 2019. *Annual report: Impact of mobile mammography and education units*. Available at: https://pinkdrive.org/annual-report-2019 [Accessed May 9, 2025].
PROJECT PURPLE. 2022. *About us*. Available at:https://www.projectpurple.org
RAMASUBBU, S. K., et al. 2021. Sociodemographic factors and quality of life in Indian cancer patients. *Indian Journal of Cancer*, 58(2), 267–273.
RATNAPRADIPA K.L., RANTA J., NAPIT K., LUMA L.B., ROBINSON T., DINKEL D., SCHABLOSKE L., WATANABE-GALLOWAY, S. 2022. Qualitative analysis of cancer care experiences among rural cancer survivors and caregivers. *The Journal of Rural Health*, 38, 876–885. https://doi.org/10.1111/jrh.12665.
RAWLA, P., SUNKARA, T., & GADUPUTI, V. 2019. Epidemiology of pancreatic cancer: Global trends, etiology and risk factors. *World Journal of Oncology*, 10(1), 10–27.
RULAND, C.M., ANDERSEN, T., JENESON, A., MOORE, S., GRIMSBØ, G.H., BØRØSUND, E., & ELLISON, M. 2013. Effects of an internet support system to assist cancer patients in reducing symptom distress: a randomized controlled trial. *Cancer Nursing,* 36(1), 6–17. https://doi.org/10.1097/NCC.0b013e31824d90d4
RYAN, C., et al. 2022. Early detection of pancreatic cancer: Challenges and opportunities. *Cancer Research*, 82(11), 2234–2242.
SAIF, M.W. 2011. Palliative care of pancreatic cancer. *JOP. Journal of the Pancreas*, 12(4), 355–357.
SCHULER, M., HORNEMANN, B., & OEHLER, C. 2019. Addressing mental health in patients with cancer: A pathway to comprehensive cancer care. *Frontiers in Oncology*, 9, 44.
SCOTT, E., JEWELL, A. 2019. Supportive care needs of people with pancreatic cancer: A literature review. *Cancer Nursing Practice*. https://doi.org/10.7748/cnp.2019.e1566
SELBY, J. V., BEAL, A. C., & FRANK, L. 2012. The Patient-Centered Outcomes Research Institute (PCORI) national priorities for research and initial research agenda. *JAMA*, 307(15), 1583–1584.

SIEGEL, R. L., MILLER, K. D., & JEMAL, A. 2020. Cancer statistics, 2020. *CA: A Cancer Journal for Clinicians*, 70(1), 7–30.

SLOAN, F. & GELBAND H. 2007. *Cancer control opportunities in low- and middle-income countries.* Washington, DC. https://doi.org/10.17226/11797

SMITH, J., et al. 2023. Advances in molecular diagnostics for pancreatic cancer. *Clinical Cancer Research*, 29(2), 400–410.

STEFAN, D. C. 2015. Cancer care in Africa: An overview of resources. *The Lancet Oncology*, 16(7), e328–e335.

STENBERG, U., RULAND, C.M. AND MIASKOWSKI, C. 2010. Review of the literature on the effects of caring for a patient with cancer. *Psycho-Oncology*, 19(10), 1013–1025.

SYMANSKI E., AN HAN H., MCCURDY S., HOPKINS L., FLORES J., HAN I., SMITH M.A., CALDWELL J., FONTENOT C., WYATT B., et al. 2023. Data to action: Community-based participatory research to address concerns about metal air pollution in overburdened neighborhoods near metal recycling facilities in Houston. *Environmental Health Perspectives*, 131, 67006. https://doi.org/10.1289/EHP11405.

TAYLOR, S. E., FALKE, R. L., SHOPTAW, S. J., & LICHTMAN, R. R. 1986. Social support, support groups, and the cancer patient. *Journal of Consulting and Clinical Psychology*, 54(5), 608–615. https://doi.org/10.1037/0022-006X.54.5.608

TEMPERO, M. A., MALAFA, M. P., AL-HAWARY, M., et al. 2020. Pancreatic Adenocarcinoma, Version 2.2020, NCCN clinical practice guidelines in oncology. *Journal of the National Comprehensive Cancer Network*, 18(1), 13–23.

TESTA, M.A. AND SIMONSON, D.C. 1996. Assessment of quality-of-life outcomes. *New England Journal of Medicine*, 334(13), 835–840.

THEOFILOU, P. 2013. Quality of life: definition and measurement. *Europe's Journal of Psychology*, 9(1): 150–152.

TUCKER, C. M., MARSISKE, M., RICE, K. G., NIELSON, J. J., & HERMAN, K. 2011. Patient-centered culturally sensitive health care: model testing and refinement. *Health Psychology*, 30(3), 342–350. https://doi.org/10.1037/a002296

UGANDA WOMEN'S CANCER SUPPORT ORGANISATION (UWOCASO). 2020. *Annual report 2020.* https://ugandacancersociety.org/wp-content/uploads/2023/09/annualreport2020.pdf

Uganda Women's Cancer Support Organisation (UWOCASO), 2023a. News. [online] Available at: https://uwocaso.org.ug/news/ [Accessed 9 May 2025].

UGANDA WOMEN'S CANCER SUPPORT ORGANISATION (UWOCASO). 2023b. *Mentally cancer free and thriving: Baseline needs assessment report.* Available at: https://uwocaso.org.ug/wp-content/uploads/2024/02/Mentally-cancer-free-Thriving-baseline-needs-assessment-report.pdf [Accessed March 3, 2025].

UICC. 2022. Uganda Women's Cancer Support Organization (UWOCASO). Available at: https://www.uicc.org/membership/uganda-womens-cancer-support-organization-uwocaso [Accessed August 5, 2024].

VITALONI, M., MICHL, P., LÖHR, M., STUNT, A., RZADKOWSKA, P., BACKMAN, E., SHARARA, L., NEVES, V., GRUENBERGER, T., & MARAVIC, Z. 2022. Pancreatic cancer from the patient perspective: The time to act is now. *Journal of Patient Experience*, 9, 2374373522111263

WALL, N. R., FULLER, R. N., MORCOS, A., & DE LEON, M. 2023. Pancreatic cancer health disparity: Pharmacologic anthropology. *Cancers*, 15(20), 5070. https://doi.org/10.3390/cancers15205070.

WALL, S., et al. 2023. Pancreatic cancer awareness and advocacy: Impact and future directions. *Journal of Cancer Education*, 38(4), 123–135.

WARNER Z.C., GILBERT-GARD K., REID B., JOSEPH W., KEPKA D., AUGUSTE P., WARNER E.L. 2023. Knowledge and awareness of colorectal cancer among a predominantly Indigenous Caribbean community. *BMC Public Health*, 23, 243. https://doi.org/10.1186/s12889-022-14810-5.

WHITEFORD, G., & OLVER, I. 2012. Meaning in illness for cancer patients: A review of the literature and a theoretical model. *International Journal of Palliative Nursing*, 18(8), 375–378.

WORLD HEALTH ORGANIZATION. 1993. *WHOQOL study protocol: The development of the World Health Organization Quality of Life Assessment Instrument (WHOQOL).* https://iris.who.int/bitstream/10665/63482/1/who_msa_mnh_psf_97.4.pdf

WORLD HEALTH ORGANIZATION (WHO). 2022. *Cancer prevention and control.* Available at: https://www.who.int

ZHU, Y., YANG, S., & LIU, Y. 2023. Psychological distress in pancreatic cancer patients: A systematic review. *Journal of Psycho-Oncology*, 31(4), 312–321.

10 Pancreatic Cancer Research and Innovation in Africa

Benny Mosoane[1], Ekene Emmanuel Nweke[2], Nnenna Elebo[2], Nolonwabo Moyakhe[1], Lesedi Nevondo[1], Jessica McIntyre[1], Dineo Disenyane[1], Bonolo Mashishi[3], Abidemi Omonisi[4,5], and Zodwa Dlamini[6]*

[1]Department of Anatomical Pathology, National Health Laboratory Services (NHLS), University of Pretoria, Pretoria, South Africa.

[2]Department of Surgery, University of Witwatersrand, Johannesburg, South Africa.

[3]Department of Medical Virology, National Health Laboratory Services (NHLS), University of Pretoria, Pretoria, South Africa.

[4]Department of Anatomic Pathology, Ekiti State University, Ado-Ekiti, Ekiti State, Nigeria.

[5]Department of Laboratory Medicine, University of Liberia, Monrovia.

[6]SAMRC Precision Oncology Research Unit (PORU), DSI/NRF SARChI Chair in Precision Oncology and Cancer Prevention (POCP), Pan African Cancer Research Institute (PACRI), University of Pretoria, Pretoria, South Africa.

*Corresponding author: Benny Mosoane

INTRODUCTION

Among solid cancers, pancreatic cancer has the lowest survival rate in Africa and is still a very deadly disease (Asombang et al., 2017). In Togo, 87% of pancreatic cancer patients die within 12.4 days of hospital admission (Bouglouga et al., 2015). Notably, over the previous 20 years, there has been a persistent increase in the incidence and death rates of pancreatic cancer, contrasting sharply with the trends observed in other common cancer types (Siegel et al., 2023). Pancreatic ductal adenocarcinoma (PDAC) is the histological diagnosis made in more than 90% of cases of pancreatic cancer, representing the most common subtype of this malignancy (Elebo et al., 2023). By 2030, PDAC is expected to rank second in terms of cancer-related mortality due to its rising incidence (Daamen et al., 2023). PDAC has taken center stage along with hepatobiliary cancers, breast cancer, and colorectal cancer as part of the cancer study conducted by the Common Epithelial Research Center (CERC) at Witwatersrand University in South Africa (Council, 2024).

Understanding the risk factors for pancreatic cancer is critical, especially in African settings where the disease typically presents at a younger age. As with many other cancers, early diagnosis is key to improving prognoses in pancreatic cancer patients, allowing for surgery with curative intent at an early stage. Obesity, chronic pancreatitis, excessive alcohol intake, tobacco use, lack of physical exercise, adoption of Western lifestyles, and early onset diabetes mellitus (DM) have been established as risk factors for pancreatic cancer according to African studies (Jonas and Brand, 2023; Nhleko et al., 2024; Pearlstein et al., 2021; Collaborators, 2019; Madela et al., 2024).

 DOI: 10.1201/9781003544401-15

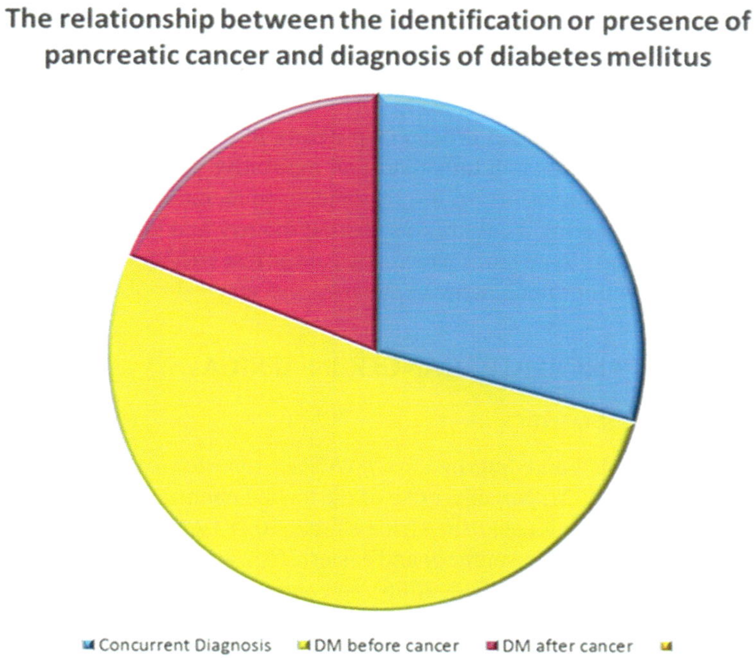

FIGURE 10.1 When was pancreatic cancer discovered and when was diabetes mellitus (DM) diagnosed? Pancreatic cancer was first discovered prior to the diagnosis of DM in 51.7% of the patients, after the pancreatic mass was detected in 19% of the patients, and both conditions were diagnosed concurrently in 29.3% of the patients. (Adapted from Kivuyo and Akoko, 2022.)

A Tanzanian study explored the association between the presence or detection of a pancreatic mass and the diagnosis of DM. The study found that all PDAC patients had experienced adult-onset diabetes. Specifically, DM was diagnosed prior to the identification of the pancreatic mass in 51.7% of patients, after the discovery of the pancreatic mass in 19% of patients, and concurrently in 29.3% of patients (Figure 10.1).

Approximately 80% of pancreatic cancer cases are due to non-inherited Kirsten rat sarcoma viral oncogene homolog (*KRAS*) somatic mutations (Norris et al., 2015; Alatise et al., 2009). Another 10–15% of pancreatic cancers are linked to known hereditary mutations and/or familial tendencies. The syndromic risk factors with the highest risks for PDAC are Peutz–Jeghers syndrome and hereditary pancreatitis, both associated with mutations in the serine/threonine kinase 11 (*STK11*), serine protease 1 (*PRSS1*), serine protease inhibitor Kazal type 1 (*SPINK1*), chymotrypsin C, cystic fibrosis transmembrane conductance regulator (*CFTR*), and cyclin-dependent kinase inhibitor 2A (*CDKN2A*) genes (Jonas and Brand, 2023).

The African population frequently presents with advanced pancreatic cancer associated with poor treatment outcomes, and the prevalence of treatment-resistant cases underscores the urgent need for exploring innovative therapeutic strategies. Identifying new biomarkers could be facilitated by stratifying pancreatic cancer into groups based on genetic, morphological, and clinical characteristics. The application of cutting-edge technologies, such as single-cell RNA sequencing and single-cell omics, could enable a more comprehensive subtyping of PDAC patients. This approach allows for a detailed characterization based on their underlying biology, prognosis, therapeutic targets, and responses to pharmacological treatments.

To optimize costs and expedite therapeutic discovery, drug repurposing has emerged as a strategy in cancer research. Elebo et al. (2024) argue that the future of treatment lies in repurposing drugs to target genetic abnormalities specific to different subtypes of pancreatic cancer. Research into the potential use of herbal medicine for chemoprevention and cancer treatment is being explored in several African countries, including Ethiopia, Tanzania, Egypt, and Uganda (Matata et al., 2018; Helene et al., 2000; Omara et al., 2020).

Research and innovation are crucial drivers of progress in the fight against pancreatic cancer. This chapter explores ongoing research initiatives and innovative approaches to pancreatic cancer management in Africa, including efforts to leverage precision medicine, artificial intelligence, and splicing technologies. By fostering a culture of scientific inquiry and knowledge exchange, Africa can contribute to global efforts to advance pancreatic cancer research and translate discoveries into clinical practice, ultimately improving patient outcomes and survival rates.

EPIDEMIOLOGY OF PANCREATIC CANCER IN AFRICA

Comparison with Global Statistics

The World Cancer Research Fund International (WCRF) reported approximately 510,992 new cases of pancreatic cancer in 2022. Notably, none of the top ten countries with the highest incidence rates of pancreatic cancer were located in Africa (Table 10.1) (WCRF, 2024). The incidence of pancreatic cancer is relatively low in southern and eastern Asia and most of Africa, with fewer than 3.5 cases per 100,000 people (Schottenfeld, 2006). In contrast, North America and Europe exhibit higher incidence rates, ranging from 11.8 to 12.5 cases per 100,000 people (Intitute, 2012, Sellam et al., 2015b).

Regional Variations within Africa

Significant regional differences in pancreatic cancer epidemiological indices, such as incidence, prevalence, and mortality, exist due to the underdiagnosis and misdiagnosis of cases in Africa. In addition, there has been slow progress in updating the African Cancer Registry Network (AFCRN)

TABLE 10.1
Data on Pancreatic Cancer from 2022 Highlights the Ten Countries with the Highest Pancreatic Cancer Mortality Rates and Incidence Rates (to Account for the Strong Correlation between Age and the Likelihood of Dying from Cancer, Age-Standardized Rates are Employed to Facilitate Comparisons among Populations with Varying Age Distributions)

World Ranking	Country	Number of Cases	Age-standardized Rates/100,000
	World	510,992	4.7
1	China	118,672	4.42
2	United States of America	60,127	8.6
3	Japan	47,627	9.8
4	Germany	21,869	8.7
5	Russian Federation	21,842	7.8
6	France	15,895	9.4
7	Italy	15,710	8.0
8	Brazil	14,670	4.7
9	India	13,661	0.96
10	United Kingdom	11,351	6.7

Source: Adapted from (WCRF, 2024).

many times due to poor coordination of data sources, a shortage of trained personnel, absence of timely population census figures, inefficient health services, and geopolitical factors such as poor procurement systems and corruption (Omonisi et al., 2020). These discrepancies frequently lead to an underestimation of the disease burden across the African continent. It is essential to conduct studies in Africa because certain epidemiologic factors may not accurately reflect the actual disease characteristics of the African population. For instance, a study from Zambia revealed that pancreatic cancer manifests at a younger age compared to the African American population in the United States (Asombang et al., 2017). The study indicated no significant statistical difference in sex presentation between the Zambian and African American populations. These findings from Zambia are comparable to those obtained at Tikur Anbessa Specialized Hospital in Addis Ababa, Ethiopia (Aga and Abdurahman, 2023).

A descriptive study was conducted in Sidi Bel Abbes, Algeria (Sellam et al., 2015b). The study reviewed the medical records of 264 patients diagnosed with pancreatic cancer between 1991 and 2013. The average age of the patients was 66.2 ± 8.7 years, suggesting a correlation between pancreatic cancer incidence and advanced age. This average age is approximately ten years younger than the data reported by the US National Cancer Institute, which indicates that pancreatic cancer incidence and mortality rates begin to rise at age 50 (Intitute, 2012). Additionally, Amri et al. (2023) have observed a higher incidence of pancreatic cancer among elderly individuals in Morocco (Amri et al., 2023).

According to Muchiri et al. (2024), the majority of patients in Kenya presenting at Kenyatta National Hospital, a tertiary-level facility, had advanced pancreatic cancer, significantly limiting their treatment options (Muchiri et al., 2024). Similarly, just 3% of Nigerian patients present with cancer that is resectable at the time of diagnosis (Alatise et al., 2010). In North African countries like Morocco, a substantial proportion of patients with advanced cancer present with elevated CA 19-9 levels and hypoalbuminemia (Bazine et al., 2018). When comparing age-specific mortality rates, it was observed that Egypt had higher rates of pancreatic cancer than the United States for individuals under the age of 20, with relative risks (RRs) of 7.7 and 4.2 for the age groups 0–15 and 15–20, respectively (Soliman et al., 2006). In contrast, the United States showed significantly higher rates than Egypt for individuals aged 40 and older, with RRs ranging from 1.8 to 80.5 for the age groups 40–45 to 75+.

Risk Factors Specific to African Populations—Genetic Factors, Environmental, and Multi-factorial Causes

A South African case–control study found that smoking and certain occupations are frequently linked to pancreatic cancer mortality (Nhleko et al., 2024). In this study, which included 23,581 controls (12,171 males and 11,410 females), it was demonstrated that, in addition to smoking, male managers and workers in craft and related trades had a significantly increased risk of pancreatic cancer mortality. Similarly, female managers, professionals, clerical support workers, and elementary occupation workers also exhibited an elevated risk. Research conducted in Europe has indicated that men employed in the building trade, miners, machinery mechanics and fitters, and motor vehicle drivers are significantly more likely to succumb to pancreatic cancer compared to women working as office clerks (Alguacil et al., 2003; Partyka et al., 2023; Alguacil et al., 2000; Luna-Sanchez et al., 2014). There are strong correlations between dose–response trends for ionizing radiation, diesel and petrol engine exhaust, asbestos, synthetic polymer dust exposure, chlorinated hydrocarbon solvents, and histologically confirmed PDAC, demonstrating the significant impact that occupational exposure to these specific agents, or combinations thereof, has on the risk of developing pancreatic cancer (Santibañez et al., 2010).

The genetic and familial components contributing to the propensity for pancreatic cancer may be underreported in Africa. In a study conducted in the South African provinces of the Free State and Northern Cape, only 2 out of 352 patients (0.56%) had a positive family history of pancreatic cancer

(Mthunzi and Noel, 2023). Studies indicate that genetic predisposition may account for up to 10% of pancreatic cancer cases, suggesting that the observed outcomes are lower than initially anticipated (Grover and Syngal, 2010; Brand and Lynch, 2000; Lynch et al., 1996; Klatte et al., 2022). According to a systematic review conducted by Pearlstein et al. (2021), there is a notable discrepancy in the volume of pancreatic cancer publications across Africa, with Egypt and Nigeria contributing the majority of these publications (Pearlstein et al., 2021). Overall, there is a paucity of experts capable of conducting hepato-pancreato-biliary research in most countries in Africa. Further investigation is essential to identify national-level factors that could account for this discrepancy.

INNOVATIVE DIAGNOSTIC TECHNIQUES

Advances in Early Detection Methods

Pancreatic cancer is globally recognized as one of the most aggressive malignancies, characterized by poor survival rates and prognosis (Sellam et al., 2015b). This is largely due to the advanced stage at which the disease is typically diagnosed, a consequence of the lack of specific symptoms and the overall inadequacy of reliable screening tools for early detection. Moreover, non-neoplastic pancreatic lesions often exhibit radiological and clinical similarities to neoplastic lesions, complicating the diagnostic process (Amri et al., 2023). These challenges are further exacerbated by age-related physiological changes, which can obscure the identification of early lesions on imaging. The biomarker CA 19-9 has emerged as the preferred diagnostic tool, offering a specificity of 82% (Khan and Thomson, 2018).

An ideal screening and early detection tool for PDAC would be particularly valuable for high-risk individuals. Such a tool would enable the identification of tumors at an early stage, when they are still amenable to surgical resection, thereby facilitating the selection of appropriate management strategies and potentially improving prognostic outcomes.

Advanced techniques such as metabolomics, genomics, epigenomics, transcriptomics, proteomics, and molecular analyses via liquid biopsy (LB) are employed for biomarker research for early diagnosis. These methodologies can detect circulating tumor cells, tumor DNA, proteins, metabolites, microRNA, and extracellular vesicles in various body fluids. Furthermore, biomarkers can be used to monitor treatment outcomes, guide therapy selection, predict tumor behavior, and differentiate between benign and malignant pancreatic tumors (Daher et al., 2024; Khan and Thomson, 2018). Research has indicated that microRNA serves as a significantly superior biomarker compared to CA 19-9 (Nesteruk et al., 2022). LB has demonstrated a 99% specificity in identifying resectable tumors (Khan and Thomson, 2018). Different types of biomarkers play distinct roles: exosomes provide information on tumor cell proliferation, metastasis, and angiogenesis; transcriptomics can reveal cell lineage; and non-coding RNA can predict tumor progression (Nakamura et al., 2019). The sensitivity and specificity of early cancer diagnosis can be enhanced by integrating biomarkers with artificial intelligence (Klett et al., 2018; Daher et al., 2024).

Given that most cancers in Africa are diagnosed at an advanced stage, there is an urgent need to adopt testing platforms such as LB. As noted in the review by Temilola et al. (2019), LB offers several advantages, including early cancer detection, disease monitoring, and the ability to track disease progression (Temilola et al., 2019). Unlike conventional tissue biopsies, LB mitigates surgical risks, reduces costs, and addresses issues related to limited access and poor patient compliance (Wang et al., 2024; Temilola et al., 2019). LB enables the detection and monitoring of cancer through the analysis of body fluids such as blood, urine, or saliva.

The adoption of LB testing in African countries has been delayed due to limited access and economic challenges. Currently, only three countries on the continent—South Africa, Kenya, and Tunisia—have made LB testing commercially available (Temilola et al., 2019; Temilola et al., 2023). A significant limitation of relying on LB at this stage is that the technology for detection and monitoring has not yet been adequately developed or validated for African populations. Only a limited number of studies focusing on various hematological and solid organ cancers have been conducted in countries such as Senegal (Zehentner et al., 2004), Cameroon (Marchio et al., 2018), The Gambia (Kirk et al., 2005; Szymańska et al., 2004), South Africa (Hanekom et al., 1999; van der Vaart et al., 2009; van der Vaart and Pretorius, 2008), Egypt (Sonbol et al., 2019; Ali et al., 2019; Iyer et al., 2010; Hosny et al., 2008; Zaher et al., 2013; Ezzat et al., 2019), and Tunisia (Hamdi et al., 2016).

Use of Artificial Intelligence and Machine Learning in Diagnosis

Artificial intelligence (AI) in diagnostic medicine has been a growing area of interest since the 1950s. Over the decades, advancements in technology have spurred increased research in the fields of machine learning (Daher et al., 2024). Much of this research in Africa has focused on the application of AI in areas such as early diagnosis, molecular subtyping, treatment response prediction, and risk prediction (Adeoye et al., 2022; Achilonu et al., 2021; Khanyile et al., 2023).

Artificial intelligence utilizes algorithms to process large volumes of data, enabling the recognition of patterns and trends (Manson et al., 2023). There are various approaches to machine learning, including supervised and unsupervised learning. Deep learning, a subset of machine learning, can manage larger and more complex datasets than basic machine learning methods (Huynh et al., 2020). In supervised learning, models can learn new functions and make predictions by using labeled input and output data. Unsupervised learning, on the other hand, can uncover new insights and address questions that may otherwise have been overlooked. AI has the potential to enhance imaging by detecting subtle changes that may be missed by the human eye or misinterpreted by the human mind. Additionally, AI can diagnose pathology by analyzing full slide imaging (Dlamini et al., 2020).

Despite the excitement surrounding these discoveries and the hope they bring to patients with pancreatic cancer, several limitations and ethical dilemmas must be considered. Cybersecurity and the risk of hacking are significant concerns as patient data loss can compromise confidentiality. Additionally, there is an issue of accountability, as AI systems cannot be held responsible for errors in diagnosis. Structural inequalities, lack of governance and regulation are major threats to the successful deployment of AI in Africa (Gwagwa et al., 2020). Bias may also be introduced depending on the data used to train AI, particularly since most AI development occurs in industrialized nations with potentially unbalanced demographic representation. This underscores the importance of engagement in AI development by African countries to ensure diverse perspectives and data incorporation.

In Kenya since 2013, Data Science Africa (DSA), which is a nonprofit organization, has been promoting the wider deployment of affordable and suitable AI solutions in Africa (Arakpogun et al., 2021). DSA has been providing a platform for medical practitioners and researchers across the continent to share information sharing on AI. DSA has organized various workshops and training sessions in Nigeria, Kenya, Tanzania, Ghana, Rwanda, South Africa, Uganda, and Ethiopia (DSA, 2024). In addition, Women in Machine Learning & Data Science (WiMLDS) promotes women's participation in AI in some countries in Africa like Algeria, Nigeria, Morocco, Uganda, Botswana, and Kenya (Arakpogun et al., 2021).

The South African public healthcare system currently faces challenges due to ineffective patient data management (Maphumulo and Bhengu, 2019). This situation contributes to a lack of transparency in handling patients' health information and results in unsecured medical records (Matlebjane and Ndayizigamiye, 2022; Matlebjane, 2022). Blockchain technology, however, offers a promising solution by enhancing the security, auditability, and reliability of healthcare records (Miya et al., 2023). These benefits make blockchain an attractive technology for managing patient health information in the South African context.

NOVEL THERAPEUTIC APPROACHES: PERSONALIZED MEDICINE AND ITS APPLICATION IN AFRICA

Key Genomic Aberrations in PDAC

PDAC can be classified into several molecular subtypes using state-of-the-art methods such as integrated genomic analysis, microarray gene expression, single-cell analyses, non-negative matrix factorization, and integrated multi-omics profiling. These molecular subtypes have been summarized in Figure 10.2 from a recent review (Elebo et al., 2023). The molecular subtypes include the classic, basal-like squamous, immunogenic, and aberrantly differentiated endocrine exocrine (ADEX).

Research has indicated that distinct molecular subtypes of PDAC may influence the clinical outcomes of the disease (Dreyer et al., 2022). Following resection, patients with the quasi-mesenchymal subtype have a worse prognosis than those with the classical molecular subtype. The quasi-mesenchymal subtype is characterized by higher expression of mesenchyme-associated genes and lower levels of GATA binding protein 6 (GATA6) (Collisson et al., 2011). The classical subtype of PDAC is distinguished by a high mutational level of *KRAS G12V*, along with strong expression of adhesion-associated and epithelial genes, including GATA6, *KRAS*, and mothers against decapentaplegic homolog 4 (*SMAD4*) (Moffitt et al., 2015).

There is a correlation between the initiation, growth, and progression of PDAC and a high degree of *KRAS* mutation (Buscail et al., 2020). Activating *KRAS* mutations, TP53 mutations are present in more than 70% of PDAC cases and are typically linked to unfavorable outcomes (Yousef et al., 2024; Elebo et al., 2023). A mutation in the *CDKN2A* gene raises the risk of pancreatic cancer. The significance of germline *CDKN2A* mutations in carcinogenesis is demonstrated by the fact that they are found in more than 3% of pancreatic cancers (Kimura et al., 2021).

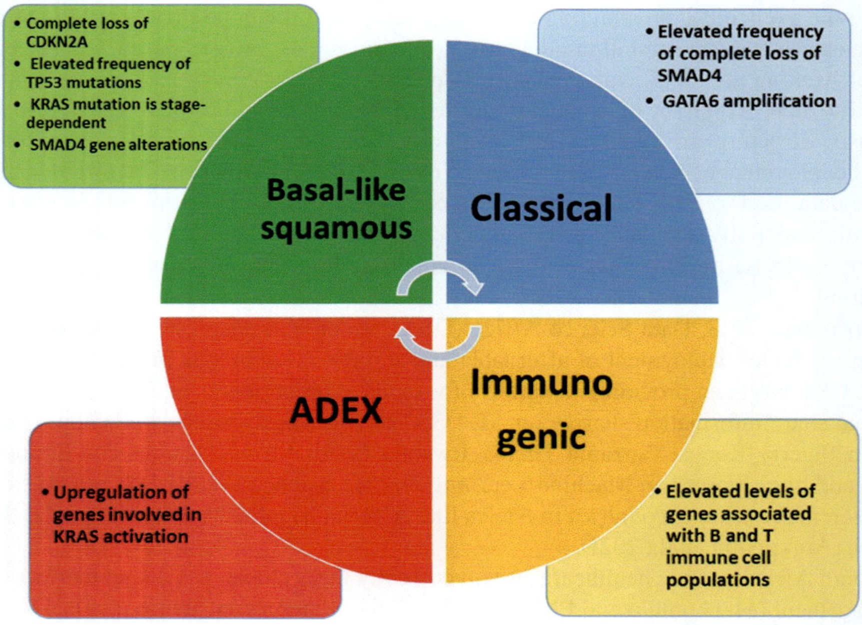

FIGURE 10.2 Molecular subtypes of pancreatic ductal adenocarcinoma (PDAC). The impact of genomic mutations for each subtype is summarized. Abbreviations: ADEX, aberrantly differentiated endocrine exocrine; CDKN2A, cyclin-dependent kinase inhibitor 2A; GATA6, GATA binding protein 6; KRAS, Kirsten rat sarcoma viral oncogene homolog; SMAD4, mothers against decapentaplegic homolog 4; TP53, tumor protein p53. (Adapted from Elebo et al., 2023.)

Several key signaling pathways are aberrant in pancreatic cancer progression. For instance, the dysregulation of the MAPK and PI3K/AKT signaling pathways occurs in the PDAC tumor microenvironment (TME) (Nweke et al., 2020). The study also indicated that early resectable PDAC is characterized by plakoglobin overexpression and activation of the insulin/IGF and MAPK pathways. The junction plakoglobin (JUP) gene encodes plakoglobin, which plays a vital role in cell adhesion as a component of desmosomes and adherens junctions (Zinkovsky and Sood, 2023). Plakoglobin forms complexes with α- or β-catenin and interacts with the cytoplasmic domains of N- or E-cadherin, thereby facilitating the connection of adherens junctions to the cytoskeleton (Knudsen and Wheelock, 1992). Recent research has revealed that plakoglobin also functions as a transcriptional factor involved in regulating WNT signaling pathways (Alaee et al., 2018).

Repurposing Drugs in PDAC Treatment

Pharmacological repurposing is defined as "the process of identifying new therapeutic use for old or existing drugs through drug redirection and therapeutic switching" (Nweke and Thimiri Govinda Raj, 2021). This approach can help avoid the tedious, costly, time-consuming, and high-risk process associated with traditional drug discovery. Repurposed medications have the ability to target one or several aberrations in targets or pathways, potentially avoiding resistance (Ntwasa and Dlamini, 2024; Elebo et al., 2023). This ability can enhance the efficacy and broaden the therapeutic applications of existing drugs, making them valuable tools in addressing complex and evolving medical conditions.

In PDAC, drug repurposing therapy focuses on the molecular subtypes and genetic modifications of cancer. As previously mentioned, different molecular alterations are used to categorize the molecular subtypes. Targeting genomic alterations in PDAC, repurposed medications such as hydroxychloroquine, chloroquine, niraparib, nitroxoline, and rucaparib are employed (Nweke and Thimiri Govinda Raj, 2021; Sarmento-Ribeiro et al., 2019; Rebelo et al., 2021; Hosein et al., 2022).

Notable drugs such as aspirin, a non-steroidal anti-inflammatory drug (NSAID), have demonstrated anti-neoplastic effects due to their ability to inhibit the prostaglandin precursors, COX-1 and COX-2 enzymes, which play a key role in regulating inflammatory processes (Elebo et al., 2023). Aspirin's acetylation inhibits the transcription factor NF-κB, which regulates the expression of genes associated with metastasis and apoptosis. A 2020 study by Qorri et al. suggests that the combination of gemcitabine, oseltamivir phosphate, and aspirin may facilitate the blockade of survival pathways that contribute to the proliferation of PDAC cell lines (Qorri et al., 2020).

The anti-neoplastic actions of metformin, an oral antidiabetic, are likely brought about by its suppression of reactive oxygen species (ROS) and the mammalian target of rapamycin (mTOR) (Amin et al., 2019). AMP-activated protein kinase (AMPK) activation and DNA damage are also linked to metformin (Algire et al., 2012). The role of metformin as a precision treatment in PDAC may be indicated due to its ability to inhibit hepatic nuclear factor gamma (HNF4G) activity (Elebo et al., 2023; Wang et al., 2021).

Vitamin D regulates cell cycle and differentiation through a poorly understood mechanism, as Cannon et al. (2016) have shown, lowering the risk of PDAC. Numerous metabolic processes, immunological reactions, and cancers are impacted by vitamin D receptors (VDR). According to a study from Ibadan University in Nigeria, the vitamin D receptor (VDR) is involved in a number of cancer biological pathways, including those in oral squamous cell carcinoma (Adisa et al., 2017).

Beta-blockers, commonly used to treat hypertension and other cardiovascular diseases, have been repurposed due to their ability to inhibit the stimulation of adrenergic receptors, which are G-protein-coupled receptors expressed by pancreatic cancer cells (Massalee and Cao, 2023; Hüttner et al., 2020). Saad et al. (2020) have indicated that long-term use of beta-blockers such as atenolol, propranolol, and carvedilol is associated with a lower incidence of pancreatic cancer (Saad et al., 2020).

Chloroquine and hydroxychloroquine, two antimalarial drugs, reduce autophagy by enhancing the efficacy of ERK inhibitors in mediating anticancer effects in PDAC driven by *KRAS* mutations (Bryant et al., 2019). Moreover, autophagy has been shown to play a critical role in the growth and progression of PDAC tumors harboring mutations in the *P53* gene (Yang et al., 2014). The reader is directed to Elebo et al.'s comprehensive work on medication repurposing (Elebo et al., 2023).

Immunotherapeutic Strategies in Pancreatic Cancer

Numerous clinical trials have identified both established and potential immunological targets for pancreatic cancer (Dodson et al., 2011; Nsingwane et al., 2020; Lenzo et al., 2021; Elebo et al., 2024). Immunomodulators, or checkpoint inhibitors, enable immune cells, including T lymphocytes, to function without restraint, thereby enhancing the immune response against cancer. Promising outcomes have been observed in the treatment of pancreatic cancer with various immune checkpoint inhibitors. African scientists are especially keen to explore new targets within the TME, such as stromal cells. Additionally, potential research areas include the application of neoantigens in immunotherapy and the use of bacterial treatments (Nsingwane et al., 2020).

Neoantigens are antigens produced by altered genes unique to tumors that arise from aberrant splicing. The immune system recognizes neoantigens as non-self, which triggers an immune response, making them significant in cancer treatment. Pancreatic cancer cells carry neoantigens that elicit an efficient immune response; however, this response is often thwarted by immune evasion, where the TME reduces the presentation of neoantigens to the immune system (Wang et al., 2016; Bailey et al., 2016). Mutations in the *KRAS* codon 12, specifically *G12V* and *G12D*, have been identified as potential neoantigens for pancreatic adenocarcinoma. These neoantigens have been shown to bind to T-cell receptors in mouse models, highlighting their potential for targeted immunotherapy (Wang et al., 2016).

Another exciting strategy gaining attention among African researchers for pancreatic cancer treatment is the emergence of cancer vaccines (Nsingwane et al., 2020). These vaccines function as adjuvants, working with the body's natural defenses to strengthen and stimulate the immune system. In this type of vaccination, cancer antigens can be delivered in various forms, including whole tumors, proteins, peptides, DNA, RNA, or antigen-pulsed dendritic cells, with the aim of stimulating an immune response specific to the targeted cancer (Reynolds, 2023).

One of the effective vaccines tested on individuals with metastatic pancreatic cancer is the pancreatic GVAX vaccine (Laheru et al., 2008). This vaccine has shown promise in eliciting an immune response against pancreatic cancer. Compared to other treatments in Europe, the peptide-based pancreatic cancer vaccine Targovax (TG01), which targets RAS mutations, advanced to phase I/II clinical trials and demonstrated notable improvement following surgical resection. It has been proposed that TG01 may enhance the efficacy of chemotherapy (Palmer et al., 2020).

ROLE OF TRADITIONAL MEDICINE AND INTEGRATIVE APPROACHES

The gold standard in the curative therapeutic treatment of pancreatic cancer remains largely surgery and adjuvant chemotherapy (Yue et al., 2017). Novel therapeutic interventions, including traditional and natural medicines, are, however, acquiring repute as a viable alternative in the treatment of solid human malignancies, especially within the African context (Gaobotse et al., 2023). The basis of the use of traditional medicine stems from extensive research into bioactive compounds that can manage and treat a myriad of illnesses, including various malignancies (Nkwe et al., 2021).

The anticancer ability of traditional medicine has been based on the cytotoxic and apoptotic properties of various African traditional plants (Ochwang'i et al., 2014). Ochwang'i et al. further expanded on this by highlighting the range of specific species of plants that have anticancer properties. This list includes *Spathodea campanulata*, *Microglossa pyrifolia*, *Prunus* africana, *Curcuma longa,* and *Catharanthus roseus* (Ochwang'i et al., 2014). The secondary metabolites from such medicinal plants have been proven to show anticancer properties through apoptotic pathways including metabolic reprogramming (Nkwe et al., 2021).

The process of developing anticancer products from medicinal plants is well elaborated by Adico et al. (2024) from Burkina Faso in their analysis of the anticancer properties in various sub-Saharan African plants. The process includes the crude and active extraction of the bioactive compounds in the whole bark, root, or leaves (Adico et al., 2024). The extraction of the bioactive compounds proceeds to the in-vitro and in-vivo investigation, which, when successful, will advance to clinical trials (Figure 10.3) (Adico et al., 2024; Nkwe et al., 2021). The effectiveness of medicinal plants largely depends on the quality and concentration of their active phytochemical components (Kumar et al., 2023).

Studies specifically assessing the anticancer properties of traditional medicine in pancreatic cancer are minimal, with even less documentation of the clinical trials that have been conducted. Adico et al. (2024) highlighted curcumin, an active component of *Curcuma longa* known for its antioxidant, anti-inflammatory, and anticancer properties, as part of the spectrum of medicinal plants that may have therapeutic effects in the treatment of pancreatic malignancies (Adico et al., 2024) .

The use of curcumin in the treatment of pancreatic malignancies has been substantiated in several studies (Guo et al., 2022; Dhillon et al., 2008; Kanai, 2014; Liu et al., 2020; Bimonte et al., 2016). Curcumin has several targets, including promoting apoptosis, reducing the expression levels of cyclin A1, CDK1 and Bcl-2 proteins, increasing caspase-9 levels, triggering cell cycle shutdown, positively regulating p21 protein expression levels by targeting the NFκB signaling pathway, and improving the generation of ROS, which results in oxidative stress in cancer cells (Adico et al., 2024).

Minimal studies have proceeded to clinical trials in the assessment of traditional medicine in pancreatic malignancies, one of those being Dhillon et al. (2008), which concluded that oral curcumin

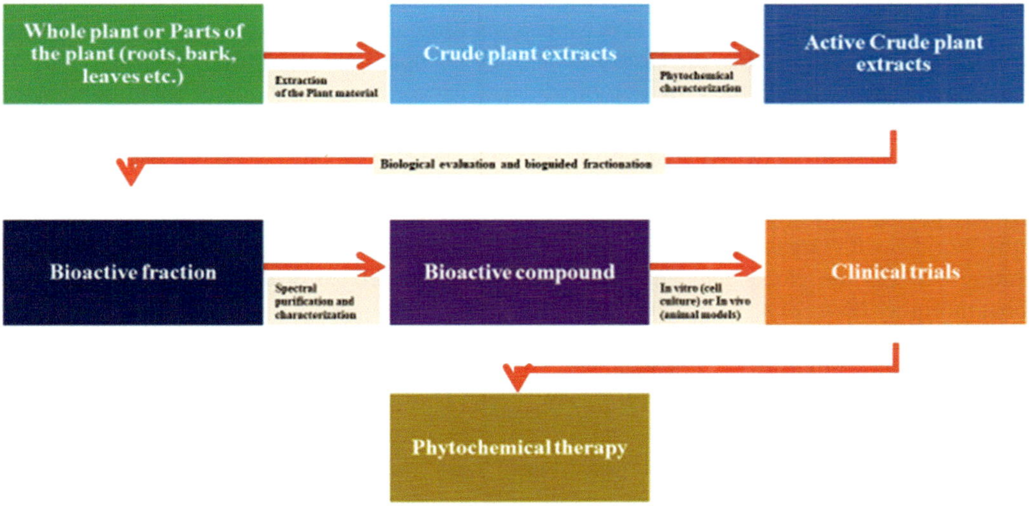

FIGURE 10.3 The process for developing anticancer products is derived from therapeutic medicinal plants. The extraction of plant material constitutes the initial step in the development of purification procedures for phytochemicals with anticancer properties. Following the extraction of the active compounds, the extract is fractionated using a suitable matrix. The resulting bioactive fractions are then analyzed using various analytical techniques, such as mass spectrometry, high-performance liquid chromatography, and nuclear magnetic resonance spectroscopy. Subsequent to the purification and characterization of bioactive fractions employing these analytical techniques, it is imperative to evaluate the anticancer efficacy of the molecules through both in vitro and in vivo studies. Furthermore, thorough assessments of pharmacokinetics, pharmacodynamics, immunogenicity, biosecurity, and potential adverse effects are performed to guarantee the development of safe and effective therapeutic agents. (Adapted from Adico et al., 2024.)

is well tolerated and has biological activity in some patients with pancreatic cancer (Dhillon et al., 2008). In support of this, there is research highlighting the molecular effects of curcumin in the suppression of proliferation and the induction of apoptosis (Li et al., 2004).

Clinical trials have demonstrated the therapeutic benefits of herbal medicine in the treatment of pancreatic malignancies (Dhillon et al., 2008; Hu et al., 2022; Triantafillidis et al., 2022). The benefits of traditional plants have so far been noted within the context of integrative medicine that includes both conventional neoadjuvant therapies combined with traditional African medicines (Bimonte et al., 2016).

Integrated medicine, including gemcitabine-based chemotherapy plus nab-paclitaxel, is currently used as the gold standard therapy for patients with advanced pancreatic cancer. However, many adverse effects are still present (Bimonte et al., 2016). The use of traditional medicine offers an opportunity for therapeutic treatment of pancreatic malignancies, while reducing the adverse effects. Further research is advocated, particularly those that go beyond theoretical investigations and provide clinically proven improvements in the management of pancreatic malignancies, especially in the African diaspora.

PANCREATIC CANCER BURDEN AND TREATMENT INNOVATIONS IN AFRICA: CHALLENGES AND LEGISLATIVE EFFORTS AND THE ROLE OF GOVERNMENT IN FACILITATING RESEARCH

Cancer constitutes a significant global public health issue, contributing to a substantial number of fatalities annually. In 2020, Africa alone reported 1,109,209 new cancer cases and 711,429 cancer-related deaths. This equates to approximately 1,949 cancer-related fatalities per day on the continent (Omotoso et al., 2023). Nigeria, with a population of 213 million, has an estimated 70 clinical oncologists (COs) treating 124,815 cancer patients (Chidebe et al., 2023). This alarming gap is expected to worsen, as 90% of Nigerian physicians express a willingness to leave the country for perceived better opportunities in Canada, the United States, and Europe (Ihua and Nsofor, 2019).

Several obstacles hinder the provision of adequate cancer care in Africa, including insufficient awareness, the high costs of treatment and research funding, a lack of expertise, and a shortage of medical professionals, particularly oncologists and specialized cancer care nurses. These limitations contribute to high cancer mortality rates and a disproportionate cancer burden in many African nations, especially in sub-Saharan Africa (SSA). For instance, in 2018, member countries of the World Health Organization (WHO) African Region reported 811,200 new cancer cases, constituting 4.5% of the global incidence, and 534,000 cancer-related deaths, representing 7.3% of global mortality (Sung et al., 2021; Omonisi et al., 2020).

Without addressing these obstacles, the prospects for Africa to overcome its cancer burden appear bleak. Projections suggest that between 2020 and 2040, the cancer burden in African nations is anticipated to exceed that of the global population (Morhason-Bello et al., 2013). An individual's risk of developing PDAC can be significantly reduced by mitigating exposure to modifiable risk factors such as alcohol and tobacco use, improving workplace safety protocols, and promoting healthy lifestyle choices. Conversely, non-modifiable factors, including age, gender, family history, inherited genetic mutations, and race/ethnicity, also play a substantial role in cancer risk. Although these factors are beyond an individual's control, awareness of their impact can encourage proactive measures to reduce modifiable risks. These measures include avoiding carcinogen exposure, obtaining vaccinations against preventable infections, and scheduling regular screenings for early cancer detection and treatment.

The incidence of cancer in South Africa, a nation of approximately 60 million people, is projected to double, with 216,336 new cases expected by 2030 (Meyer et al., 2024). Despite this looming increase, the country lacks a comprehensive national act addressing cancer, as well as adequate funding for coordinated prevention and control efforts. Moreover, there are deficiencies in policies and legal frameworks specifically targeting the disease. However, elements of legislation supporting cancer prevention and control are derived from Section 27 and other guarantees found in the Bill of Rights of the South African Constitution (1996) (Ngwena, 2003; Sarkin, 2000).

Kenya has become the first nation in Africa to establish a national cancer statute aligned with international guidelines (Kenya, 2012). This act is integrated with national cancer strategy frameworks and the National Cancer Institute (NCI), supporting both a National Health Insurance (NHI) program and the constitutional right to health. The passage of this legislation was facilitated by contributions from patient advocacy organizations and civil society members.

In 2016, Uganda enacted a special law establishing the Uganda Cancer Institute. This act aims to create an independent institution dedicated to managing cancer and related diseases within Uganda. Its objectives include establishing a state-of-the-art cancer institute that meets international standards, addressing the challenges in cancer treatment, reducing the need for Ugandan cancer patients to seek treatment abroad, ensuring timely access to specialized cancer medication, and enhancing the capacity of Ugandan oncologists to manage complex cancer conditions (Uganda Cancer Institute Act, 2017). Subsequently, in 2017, the Nigerian Federal Government established the National Institute for Cancer Research and Treatment (NICRAT) to promote and guide scientific advancements in cancer prevention, treatment, and care, among other objectives (Treatment, 2024). It has taken about seven years between when the bill that established the institute was signed into law and when the institute was fully established to commence operations (Malami et al., 2023).

To address pancreatic cancer, nations such as Zimbabwe are adopting advanced surgical techniques. In 2023, Chitungwiza Central Hospital, a public healthcare facility, performed the first laparoscopic-assisted Whipple procedure, or pancreaticoduodenectomy (Chingwere, 2023). As reported by Seward et al. (2022) and O'Brien et al. (2023), Zimbabwe aims to provide top-tier healthcare services (Seward et al., 2022; O'Brien et al., 2023). The country's commitment to universal health coverage (UHC) ensures that all individuals have access to a comprehensive range of high-quality healthcare services whenever and wherever needed, without facing financial hardship (Chimbwanda, 2023).

The multimodal treatment of PDAC in Morocco demonstrates survival benefits comparable to those observed in Asian and Caucasian populations (Bazine et al., 2018). In Tanzania, Kivuyo and Akoko (2022) observed that many PDAC patients are inadequately staged, resulting in variations in treatment approaches (Kivuyo and Akoko, 2022). Standardizing patient workups is crucial to ensuring equitable access to curative interventions and improving survival outcomes.

PUBLIC AWARENESS AND EDUCATION: AWARENESS PROGRAMS AND CAMPAIGNS

A higher prevalence of delayed pancreatic cancer diagnosis has been observed in a group of young people from North Africa. An Algerian study reported that 20% of patients with pancreatic cancer consumed alcohol, and 32% smoked cigarettes, both of which are potential risk factors for the disease (Sellam et al., 2015a). Numerous published studies have indicated that smokers have approximately twice the risk of developing pancreatic cancer compared to nonsmokers (Pandol et al., 2012; Subhan et al., 2022; Bogumil et al., 2024). If sociocultural norms, negligence, or lack of awareness are not addressed, patients with pancreatic cancer may have poor survival outcomes. Therefore, it is crucial to raise awareness among healthcare professionals and citizens in African nations.

At the conclusion of the high-level advocacy event on cancer in Africa, held in Niamey, Niger, alongside the African Union summit, an appeal was made to African nations to incorporate basic cancer interventions, such as addressing risk factors, early diagnosis, and screening, at the primary healthcare level to promote greater equity. Dr. Matshidiso Moeti, the regional director of WHO for Africa, emphasized this call while addressing attendees. A collective request was also made to increase taxes on alcohol, tobacco, and other items that contribute to cancer risk (World Health Oganization, 2019).

Every November, the world observes Pancreatic Cancer Awareness Month. However, pancreatic cancer-related media activity has not gained significant traction in Africa. There are few African Facebook pages that host events and mark awareness related to pancreatic cancer (Mac'yong, 2020; Africa, 2022; Centre, 2021; Zambia, 2022). The most recent updates to the pancreatic cancer fact sheet on the Cancer Association of South Africa (CANSA) website were made in 2021 (Herbst, 2021). Apart from CANSA in South Africa, there is a notable lack of government-supported information sources that raise public awareness and education about pancreatic cancer, as well as awareness initiatives and programs, in other African nations.

CHALLENGES AND LIMITATIONS IN AFRICA

With the projected increase in the incidence of PDAC expected to become the second leading cause of cancer-related mortality by 2030, African countries are confronted with the challenge of developing population-specific interventions. Currently, research efforts on the continent are fragmented, with research centers both within and across countries often working in isolation. In Africa, the fight against pancreatic cancer is hindered by a lack of health infrastructure, including limited molecular laboratories and platforms for advanced pancreatic cancer research. Additional challenges include insufficient funding, a shortage of experts in hepato-pancreato-biliary research, lack of political will, inadequate regional collaboration, and limited information sharing on pancreatic cancer.

Although there are few centers of excellence in research and innovation related to pancreatic cancer across Africa, these initiatives frequently fail to translate into practical, community-oriented solutions. The epidemiologic data and risk factors of pancreatic cancer remain poorly understood at regional and national levels, largely due to the lack of comprehensive research.

Splicing alterations in PDAC constitute an emerging molecular landscape with significant translational potential, yet this area of research has not garnered substantial attention within the African scientific community. It is encouraging that African countries are increasingly incorporating AI into healthcare, which holds promise for advancing research and innovation on the continent. However, there is a slow adoption of the continent's rich flora, leading to missed opportunities in harnessing traditional medicines to combat this deadly cancer. Moreover, there is generally insufficient funding for coordinated cancer prevention and control efforts, further impeding progress in addressing this escalating health challenge.

CONCLUSION

Despite pancreatic cancer being a highly fatal malignancy with the worst survival rate among solid malignancies in Africa, its incidence on the continent is relatively low, with fewer than 3.5 cases per 100,000 people. Unfortunately, the majority of patients present with advanced pancreatic cancer, significantly limiting their treatment options. Only 3% of Nigerian patients have resectable cancer at the time of diagnosis. In Africa, the risk factors for the disease include excessive alcohol intake, smoking, obesity, occupational exposures, and underreported genetic mutations.

Innovative diagnostic techniques that enable early detection of pancreatic cancer and facilitate personalized therapy are gaining significant interest in the African research landscape. The adoption of artificial intelligence is expected to enhance early diagnosis, molecular subtyping, treatment response prediction, and risk prediction. Molecular subtyping of pancreatic cancer has the potential to positively influence clinical outcomes.

In Africa, drug repurposing, which avoids the excessive costs, lengthy timeframes, and high-risk processes associated with traditional drug discovery, broadens the therapeutic applications of existing drugs. This approach makes them valuable tools in addressing complex and evolving cancers. Research indicates that some benefits from nonsteroidal anti-inflammatory drugs (NSAIDs), antidiabetic drugs, antimalarials, beta-blockers, and other common medication may facilitate the suppression of survival pathways necessary for pancreatic cancer cell lines. Cutting-edge immunotherapeutic strategies and cancer vaccines have also garnered interest among African researchers. Encouragingly, traditional and integrative approaches are emerging as viable alternatives in pancreatic cancer management within the African continent.

African countries are adopting legislative standards that facilitate cancer care and management. Mitigating exposure to modifiable risk factors for pancreatic cancer, supported by the WHO Africa Region, is crucial. Awareness of the impact of non-modifiable risk factors, such as genetics, can encourage proactive measures to reduce modifiable risks. However, awareness programs and campaigns addressing pancreatic cancer remain limited at the community level.

REFERENCES

ACHILONU, O. J., FABIAN, J., BEBINGTON, B., SINGH, E., NIMAKO, G., EIJKEMANS, R. & MUSENGE, E. 2021. Use of machine learning and statistical algorithms to predict hospital length of stay following colorectal cancer resection: A South African pilot study. *Front Oncol,* 11, 644045.

ADEOYE, J., AKINSHIPO, A., THOMSON, P. & SU, Y. X. 2022. Artificial intelligence-based prediction for cancer-related outcomes in Africa: status and potential refinements. *J Glob Health,* 12, 03017.

ADICO, M. D. W., BAYALA, B., BUNAY, J., BARON, S., SIMPORE, J. & LOBACCARO, J.-M. A. 2024. Contribution of Sub-Saharan African medicinal plants to cancer research: scientific basis 2013–2023. *Pharmacol Res,* 202, 107138.

ADISA, A., UDEABOR, S., OGUN, G., ORLOWSKA, A., SADER, R. & GHANAATI, S. 2017. Immunohistochemical expression of vitamin-D receptor in oral and skin squamous cell carcinoma of a Black African subpopulation. *Med J Zambia,* 44, 64–68.

AFRICA, D. S. 2024. *DSA events.* Data Science Africa. Available: https://www.datascienceafrica.org/ [Accessed September 19, 2024].

AFRICA, P. C. A. S. 2022. *Pancreatic cancer awareness South Africa.* Available at: https://web.facebook.com/PanCanSA/?_rdc=1&_rdr 2024 [Accessed July 2024].

DINAGDE, T.A. & ABUBEKER, Z. 2024. Surgical management of pancreatic cancer in Tikur Anbessa Specialized Hospital, Addis Ababa, Ethiopia: a 5 years retrospective descriptive study. *BMC Surg.* Aug 6; 24(1), 223. doi: 10.1186/s12893-024-02503-2. PMID: 39103810; PMCID: PMC11302827.

ALAEE, M., NOOL, K. & PASDAR, M. 2018. Plakoglobin restores tumor suppressor activity of p53(R175H) mutant by sequestering the oncogenic potential of β-catenin. *Cancer Sci,* 109, 1876–88.

ALATISE, O., NDUBUBA, D., OJO, O., AGBAKWURU, E., ADEKANLE, O. & AROWOLO, O. 2009. Pancreatic cancer in Nigeria: past, present and future. *Nigeria J Gastroenterol Hepatol*, 1(2).

ALATISE, O. I., LAWAL, O. O. & TOPE, O. O. 2010. Management challenges of pancreatic cancer in a resource scarce setting. *Africa J Health Sci,* 17, 36–41.

ALGIRE, C., MOISEEVA, O., DESCHÊNES-SIMARD, X., AMREIN, L., PETRUCCELLI, L., BIRMAN, E., VIOLLET, B., FERBEYRE, G. & POLLAK, M. N. 2012. Metformin reduces endogenous reactive oxygen species and associated DNA damage. *Cancer Prev Res (Phila),* 5, 536–43.

ALGUACIL, J., POLLÁN, M. & GUSTAVSSON, P. 2003. Occupations with increased risk of pancreatic cancer in the Swedish population. *Occup Environ Med,* 60, 570–6.

ALGUACIL, J., PORTA, M., BENAVIDES, F. G., MALATS, N., KOGEVINAS, M., FERNÁNDEZ, E., CARRATO, A., RIFÀ, J., GUARNER, L. & GROUP, F. T. P. I. S. 2000. Occupation and pancreatic cancer in Spain: a case-control study based on job titles. *Int J Epidemiol,* 29, 1004–13.

ALI, L. H., HIGAZI, A. M., MONESS, H. M., FARAG, N. M., SAAD, Z. M., MOUKAREB, H. A., SOLIMAN, W., EL SAGHEER, G., ABD EL HAMID, S. R. & ABDL HAMID, H. 2019. Clinical significances and diagnostic utilities of both miR-215 and squamous cell carcinoma antigen-IgM versus alpha-fetoprotein in Egyptian patients with hepatitis C virus-induced hepatocellular carcinoma. *Clin Exp Gastroenterol,* 12, 51–66.

AMIN, S., LUX, A. & O'CALLAGHAN, F. 2019. The journey of metformin from glycaemic control to mTOR inhibition and the suppression of tumour growth. *Br J Clin Pharmacol,* 85, 37–46.

AMRI, F., KOULALI, H., JABI, R., ZAZOUR, A., BOUZIANE, M., ISMAILI, Z. & KHARRASSE, G. 2023. Pancreatic cancer: experience from an emerging country in North Africa. *J Cancer Res Clin Oncol,* 149, 14297–302.

ARAKPOGUN, E. O., ELSAHN, Z., OLAN, F. & ELSAHN, F. 2021. Artificial intelligence in Africa: Challenges and opportunities. In *The fourth industrial revolution: implementation of artificial intelligence for growing business success*, pp. 375–388.

ASOMBANG, A. W., MADSEN, R., SIMUYANDI, M., PHIRI, G., BECHTOLD, M., IBDAH, J., LISHIMPI, K. & BANDA, L. 2017. Pancreatic cancer: patterns in a low-to middle-income Ppopulation, Zambia. *Med J Zambia,* 44**,** 212–17.

BAILEY, P., CHANG, D. K., FORGET, M. A., LUCAS, F. A., ALVAREZ, H. A., HAYMAKER, C., CHATTOPADHYAY, C., KIM, S. H., EKMEKCIOGLU, S., GRIMM, E. A., BIANKIN, A. V., HWU, P., MAITRA, A. & ROSZIK, J. 2016. Exploiting the neoantigen landscape for immunotherapy of pancreatic ductal adenocarcinoma. *Sci Rep,* 6**,** 35848.

BAZINE, A., FETOHI, M., TORES, M., TANZ, R., TAHIRI, M. E. H., CHOHO, A. & ICHOU, M. 2018. Pancreatic ductal adenocarcinoma in a Moroccan population: analysis of six years experience. *Int Surg J,* 5**,** 1628–32.

BIMONTE, S., BARBIERI, A., LEONGITO, M., PICCIRILLO, M., GIUDICE, A., PIVONELLO, C., DE ANGELIS, C., GRANATA, V., PALAIA, R. & IZZO, F. 2016. Curcumin antiCancer studies in pancreatic cancer. *Nutrients,* July 8(7), 433. doi: 10.3390/nu8070433. PMID: 27438851; PMCID: PMC4963909.

BOGUMIL, D., STRAM, D., PRESTON, D. L., PANDOL, S. J., WU, A. H., MCKEAN-COWDIN, R., CONTI, D. V. & SETIAWAN, V. W. 2024. Excess pancreatic cancer risk due to smoking and modifying effect of quitting smoking: the Multiethnic Cohort Study. *Cancer Causes Control,* 35**,** 541–8.

BOUGLOUGA, O., LAWSON-ANANISSOH, L. M., BAGNY, A., KAAGA, L. & REDAH, D. 2015. [Pancreatic cancer: Epidemiological, clinical, and management aspects in the department of hepatogastroenterology at the Lome Campus teaching hospital (Togo)]. *Med Sante Trop,* 25**,** 323–6.

BRAND, R. E. & LYNCH, H. T. 2000. Hereditary pancreatic adenocarcinoma. A clinical perspective. *Med Clin North Am,* 84**,** 665–75.

BRYANT, K. L., STALNECKER, C. A., ZEITOUNI, D., KLOMP, J. E., PENG, S., TIKUNOV, A. P., GUNDA, V., PIEROBON, M., WATERS, A. M., GEORGE, S. D., TOMAR, G., PAPKE, B., HOBBS, G. A., YAN, L., HAYES, T. K., DIEHL, J. N., GOODE, G. D., CHAIKA, N. V., WANG, Y., ZHANG, G. F., WITKIEWICZ, A. K., KNUDSEN, E. S., PETRICOIN, E. F., 3RD, SINGH, P. K., MACDONALD, J. M., TRAN, N. L., LYSSIOTIS, C. A., YING, H., KIMMELMAN, A. C., COX, A. D. & DER, C. J. 2019. Combination of ERK and autophagy inhibition as a treatment approach for pancreatic cancer. *Nat Med,* 25**,** 628–40.

BUSCAIL, L., BOURNET, B. & CORDELIER, P. 2020. Role of oncogenic KRAS in the diagnosis, prognosis and treatment of pancreatic cancer. *Nat Rev Gastroenterol Hepatol,* 17**,** 153–68.

CANNON, T.L., FORD, J., & HESTER, D., TRUMP, D.L. 2016. The incidental use of high-dose vitamin D3 in pancreatic cancer. *Case Rep Pancreat Cancer.* May 1; 2(1), 32–35. doi: 10.1089/crpc.2016.0003. PMID: 30631812; PMCID: PMC6319684.

CENTRE, T. N. O. 2021. *Pancreatic cancer awareness.* Available: https://web.facebook.com/photo.php?fbid=4558538554237914&__hwl=1&_rdc=1&_rdr [Accessed July 21, 2024].

CHIDEBE, R. C. W., ORJIAKOR, T. C., LASEBIKAN, N., JOSEPH, A., TOLAND, S. & SIMONS, A. 2023. Brain drain in cancer care: the shrinking clinical oncology workforce in Nigeria. *JCO Glob Oncol,* 9**,** e2300257.

CHIMBWANDA, T. 2023. *Zimbabwe launches health resilience fund and strategic frameworks to bolster universal health coverage.* World Health Organization - Zimbabwe. Available: https://www.afro.who.int/countries/zimbabwe/news/zimbabwe-launches-health-resilience-fund-and-strategic-frameworks-bolster-universal-health-coverage#:~:text=Harare%2C%20Zimbabwe%20%E2%80%93%20Universal%20Health%20Coverage,hardship%20to%20pay%20for%20it. [Accessed July 14, 2024].

CHINGWERE, M. 2023. *Surgeons perform delicate pancreatic cancer op.* The Herald Available: https://www.herald.co.zw/surgeons-perform-delicate-pancreatic-cancer-op/ [Accessed July 2024].

COLLABORATORS, G. P. C. 2019. The global, regional, and national burden of pancreatic cancer and its attributable risk factors in 195 countries and territories, 1990-2017: a systematic analysis for the Global Burden of Disease Study 2017. *Lancet Gastroenterol Hepatol,* 4**,** 934–47.

COLLISSON, E. A., SADANANDAM, A., OLSON, P., GIBB, W. J., TRUITT, M., GU, S., COOC, J., WEINKLE, J., KIM, G. E., JAKKULA, L., FEILER, H. S., KO, A. H., OLSHEN, A. B., DANENBERG, K. L., TEMPERO, M. A., SPELLMAN, P. T., HANAHAN, D. & GRAY, J. W. 2011. Subtypes of pancreatic ductal adenocarcinoma and their differing responses to therapy. *Nat Med,* 17**,** 500–3.

COUNCIL, S. A. M. R. 2024. *Research centres.* Available: https://www.samrc.ac.za/research/centre-and-platforms-office/research-centres [Accessed July 5, 2024].

DAAMEN, L. A., MOLENAAR, I. Q. & GROOT, V. P. 2023. Recent advances and future challenges in pancreatic cancer care: early detection, liquid biopsies, precision medicine and artificial intelligence. *J Clin Med,* 12, 7485.

DAHER, H., PUNCHAYIL, S. A., ISMAIL, A. A. E., FERNANDES, R. R., JACOB, J., ALGAZZAR, M. H. & MANSOUR, M. 2024. Advancements in pancreatic cancer detection: integrating biomarkers, imaging technologies, and machine learning for early diagnosis. *Cureus,* 16, e56583.

DHILLON, N., AGGARWAL, B. B., NEWMAN, R. A., WOLFF, R. A., KUNNUMAKKARA, A. B., ABBRUZZESE, J. L., NG, C. S., BADMAEV, V. & KURZROCK, R. 2008. Phase II trial of curcumin in patients with advanced pancreatic cancer. *Clin Cancer Res,* 14, 4491–9.

DLAMINI, Z., FRANCIES, F. Z., HULL, R. & MARIMA, R. 2020. Artificial intelligence (AI) and big data in cancer and precision oncology. *Comput Struct Biotechnol J,* 18, 2300–11.

DODSON, L. F., HAWKINS, W. G. & GOEDEGEBUURE, P. 2011. Potential targets for pancreatic cancer immunotherapeutics. *Immunotherapy,* 3, 517–37.

DREYER, S. B., UPSTILL-GODDARD, R., LEGRINI, A., BIANKIN, A. V., JAMIESON, N. B., CHANG, D. K., JAMIESON, N. B. & CHANG, D. K. 2022. Genomic and molecular analyses identify molecular subtypes of pancreatic cancer recurrence. *Gastroenterology,* 162, 320–4.e4.

ELEBO, N., ABDEL-SHAFY, E. A., CACCIATORE, S. & NWEKE, E. E. 2023. Exploiting the molecular subtypes and genetic landscape in pancreatic cancer: the quest to find effective drugs. *Front Genet,* 14, 1170571.

ELEBO, N., ABDEL-SHAFY, E. A., OMOSHORO-JONES, J. A. O., NSINGWANE, Z., HUSSEIN, A. A. A., SMITH, M., CANDY, G., CACCIATORE, S., FRU, P. & NWEKE, E. E. 2024. Comparative immune profiling of pancreatic ductal adenocarcinoma progression among South African patients. *BMC Cancer,* 24, 809.

EZZAT, W. M., AMR, K. S., ELHOSARY, Y. A., HEGAZY, A. E., FAHIM, H. H., ELTAWEEL, N. H. & KAMEL, R. R. 2019. Detection of DNA methylated microRNAs in hepatocellular carcinoma. *Gene,* 702, 153–7.

GAOBOTSE, G., VENKATARAMAN, S., BROWN, P. D., MASISI, K., KWAPE, T. E., NKWE, D. O., RANTONG, G. & MAKHZOUM, A. 2023. The use of African medicinal plants in cancer management. *Front Pharmacol,* 14, 1122388.

GROVER, S. & SYNGAL, S. 2010. Hereditary pancreatic cancer. *Gastroenterology,* 139, 1076–80, 1080.e1–2.

GUO, W., DING, Y., PU, C., WANG, Z., DENG, W. & JIN, X. 2022. Curcumin inhibits pancreatic cancer cell proliferation by regulating Beclin1 expression and inhibiting the hypoxia-inducible factor-1α-mediated glycolytic pathway. *J Gastrointest Oncol,* 13, 3254–3262.

GWAGWA, A., KRAEMER-MBULA, E., RIZK, N., RUTENBERG, I. & DE BEER, J. 2020. Artificial intelligence (AI) deployments in Africa: benefits, challenges and policy dimensions. *African J Inform Commun,* 26, 1–28.

HAMDI, K., BLANCATO, J., GOERLITZ, D., ISLAM, M., NEILI, B., ABIDI, A., GAT, A., AYED, F. B., CHIVI, S., LOFFREDO, C., JILLSON, I., ELGAAIED, A. B. & MARRAKCHI, R. 2016. Circulating cell-free miRNA expression and its association with clinicopathologic features in inflammatory and non- inflammatory breast cancer. *Asian Pac J Cancer Prev,* 17, 1801–10.

HANEKOM, G. S., STUBBINGS, H. M., JOHNSON, C. A. & KIDSON, S. H. 1999. The detection of circulating melanoma cells correlates with tumour thickness and ulceration but is not predictive of metastasis for patients with primary melanoma. *Melanoma Res,* 9, 465–73.

HELENE, T., SERGE, F., NGADJUI, B. T., ETIENNE, D. & ABEGAZ, B. M. 2000. Phenolic metabolites from the seeds of Canarium schweinfurthii. *Bull Chem Soc Ethiop,* 14, 155–9.

HERBST, M. C. 2021. *Fact sheet on pancreatic cancer.* Cancer Association of South Africa (CANSA) Available: https://cansa.org.za/files/2021/04/Fact-Sheet-on-Pancreatic-Cancer-NCR-2017-web-April-2021.pdf [Accessed July 21, 2024].

HOSEIN, A. N., DOUGAN, S. K., AGUIRRE, A. J. & MAITRA, A. 2022. Translational advances in pancreatic ductal adenocarcinoma therapy. *Nature Cancer,* 3, 272–86.

HOSNY, G., FARAHAT, N., TAYEL, H. & HAINAUT, P. 2008. Ser-249 TP53 and CTNNB1 mutations in circulating free DNA of Egyptian patients with hepatocellular carcinoma versus chronic liver diseases. *Cancer Lett,* 264, 201–8.

HU, J., JIANG, J., LIU, R., CHENG, M., ZHU, G., HE, S., SHI, B., ZHAO, Y., HE, Z., YU, H., ZHANG, X., ZHENG, H. & HUA, B. 2022. Clinical efficacy and safety of Traditionaltraditional medicine preparations combined with chemotherapy for advanced pancreatic cancer: a systematic review and meta-analysis. *Front Oncol,* 12, 828450.

HÜTTNER, F. J., ROOMAN, I., BOUCHE, G., KNEBEL, P., HÜSING, J., MIHALJEVIC, A. L., HACKERT, T., STROBEL, O., BÜCHLER, M. W. & DIENER, M. K. 2020. Pancreatic resection with perioperative drug repurposing of propranolol and etodolac: trial protocol of the phase-II randomised placebo controlled PROSPER trial. *BMJ Open,* 10**,** e040406.

HUYNH, E., HOSNY, A., GUTHIER, C., BITTERMAN, D. S., PETIT, S. F., HAAS-KOGAN, D. A., KANN, B., AERTS, H. J. & MAK, R. H. 2020. Artificial intelligence in radiation oncology. *Nat Rev Clin Oncol,* 17**,** 771–81.

IHUA, B. & NSOFOR, I. 2019. Emigration of Nigerian medical doctors survey report. Technical Report for NOI Polls and Nigeria Health Watch, Abuja, Nigeria.

INTERNATIONAL, W. C. R. F. 2024. *Pancreatic cancer statistics.* Available: https://www.wcrf.org/cancer-trends/pancreatic-cancer-statistics/ [Accessed July 21, 2024].

INTITUTE, N. C. 2012. *DevCan - Probability of developing or dying of cancer, version 6.6.1.* Available: https://surveillance.cancer.gov/ [Accessed July 21, 2024].

IYER, P., ZEKRI, A. R., HUNG, C. W., SCHIEFELBEIN, E., ISMAIL, K., HABLAS, A., SEIFELDIN, I. A. & SOLIMAN, A. S. 2010. Concordance of DNA methylation pattern in plasma and tumor DNA of Egyptian hepatocellular carcinoma patients. *Exp Mol Pathol,* 88**,** 107–11.

JONAS, E. & BRAND, M. 2023. Early detection of pancreatic adenocarcinoma. *South African J Surg,* 61**,** 2–4.

KANAI, M. 2014. Therapeutic applications of curcumin for patients with pancreatic cancer. *World J Gastroenterol,* 20**,** 9384–91.

KENYA, L. O. 2012. *Cancer prevention and control.* Available: http://guidelines.health.go.ke:8000/media/CancerPreventionandControlAct15of2012.pdf [Accessed July 13, 2024].

KHAN, Z. & THOMSON, S. 2018. New technology in the management of pancreatic cancer. *South African Gastroenterol Rev,* 16**,** 13–17.

KHANYILE, R., MARIMA, R., MBEJE, M., MUTAMBIRWA, S., MONTWEDI, D. & DLAMINI, Z. 2023. AI tools offering cancer clinical applications for risk predictor, early detection, diagnosis, and accurate prognosis: perspectives in personalised care. In DLAMINI, Z. (ed.) *Artificial* intelligence *and* precision oncology: bridging cancer research *and* clinical decision *support* (pp. 293–312). Cham: Springer.

KIMURA, H., KLEIN, A. P., HRUBAN, R. H. & ROBERTS, N. J. 2021. The role of inherited pathogenic CDKN2A variants in susceptibility to pancreatic cancer. *Pancreas,* 50**,** 1123–30.

KIRK, G. D., LESI, O. A., MENDY, M., SZYMAÑSKA, K., WHITTLE, H., GOEDERT, J. J., HAINAUT, P. & MONTESANO, R. 2005. 249(ser) TP53 mutation in plasma DNA, hepatitis B viral infection, and risk of hepatocellular carcinoma. *Oncogene,* 24**,** 5858–67.

KIVUYO, N. E. & AKOKO, L. O. 2022. Patients' characteristics, diagnosis, and management of pancreatic masses in low resource settings. *Ann Pancr Cancer,* 5.

KLATTE, D. C. F., WALLACE, M. B., LÖHR, M., BRUNO, M. J. & VAN LEERDAM, M. E. 2022. Hereditary pancreatic cancer. *Best Pract Res Clin Gastroenterol,* 58–59**,** 101783.

KLETT, H., FUELLGRAF, H., LEVIT-ZERDOUN, E., HUSSUNG, S., KOWAR, S., KÜSTERS, S., BRONSERT, P., WERNER, M., WITTEL, U., FRITSCH, R., BUSCH, H. & BOERRIES, M. 2018. Identification and validation of a diagnostic and prognostic multi-gene biomarker panel for pancreatic ductal adenocarcinoma. *Front Genet,* 9**,** 108.

KNUDSEN, K. A. & WHEELOCK, M. J. 1992. Plakoglobin, or an 83-kD homologue distinct from beta-catenin, interacts with E-cadherin and N-cadherin. *J Cell Biol,* 118**,** 671–9.

KUMAR, A., P, N., KUMAR, M., JOSE, A., TOMER, V., OZ, E., PROESTOS, C., ZENG, M., ELOBEID, T., K, S. & OZ, F. 2023. Major phytochemicals: recent advances in health benefits and extraction method. *Molecules,* 28(2), 887. 10.3390/molecules28020887

LAHERU, D., LUTZ, E., BURKE, J., BIEDRZYCKI, B., SOLT, S., ONNERS, B., TARTAKOVSKY, I., NEMUNAITIS, J., LE, D., SUGAR, E., HEGE, K. & JAFFEE, E. 2008. Allogeneic granulocyte macrophage colony-stimulating factor–secreting tumor immunotherapy alone or in sequence with cyclophosphamide for metastatic pancreatic cancer: a pilot study of safety, feasibility, and immune activation. *Clin Cancer Res,* 14**,** 1455–1463.

LENZO, F. L., KATO, S., PABLA, S., DEPIETRO, P., NESLINE, M. K., CONROY, J. M., BURGHER, B., GLENN, S. T., KUVSHINOFF, B., KURZROCK, R. & MORRISON, C. 2021. Immune profiling and immunotherapeutic targets in pancreatic cancer. *Ann Transl Med,* 9**,** 119.

LI, L., AGGARWAL, B. B., SHISHODIA, S., ABBRUZZESE, J. & KURZROCK, R. 2004. Nuclear factor-kappaB and IkappaB kinase are constitutively active in human pancreatic cells, and their down-regulation by curcumin (diferuloylmethane) is associated with the suppression of proliferation and the induction of apoptosis. *Cancer,* 101**,** 2351–62.

LIU, P., YING, Q., LIU, H., YU, S. Q., BU, L. P., SHAO, L. & LI, X. Y. 2020. Curcumin enhances anti-cancer efficacy of either gemcitabine or docetaxel on pancreatic cancer cells. *Oncol Rep,* 44, 1393–1402.

LUNA-SANCHEZ, S., LOBETO MARTÍNEZ, M. R. & GARAY MUÑOZ, J. S. 2014. Occupational exposure to chlorinated hydrocarbons and pancreatic cancer: review of recent literature. *Medicina y Seguridad del Trabajo* 60(235), 420–433.

LYNCH, H. T., SMYRK, T., KERN, S. E., HRUBAN, R. H., LIGHTDALE, C. J., LEMON, S. J., LYNCH, J. F., FUSARO, L. R., FUSARO, R. M. & GHADIRIAN, P. 1996. Familial pancreatic cancer: a review. *Semin Oncol,* 23, 251–75.

MAC'YONG, N. 2020. *Pancreatic cancer awereness walk, running and cycling.* Available: https://web.facebook.com/events/ratanda-stadium/pancreatic-cancer-awareness-walk-run-and-cycling/1267556416915757/?_rdc=1&_rdr [Accessed July 21, 2024].

MADELA, F., SITHOLE, M., NTANZI, N., CHILIZA, K., MWAZHA, A., THOMSON, S. & ALDOUS, C. 2024. Correlation between chronic pancreatitis and pancreatic cancer at a Central Hospital in KwaZulu-Natal, South Africa. *HPB,* 26, S347.

MALAMI, S. A., ALIYU, U. M., NWOKWU, U. E. & MOHAMMED, W. U. 2023. Abstract A124: NICRAT: the institute pioneering a new approach to cancer research and treatment in sub-Saharan Africa. *Cancer Epidemiol Biomarkers Prev,* 32, A124.

MANSON, E. N., HASFORD, F., TRAUERNICHT, C., IGE, T. A., INKOOM, S., INYANG, S., SAMBA, O., KHELASSI-TOUTAOUI, N., LAZARUS, G., SOSU, E. K., POKOO-AIKINS, M. & STOEVA, M. 2023. Africa's readiness for artificial intelligence in clinical radiotherapy delivery: medical physicists to lead the way. *Physica Medica,* 113, 102653.

MAPHUMULO, W. T. & BHENGU, B. R. 2019. Challenges of quality improvement in the healthcare of South Africa post-apartheid: a critical review. *Curationis,* 42, e1–9.

MARCHIO, A., AMOUGOU ATSAMA, M., BÉRÉ, A., KOMAS, N. P., NOAH NOAH, D., ATANGANA, P. J. A., CAMENGO-POLICE, S. M., NJOUOM, R., BEKONDI, C. & PINEAU, P. 2018. Droplet digital PCR detects high rate of TP53 R249S mutants in cell-free DNA of middle African patients with hepatocellular carcinoma. *Clin Exp Med,* 18, 421–31.

MASSALEE, R. & CAO, X. 2023. Repurposing beta-blockers for combinatory cancer treatment: effects on conventional and immune therapies. *Front Pharmacol,* 14.

MATATA, D. Z., NGASSAPA, O. D., MACHUMI, F. & MOSHI, M. J. 2018. Screening of plants used as traditional anticancer remedies in Mkuranga and same districts, Tanzania, using brine shrimp toxicity bioassay. *Evid Based Complement Alternat Med,* 2018, 3034612.

MATLEBJANE, D. A. 2022. *Adoption of blockchain to enhance patients' health information management,* Johannesburg: University of Johannesburg, South Africa.

MATLEBJANE, D. A. & NDAYIZIGAMIYE, P. 2022. Antecedents of the adoption of blockchain to enhance patients' health information management in South Africa. *South African J Inform Manag,* 24, 1–9.

MEYER, S., HARRIES, J., TORODE, J. & BALDWIN-RAGAVEN, L. 2024. Harnessing the law to advance equitable cancer care in South Africa: exploring the feasibility, desirability and added value of a dedicated national cancer act. *Ecancermedicalscience,* 18, 1658.

MIYA, T. V., MOSOANE, B., LOLAS, G. & DLAMINI, Z. 2023. Healthcare transformation using blockchain technology in the era of society 5.0. In DLAMINI, Z. (ed.) *Society 5.0 and* next generation healthcare: patient-focused *and* technology-assisted precision *therapies* (pp. 223–248). Cham: Springer.

MOFFITT, R. A., MARAYATI, R., FLATE, E. L., VOLMAR, K. E., LOEZA, S. G., HOADLEY, K. A., RASHID, N. U., WILLIAMS, L. A., EATON, S. C., CHUNG, A. H., SMYLA, J. K., ANDERSON, J. M., KIM, H. J., BENTREM, D. J., TALAMONTI, M. S., IACOBUZIO-DONAHUE, C. A., HOLLINGSWORTH, M. A. & YEH, J. J. 2015. Virtual microdissection identifies distinct tumor- and stroma-specific subtypes of pancreatic ductal adenocarcinoma. *Nat Genet,* 47, 1168–78.

MORHASON-BELLO, I. O., ODEDINA, F., REBBECK, T. R., HARFORD, J., DANGOU, J. M., DENNY, L. & ADEWOLE, I. F. 2013. Challenges and opportunities in cancer control in Africa: a perspective from the African Organisation for Research and Training in Cancer. *Lancet Oncol,* 14, e142–51.

MTHUNZI, R. & NOEL, C. 2023. Geographic distribution of pancreaticobiliary malignancy in central South Africa presenting to the Universitas Academic Hospital Complex. *South African J Surg,* 61, 7–11.

MUCHIRI, S. W., OGUTU, E. O., OYIRO, P. O. & AABAKKEN, L. 2024. Clinicopathologic characteristics and treatment outcomes of pancreatic cancer patients at a tertiary referral hospital in Kenya. *Ecancermedicalscience,* 18, 1682.

NAKAMURA, S., SADAKARI, Y., OHTSUKA, T., OKAYAMA, T., NAKASHIMA, Y., GOTOH, Y., SAEKI, K., MORI, Y., NAKATA, K., MIYASAKA, Y., ONISHI, H., ODA, Y., GOGGINS, M. & NAKAMURA, M. 2019. Pancreatic juice exosomal microRNAs as biomarkers for detection of pancreatic ductal adenocarcinoma. *Ann Surg Oncol,* 26**,** 2104–11.

NESTERUK, K., LEVINK, I. J. M., DE VRIES, E., VISSER, I. J., PEPPELENBOSCH, M. P., CAHEN, D. L., FUHLER, G. M. & BRUNO, M. J. 2022. Extracellular vesicle-derived microRNAs in pancreatic juice as biomarkers for detection of pancreatic ductal adenocarcinoma. *Pancreatology,* 22**,** 626–35.

NGWENA, C. 2003. Access to health care services as a justiciable socio-economic right under the South African constitution. *Med Law Int,* 6**,** 13–23.

NHLEKO, M., EDOKA, I. & MUSENGE, E. 2024. Pancreatic cancer mortality in South Africa: a case-control study. *South African Med Journal,* 114**,** 27–32.

NKWE, D. O., LOTSHWAO, B., RANTONG, G., MATSHWELE, J., KWAPE, T. E., MASISI, K., GAOBOTSE, G., HEFFERON, K. & MAKHZOUM, A. 2021. Anticancer mechanisms of bioactive compounds from solanaceae: an update. *Cancers,* 13(19), 4989.

NORRIS, A. L., ROBERTS, N. J., JONES, S., WHEELAN, S. J., PAPADOPOULOS, N., VOGELSTEIN, B., KINZLER, K. W., HRUBAN, R. H., KLEIN, A. P. & ESHLEMAN, J. R. 2015. Familial and sporadic pancreatic cancer share the same molecular pathogenesis. *Familial Cancer,* 14**,** 95–103.

NSINGWANE, Z., CANDY, G., DEVAR, J., OMOSHORO-JONES, J., SMITH, M. & NWEKE, E. 2020. Immunotherapeutic strategies in pancreatic ductal adenocarcinoma (PDAC): current perspectives and future prospects. *Mol Biol Rep,* 47**,** 6269–80.

NTWASA, M. & DLAMINI, Z. 2024. Editorial: molecular targets for anticancer drug discovery and development. *Front Genet,* 15**,** 1374867.

NWEKE, E., NTWASA, M., BRAND, M., DEVAR, J., SMITH, M. & CANDY, G. 2020. Increased expression of plakoglobin is associated with upregulated MAPK and PI3K/AKT signalling pathways in early resectable pancreatic ductal adenocarcinoma. *Oncol Lett,* 19**,** 4133–41.

NWEKE, E. E. & THIMIRI GOVINDA RAJ, D. B. 2021. Drug sensitivity and drug repurposing platform for cancer precision medicine. *Adv Exp Med Biol.* 1326, 47–53. doi: 10.1007/5584_2021_622. PMID: 33629259.

O'BRIEN, N., LI, E., CHAIBVA, C. N., GOMEZ BRAVO, R., KOVACEVIC, L., KWAME AYISI-BOATENG, N., LOUNSBURY, O., NWABUFO, N. F. F., SENKYIRE, E. K., SERAFINI, A., SURAFEL ABAY, E., VAN DE VIJVER, S., WANJALA, M., WANGARI, M. C., MOOSA, S. & NEVES, A. L. 2023. Strengths, weaknesses, opportunities, and threats analysis of the use of digital health technologies in primary health care in the Sub-Saharan African region: qualitative study. *J Med Internet Res,* 25**,** e45224.

OCHWANG'I, D. O., KIMWELE, C. N., ODUMA, J. A., GATHUMBI, P. K., MBARIA, J. M. & KIAMA, S. G. 2014. Medicinal plants used in treatment and management of cancer in Kakamega County, Kenya. *J Ethnopharmacol,* 151**,** 1040–55.

OMARA, T., KIPROP, A. K., RAMKAT, R. C., CHERUTOI, J., KAGOYA, S., MORAA NYANGENA, D., AZEZE TEBO, T., NTEZIYAREMYE, P., NYAMBURA KARANJA, L., JEPCHIRCHIR, A., MAIYO, A., JEMATIA KIPTUI, B., MBABAZI, I., KIWANUKA NAKIGULI, C., NAKABUYE, B. V. & CHEPKEMOI KOSKE, M. 2020. Medicinal plants used in traditional management of cancer in Uganda: a review of ethnobotanical surveys, phytochemistry, and anticancer studies. *Evid Based Complement Alternat Med,* 2020**,** 3529081.

OMONISI, A. E., LIU, B. & PARKIN, D. M. 2020. Population-based cancer registration in Sub-Saharan Africa: its role in research and cancer control. *JCO Glob Oncol,* 6**,** 1721–8.

OMOTOSO, O., TEIBO, J. O., ATIBA, F. A., OLADIMEJI, T., PAIMO, O. K., ATAYA, F. S., BATIHA, G. E. & ALEXIOU, A. 2023. Addressing cancer care inequities in sub-Saharan Africa: current challenges and proposed solutions. *Int J Equity Health,* 22**,** 189.

PALMER, D. H., VALLE, J. W., MA, Y. T., FALUYI, O., NEOPTOLEMOS, J. P., JENSEN GJERTSEN, T., IVERSEN, B., AMUND ERIKSEN, J., MØLLER, A.-S., AKSNES, A.-K., MILLER, R. & DUELAND, S. 2020. TG01/GM-CSF and adjuvant gemcitabine in patients with resected RAS-mutant adenocarcinoma of the pancreas (CT TG01-01): a single-arm, phase 1/2 trial. *Br J Cancer,* 122**,** 971–7.

PANDOL, S. J., APTE, M. V., WILSON, J. S., GUKOVSKAYA, A. S. & EDDERKAOUI, M. 2012. The burning question: why is smoking a risk factor for pancreatic cancer? *Pancreatology,* 12**,** 344–9.

PARTYKA, O., PAJEWSKA, M., KWAŚNIEWSKA, D., CZERW, A., DEPTAŁA, A., BUDZIK, M., CIPORA, E., GĄSKA, I., GAZDOWICZ, L., MIELNIK, A., SYGIT, K., SYGIT, M., KRZYCH-FAŁTA, E., SCHNEIDER-MATYKA, D., GROCHANS, S., CYBULSKA, A. M., DROBNIK, J., BANDURSKA,

E., CIEĆKO, W., RATAJCZAK, P., KAMECKA, K., MARCZAK, M. & KOZŁOWSKI, R. 2023. Overview of pancreatic cancer epidemiology in Europe and recommendations for screening in high-risk populations. *Cancers,* 15(14), 3634.

PEARLSTEIN, E. F., TAZINKENG, N. N., SAWYER, K., CHUKWUDIKE, E. S., BAINS, K., ASANTE, C., IBRAHIM OMAR SUNDI, P. R., CHISHINGA, N. & ASOMBANG, A. W. 2021. S61. Epidemiology of pancreatic cancer in Africa: a systematic review. *Am J Gastroenterol,* 116, S26–7.

QORRI, B., HARLESS, W. & SZEWCZUK, M. R. 2020. Novel molecular mechanism of aspirin and celecoxib targeting mammalian neuraminidase-1 impedes epidermal growth factor receptor signaling axis and induces apoptosis in pancreatic cancer cells. *Drug Des Devel Ther,* 14, 4149–67.

REBELO, R., POLÓNIA, B., SANTOS, L. L., VASCONCELOS, M. H. & XAVIER, C. P. R. 2021. Drug repurposing opportunities in pancreatic ductal adenocarcinoma. *Pharmaceuticals,* 14 (3), 280. doi: 10.3390/ph14030280

REYNOLDS, S. 2023. *An mRNA vaccine to treat pancreatic cancer.* National Health Institute. Available: https://www.nih.gov/news-events/nih-research-matters/mrna-vaccine-treat-pancreatic-cancer [Accessed August 8, 2024].

SAAD, A., GOLDSTEIN, J., MARGALIT, O., SHACHAM-SHMUELI, E., LAWRENCE, Y. R., YANG, Y. X., REISS, K. A., GOLAN, T., MAMTANI, R., HALPERN, N., ADERKA, D., MOUALLEM, M., GOLDSTEIN, A., GIANTONIO, B. & BOURSI, B. 2020. Assessing the effects of beta-blockers on pancreatic cancer risk: a nested case-control study. *Pharmacoepidemiol Drug Saf,* 29, 599–604.

SANTIBAÑEZ, M., VIOQUE, J., ALGUACIL, J., DE LA HERA, M. G., MORENO-OSSET, E., CARRATO, A., PORTA, M. & KAUPPINEN, T. 2010. Occupational exposures and risk of pancreatic cancer. *Eur J Epidemiol,* 25, 721–30.

SARKIN, J. 2000. A review of health and human rights after five years of democracy in South Africa. *Med Law,* 19, 287–307.

SARMENTO-RIBEIRO, A. B., SCORILAS, A., GONÇALVES, A. C., EFFERTH, T. & TROUGAKOS, I. P. 2019. The emergence of drug resistance to targeted cancer therapies: clinical evidence. *Drug Resist Updat,* 47, 100646.

SCHOTTENFELD, J. F. 2006. *Cancer Epidemiology and Prevention*, Oxford: Oxford University Press.

SELLAM, F., HARIR, N., KHALED, M. B., MRABENT, N. M., SALAH, R., BENCHOUK, A. & DIAF, M. 2015a. Delayed diagnosis of pancreatic cancer reported as more common in a population of North African young adults. *J Gastroint Oncol,* 6, 505–10.

SELLAM, F., MRABENT, N. M., HARIR, N., KHALED, M. B., SALAH, R., DIAF, M. & TOU, A. 2015b. Pancreatic cancer in a Northern African population: a retrospective analysis spanning two decades. *JOP J Pancreas,* 16, 444–8.

SEWARD, N., HANLON, C., ABDELLA, A., ABRAHAMS, Z., ALEM, A., ARAYA, R., BACHMANN, M., BEKELE, A., BOGALE, B., BRIMA, N., CHIBANDA, D., CURRAN, R., DAVIES, J., BEYENE, A., FAIRALL, L., FARRANT, L., FRISSA, S., GALLAGHER, J., GAO, W., GWYTHER, L., HARDING, R., KARTHA, M. R., LEATHER, A., LUND, C., MARX, M., NKHOMA, K., MURDOCH, J., PETERSEN, I., PETRUS, R., VAN RENSBURG, A., SANDALL, J., SEVDALIS, N., SHEENAN, A., TADESSE, A., THORNICROFT, G., VERHEY, R., WILLOTT, C. & PRINCE, M. 2022. HeAlth System StrEngThening in four sub-Saharan African countries (ASSET) to achieve high-quality, evidence-informed surgical, maternal and newborn, and primary care: protocol for pre-implementation phase studies. *Glob Health Action,* 15, 1987044.

SIEGEL, R. L., MILLER, K. D., WAGLE, N. S. & JEMAL, A. 2023. Cancer statistics, 2023. *CA Cancer J Clin,* 73 (1), 17–48. doi: 10.3322/caac.21763.

SOLIMAN, A. S., ZHANG, Q., SALEH, T., ZARZOUR, A., SELIM, M., ABDEL-FATTAH, M. & ABBRUZZESE, J. L. 2006. Pancreatic cancer mortality in Egypt: comparison to the United States pancreatic cancer mortality rates. *Cancer Detect Prevent,* 30, 473–9.

SONBOL, A. A., EL-GAMMAL, A. S., ABO EL-YAZEED, A. S. & MONTASER, L. M. 2019. Role of circulating endothelial progenitor cells in patients with breast cancer. *Menoufia Med J,* 32, 770–6.

SUBHAN, M., SAJI PAREL, N., KRISHNA, P. V., GUPTA, A., UTHAYASEELAN, K., UTHAYASEELAN, K. & KADARI, M. 2022. Smoking and pancreatic cancer: smoking patterns, tobacco type, and dose-response relationship. *Cureus,* 14, e26009.

SUNG, H., FERLAY, J., SIEGEL, R. L., LAVERSANNE, M., SOERJOMATARAM, I., JEMAL, A. & BRAY, F. 2021. Global Cancer Statistics 2020: GLOBOCAN estimates of incidence and mortality worldwide for 36 cancers in 185 countries. *CA Cancer J Clin,* 71, 209–49.

SZYMAŃSKA, K., LESI, O. A., KIRK, G. D., SAM, O., TANIERE, P., SCOAZEC, J. Y., MENDY, M., FRIESEN, M. D., WHITTLE, H., MONTESANO, R. & HAINAUT, P. 2004. Ser-249TP53 mutation in tumour and plasma DNA of hepatocellular carcinoma patients from a high incidence area in the Gambia, West Africa. *Int J Cancer,* 110**,** 374–9.

TEMILOLA, D. O., ADEOLA, H. A., GROBBELAAR, J. & CHETTY, M. 2023. Liquid biopsy in head and neck cancer: its present state and future role in Africa. *Cells,* 12**,** 2663.

TEMILOLA, D. O., WIUM, M., COULIDIATI, T. H., ADEOLA, H. A., CARBONE, G. M., CATAPANO, C. V. & ZERBINI, L. F. 2019. The prospect and challenges to the flow of liquid biopsy in Africa. *Cells,* 8(8), 862. doi: 10.3390/cells8080862.

TREATMENT, N. F. G. E. T. N. I. F. C. R. A. 2024. *About us*. National Institute for Cancer Research and Treatment (NICRAT). Available: https://www.nicrat.gov.ng/#:~:text=The%20National%20Institute%20for%20Cancer%20Research%20and%20Treatment%20(NICRAT)%20was,Treatment%20and%20Control%20in%20Nigeria. [Accessed August 16, 2024].

TRIANTAFILLIDIS, J. K., TRIANTAFYLLIDI, E., SIDERIS, M., PITTARAS, T. & PAPALOIS, A. E. 2022. Herbals and plants in the treatment of pancreatic cancer: a systematic review of experimental and clinical studies. *Nutrients,* 14(3), 619. doi: 10.3390/nu14030619.

UGANDA CANCER INSTITUTE ACT, A. O. 2017. *Uganda Cancer Institute Act, 2016*. Uganda. Available: https://ulii.org/akn/ug/act/2016/19/eng@2016-11-29 [Accessed July 13, 2024].

VAN DER VAART, M. & PRETORIUS, P. J. 2008. A method for characterization of total circulating DNA. *Ann N Y Acad Sci,* 1137**,** 92–7.

VAN DER VAART, M., SEMENOV, D. V., KULIGINA, E. V., RICHTER, V. A. & PRETORIUS, P. J. 2009. Characterisation of circulating DNA by parallel tagged sequencing on the 454 platform. *Clin Chim Acta,* 409**,** 21–7.

WANG, C., ZHANG, T., LIAO, Q., DAI, M., GUO, J., YANG, X., TAN, W., LIN, D., WU, C. & ZHAO, Y. 2021. Metformin inhibits pancreatic cancer metastasis caused by SMAD4 deficiency and consequent HNF4G upregulation. *Protein Cell,* 12**,** 128–44.

WANG, H., ZHANG, Y., ZHANG, H., CAO, H., MAO, J., CHEN, X., WANG, L., ZHANG, N., LUO, P., XUE, J., QI, X., DONG, X., LIU, G. & CHENG, Q. 2024. Liquid biopsy for human cancer: cancer screening, monitoring, and treatment. *MedComm,* 5**,** e564.

WANG, Q. J., YU, Z., GRIFFITH, K., HANADA, K., RESTIFO, N. P. & YANG, J. C. 2016. Identification of T-cell receptors targeting KRAS-mutated human tumors. *Cancer Immunol Res,* 4**,** 204–14.

WORLD HEALTH OGANISATION, A. R. 2019. *Greater public awareness, stronger collaboration, early detection and screening are critical to fight cancer in Africa*. Available: https://www.afro.who.int/news/greater-public-awareness-stronger-collaboration-early-detection-and-screening-are-critical [Accessed July 21, 2024].

YANG, A., RAJESHKUMAR, N. V., WANG, X., YABUUCHI, S., ALEXANDER, B. M., CHU, G. C., VON HOFF, D. D., MAITRA, A. & KIMMELMAN, A. C. 2014. Autophagy is critical for pancreatic tumor growth and progression in tumors with p53 alterations. *Cancer Discov,* 4**,** 905–13.

YOUSEF, A., YOUSEF, M., CHOWDHURY, S., ABDILLEH, K., KNAFL, M., EDELKAMP, P., ALFARO-MUNOZ, K., CHACKO, R., PETERSON, J., SMAGLO, B. G., WOLFF, R. A., PANT, S., LEE, M. S., WILLIS, J., OVERMAN, M., DOSS, S., MATRISIAN, L., HURD, M. W., SNYDER, R., KATZ, M. H. G., WANG, H., MAITRA, A., SHEN, J. P. & ZHAO, D. 2024. Impact of KRAS mutations and co-mutations on clinical outcomes in pancreatic ductal adenocarcinoma. *NPJ Precis Oncol,* 8**,** 27.

YUE, Q., GAO, G., ZOU, G., YU, H. & ZHENG, X. 2017. Natural products as adjunctive treatment for pancreatic cancer: recent trends and advancements. *Biomed Res Int,* 2017**,** 8412508.

ZAHER, E. R., ANWAR, M. M., KOHAIL, H. M., EL-ZOGHBY, S. M. & ABO-EL-ENEEN, M. S. 2013. Cell-free DNA concentration and integrity as a screening tool for cancer. *Indian J Cancer,* 50**,** 175–83.

ZAMBIA, T. 2022. *Today is world pancreatic cancer day.* Available: https://web.facebook.com/teledoctorzambia/posts/3335557560048697/?_rdc=1&_rdr [Accessed July 21, 2024].

ZEHENTNER, B. K., PERSING, D. H., DEME, A., TOURE, P., HAWES, S. E., BROOKS, L., FENG, Q., HAYES, D. C., CRITICHLOW, C. W., HOUGHTON, R. L. & KIVIAT, N. B. 2004. Mammaglobin as a novel breast cancer biomarker: multigene reverse transcription-PCR assay and sandwich ELISA. *Clin Chem,* 50**,** 2069–76.

ZINKOVSKY, D. & SOOD, M. R. 2023. Isolated JUP plakoglobin gene mutation with left ventricular fibrosis in familial arrhythmogenic right ventricular cardiomyopathy. *J Cardiovasc Electrophysiol,* 34**,** 2112–21.

11 Building Partnerships for Progress

Thabiso Victor Miya[1*], *Godfrey Grech*[2], *and Zodwa Dlamini*[1]

[1]SAMRC Precision Oncology Research Unit (PORU), DSI/NRF SARChI Chair in Precision Oncology and Cancer Prevention (POCP), Pan African Cancer Research Institute (PACRI), University of Pretoria, Pretoria, South Africa.

[2]Department of Pathology, Faculty of Medicine and Surgery, University of Malta, Malta.

*Corresponding author: Thabiso Victor Miya

COLLABORATION AND INTEGRATION: THE FUTURE OF 21ST-CENTURY HEALTHCARE

The 21st century has seen a rise in domestic and global partnerships, collaborations, and integration of organizations within the healthcare ecosystem. Groups of healthcare providers are now the norm, making solo practitioners increasingly uncommon. Through cooperative agreements and mergers and acquisitions, healthcare systems are extending their reach not only within their local area but also nationally and beyond. In addition to being team-based, research teams may now consist of personnel from market-sharing firms. An interdisciplinary approach has been adopted in medical research, bringing together academia, healthcare, industry, and NGOs. Together, payers, businesses, professional member associations, and patient advocacy groups are now more willing to cooperate and lobby on common health policy priorities. This scenario sums up the current state of cancer care (Nevidjon, 2016).

Since the isolated method of addressing societal issues in the past was ineffective, collaborations between governmental, nonprofit, and commercial sectors are required to address issues in the 21st century (Rosenbaum et al., 2013). Organizational competition has not vanished. Nonetheless, there is a growing understanding of how interconnected the healthcare industry is and that new levels of cooperation and skillsets will be necessary to achieve the triple purpose of healthcare reform. Since real-time connectivity is available around the clock, every day of the week, both individuals and companies may collaborate seamlessly at any time and from any location. Global partnership prospects have also risen as a result of this connectivity. Numerous important variables, such as the complexity of the healthcare environment, the demands in the political and economic climate, knowledge specialization, technology, and resource availability, will continue to drive companies to collaborate, coordinate, and integrate (O'Neil, 2013). However, among the professional and advocacy member associations, redundancy and duplication are not only present but have been growing. In addition to having representation from the same agency or organization in each coalition, there is overlap and duplication among coalitions. How these relationships might alter in the future to mobilize and utilize available resources and provide better results is a question. All sides may profit from a partnership, but they can also be difficult to establish and manage (Nevidjon, 2016).

There are numerous instances of effective team-based research and care delivery strategies in the context of cancer care. Nonetheless, several Institute of Medicine (IOM) studies note that effective interprofessional teams are required to address the challenging clinical problems that lie ahead.

DOI: 10.1201/9781003544401-16

These difficulties include the patient's ongoing financial burden, the pipeline of new diagnostic and therapeutic breakthroughs, and the future workforce. Cancer nurses play a crucial role in helping healthcare providers solve challenging problems, but it is possible that they do not currently have as much of an impact on these partnerships to contribute to the future. Partnerships start on an individual basis and then spread to businesses and professions (Nevidjon, 2016). Thus, this chapter discusses and evaluates the significance of collaborations across governments, communities, healthcare organizations, research institutions, and other stakeholders in improving pancreatic cancer awareness, prevention, diagnosis, and outcomes. It also examines the hurdles of forging these partnerships, as well as prospective cancer-related research and technological advancements.

BUILDING EFFECTIVE AND SUSTAINABLE PARTNERSHIPS IN PANCREATIC CANCER

The Importance of Collaboration and Partnership

Pancreatic cancer is distinguished by its aggressiveness, late-stage diagnosis, scarcity of available treatments, and low odds of survival (Sun et al., 2020; Liu et al., 2021). However, the pancreatic cancer burden varies among populations (Martin and Snyder, 2023; Katona et al., 2023; Brooks et al., 2023; Halder et al., 2023). Thus, pancreatic cancer induces substantial health disparities, and improving outcomes and lowering disparities will require a multidisciplinary approach. Pancreatic cancer-related health disparities can take many different forms, including unequal access to healthcare, discrepancies in disease awareness and screening, variations in treatment outcomes, and differing rates of disease prevalence and mortality among diverse population groups (Behar-Horenstein et al., 2020; Xu et al., 2019).

Addressing pancreatic cancer health inequities necessitates collaborations among multiple disciplines (Figure 11.1). There are several obstacles and future directions in this field. Precision medicine developments could lead to better pancreatic cancer results. Working together, scientists, oncologists, and geneticists can help find and confirm more biomarkers linked to pancreatic cancer risk, prognosis, and therapy response (El Kaoutari et al., 2023; Dhani et al., 2023; Zhao et al., 2023). These kinds of partnerships can help with the creation of individualized treatment plans and focused therapies for certain individuals. To put new ideas into practice that increase access to high-quality care and reduce disparities, researchers and healthcare systems must work together (Roth

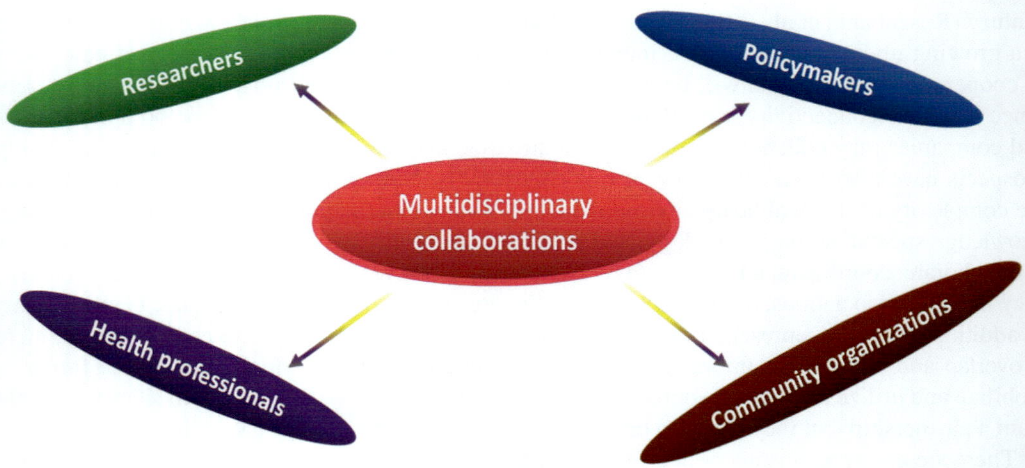

FIGURE 11.1 Various disciplines need to collaborate to address pancreatic cancer health inequities. These include public health professionals, researchers, community organizations, and policymakers.

et al., 2022; Wang et al., 2022). This involves programs like community health worker interventions, telehealth services, and patient navigation. Healthcare systems and researchers can ensure the efficacy of these interventions in a variety of groups by collaborating to implement and assess them. Outreach and community involvement programs can be supported by interdisciplinary partnerships between researchers, public health specialists, and community organizations (Wall et al., 2023).

Advancing pancreatic cancer Interventions through Multidisciplinary Collaboration

Through cooperation, culturally suitable interventions and educational campaigns can be developed by considering the requirements, beliefs, and preferences of the community. These initiatives can help remove obstacles to screening, treatment, and early diagnosis (Raoof et al., 2022), as well as advance health literacy and increase public knowledge of pancreatic cancer disparities. Advocating for policy changes that address pancreatic cancer disparities requires multidisciplinary collaborations between researchers and policymakers. Working together can entail carrying out research, sharing the results, influencing evidence-based policy, and promoting more financing for treatment, prevention, and research (Bledsoe and Grizzle, 2013; Howell et al., 2013). Engaging with legislators can identify and implement legislation that promotes fair access to healthcare. To share and integrate data to better understand pancreatic cancer discrepancies, collaboration between researchers, clinicians, and health systems is crucial. Collaborating to share resources and data enables researchers to perform thorough analyses and spot trends associated with discrepancies. As well as facilitating the creation of registries to track and assess initiatives aimed at reducing health disparities, multidisciplinary cooperation can aid in the development of standardized data collection techniques. It is essential to set up efficient platforms and communication channels for knowledge sharing and teamwork. It is frequently necessary to allocate resources, such as money, personnel, and infrastructure, to facilitate collaboration. It might be difficult to secure and distribute resources among cooperating companies fairly.

For interdisciplinary cooperation to be successful, funding for research, intervention implementation, and sustainability must be guaranteed. Disciplines may differ in their institutional cultures, methods, and priorities. Respect for one another, a readiness to cooperate, and mutual understanding are necessary to overcome disciplinary silos and close these gaps (Wall et al., 2023). Understanding the pros and cons of collaborations and partnerships enables leaders to make informed judgments about these opportunities (Nevidjon, 2016) (Table 11.1). Developing connections and encouraging a

TABLE 11.1
Pros and Cons of Partnerships among Various Organizations and Stakeholders as Stipulated by Nevidjon (2016)

Pros	Cons
Shared resources	Restricted reach or compromised goals
Combination of each organization's unique strengths and areas of competence	It takes time to establish a relationship and foster trust
New viewpoints that lead to fresh ideas or answers to difficult issues	Inefficiency if everyone fails to carry out their obligations
May provide certain participating organizations with tax benefits	Business expansion might be hindered by agreements including conflicts of interest
Improved financial availability	Resources constrained by organizational versus partnership priorities
Increased sway over policies	Loss of adaptability to act quickly
Improved standing within the company	Credit might not be distributed fairly

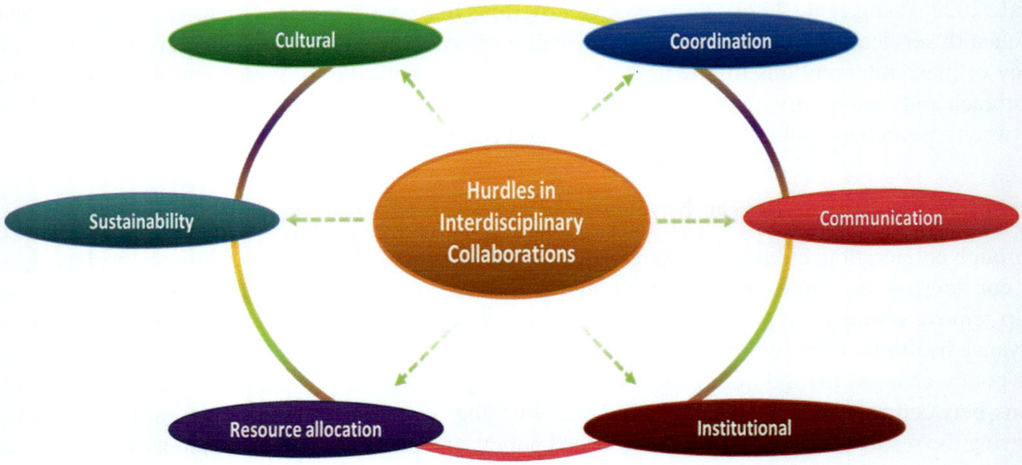

FIGURE 11.2 Hurdles and difficulties in interdisciplinary collaborations include coordination and communication, resource allocation, institutional and cultural hurdles, and sustainability.

collaborative spirit among involved parties can assist in removing these obstacles. It can be difficult to keep up long-term partnerships and to continue interventions after research programs. For multidisciplinary collaborations to be successful over the long term, sustainability methods involving funding, training, and capacity building must be developed along with relationships with community organizations (Wall et al., 2023).

Challenges in Interdisciplinary Partnerships

Several hurdles and difficulties may arise during interdisciplinary partnerships (Figure 11.2). Coordination and good communication between various stakeholders are essential for effective collaboration. It can be difficult to overcome corporate hierarchies, language obstacles, and differences in expertise. It is essential to set up efficient platforms and communication channels for knowledge sharing and teamwork. It is frequently necessary to allocate resources, such as money, personnel, and infrastructure, to facilitate collaboration. It might be difficult to secure and distribute resources among cooperating companies fairly. For interdisciplinary cooperation to be successful, funding for research, intervention implementation, and sustainability must be guaranteed. Disciplines may differ in their institutional cultures, methods, and priorities. Respect for one another, a readiness to cooperate, and mutual understanding are necessary to overcome disciplinary silos and close these gaps (Wall et al., 2023).

Developing connections and encouraging a collaborative spirit among involved parties can assist in removing these obstacles. It can be difficult to keep up long-term partnerships and to continue interventions after research programs. For multidisciplinary collaborations to be successful over the long term, sustainability methods involving funding, training, and capacity building must be developed along with relationships with community organizations. To sum up, multidisciplinary teamwork is essential to resolving health inequities related to pancreatic cancer. Precision medicine, innovations in the health system, community involvement, policy lobbying, and data integration are some of the future directions. Successful collaborations must overcome obstacles to sustainability, institutional impediments, communication, and resource allocation. Stakeholders can improve outcomes for all groups by collaborating to create comprehensive policies, carry out efficient treatments, and reduce pancreatic cancer discrepancies (Wall et al., 2023).

CASE STUDIES OF SUCCESSFUL COLLABORATIONS

Community-Based Participatory research

Community-based participatory research (CBPR) involves the active participation of community members, organizations, and researchers at all phases. CBPR has been effectively used in pancreatic cancer research, addressing health inequities, improving the understanding of the disease, and creating beneficial therapies. The following case studies and best practices highlight the role that CBPR plays in pancreatic cancer. To address health inequities, including pancreatic cancer, in Detroit, community organizations, academic institutions, and researchers have partnered to form the Detroit Urban Research Center (DURC). DURC involved the community in the planning and execution of studies, community outreach initiatives, and policy advocacy campaigns through CBPR (Heisler et al., 2019; Izumi et al., 2010). In addition to increasing community involvement in research, this collaborative strategy facilitated community-driven research goals and produced culturally appropriate pancreatic cancer preventive and treatment approaches. CBPR has been employed to investigate and resolve pancreatic cancer disparities in Latino communities. Researchers identified cultural attitudes, care barriers, and information requirements linked to pancreatic cancer by working with community organizations, health facilities, and individuals of Latino origin (Pulte et al., 2012; Bogumil et al., 2020). To increase Latino populations' access to pancreatic cancer prevention and treatment services, culturally relevant educational materials, community outreach initiatives, and navigation services were developed using the results of these studies (Huang et al., 2021a).

A joint effort called the Appalachian Community Cancer Network (ACCN) uses CBPR to address cancer disparities in the Appalachian region of the United States, where rates of pancreatic cancer incidence and death are high. The ACCN works with researchers, healthcare professionals, and community members to identify needs, create treatments, and assess their efficacy. The ACCN has reduced pancreatic cancer disparities in the Appalachian area by implementing screening programs, community education campaigns, and survivorship support services through CBPR (Zoellner et al., 2021). For those with pancreatic cancer, CBPR has been crucial in the creation of programs for survivorship. The development and implementation of support services, such as peer support groups, training on survivorship education, and resource access, involved the active participation of community members and survivors (Youn et al., 2013; Yang et al., 2018; Arthur et al., 2016). These initiatives increase the quality of life and improve survivorship outcomes by attending to the specific physical, emotional, and psychosocial needs of pancreatic cancer survivors (Wall et al., 2023).

The best practices in CBPR for pancreatic cancer research are:

- *Including the community from the beginning*: Including the community in the stages of study design, implementation, and dissemination guarantees that the findings are relevant to the needs and goals of the community.
- *Developing connections based on trust*: Successful CBPR depends on researchers and community partners developing a relationship based on trust. Collaboration and trust are facilitated by open and honest communication, respect for one another, and acknowledgment of the knowledge and contributions of all parties involved.
- *Including cultural competence*: It is critical to identify and address the cultural determinants of health that affect health behaviors, attitudes, and access to care. To guarantee that interventions are culturally appropriate, cultural competence training for researchers and the incorporation of cultural viewpoints into study protocols are essential.
- *Equitable power and resource allocation*: Both are necessary for community partners and researchers to implement CBPR. A sense of ownership and sustainability is fostered when community members are meaningfully involved in capacity building, resource allocation, and decision-making processes.

- *Building capacity*: To improve community members' research abilities and knowledge, CBPR should involve capacity-building initiatives. This enables community partners to maintain interventions after the research project is finished, bolster their advocacy activities, and actively participate in the research process.

In addressing health inequities, encouraging community engagement, and creating interventions that are sensitive to community needs, CBPR in pancreatic cancer research has proven to be effective. CBPR will encourage more impactful and equitable approaches to pancreatic cancer treatment, prevention, and support by utilizing the knowledge of both researchers and community participants (Wall et al., 2023).

FUTURE DIRECTIONS AND RECOMMENDATIONS

Future Research

The higher incidence of pancreatic cancer in developed nations could be related to certain lifestyle variables. Large-scale prospective studies are required to gain a better understanding of the etiology of pancreatic cancer and the risk factors associated with it. Individuals who have a family history of pancreatic ductal adenocarcinoma may be candidates for screening. Nevertheless, there is disagreement over the ideal imaging method, ideal age, and appropriate interval for screening. To better understand the course of this disease and to facilitate the development of efficient screening and treatment techniques, more retrospective and prospective studies that track these patients with familial pancreatic cancer over time are needed (McGuigan et al., 2018).

It is known that pancreatic intraepithelial neoplasia (PanIN), mucinous cystic neoplasms (MCN), and intraductal papillary mucinous neoplasms (IPMN) are precursors of pancreatic adenocarcinoma (Esposito et al., 2014). Early detection of these lesions in patients and the creation of a suitable follow-up plan will allow for early treatment in high-risk individuals while averting needless surgery for low-risk lesions. This can be accomplished by carrying out sizable prospective and retrospective studies that monitor these patient groups over extended lengths of time, allowing for a deeper comprehension of the illness process. Additionally, it will be possible to identify the risk factors linked to these pre-malignant disorders, which creates the opportunity for target populations to be checked for these conditions. Some patients have shown an improvement in survival since the advent of neoadjuvant therapy, whereas others who had previously treatable disease have developed untreatable disease as a result of their treatment. To determine which patients will benefit most from this strategy, more randomized studies are necessary. The identification of new biomarkers may aid in the process of making decisions and allow for personalized treatment and precision medicine (McGuigan et al., 2018). The shortcomings of the pancreatic cancer treatment plan currently in use highlight the need for fresh approaches to study to make possible discoveries. Oncolytic virus treatment and gene editing technology are two novel therapeutic methods that have shown promise in several pre-clinical and early-phase clinical trials (Chu et al., 2012; Yamamoto et al., 2014). An assessment of various therapeutic approaches by Rouanet et al. offers a great summary of the state of these investigational treatments at the moment (Rouanet et al., 2017).

The cornerstone of effective pancreatic cancer treatment is still surgical resection. Although vascular resection makes it possible to attain clear margins, it is unclear whether this improves survival. Additional retrospective research identifying patients who have received this treatment and the results related to it will contribute to the body of knowledge and aid in the development of future guidelines (McGuigan et al., 2018). The aging of the population and the anticipated growth in the frequency of numerous risk factors for this nearly always fatal cancer will only contribute to the increase in the worldwide burden of pancreatic cancer (Pourshams et al., 2019). Policymakers must get ready for the ongoing increase in the incidence of pancreatic cancer (Klein, 2019). Pancreatic cancer treatment based on medication and surgery should be given high priority in policy agendas and clinical standards given the rising trends in incidence and mortality, especially for industrialized countries. This would help to lower the disease's related mortality (Huang et al., 2021b).

Innovations and Emerging Trends

Innovations in Technology and Access

Innovations in technology have the potential to significantly improve access to information, diagnosis, treatment, and supportive care, thereby mitigating health inequities associated with pancreatic cancer. Prioritizing and addressing the issues around access to and future orientations of technological developments in addressing pancreatic cancer health disparities is imperative. For those living in underdeveloped areas, telemedicine and virtual care technology can improve their access to specialist healthcare services (Portman et al., 2021; Katz et al., 2016). Policymakers, software developers, and healthcare professionals working together can make it easier to put telemedicine platforms in place that support multidisciplinary tumor boards, follow-up treatment, and remote consultations (Satturwar et al., 2020). However, to guarantee that everyone has fair access to virtual care, issues including internet connectivity, digital device access, and privacy concerns must be resolved. Researchers, technology developers, and healthcare professionals can work together to create mobile health applications (apps) that enable users to monitor their symptoms, take their medication, access learning materials, and establish connections with support systems (Gustavell et al., 2020; Gustavell et al., 2019). Those who have trouble making frequent in-person visits or have restricted access to healthcare institutions may find these apps especially helpful. For app adoption and effectiveness to be equitable, user-friendly, culturally aware, and language-accessible apps must be made available (Wall et al., 2023).

The potential of artificial intelligence (AI) and data analytics can be leveraged by healthcare practitioners, researchers, and data scientists working together to enhance pancreatic cancer outcomes. AI algorithms can help with response monitoring, treatment planning, risk prediction, and early diagnosis (Jiang et al., 2023; Tovar et al., 2023). However, there are obstacles to overcome, such as the lack of representative and varied datasets, the need to address biases in AI algorithms, and the need to integrate AI technologies into clinical practice equitably and ethically. The establishment of health information exchange networks can be facilitated by partnerships between technology developers, healthcare institutions, and legislators (Gagliardi et al., 2016). These networks provide seamless care coordination and reduce the burden of transferring medical information on patients by facilitating the rapid and safe interchange of patient data between healthcare providers. Fair access to thorough patient information depends on resolving data-sharing issues and ensuring interoperability and data privacy. Wearable technology (Smokovski et al., 2024), including fitness trackers and smartwatches, makes it possible to continuously monitor a patient's vital signs, level of physical activity, and symptom development. These devices can be used in conjunction with research, development teams, and healthcare practitioners to give real-time data for patient self-management, individualized therapy monitoring, and early detection. Ensuring wearable device accuracy, dependability, and cost, as well as resolving inequalities in access to and use of these technologies, present challenges (Low et al., 2021; Cos et al., 2021).

If individuals encounter obstacles in utilizing technology or lack digital literacy, technological advancements may worsen health inequalities. To close the digital divide, healthcare providers, community organizations, and legislators must work together to offer digital literacy training, guarantee that internet access is affordable and readily available, and customize technological solutions to meet the needs and preferences of various demographics. To create rules and guidelines that guarantee the moral application of technology in pancreatic cancer care, legislators must work with researchers, technology developers, and politicians. This entails dealing with concerns about security, privacy, informed consent, and transparency in the gathering and use of data. Collaboration between many stakeholders, such as healthcare providers, technology developers, researchers, legislators, and patient advocacy groups, is necessary to tackle issues about technological advancements and access. To ensure that technological advances are culturally sensitive, user-friendly, and effective across diverse groups, efforts should be directed at encouraging equity in technology development and implementation, resolving access hurdles, and addressing access issues. To sum up, advancements in technology hold promises for enhancing both the quality and accessibility of pancreatic cancer treatment. Leveraging telemedicine, wearable technology, AI, mobile health apps, health information exchange, and remote monitoring are among the future directions that need to be explored (Nipp et al., 2022). To guarantee an equitable and efficient use of technology in resolving pancreatic cancer health inequities, however, issues with access, health literacy, data privacy, and regulatory concerns must be addressed (Wall et al., 2023).

Precision Medicine and Personalized Care

A major benefit of precision medicine is its ability to reduce health inequities and improve pancreatic cancer outcomes by customizing medical therapies for each patient based on their specific characteristics. Precision medicine and tailored care are still in their infancy as approaches to reducing health inequities associated with pancreatic cancer. Thanks to developments in genomic profiling, the precise genetic mutations and changes that induce pancreatic cancer have been identified (Pedrosa et al., 2023; Tsai et al., 2023). Researchers, oncologists, and geneticists' collaborations can improve our comprehension of pancreatic cancer's genomic landscape and pave the way for the creation of targeted treatments. By guaranteeing that patients receive medication that is most likely to be beneficial for their particular cancer subtype, personalized therapy techniques based on individual genetic profiles can enhance treatment outcomes and reduce inequities. Liquid biopsies have demonstrated potential for the early identification and monitoring of pancreatic cancer. These procedures entail the study of circulating tumor DNA or other biomarkers in body fluids. Research, clinical, and diagnostic specialists working together can progress liquid biopsy test development and validation for early screening and detection (Yang et al., 2023; Yu et al., 2023).

Given that pancreatic cancer is frequently identified at an advanced stage, early detection is essential for improving outcomes. Using individualized screening techniques that consider a person's genetic predisposition and risk factors can assist in identifying people who are at high risk and allow for prompt interventions. Researchers and physicians working together can find predictive biomarkers that show how likely a patient is to respond to a certain treatment (Heiselman et al., 2023; Vannini et al., 2023). Personalized treatment regimens can be established to maximize therapeutic responses by knowing the molecular and genetic properties of pancreatic cancers. This strategy can reduce toxicities, cut down on pointless therapies, and enhance patient outcomes for a variety of patient populations. To obtain a thorough understanding of pancreatic cancer biology, collaboration between researchers and bioinformatics specialists can help integrate multi-omics data, such as genomics, transcriptomics, proteomics, and metabolomics (Kane et al., 2021; Turanli et al., 2021; Xu et al., 2021).

The integrated analysis of multi-omics data can yield novel therapeutic targets, personalized treatment strategies, and an understanding of the underlying mechanisms. Multidisciplinary teams can utilize the promise of AI and big data analytics to provide new opportunities for personalized treatment. One of the biggest obstacles to putting personalized care and precision medicine into practice is ensuring that different groups have equitable access and can afford them (De Pastena et al., 2021; Chen et al., 2018). To address these issues, cooperation between legislators, medical professionals, patient activists, and insurance companies is required. The main goals should be to lower obstacles to care, guarantee insurance for genetic testing and focused treatments, and encourage patient education and knowledge on the advantages and accessibility of tailored care alternatives. Significant ethical and privacy issues are brought up by precision medicine, especially regarding the gathering, storing, and sharing of personal genomic data (Reddy et al., 2020; Peota, 2012).

Establishing strong frameworks and rules to address these challenges requires collaboration between researchers, bioethicists, and legal professionals. Upholding patient autonomy, guaranteeing informed consent, and safeguarding patient confidentiality are fundamental to preserving confidence and advancing fair and impartial access to customized healthcare. It might be difficult to incorporate precision medicine techniques into everyday clinical practice. It takes cooperation between researchers, healthcare providers, and health systems to create recommendations, set up clinical decision-support technologies, and give medical professionals the required education and training. Collaboration can also help ensure that individualized treatment becomes the norm by facilitating the translation of research findings into practicable clinical interventions. To sum up, individualized treatment and precision medicine have the power to transform pancreatic cancer management and reduce health inequities. Soon, there will be advancements in liquid biopsies, targeted therapeutics, genetic profiling, and the integration of multi-omics data (Wall et al., 2023).

Policy and Advocacy

To alleviate pancreatic cancer health inequities, policy changes and an emphasis on health equity are essential. In the future, efforts will be directed at securing more funds for pancreatic cancer-specific research, with an emphasis on delving into the causes, risk factors, and underlying mechanisms of the illness. Policymakers, patient advocacy organizations, and academics working together can affect funding priorities and guarantee that pancreatic cancer research is adequately funded. More financing for research can result in improvements in targeted medicines, early identification, and available treatment alternatives. Enhancing access to and use of pancreatic cancer screening techniques should be the goal of policy initiatives, especially for high-risk populations. Working together, legislators, medical professionals, and professional associations can produce guidelines that support routine screening for people with established risk factors, such as a family history of pancreatic cancer or particular genetic mutations. Increasing knowledge of warning indicators and symptoms can also help with early detection. All people with pancreatic cancer should have better access to high-quality care as the primary goal of policy changes. This involves making certain that medical centers and practitioners in underprivileged regions have the tools and knowledge required to properly identify and manage pancreatic cancer (Wall et al., 2023).

Policymakers, healthcare institutions, and community organizations working together can identify access hurdles and create plans to remove them, such as increasing healthcare coverage, enhancing transit alternatives, and putting telemedicine services into place. The discrepancies in health insurance that lead to unequal access to pancreatic cancer care should be addressed by policy changes. Collaboration among legislators, insurance companies, and patient advocacy organizations can advance policies that offer comprehensive coverage for diagnostic testing, treatment alternatives, screening, and supportive care services. To provide equal access to essential care, efforts should concentrate on lowering financial obstacles, including high out-of-pocket expenses and pre-existing condition exclusions. To reduce pancreatic cancer disparities, policy improvements should highlight the significance of culturally competent care. Policymakers, healthcare providers, and community organizations can collaborate to support programs that remove cultural barriers to care, guarantee linguistic access to services, and encourage diversity and inclusion in the healthcare workforce. Policies ought to support the creation of training initiatives that improve the cultural competence of healthcare professionals, empowering them to provide patients with individualized treatment that honors their values, beliefs, and preferences. Prioritizing patient education programs and health literacy activities can help policymakers better understand pancreatic cancer prevention, early detection, treatment options, and survivability (Wall et al., 2023).

Policies that support easily understood health information can be developed and put into action through cooperation between legislators, healthcare professionals, and patient advocacy groups. These policies can include educational materials, online resources, and community-based seminars. Reaching varied groups should be the main goal, considering factors like language, literacy, and culture. It is necessary to address the social determinants of health that lead to unequal outcomes to resolve pancreatic cancer disparities. Policy changes should focus on more general issues, including housing, work, education, and socioeconomic status, since these affect health behaviors and healthcare access. Collaboration among public health professionals, legislators, and community organizations can advance social determinants of health policies, which will ultimately increase health equity (Wall et al., 2023).

CONCLUSION

There will undoubtedly be a rise in pancreatic cancer throughout the next 10–20 years. Nevertheless, early cancer prevention through knowledge of the risk factors for pancreatic cancer is an affordable and efficient way to prevent pancreatic cancer in the first place, given the disease's high death rate and challenging early diagnosis. Given the unchangeable elements that contribute to pancreatic cancer, we can screen the population at risk for the disease and offer trustworthy screening methods and sensible diagnostic suggestions for an early diagnosis. By researching the modifiable risk factors associated with pancreatic cancer, we may be able to prevent the disease at an earlier stage and potentially even block it in its early stages, greatly lowering the incidence of the disease. An effective tobacco-control policy, suggestions for healthier lifestyles, and expanding the reach of screening, education, and vaccination programs should all be part of a global strategy for pancreatic cancer prevention and control. This will help to increase public awareness of the importance of taking preventative measures (Hu et al., 2021).

REFERENCES

ARTHUR, A. E., DELK, A., DEMARK-WAHNEFRIED, W., CHRISTEIN, J. D., CONTRERAS, C., POSEY, J. A., VICKERS, S., OSTER, R. & ROGERS, L. Q. 2016. Pancreatic cancer survivors' preferences, barriers, and facilitators related to physical activity and diet interventions. *Journal of Cancer Survivorship,* 10**,** 981–989.

BEHAR-HORENSTEIN, L., WARREN, R. C., SETIAWAN, V. W., PERKINS, C. & SCHMITTGEN, T. D. 2020. Enhancing African American participation in biospecimens: a case in point for pancreatic cancer. *Cancer Health Disparities,* 29, 1–7.

BLEDSOE, M. J. & GRIZZLE, W. E. 2013. Use of human specimens in research: The evolving United States regulatory, policy, and scientific landscape. *Diagnostic Histopathology,* 19**,** 322–330.

BOGUMIL, D., CONTI, D. V., SHENG, X., XIA, L., SHU, X.-O., PANDOL, S. J., BLOT, W. J., ZHENG, W., LE MARCHAND, L. & HAIMAN, C. A. 2020. Replication and genetic risk score analysis for pancreatic cancer in a diverse multiethnic population. *Cancer Epidemiology, Biomarkers & Prevention,* 29**,** 2686–2692.

BROOKS, G. A., TOMAINO, M. R., RAMKUMAR, N., WANG, Q., KAPADIA, N. S., O'MALLEY, A. J., WONG, S. L., LOEHRER, A. P. & TOSTESON, A. N. 2023. Association of rurality, socioeconomic status, and race with pancreatic cancer surgical treatment and survival. *JNCI: Journal of the National Cancer Institute,* 115**,** 1171–1178.

CHEN, Q., MERATH, K., BAGANTE, F., AKGUL, O., DILLHOFF, M., CLOYD, J. & PAWLIK, T. M. 2018. A comparison of open and minimally invasive surgery for hepatic and pancreatic resections among the medicare population. *Journal of Gastrointestinal Surgery,* 22**,** 2088–2096.

CHU, Q. D., SUN, G., POPE, M., LURAGUIZ, N., CURIEL, D. T., KIM, R., LI, B. D. & MATHIS, J. M. 2012. Virotherapy using a novel chimeric oncolytic adenovirus prolongs survival in a human pancreatic cancer xenograft model. *Surgery,* 152**,** 441–448.

COS, H., LI, D., WILLIAMS, G., CHININIS, J., DAI, R., ZHANG, J., SRIVASTAVA, R., RAPER, L., SANFORD, D. & HAWKINS, W. 2021. Predicting outcomes in patients undergoing pancreatectomy using wearable technology and machine learning: prospective cohort study. *Journal of Medical Internet Research,* 23**,** e23595.

DE PASTENA, M., ESPOSITO, A., PAIELLA, S., SURCI, N., MONTAGNINI, G., MARCHEGIANI, G., MALLEO, G., SECCHETTIN, E., CASETTI, L. & RICCI, C. 2021. Cost-effectiveness and quality of life analysis of laparoscopic and robotic distal pancreatectomy: A propensity score-matched study. *Surgical Endoscopy,* 35**,** 1420–1428.

DHANI, H., HINESTROSA, J. P., IZAGUIRRE-CARBONELL, J., BALCER, H. I., KURZROCK, R. & BILLINGS, P. R. 2023. Case report: Early detection of pancreatic pre-cancer lesion in multimodal approach with exosome liquid biopsy. *Frontiers in Oncology,* 13**,** 1170513.

EL KAOUTARI, A., FRAUNHOFFER, N. A., AUDEBERT, S., CAMOIN, L., BERTHOIS, Y., GAYET, O., ROQUES, J., BIGONNET, M., BONGRAIN, C. & CICCOLINI, J. 2023. Pancreatic ductal adenocarcinoma ubiquitination profiling reveals specific prognostic and theranostic markers. *EBioMedicine,* 92, 1–16.

ESPOSITO, I., KONUKIEWITZ, B., SCHLITTER, A. M. & KLÖPPEL, G. 2014. Pathology of pancreatic ductal adenocarcinoma: facts, challenges and future developments. *World Journal of Gastroenterology: WJG,* 20, 13833.

GAGLIARDI, A. R., SOONG, D. & GALLINGER, S. 2016. Identifying factors influencing pancreatic cancer management to inform quality improvement efforts and future research: A scoping systematic review. *Pancreas,* 45, 161–166.

GUSTAVELL, T., SUNDBERG, K. & LANGIUS-EKLÖF, A. 2020. Using an interactive app for symptom reporting and management following pancreatic cancer surgery to facilitate person-centered care: descriptive study. *JMIR mHealth and uHealth,* 8, e17855.

GUSTAVELL, T., SUNDBERG, K., SEGERSVÄRD, R., WENGSTRÖM, Y. & LANGIUS-EKLÖF, A. 2019. Decreased symptom burden following surgery due to support from an interactive app for symptom management for patients with pancreatic and periampullary cancer. *Acta Oncologica,* 58, 1307–1314.

HALDER, R., VEERAVELLI, S., CHENG, C., ESTRADA-MENDIZABAL, R. J. & RECIO-BOILES, A. 2023. Health disparities in presentation, treatment, genomic testing, and outcomes of pancreatic cancer in hispanic and non-hispanic patients. *Journal of Racial and Ethnic Health Disparities,* 10, 3131–3139.

HEISELMAN, J. S., ECKER, B. L., LANGDON-EMBRY, L., O'REILLY, E. M., MIGA, M. I., JARNAGIN, W. R., DO, R. K., HORVAT, N., WEI, A. C. & CHAKRABORTY, J. 2023. Registration-based biomarkers for neoadjuvant treatment response of pancreatic cancer via longitudinal image registration. *Journal of Medical Imaging,* 10, 036002.

HEISLER, M., LAPIDOS, A., HENDERSON, J., GUZMAN, R. M., WOLFE, J., MEYER, P., LAW, D., KIEFFER, E. C., ERNST, C. & DJELAJ, V. 2019. Study protocol for a Community Health Worker (CHW)-led comprehensive neighborhood-focused program for medicaid enrollees in detroit. *Contemporary Clinical Trials Communications,* 16, 100456.

HOWELL, L. A., SINICROPE, P. S., BROCKMAN, T. A., PATTEN, C. A., DECKER, P. A., EHLERS, S. L., NADEAU, A., RABE, K. G., BREITKOPF, C. R. & PETERSEN, G. M. 2013. Receptivity and preferences of pancreatic cancer family members for participating in lifestyle programs to reduce cancer risk. *Hereditary Cancer in Clinical Practice,* 11, 1–9.

HU, J.-X., ZHAO, C.-F., CHEN, W.-B., LIU, Q.-C., LI, Q.-W., LIN, Y.-Y. & GAO, F. 2021. Pancreatic cancer: A review of epidemiology, trend, and risk factors. *World Journal of Gastroenterology,* 27, 4298.

HUANG, B. Z., WANG, S., BOGUMIL, D., WILKENS, L. R., WU, L., BLOT, W. J., ZHENG, W., SHU, X. O., PANDOL, S. J. & LE MARCHAND, L. 2021a. Red meat consumption, cooking mutagens, NAT1/2 genotypes and pancreatic cancer risk in two ethnically diverse prospective cohorts. *International Journal of Cancer,* 149, 811–819.

HUANG, J., LOK, V., NGAI, C. H., ZHANG, L., YUAN, J., LAO, X. Q., NG, K., CHONG, C., ZHENG, Z.-J. & WONG, M. C. 2021b. Worldwide burden of, risk factors for, and trends in pancreatic cancer. *Gastroenterology,* 160, 744–754.

IZUMI, B. T., SCHULZ, A. J., ISRAEL, B. A., REYES, A. G., MARTIN, J., LICHTENSTEIN, R. L., WILSON, C. & SAND, S. L. 2010. The one-pager: A practical policy advocacy tool for translating community-based participatory research into action. *Progress in Community Health Partnerships: Research, Education, and Action,* 4, 141–147.

JIANG, J., CHAO, W.-L., CULP, S. & KRISHNA, S. G. 2023. Artificial intelligence in the diagnosis and treatment of pancreatic cystic lesions and adenocarcinoma. *Cancers,* 15, 2410.

KANE, L. E., MELLOTTE, G. S., CONLON, K. C., RYAN, B. M. & MAHER, S. G. 2021. Multi-omic biomarkers as potential tools for the characterisation of pancreatic cystic lesions and cancer: Innovative patient data integration. *Cancers,* 13, 769.

KATONA, B. W., KLUTE, K., BRAND, R. E., EVERETT, J. N., FARRELL, J. J., HAWTHORNE, K., KAUL, V., KUPFER, S. S., PAIELLA, S. & SIMEONE, D. M. 2023. Racial, ethnic, and sex-based disparities among high-risk individuals undergoing pancreatic cancer surveillance. *Cancer Prevention Research,* 16, 343–352.

KATZ, M. H., SLACK, R., BRUNO, M., MCMILLAN, J., FLEMING, J. B., LEE, J. E., BEDNARSKI, B., PAPADOPOULOS, J. & MATIN, S. F. 2016. Outpatient virtual clinical encounters after complex surgery for cancer: A prospective pilot study of "TeleDischarge". *Journal of Surgical Research,* 202, 196–203.

KLEIN, A. P. 2019. Pancreatic cancer: A growing burden. *The Lancet Gastroenterology & Hepatology,* 4, 895–896.

LIU, X., LI, Z. & WANG, Y. 2021. Advances in targeted therapy and immunotherapy for pancreatic cancer. *Advanced Biology,* 5, 1900236.

LOW, C. A., LI, M., VEGA, J., DURICA, K. C., FERREIRA, D., TAM, V., HOGG, M., ZEH III, H., DORYAB, A. & DEY, A. K. 2021. Digital biomarkers of symptom burden self-reported by perioperative patients undergoing pancreatic surgery: prospective longitudinal study. *JMIR Cancer,* 7, e27975.

MARTIN, A. N. & SNYDER, R. A. 2023. Racial disparities in pancreatic cancer clinical trials: Defining the problem and identifying solutions. *Advances in Cancer Research* 159, 185–201.

MCGUIGAN, A., KELLY, P., TURKINGTON, R. C., JONES, C., COLEMAN, H. G. & MCCAIN, R. S. 2018. Pancreatic cancer: A review of clinical diagnosis, epidemiology, treatment and outcomes. *World Journal of Gastroenterology,* 24, 4846.

NEVIDJON, B. 2016. Partnerships of the future. *Seminars in Oncology Nursing*, 23, 164–171.

NIPP, R. D., GAUFBERG, E., VYAS, C., AZOBA, C., QIAN, C. L., JAGGERS, J., WEEKES, C. D., ALLEN, J. N., ROELAND, E. J. & PARIKH, A. R. 2022. Supportive oncology care at home intervention for patients with pancreatic cancer. *JCO Oncology Practice,* 18, e1587–e1593.

O'NEIL, E. H. 2013. *Collaboration—For Whose Benefit?* Thorofare, NJ: SLACK Incorporated.

PEDROSA, L., ARAUJO, I., CUATRECASAS, M., SOY, G., LOPEZ, S., MAUREL, J., SANCHEZ-MONTES, C., MONTIRONI, C., SAURI, T. & SENDINO, O. 2023. Targeted transcriptomic analysis of pancreatic adenocarcinoma in EUS-FNA samples by NanoString technology. *Frontiers in Molecular Biosciences,* 10, 1161893.

PEOTA, C. 2012. A question of consent. *Minnesota Medicine,* 95, 7–7.

PORTMAN, D. G., THIRLWELL, S., DONOVAN, K. A. & ELLINGTON, L. 2021. Virtual teaming: Leveraging team science sense-making during COVID-19. *Journal of Patient Experience,* 8, 2374373521996945.

POURSHAMS, A., SEPANLOU, S.G., IKUTA, K.S., BISIGNANO, C., SAFIRI, S., ROSHANDEL, G., SHARIF, M., KHATIBIAN, M., FITZMAURICE, C., NIXON, M.R. & ABBASI, N. 2019. The global, regional, and national burden of pancreatic cancer and its attributable risk factors in 195 countries and territories, 1990–2017: A systematic analysis for the Global Burden of Disease Study 2017. *The Lancet*, 4(12), 934–947.

PULTE, D., REDANIEL, M., BRENNER, H. & JEFFREYS, M. 2012. Changes in survival by ethnicity of patients with cancer between 1992–1996 and 2002–2006: Is the discrepancy decreasing? *Annals of Oncology,* 23, 2428–2434.

RAOOF, S., LEE, R. J., JAJOO, K., MANCIAS, J. D., REBBECK, T. R. & SKATES, S. J. 2022. Multicancer early detection technologies: A review informed by past cancer screening studies. *Cancer Epidemiology, Biomarkers & Prevention,* 31, 1139–1145.

REDDY, A., AMARNANI, A., CHEN, M., DYNES, S., FLORES, B., MOSHCHINSKY, A., LEE, Y. J., KURBATOV, V., SHAPIRA, I. & VIGNESH, S. 2020. Privacy concerns about personal health information and fear of unintended use of biospecimens impact donations by African American patients. *Journal of Cancer Education,* 35, 522–529.

ROSENBAUM, L., VAN BUREN, E. & MENNEL, J. 2013. Partnerships for the future. *Business Trends*, p. 28. http://www2.deloitte.com/content/dam/Deloitte/ie/Documents/Finance/CFO/2013- SO-Business -Trends_vFINAL.pdf. Accessed: October 29, 2024.

ROTH, G. S., FAYET, Y., BENMAMECHE-MEDJAHED, S., DUCIMETIÈRE, F., CHARRETON, A., CROPET, C., CHABAUD, S., MARION-AUDIBERT, A.-M., BERTHELET, O. & WALTER, T. 2022. Structural and socio-spatial determinants influencing care and survival of patients with a pancreatic adenocarcinoma: Results of the PANDAURA cohort. *Cancers,* 14, 5413.

ROUANET, M., LEBRIN, M., GROSS, F., BOURNET, B., CORDELIER, P. & BUSCAIL, L. 2017. Gene therapy for pancreatic cancer: Specificity, issues and hopes. *International Journal of Molecular Sciences,* 18, 1231.

SATTURWAR, S., MONACO, S. E., XING, J. & PANTANOWITZ, L. 2020. The utility of cell blocks for international cytopathology teleconsultation by whole slide imaging. *Cytopathology,* 31, 419–425.

SMOKOVSKI, I., STEINLE, N., BEHNKE, A., BHASKAR, S. M., GRECH, G., RICHTER, K., NIKLEWSKI, G., BIRKENBIHL, C., PARINI, P. & ANDREWS, R. J. 2024. Digital biomarkers: 3PM approach revolutionizing chronic disease management—EPMA 2024 position. *EPMA Journal*, 15, 149–162.

SUN, J., RUSSELL, C. C., SCARLETT, C. J. & MCCLUSKEY, A. 2020. Small molecule inhibitors in pancreatic cancer. *RSC Medicinal Chemistry,* 11, 164–183.

TOVAR, D. R., ROSENTHAL, M. H., MAITRA, A. & KOAY, E. J. 2023. Potential of artificial intelligence in the risk stratification for and early detection of pancreatic cancer. *Artificial Intelligence Surgery,* 3, 14.

TSAI, Y. S., CHAREDDY, Y. S., PRICE, B. A., PARKER, J. S. & PECOT, C. V. 2023. An integrated model for predicting KRAS dependency. *PLoS Computational Biology,* 19, e1011095.

TURANLI, B., YILDIRIM, E., GULFIDAN, G., ARGA, K. Y. & SINHA, R. 2021. Current state of "omics" biomarkers in pancreatic cancer. *Journal of Personalized Medicine,* 11, 127.

VANNINI, I., ROSSI, T., MELLONI, M., VALGIUSTI, M., URBINI, M., PASSARDI, A., BARTOLINI, G., GALLIO, C., AZZALI, I. & BANDINI, S. 2023. Analysis of EVs from patients with advanced pancreatic cancer identifies antigens and miRNAs with predictive value. *Molecular Therapy Methods & Clinical Development,* 29, 473–482.

WALL, N. R., FULLER, R. N., MORCOS, A. & DE LEON, M. 2023. Pancreatic cancer health disparity: Pharmacologic anthropology. *Cancers,* 15, 5070.

WANG, F., SHU, X., PAL, T., BERLIN, J., NGUYEN, S. M., ZHENG, W., BAILEY, C. E. & SHU, X.-O. 2022. Racial/ethnic disparities in mortality related to access to care for major cancers in the United States. *Cancers,* 14, 3390.

XU, C., WANG, Y., YANG, H., HOU, J., SUN, L., ZHANG, X., CAO, X., HOU, Y., WANG, L. & CAI, Q. 2019. Association between cancer incidence and mortality in web-based data in China: Infodemiology study. *Journal of Medical Internet Research,* 21, e10677.

XU, D., WANG, Y., LIU, X., ZHOU, K., WU, J., CHEN, J., CHEN, C., CHEN, L. & ZHENG, J. 2021. Development and clinical validation of a novel 9-gene prognostic model based on multi-omics in pancreatic adenocarcinoma. *Pharmacological Research,* 164, 105370.

YAMAMOTO, Y., HIRAOKA, N., GOTO, N., RIN, Y., MIURA, K., NARUMI, K., UCHIDA, H., TAGAWA, M. & AOKI, K. 2014. A targeting ligand enhances infectivity and cytotoxicity of an oncolytic adenovirus in human pancreatic cancer tissues. *Journal of Controlled Release,* 192, 284–293.

YANG, H., LI, W., REN, L., YANG, Y., ZHANG, Y., GE, B., LI, S., ZHENG, X., LIU, J. & ZHANG, S. 2023. Progress on diagnostic and prognostic markers of pancreatic cancer. *Oncology Research,* 31, 83.

YANG, H. T., WU, M. C. & SHUN, S. C. 2018. Care plan for resuming the physical activity of patients with pancreatic cancer and diabetes after surgery. *Hu li za zhi The Journal of Nursing,* 65, 104–111.

YOUN, P., LI, H., MILANO, M., STOVALL, M., CONSTINE, L. & TRAVIS, L. 2013. Long-term survival among Hodgkin's lymphoma patients with gastrointestinal cancer: a population-based study. *Annals of Oncology,* 24, 202–208.

YU, Z., YANG, Y., FANG, W., HU, P., LIU, Y. & SHI, J. 2023. Dual tumor exosome biomarker co-recognitions based nanoliquid biopsy for the accurate early diagnosis of pancreatic cancer. *ACS Nano,* 17, 11384–11395.

ZHAO, Y., TANG, J., JIANG, K., LIU, S.-Y., AICHER, A. & HEESCHEN, C. 2023. Liquid biopsy in pancreatic cancer–Current perspective and future outlook. *Biochimica et Biophysica Acta (BBA)-Reviews on Cancer,* 1878, 188868.

ZOELLNER, J. M., PORTER, K. J., BROCK, D.-J. P., MITCHELL, E. M. K., CHAPMAN JR, H., CLARKSTON, D., COHN, W., HAUSER, L., MORRIS, D. W. & RAMEY, S. Y. 2021. Advancing engagement and capacity for rural cancer control: A mixed-methods case study of a Community-Academic Advisory Board in the Appalachia region of Southwest Virginia. *Research Involvement and Engagement,* 7, 44.

12 Conclusion
A Roadmap for the Future of Pancreatic Cancer Care in Africa and beyond

Zodwa Dlamini[1], Richard Khanyile[1,2], Thulo Molefi[1,2], Tebogo Marutha[1], Nkhensani Chauke-Malinga[3], and Rodney Hull[1]*

[1]SAMRC Precision Oncology Research Unit (PORU), DSI/NRF SARChI Chair in Precision Oncology and Cancer Prevention (POCP), Pan African Cancer Research Institute (PACRI), University of Pretoria, Pretoria, South Africa.

[2]Department of Medical Oncology, Faculty of Health Sciences, Steve Biko Academic Hospital, University of Pretoria, Pretoria, South Africa.

[3]Papillon Plastic Surgery, Linksfield West, Johannesburg, South Africa.

*Corresponding author: Zodwa Dlamini

CHARTING THE PATH FORWARD: INTEGRATING GLOBAL INSIGHTS WITH AFRICAN REALITIES

As we conclude this exploration into pancreatic cancer care, it is essential to reflect on the journey and the insights gained throughout this book. From the fundamental challenges of diagnosis and treatment to the innovative solutions emerging from both global and African contexts, our examination has unveiled the multifaceted nature of pancreatic cancer and the diverse strategies needed to address it. This concluding chapter aims to weave together these threads, providing a cohesive vision for the future of pancreatic cancer care. By integrating lessons learned, addressing persistent barriers, and envisioning a collaborative path forward, we aim to chart a course that not only advances scientific knowledge but also translates into tangible improvements in patient outcomes, particularly within the African context.

A GLOBAL IMPERATIVE WITH AFRICAN PERSPECTIVES

Pancreatic cancer remains one of the most challenging cancers to diagnose and treat, with a dismal prognosis and an increasing incidence worldwide. The complexity of this disease, characterized by late-stage diagnosis, rapid progression, and limited therapeutic options, necessitates a global response. This book, *Understanding and Conquering Pancreatic Cancer: Global Solutions with African Perspectives*, has meticulously examined the multifaceted nature of pancreatic cancer, emphasizing the need for a nuanced approach that incorporates both global insights and African-specific strategies (World Health Organization, 2020).

Africa presents unique challenges in addressing pancreatic cancer, including limited healthcare infrastructure, low cancer awareness, and significant disparities in access to care (Adeloye et al., 2019). However, it also offers unique opportunities for innovation, particularly in leveraging community-based approaches, adopting low-cost technologies, and integrating traditional knowledge

 DOI: 10.1201/9781003544401-17

systems with modern medical practices (Dlamini & Mbewu, 2023). As we move forward, it is essential to build on the foundations laid in this book to develop and implement a comprehensive roadmap for pancreatic cancer care in Africa and beyond.

THE URGENCY OF EARLY DETECTION AND PRECISION MEDICINE

One of the most pressing needs in the fight against pancreatic cancer is the development of reliable early detection methods (gray circle Figure 12.1). As highlighted in Chapter 5, early detection is critical to improving survival rates. Yet, it remains a significant challenge due to the asymptomatic nature of early-stage pancreatic cancer and the lack of specific screening tools (Zhou et al., 2022). Advances in imaging technologies, such as contrast-enhanced endoscopic ultrasound (EUS) and diffusion-weighted magnetic resonance imaging (MRI), offer promising avenues for early detection (Tummala et al., 2020). Moreover, the integration of artificial intelligence (AI) and machine learning algorithms into these diagnostic tools could revolutionize early detection by enabling the analysis of complex data sets and the identification of subtle patterns indicative of pancreatic cancer (Liu et al., 2021).

Precision medicine, as discussed in Chapter 4, represents a transformative approach to cancer treatment, offering the potential for tailored therapies that target the unique genetic and molecular profiles of individual tumors (Collisson et al., 2020). However, the implementation of precision medicine in Africa faces significant challenges, including limited access to advanced genomic technologies, high costs, and a lack of trained personnel (Muzammil et al., 2019). To overcome these barriers, there is a need for investment in capacity-building initiatives, collaboration with international research institutions, and the development of cost-effective precision medicine approaches that are accessible to African populations (Olopade, 2021).

ADDRESSING BARRIERS TO CARE: INFRASTRUCTURE, ACCESS, AND EQUITY

Chapter 8 of this book delves into the numerous barriers that hinder access to care for pancreatic cancer patients in Africa, including geographical, financial, and cultural obstacles (Jemal et al., 2020). These barriers contribute to the late-stage presentation of the disease, limited treatment options, and poor survival outcomes (Parkin et al., 2020). Improving access to care requires a multifaceted approach that addresses the root causes of these barriers (green circle in Figure 12.1).

Infrastructure development: Strengthening healthcare infrastructure is paramount to improving pancreatic cancer care in Africa. This includes expanding the availability of diagnostic facilities, establishing specialized cancer centers, and enhancing the capacity of healthcare workers through targeted training programs (Olopade, 2021). Furthermore, the development of telemedicine platforms can bridge the gap between remote areas and urban centers, ensuring that patients in underserved regions have access to expert consultations and follow-up care (Smith & Mistry, 2021).

Financial accessibility: The high cost of cancer treatment is a significant barrier to care in Africa. Policymakers must prioritize the development of funding mechanisms that reduce the financial burden on patients, such as subsidized treatment programs, government-funded cancer care initiatives, and partnerships with non-governmental organizations (NGOs) and international donors (Adeloye et al., 2019). Additionally, the inclusion of pancreatic cancer in national health insurance schemes could significantly improve access to care for low-income populations (Brouwer et al., 2020).

Equity in healthcare: Achieving equity in healthcare is essential for ensuring that all patients, regardless of their socio-economic status, have access to the same quality of care. This includes addressing disparities in healthcare delivery, improving health literacy, and reducing cancer-related stigma through public education campaigns (Parkin et al., 2020). Community engagement and the involvement of patient advocacy groups are critical to creating a supportive environment that empowers patients and their families to actively participate in their care (Dlamini & Mbewu, 2023).

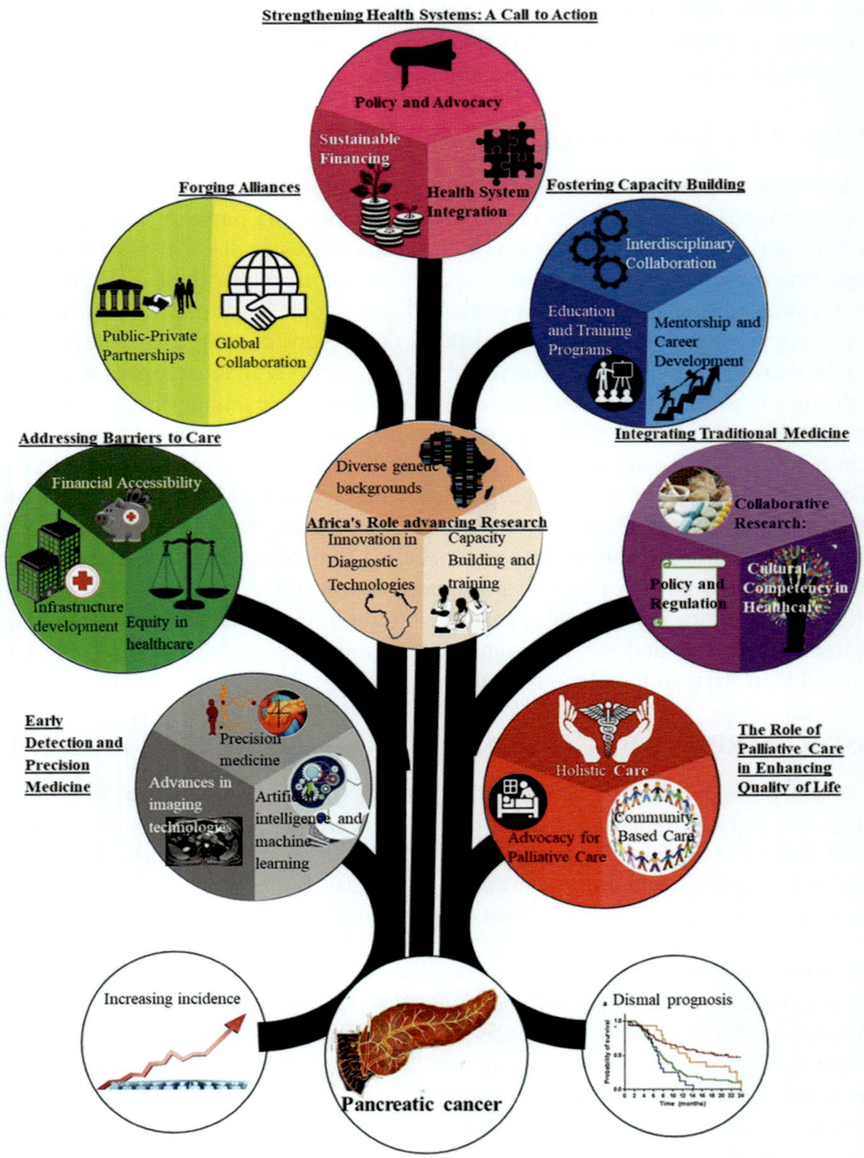

FIGURE 12.1 The future of pancreatic cancer care. Worldwide, pancreatic cancer incidence is on the rise, and combined with the poor prognosis for most patients, this makes pancreatic cancer a major health concern. This poor prognosis is exacerbated by the fact that pancreatic cancer is generally diagnosed when the disease is in its later stages. This is why diagnosing the disease at its early stages is important. In this regard, the development of new imaging technologies and the use of artificial intelligence in combination with personalized medicine will play a vital role in the early detection of the disease, resulting in a better prognosis. However, the lack of highly effective treatments and the nature of the disease mean that palliative care will remain an important feature in the care of pancreatic cancer patients. Holistic approaches and involving the community in this process are important factors for effective palliative care.

ADVANCING RESEARCH AND INNOVATION: AFRICA'S ROLE IN THE GLOBAL EFFORT

Africa's contribution to the global fight against pancreatic cancer is critical, particularly in the realm of research and innovation. As discussed in Chapter 10, the continent has unique strengths that can be leveraged to advance the understanding of pancreatic cancer and develop novel therapeutic approaches (Olopade, 2021).

Genomic research and precision medicine: African populations have diverse genetic backgrounds that provide valuable insights into the molecular mechanisms of pancreatic cancer. By conducting large-scale genomic studies in African populations, researchers can identify unique genetic mutations and biomarkers that may lead to the development of targeted therapies (Muzammil et al., 2019). Furthermore, collaboration between African research institutions and international partners can facilitate knowledge exchange and the transfer of advanced technologies (Smith & Mistry, 2021).

Innovation in diagnostic technologies: Africa has the potential to become a hub for innovation in low-cost diagnostic technologies. For example, the development of portable imaging devices and point-of-care diagnostic tests could revolutionize early detection in resource-limited settings (Liu et al., 2021). Additionally, the integration of AI into diagnostic tools offers a promising avenue for improving the accuracy and efficiency of pancreatic cancer diagnosis (Tummala et al., 2020).

Capacity building and training: Investing in the training of healthcare professionals is essential for advancing pancreatic cancer care in Africa. This includes the establishment of specialized training programs in oncology, genomics, and palliative care, as well as the promotion of interdisciplinary collaboration among healthcare providers, researchers, and policymakers (Olopade, 2021). Building a robust research infrastructure that supports the development of African-led research initiatives is also crucial for ensuring that the continent plays a leading role in the global fight against pancreatic cancer (Dlamini & Mbewu, 2023).

THE ROLE OF PALLIATIVE CARE IN ENHANCING QUALITY OF LIFE

Palliative care, as explored in Chapter 7, is a vital component of pancreatic cancer care, particularly given the aggressive nature of the disease and the high likelihood of late-stage diagnosis (Jemal et al., 2020). The integration of palliative care into the cancer care continuum is essential for addressing the physical, emotional, and psychological needs of patients and their families (red circle in Figure 12.1) (Smith & Mistry, 2021).

Holistic care approach: Palliative care should be tailored to the individual needs of patients, focusing not only on symptom management but also on providing emotional support and enhancing the quality of life (Brouwer et al., 2020). This includes addressing the psychological distress associated with a pancreatic cancer diagnosis, managing pain and other symptoms, and providing spiritual care (Olopade, 2021). Healthcare providers should be trained in palliative care principles and practices to ensure that patients receive comprehensive, compassionate care throughout their cancer journey (Smith & Mistry, 2021).

Community-based palliative care: In resource-limited settings, community-based palliative care programs can play a critical role in providing support to patients and their families. These programs can be implemented through partnerships with local NGOs, faith-based organizations, and community health workers (Jemal et al., 2020). By engaging the community in the delivery of palliative care, healthcare providers can ensure that patients receive care that is culturally appropriate and accessible (Adeloye et al., 2019).

Advocacy for palliative care: Advocacy efforts are needed to raise awareness about the importance of palliative care and to secure the necessary funding and resources for its implementation. Policymakers should prioritize the integration of palliative care into national cancer control plans and healthcare systems (Dlamini & Mbewu, 2023). Additionally, public education campaigns can help to reduce the stigma associated with palliative care and encourage patients to seek support early in their disease trajectory (Parkin et al., 2020).

FORGING ALLIANCES: THE POWER OF COLLABORATION AND PARTNERSHIP

Chapter 11 emphasizes the importance of collaboration and partnership in driving progress in pancreatic cancer care. The challenges posed by pancreatic cancer are too complex to be addressed by any single entity; instead, a coordinated effort involving governments, healthcare organizations, research institutions, industry partners, and advocacy groups is required (Collisson et al., 2020) (yellow circle in Figure 12.1).

Public–private partnerships: Public–private partnerships can play a crucial role in advancing pancreatic cancer care by pooling resources, expertise, and technologies (Brouwer et al., 2020). These partnerships can facilitate the development of innovative diagnostic and therapeutic solutions, as well as the implementation of large-scale public health initiatives aimed at raising awareness and improving access to care (Olopade, 2021).

Global collaboration: Africa's involvement in global cancer research networks is essential to ensure that the continent benefits from the latest scientific advancements. By participating in international research consortia, African researchers can contribute to the global knowledge base on pancreatic cancer and gain access to cutting-edge technologies and methodologies (Smith & Mistry, 2021). Additionally, global collaboration can help to standardize treatment protocols and improve the quality of care across different regions (Dlamini & Mbewu, 2023).

FOSTERING CAPACITY BUILDING: EMPOWERING THE NEXT GENERATION

The future of pancreatic cancer care in Africa hinges on the development of a robust healthcare workforce that is well-equipped to tackle the unique challenges of this disease. As highlighted in Chapter 9, capacity building through education, training, and mentorship is essential for ensuring that healthcare professionals, researchers, and policymakers have the necessary skills and knowledge to advance pancreatic cancer care on the continent (Olopade, 2021) (blue circle in Figure 12.1).

Education and training programs: There is a critical need for the establishment of specialized education and training programs in oncology, surgery, pathology, and palliative care across African medical schools and universities (Smith & Mistry, 2021). These programs should be designed to address the specific needs of African populations, considering local disease patterns, cultural contexts, and resource limitations (Jemal et al., 2020). Furthermore, partnerships with international institutions can facilitate the exchange of knowledge and expertise, enabling African healthcare professionals to stay abreast of the latest developments in pancreatic cancer care (Adeloye et al., 2019).

Mentorship and career development: Mentorship programs are vital for nurturing the next generation of African cancer researchers and clinicians (Muzammil et al., 2019). By connecting young professionals with experienced mentors, these programs can provide guidance, support, and opportunities for career development. Additionally, creating pathways for professional advancement and recognizing excellence in the field of oncology can help retain talent and motivate healthcare workers to pursue careers in pancreatic cancer care (Dlamini & Mbewu, 2023).

Interdisciplinary collaboration: Pancreatic cancer care requires a multidisciplinary approach that brings together experts from various fields, including surgery, oncology, radiology, pathology, and palliative care (Collisson et al., 2020). Encouraging interdisciplinary collaboration within healthcare institutions and research centers can lead to more comprehensive and effective treatment strategies. Moreover, fostering a culture of teamwork and shared learning can enhance the overall quality of care provided to patients (Smith & Mistry, 2021).

INTEGRATING TRADITIONAL MEDICINE: A HOLISTIC APPROACH TO CANCER CARE

Chapter 6 of this book explores the potential role of traditional medicine in the management of pancreatic cancer, particularly in the African context where traditional healers are often the first point of contact for patients seeking healthcare (Dlamini & Mbewu, 2023). Integrating traditional medicine with conventional cancer treatment offers a holistic approach that respects cultural practices while ensuring that patients receive evidence-based care (purple circle in Figure 12.1).

New treatment options may be found by researching traditional remedies, and these can be used in combination with modern treatments to improve the prognosis for patients with pancreatic cancer. Africa as a continent has many unique challenges and barriers to pancreatic cancer care. These include the need to develop healthcare infrastructure, address healthcare inequalities, and financial constraints. However, Africa also presents unique opportunities for research and innovation due to the genetic diversity of African populations, the requirement for the training of new healthcare workers which would result in capacity building, and the need for new diagnostic tools and procedures which would drive the development of low-cost diagnostics. Capacity building is important for the care of patients with pancreatic cancer and requires mentorship, education, and training programs to develop the next generation of healthcare workers in many disciplines. This is important as a multidisciplinary approach has been shown to be the best way to care for pancreatic cancer patients. Collaborations between the public and private sectors globally are also important in driving research into pancreatic cancer care and treatment and will help to provide a way forward for the care of pancreatic cancer patients through sustainable funding. Integration of healthcare is also of equal importance in the future; the integration of healthcare will provide a more organized and comprehensive strategy to care for pancreatic cancer patients.

Collaborative research: Collaborative research between traditional healers and biomedical scientists can help to identify and validate the efficacy of traditional remedies used in the treatment of pancreatic cancer (Parkin et al., 2020). Such research can also provide insights into the potential for developing new therapeutic agents derived from medicinal plants and other natural products (Adeloye et al., 2019). Establishing frameworks for ethical and scientific collaboration is crucial to ensure that traditional knowledge is respected and protected while contributing to the advancement of cancer care (Smith & Mistry, 2021).

Cultural competency in healthcare: Healthcare providers must be trained in cultural competency to effectively engage with patients who use traditional medicine (Olopade, 2021). This includes understanding the cultural significance of healers and traditional healing practices, communicating respectfully with traditional healers, and incorporating patients' beliefs and preferences into their care plans (Dlamini & Mbewu, 2023). By bridging the gap between traditional and modern medicine, healthcare providers can offer more culturally appropriate and patient-centered care (Jemal et al., 2020).

Policy and regulation: Governments and regulatory bodies should develop policies that recognize and regulate the practice of traditional medicine within the broader healthcare system (Collisson et al., 2020). This includes establishing standards for the safety and efficacy of traditional remedies, providing training for traditional healers, and integrating traditional medicine into national cancer control programs (Brouwer et al., 2020). Such policies can help ensure that patients receive safe, effective, and culturally sensitive care (Parkin et al., 2020).

STRENGTHENING HEALTH SYSTEMS: A CALL TO ACTION

The overarching goal of this book is to provide a comprehensive understanding of pancreatic cancer and to propose actionable strategies for improving its management, particularly in Africa. However, achieving these goals requires a concerted effort to strengthen health systems at every level (Dlamini & Mbewu, 2023) (pink circle in Figure 12.1).

Policy and advocacy: Effective policymaking and advocacy are essential for addressing the systemic challenges that hinder pancreatic cancer care in Africa (Olopade, 2021). Governments must prioritize cancer care within their national health agendas, allocating sufficient resources for infrastructure development, training, and research (Brouwer et al., 2020). Advocacy efforts should focus on raising awareness about pancreatic cancer, reducing stigma, and mobilizing communities to demand better access to care (Smith & Mistry, 2021).

Health system integration: Pancreatic cancer care must be integrated into the broader health system to ensure that patients receive coordinated and comprehensive care (Collisson et al., 2020). This includes linking primary care providers with specialized cancer centers, developing referral pathways, and ensuring that patients have access to the full spectrum of cancer services, from prevention and early detection to treatment and palliative care (Adeloye et al., 2019). Health system integration also involves the use of data and information systems to monitor patient outcomes, track disease patterns, and inform policy decisions (Muzammil et al., 2019).

Sustainable financing: Sustainable financing is critical for ensuring the long-term viability of pancreatic cancer care programs (Parkin et al., 2020). Governments should explore innovative financing mechanisms, such as public–private partnerships, health insurance schemes, and international funding, to support cancer care initiatives (Jemal et al., 2020). Additionally, financial incentives should be provided to encourage investment in cancer research, infrastructure development, and capacity building (Smith & Mistry, 2021).

CONCLUSION: TOWARD A FUTURE OF HOPE AND PROGRESS

An illustration of the various factors forming a roadmap for the care of pancreatic cancer in Africa and the rest of the world is given in Figure 12.1. The importance of formulating a strategy for the care of pancreatic cancer is due to the increasing incidence of pancreatic cancer and the dismal prognosis for pancreatic cancer patients. The fight against pancreatic cancer is one of the most formidable challenges in global health, but it is a challenge that can be met with determination, innovation, and collaboration. By integrating global solutions with African perspectives, we can create a roadmap for the future that addresses the unique needs of African populations while contributing to the global effort to conquer pancreatic cancer (Dlamini & Mbewu, 2023).

This book has highlighted the importance of early detection, precision medicine, and equitable access to care, while also emphasizing the critical role of research, capacity building, and the integration of traditional medicine. It has underscored the need for collaboration at all levels, from local communities to global networks, and the necessity of strong health systems that can support the delivery of high-quality cancer care (Olopade, 2021).

As we move forward, let us be guided by the principles of equity, innovation, and compassion. Let us work together to build a future where pancreatic cancer is no longer a death sentence, but a disease that can be effectively managed and ultimately cured. And let us remember that every step we take toward this goal brings us closer to a world where all people, regardless of where they live, have the opportunity to live healthy, fulfilling lives (Dlamini & Mbewu, 2023).

REFERENCES

ADELOYE, D., DAVID, R. A., ADEREMI, A. V., ISEOLORUNKANMI, A., OYEDOKUN, A., IWEALA, E. E. J., ... & AUTA, A. 2019. Health workforce and governance: The crisis in Nigeria. *Human Resources for Health*, *17*(1), 79.

BROUWER, E. D., VAN BAARLE, D., SIEBERS, R., LEENSTRA, T., & VAN ESCH, B. 2020. Cancer care in Sub-Saharan Africa: Challenges and opportunities. *European Journal of Cancer Care*, *29*(6), e13250.

COLLISSON, E. A., BAILEY, P., CHANG, D. K., & BIANKIN, A. V. 2020. Molecular subtypes of pancreatic cancer. *Nature Reviews Gastroenterology & Hepatology*, *16*(4), 207–220.

DLAMINI, Z., & MBEWU, A. 2023. *Understanding and Conquering Pancreatic Cancer: Global Solutions with African Perspectives*. Pretoria: Pan African Cancer Research Institute.

JEMAL, A., BRAY, F., CENTER, M. M., FERLAY, J., WARD, E., & FORMAN, D. 2020. Global cancer statistics. *CA: A Cancer Journal for Clinicians*, *70*(1), 5–30.

LIU, X., LI, X., ZHANG, J., JIN, J., LI, L., & ZHENG, L. 2021. Artificial intelligence in the diagnosis of pancreatic cancer: a review. *Artificial Intelligence in Medicine*, *112*, 102039.

MUZAMMIL, A., OLOPADE, O. I., & HUO, D. 2019. Genomic insights into disparities in pancreatic cancer outcomes among African populations. *Journal of the National Cancer Institute*, *111*(8), 794–801.

OLOPADE, O. I. 2021. Addressing cancer health disparities: a global perspective. *The Lancet Oncology*, *22*(10), 1375–1386.

PARKIN, D. M., BRAY, F., FERLAY, J., & PISANI, P. 2020. Global cancer statistics, 2020. *CA: A Cancer Journal for Clinicians*, *55*(2), 74–108.

SMITH, R. D., & MISTRY, P. 2021. Health systems strengthening for universal health coverage: Learning from the past to build the future. *Global Health*, *17*(1), 73.

TUMMALA, P., JUNAIDI, O., AGARWAL, B. 2020. Imaging of pancreatic cancer: Recent advances. *Current Gastroenterology Reports*, *22*(10), 49.

World Health Organization. (2020). *Global Cancer Report 2020*. Geneva: World Health Organization.

Zhou, B., Xu, J. W., Cheng, Y. G., Gao, J. Y., Hu, S. Y., Wang, L., & Zhang, G. Y. (2022). Early detection of pancreatic cancer: Where are we now and where are we going? *International Journal of Cancer*, *150*(4), 1046–1058.

Index

For Product Safety Concerns and Information please contact our
EU representative GPSR@taylorandfrancis.com Taylor & Francis
Verlag GmbH, Kaufingerstraße 24, 80331 München, Germany